Springer Series in

MATERIALS SCIENCE 144

Springer Series in

MATERIALS SCIENCE

Editors: R. Hull C. Jagadish R.M. Osgood, Jr. J. Parisi Z. Wang H. Warlimont

The Springer Series in Materials Science covers the complete spectrum of materials physics, including fundamental principles, physical properties, materials theory and design. Recognizing the increasing importance of materials science in future device technologies, the book titles in this series reflect the state-of-the-art in understanding and controlling the structure and properties of all important classes of materials.

Please view available titles in *Springer Series in Materials Science* on series homepage http://www.springer.com/series/856

Jacek Kossut
Jan A. Gaj
Editors

Introduction to the Physics of Diluted Magnetic Semiconductors

With 236 Figures

Springer

Editors

Professor Jacek Kossut
Institute of Physics, Polish Academy of Sciences
Al. Lotnikow 32/46, 02-668 Warszawa, Poland
E-mail: jacek.kossut@ifpan.edu.pl

Professor Jan A. Gaj
Warsaw University, Institute of Experimental Physics, Faculty of Physics
ul. Hoża 69, 00-681 Warszawa, Poland
E-mail: jan.gaj@fuw.edu.pl

Springer Series in Materials Science ISSN 0933-033X
ISBN 978-3-642-15855-1 e-ISBN 978-3-642-15856-8
DOI 10.1007/978-3-642-15856-8
Springer Heidelberg Dordrecht London New York

Cover design: eStudio Calamar Steinen

Printed on acid-free paper

Springer is part of Springer Science+Business Media (www.springer.com)

Preface

Diluted magnetic semiconductors, or semimagnetic semiconductors, seemed for a while to be one of those research topics whose glory (i.e., the period of most extensive research) belonged already to the past. This particularly applied to "traditional" diluted magnetic semiconductors, i.e., substitutional alloys of either II–VI or IV–VI semiconductors with transition metal ions. Fortunately, a discovery, in the beginning of the nineties [1, 2], of ferromagnetic ordering in III–V DMSs with critical temperatures reaching $\sim$170 K has renewed and greatly intensified an interest in those materials. This was, at least partially, related to expectations that their Curie temperatures can be relatively easily brought to room temperature range through a clearly delineated path and, partially, due to the great successes, also commercial, of metallic version of spintronics, which earned its founders the Nobel Prize in 2007. The semiconductor version of spintronics has attracted researchers also because of hopes to engage it in efforts to construct quantum information processing devices. While these hopes and expectations are not fully realized yet, the effort is going on.

As a good example of recent achievements, new results on quantum dots containing a single magnetic ion should be mentioned. A great progress has been achieved in studies of excitonic states in such quantum dots, so far limited to InAs/GaAs [3,4] and CdTe/ZnTe [5, 6] material systems and to Manganese as the magnetic ion. Furthermore, in the II–VI QDs, first results on the optical control of the Mn spin states have been experimentally demonstrated [7–9] and theoretically analyzed [10]; the studies of Mn spin dynamics and control in III–V QDs will certainly follow. Similarly, a considerable effort is directed toward growth and characterization of another "fashionable" objects of nanotechnology, namely nanorods mostly grown by vapor–liquid–solid technique, composed, e.g., of ZnMnTe. The research on DMS nanorods and structures made of those objects is currently very actively developed, see, e.g., [11].

With such a tremendous interest worldwide, of course, there appeared several books with excellent reviews on physics and potential applications of diluted magnetic semiconductors [12–18] (including skeptical reviews, [19,20]). Also, the "old" materials such as II–VIs with transition metal ions are already covered extensively in several books, data collections etc. [21–29]. Therefore, it seems that there is little room for yet another book on a similar subject. However, the collection of previous articles seemed to share one feature: they were devoted to, at the time of their appearance, most recent achievements in the field. On the other hand, a systematic, pedagogical approach to the basics of physics of diluted magnetic semiconductors was, in our opinion, conspicuously missing. This is particularly true in the case of low-dimensional structures composed of diluted magnetic semiconductors. This book intends to fill this gap, without dropping entirely an ambition to speak about the recent achievements and research directions, such as time-resolved investigations of spin dynamics. Whether or not we succeeded in reaching this goal remains to be judged by the readers.

Since both editors of this volume consider experiment to be a driving force in the area of present interest, it was our intention to start with chapters that stressed the experimental facts and issues with theoretical explanations and ideas to follow.

The history of diluted magnetic semiconductors can be traced down to research on ferromagnetic spinels with Cr and Eu chalcogenides [30] mostly at IBM Research Center roughly in the sixties of the previous century. One must note early theoretical ideas concerning magnetic polarons and spin molecules, as well as spin-dependent scattering and indirect coupling via carriers. These can be viewed as precursors of the field of diluted magnetic semiconductors. However, it was only when magnetic components (Manganese ions, in most of the cases) were started to be introduced into relatively simple semiconductor matrices, such as CdTe, that the field began to gain impetus in mid-seventies of the twentieth century. Motivations to introduce Manganese was certainly different: it was hoped that (Hg,Mn)Te would be mechanically more durable than (Hg,Cd)Te – the material for infrared detectors made of narrow gap semiconductors, or it was hoped that electron resonances would become detectable in the internal effective field (rather than in the external magnetic filed), or that addition of Mn to HgTe would result in particularly high electron mobility in such ternary materials (as found in early work of Morissy in his Ph.D. thesis in Oxford). Furdyna hoped to detect spin resonance of Mn in highly conductive HgTe exploiting transmission of helicon waves in a magnetic field [31]. Not all of those ideas survived the test of time. But, without doubt, it was discovery of very pronounced magnetooptical activity of these materials, particularly of the giant Faraday rotation (whose origin was traced down to giant Zeeman splitting) of excitonic states in CdMnTe [32, 33] that sparked a broader interest. It was during the ICPS in Edinburgh in 1978 that the first invited talk on diluted magnetic semiconductors was given [34].

As mentioned, there are several review volumes or chapters devoted to the early work on physics of diluted magnetic semiconductors. Without attempting to be exhaustive, let us mention additionally [35] and references therein. We are also aware that while this volume is being prepared, a team of authors just completed preparation of a comprehensive book describing the spintronics-related physics of ferromagnetic III–V-based diluted magnetic semiconductors that are most popular these days [15]. Partly for this reason, this book, apart from attempting to have a pedagogical character, concentrates on "orthodox" II–VI diluted magnetic semiconductors since we believe that all basic ideas can be introduced using examples from that thoroughly studied group of materials. We hope also to reduce duplication of the material with our choice. A certain emphasis is put on low-dimensional structures, quantum wells and quantum dots, and on phenomena that only recently started to be accessible experimentally, e.g., spin relaxation and coherence.

With all these remarks in mind, the structure of the volume is the following: we begin with what we consider a most rudimentary Chap. 1 which may serve as an introduction to the field suitable for students just entering the field. We present there the simplest mean field approach that provides an acceptable description in many cases and was the first that was historically used. The next Chap. 2, by Pacuski, deals with those materials – semiconductors with wide gap between the valence band and the conduction band – where apart from apparent similarities such a simple mean field/virtual crystal approximation evidently does not work. Merkulov and Rodina in the next Chap. 3 to provide the reader with a very general and thorough theoretical considerations concerning the very origins of the specificity of diluted magnetic semiconductors, namely the spin-dependent coupling between the valence and conduction band carriers and electrons localized on half-filled shell of the "magnetic" ions, such as Manganese. In particular, an attention is paid to the role of dimensionality in the physics of this interaction. In the next Chap. 4, Furdyna and coworkers introduce us to the realm of experiments on low-dimensional structures such as quantum wells, superlattices (in particular, spin superlattices), etc. This is followed by a special Chap. 5 by Henneberger and Puls devoted to quantum dots that are either embedded in the magnetic environment or are magnetic themselves. In the spirit of the book (experiments first, theories – later) Chap. 6 contributed by Hawrylak gives a thorough discussion of the $sp{-}d$ coupling in the specific case of the quantum dots paying attention to many body aspects of the problem. When talking about dots, it is impossible not to mention magnetic polarons, localized magnetic polarons in particular, which constitute the topic of Chap. 7 by Yakovlev and Ossau. Temporal behavior of various degrees of freedom are discussed in next two Chaps. 8 and 9, first, by Yakovlev and Merkulov, discusses the relaxation of spins in diluted magnetic semiconductors, while the second, by Crooker, concentrates on the coherence of precessing spins in the DMS structures, something that may be of importance in quantum information processing in diluted magnetic semiconductor spintronic qubits. Many body aspects of optical properties (Chap. 10 by Perez and Kossacki) and charge transport (Chap. 11 by Jaroszyński) are covered next. Finally, Giebultowicz and Kępa (Chap. 12) provide us with a detailed discussion of the interlayer coupling in diluted magnetic superlattices with non-magnetic interlayers as seen by neutron scattering experiments.

As mentioned, there is some degree of topical overlap between the chapters. We found it very difficult (although we did try) to keep the notation exactly the same in the whole book (e.g., the exchange constant α needs to have, sometimes, additional super or subscripts, and sometimes it does not, when the approach is made simpler). Although this may be viewed as incompatible with an intended pedagogical spirit of the present volume, we tend to agree that that some variations of notation among the chapters is possible, provided that they are explained properly in the text, since they reflect the authors' varying views concerning the importance of particular aspects of the physics in question.

As we saw in the past, the motivations to study a fascinating class of materials, such as DMSs, might vary in time. At present, it is spintronics and quantum computation that are the driving forces. It is, thus, not unreasonable to expect that new stimuli will appear also in the future, e.g., located at the interface of spin physics and life sciences. We think, therefore, that an effort to study DMSs represents a good investment. We hope that the present collection of chapters will help the reader in this task.

Warsaw
September 2010

Jan Gaj
Jacek Kossut

References

1. H. Munekata et al., Phys. Rev. Lett. **63**, 1849 (1989)
2. H. Ohno et al., Phys. Rev. Lett. **68**, 2662 (1992)
3. O. Krebs et al., Phys. Rev. B **80**, 165315 (2009)
4. J. van Bree et al., Phys. Rev. B **78**, 165414 (2008)
5. Y. Leger et al., Phys. Rev. Lett. **97**, 107401 (2006)
6. L. Besombes et al., Phys. Rev. B **78**, 125324 (2008)
7. C. Le Gall et al., Phys. Rev. Lett. **102**, 127402 (2009)
8. M. Goryca et al., Phys. Rev. Lett. **103**, 087401 (2009)
9. C. Le Gall et al., Phys. Rev. B **81**, 245315 (2010)
10. D.E. Reiter, T. Kuhn, V.M. Axt, Phys. Rev. Lett. **102**, 177403 (2009)
11. W. Zaleszczyk et al., Nano Lett. **8**, 4061 (2008)
12. D.D. Awschalom, D. Loss, N. Samarth (eds.), *Semiconductor Spintronics and Quantum Computations* (Springer, Berlin, 2002)
13. D.D. Awschalom, R.A. Buhrman, J.M. Daughton, S. von Molnár, M.L. Roukes (eds.), *Spin Electronics* (Kluver, Dordrecht, 2004)
14. M. Ziese, J. Thornton (eds.), *Spin Electronics* (Springer, Berlin, 2001)
15. T. Dietl, D.D. Awschalom, M. Kamińska, H. Ohno (eds.), *Spintronics*, in *Semiconductors and Semimetals*, vol 82 (Elsevier, Amsterdam, 2008)
16. J. Fabian, I. Zutic, S. Das Sarma, Rev. Mod. Phys. **76**, 323 (2004)
17. J. Fabian, A. Matos-Abiague, C. Ertler, P. Stano, I. Zutic, Acta Phys. Slovaca, **4**, 565 (2007)
18. see also special issue of *Semiconductor Science and Technology*, ed. by H. Ohno, vol 17, issue 4, 2002
19. M.I. Dyakonov, in *Future Trends in Microelectronics: The Nano, The Giga, and the Ultra*, eds. by S. Luryi, J. Xu, A. Zaslavski (Wiley, NY, USA, 2004, cond-mat/0411672), p. 157
20. M.I. Dyakonov, in *Future Trends in Microelecronics: The Nano Millenium*, eds. by S. Luryi, J. Xu, A. Zaslavski (Wiley, NY, USA, 2002, cond-mat/0110326), p. 307
21. R.R. Gałązka, J. Kossut, in *Landolt-Börnstein*, ed. by O. Madelung, vol 17b (Springer, Berlin, 1982)
22. R.R. Gałązka, J. Kossut, T. Story, in *Landolt-Börnstein*, vol 41 (Springer, Berlin, 1999), p. 650
23. J. Kossut, W. Dobrowolski, in *Handbook of Magnetic Materials*, ed. by K.H.J. Buschow, vol 7 (North Holland, Amsterrdam, 1993), p. 231
24. J. Kossut, W. Dobrowolski, T. Story, in *Handbook of Magnetic Materials*, ed. by K.H.J. Buschow, vol 15 (North Holland, Amsterdam, 2003), p. 289
25. J.K. Furdyna, J. Kossut (eds.), vol 25 of *Semiconductors and Semimetals* (Academic Press, Boston, 1988) (Russian translation: Mir, Moskva, 1992)
26. J.K. Furdyna, J. Appl. Phys. **64**, R29 (1988)

27. T. Dietl, in *Handbook of Semiconductors*, eds. by T.S. Moss, S. Mahajan (Elsevier, Amsterdam, 1997), p. 1521
28. J. Kossut, W. Dobrowolski, in *Narrow Gap II–VI Compounds for Optoelectronic and Electromagnetic Applications*, ed. by P. Capper (Chapman and Hall, London, 1997), p. 401
29. M. Jain (ed.), *Diluted Magnetic Semiconductors* (World Scientific, Singapore, 1991)
30. See special volume of IBM J. Res. Develop. **14**, 3 (1970)
31. J. Furdyna, R.T. Holm, Phys. Rev **15**, 844 (1977)
32. J.A. Gaj, R.R. Gałązka, M. Nawrocki, Solid State Commun. **25**, 193 (1978)
33. A.V. Komarov, S.M. Ryabchenko, O.V. Terletskii, I.I. Zheru, R.D. Ivanchuk, Pis'ma v Zh. Eksp. Teor. Fiz. **73**, 608 (1977) (Sov. Phys. JETP **46**, 318)
34. R.R. Gałązka, in *Proc. 14th Internat. Conf. Phys. Semicond, Edinburgh, 1978*, ed. by B.L.H. Wilson, (Internat. Conf. Series vol 43), (Institute of Physics, London, 1978), p. 133
35. N.B. Brandt, V.V. Moshchalkov, Adv. Phys. **33**, 193 (1984)

Contents

Contributors

Scott A. Crooker National High Magnetic Field Laboratory, Los Alamos National Laboratory, Los Alamos, NM 87545, USA, crooker@lanl.gov

Margaret Dobrowolska Department of Physics, University of Notre Dame, Notre Dame, IN 46556, USA, mdobrowo@nd.edu

Jacek K. Furdyna Department of Physics, University of Notre Dame, Notre Dame, IN 46556, USA, Jacek.K.Furdyna.1@nd.edu

Jan A. Gaj Institute of Experimental Physics, University of Warsaw, Hoża 69 00-681, Warsaw, Poland, jan.gaj@fuw.edu.pl

Tomasz M. Giebultowicz Physics Department, Oregon State University, Corvallis, OR 97331, USA, tgiebult@physics.oregonstate.edu

Pawel Hawrylak Quantum Theory Group, Institute for Microstructural Sciences, National Research Council of Canada, Ottawa, ON, Canada K1A 0R6, pawel.hawrylak@nrc-cnrc.gc.ca

Fritz Henneberger Institute of Physics, Humboldt-University, 12489 Berlin, Germany, fh@physik.hu-berlin.de

Jan Jaroszynski National High Magnetic Field Laboratory, 1800 East Paul Dirac Drive, Tallahassee, FL 32310, USA, jaroszy@magnet.fsu.edu

Henryk Kępa Institute of Experimental Physics, Faculty of Physics, University of Warsaw, Hoża 69 00-681, Warsaw, Poland
and
Physics Department, Oregon State University, Corvallis, OR 97331, USA, Henryk.Kepa@fuw.edu.pl

Piotr Kossacki Institute of Experimental Physics, Faculty of Physics, University of Warsaw, Hoża 69 00-681 Warsaw, Poland, Piotr.Kossacki@fuw.edu.pl

Jacek Kossut Institute of Physics, Polish Academy of Sciences, Al. Lotników 32/46, 02-668 Warsaw, Poland, jacek.kossut@ifpan.edu.pl

Sanhoon Lee Department of Physics, Korea University, Seoul, 136-701, Korea, slee3@nd.edu

Xinyu Liu Department of Physics, University of Notre Dame, Notre Dame, IN 46556, USA, xliu2@nd.edu

Igor A. Merkulov Ioffe Physico-Technical Institute, Russian Academy of Sciences, 194021 St. Petersburg, Russia
and
Condensed Matter Sciences Division, Oak Ridge National Laboratory, Oak Ridge, TN 37831-6016, USA, imp@mail.ornl.gov

Wolfgang Ossau Experimentelle Physik 3, Universität Würzburg, 94074 Würzburg, Germany, Wolfgang.Ossau@physik.uni-wuerzburg.de

Wojciech Pacuski Institute of Experimental Physics, University of Warsaw, Hoża 69 00-681 Warsaw, Poland, Wojciech.Pacuski@fuw.edu.pl

Florent Perez Institute des NanoSciences de Paris, CNRS Université Paris 6, France, perez@insp.jussieu.fr

Joachim Puls Institute of Physics, Humboldt-University, 12489 Berlin, Germany, puls@physik.hu-berlin.de

Anna V. Rodina Ioffe Physico-Technical Institute, Russian Academy of Sciences, 194021 St. Petersburg, Russia, anna.rodina@mail.ioffe.ru

Tomasz Wojtowicz Institute of Physics, Polish Academy of Sciences, Al. Lotników 32/46, 02-668, Warsaw, Poland, Tomasz.Wojtowicz@ifpan.edu.pl

Dmitri R. Yakovlev Experimentelle Physik 2, Technische Universität Dortmund, 44221 Dortmund, Germany
and
Ioffe Physico-Technical Institute, Russian Academy of Sciences, 194021 St. Petersburg, Russia, dmitri.yakovlev@tu-dortmund.de

Chapter 1
Basic Consequences of $sp-d$ and $d-d$ Interactions in DMS

Jan A. Gaj and Jacek Kossut

Abstract In this introductory chapter, we describe the basic features of diluted magnetic semiconductors. We focus on giant Zeeman splitting of excitons and related giant Faraday rotation. We show that the spin splitting is proportional to the magnetization and develop a simple model, making use of virtual crystal and mean field approximations, describing the energy states in the presence of magnetic field in diluted magnetic semiconductors having moderate and narrow gaps between the valence and conduction bands. We discuss limitations of this model description and mention cases, which make a more refined theories necessary. We introduce and discuss $sp-d$ exchange interaction constants, show the trends existing in their values in the family of II–VI DMSs. We finally discuss the $d-d$ exchange interaction, which is responsible for magnetic properties of the materials in question and show how to conveniently parametrize the magnetization.

1.1 Introduction

This chapter presents the most important experimental features that distinguish a diluted magnetic semiconductor from materials containing no magnetic ions. To achieve this goal, we select the most typical material as an example that we shall use to illustrate our considerations here, referring to other diluted magnetic semiconductors only rather sporadically. With this intention, let us define a dilute magnetic semiconductor as a substitutional ternary alloy of II–VI semiconductor compounds, such as CdTe, where part of the cations in the Cd sublattice are replaced (at random) by transition metal ions. Why do II–VI compounds represent the simple case?

J.A. Gaj (✉)
Institute of Experimental Physics, Faculty of Physics, University of Warsaw, Hoża 69 00-681, Warsaw, Poland
e-mail: jan.gaj@fuw.edu.pl

J. Kossut
Institute of Physics, Polish Academy of Sciences, Al. Lotników 32/46, 02-668 Warsaw, Poland
e-mail: jacek.kossut@ifpan.edu.pl

J. Kossut and J.A. Gaj (eds.), *Introduction to the Physics of Diluted Magnetic Semiconductors*, Springer Series in Materials Science 144,
DOI 10.1007/978-3-642-15856-8_1, © Springer-Verlag Berlin Heidelberg 2010

Namely, transition metal ions may, and often do, form centers that are neutral with respect to the background crystalline lattice, i.e., there are no excess electric charges localized on them. This is in contrast to III–V-based diluted magnetic semiconductors, where a typical situation is that the doping transition metal ion is also a charged center. Why CdTe, in particular? First its lattice structure is that of zinc blende so that complications of the wurtzite or rocksalt case can be left for more in-depth going parts of the book. Second, the band gap of CdTe is intermediate in width, so, again, complications related to the narrow gap semiconductors do not have to be accounted for (we shall mention them here only relatively briefly) and neither do anomalies associated with the extremely wide band gap situation have to be worried about (those are left for the next chapter).

Again, limiting ourselves to simple cases, we shall, unless otherwise stated, think about Manganese when saying "a transition metal ion". The simplicity of Manganese stems from the fact that its electronic d-shell is half-filled, the orbital momentum of the d-shell vanishes while the spin has a maximum, for transition metal series, possible value of 5/2, and the ground state of d-electrons represents a spin singlet. Thus, the archetypal diluted magnetic semiconductor is a ternary $Cd_{1-x}Mn_xTe$, where x represents the molar fraction of Mn ions that replace Cd ions. Again, to keep things as simple as possible we shall here envisage then that the d-electrons of manganese are considerably more localized spatially that the s- and p-electrons (that are assumed to contribute to the conduction and valence bands). Their interaction with the delocalized electrons is relatively weak (see further chapters in this volume) which substantiates, in turn, a picture in which these two electronic systems – d-electrons of Manganese and the delocalized s- and p-electrons – may be considered separately, to a certain extent. The weak interaction between these two electronic subsystems may be treated effectively as a spin dependent perturbation, within a combination of virtual crystal approximation and mean field approach. It is amazing how many experimental observations can be quantitatively understood with this extremely simplified view. Let us begin our encounter with diluted magnetic semiconductors by describing in a rather detailed fashion the optical phenomenon – the Faraday rotation – which turned out to be the most convincing, if not the first, experimental indication that the spin-dependent coupling between the local moments of d-electrons of Manganese and delocalized states forming the top of the valence and the bottom of the conduction bands (1) exists, (2) is strong enough to be easily measurable, and (3) leads to interesting consequences.

1.2 Giant Faraday and Zeeman Effects

1.2.1 Giant Faraday Effect and its Origin

1.2.1.1 Experimental Faraday Rotation Measurements

From the very beginning of the studies of Diluted Magnetic Semiconductors, the Faraday effect, i.e., a rotation of polarization plane of light propagating in a medium

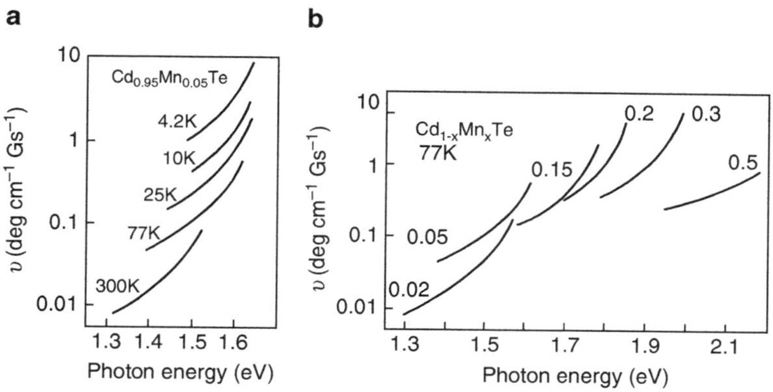

Fig. 1.1 Faraday rotation spectra of $Cd_{1-x}Mn_xTe$ alloys measured at indicated temperatures (**a**) on bulk samples with indicated Mn mole fraction x values (**b**). Reprinted with permission from [1]

along a magnetic field, attracted attention by its unusual magnitude and temperature dependence. The first Diluted Magnetic Semiconductor to reveal these striking properties was (Cd,Mn)Te [1, 2]. The Faraday rotation spectra measured in $Cd_{0.95}Mn_{0.05}Te$ at different temperatures are shown in Fig. 1.1a and those measured at 77 K in $Cd_{1-x}Mn_xTe$ of different Mn mole fraction x values are shown in Fig. 1.1b.

The effect shown in Fig. 1.1 was measured in bulk samples with thickness of order of 1 mm in the spectral region of transparency, i.e., below the energy gap. The two striking features distinguishing (Cd,Mn)Te from semiconductors without magnetic ions are well visible: a strong temperature dependence (more than an order of magnitude between 4.2 K and 300 K) and magnitude of the effect exceeding those measured for the CdTe host crystal by more than two orders of magnitude at sufficiently low temperature. The large rotation angles made the measurements very simple: it was enough to measure a transmission spectrum in magnetic field of a sample placed between crossed polarizers. An oscillatory spectrum was obtained (Fig. 1.2), with intensity minima corresponding to rotation by multiples of π. To determine the origin of the angular scale, it was enough to measure the transmitted light intensity versus magnetic field at a selected wavelength.

To analyze the physical mechanisms behind the observed features, we shall start by recalling the basic empirical description of the Faraday effect.

1.2.1.2 Empirical Description of the Faraday Effect

As mentioned above, by the Faraday effect we understand rotation of polarization plane of light propagating along a magnetic field (Faraday configuration). In the magnetic field, the eigenmodes of the light propagation correspond to right (σ^+) and left (σ^-) circular polarizations. A linearly polarized light can be decomposed into these two circular eigenmodes (Fig. 1.3), which acquire a phase difference after traveling a distance in the medium.

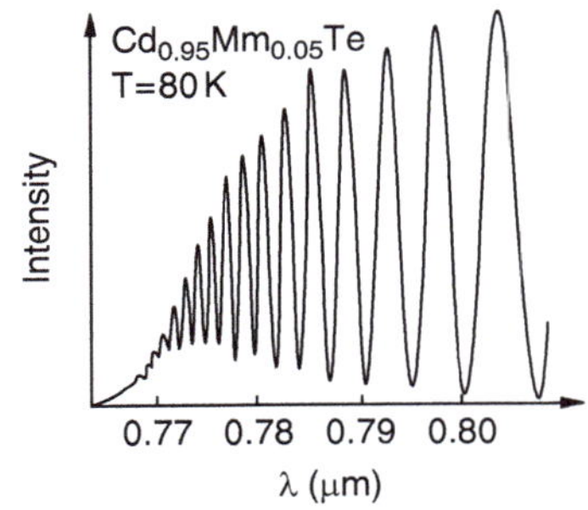

Fig. 1.2 Faraday rotation of $Cd_{1-x}Mn_xTe$ alloys can be determined by simple measurements of transmitted light intensity vs. photon energy in a given magnetic field

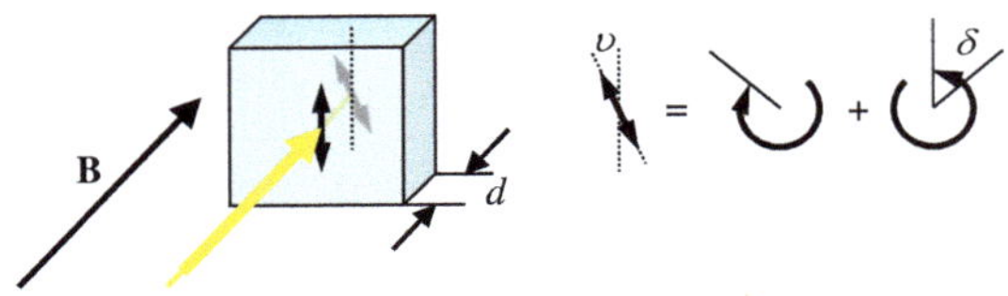

Fig. 1.3 Explanation of Faraday rotation by decomposition of linearly polarized light into circularly polarized eigenmodes

Recomposing the two phase-shifted components we obtain a rotation angle ϑ, which is twice smaller than their phase shift δ, which, in turn, is proportional to the difference between their refraction indices and to the path d traveled by the light in the sample

$$\vartheta(\omega) = \frac{\omega}{2c} d(n_+(\omega) - n_-(\omega)),$$

where ω denotes circular frequency and c speed of light.

Usually, we introduce **Verdet constant**, i.e., rotation angle per unit thickness of the sample

$$\vartheta(\omega)/d = \frac{\omega}{2c}(n_+(\omega) - n_-(\omega)). \tag{1.1}$$

1.2.1.3 Sources of the Faraday Rotation

Sources of refraction index dispersion are usually found in optical transitions corresponding to light absorption (extinction) at certain frequencies. According to (1.1), we can thus look for sources of the Faraday rotation in polarization-dependent absorption, originating from optical transitions which can be affected by the magnetic field. In the case of (Cd,Mn)Te, a strong dispersion of the the Faraday rotation when the photon energy approaches the energy gap suggests an interband origin of the observed effect. This hypothesis can be verified by studying magnetooptical Kerr effect (MOKE), i.e., polarization rotation of the reflected light in the magnetic field. This has been done in [1], where the energy position of a structure observed in MOKE was shown to coincide with that of the excitonic reflectivity structure (Fig. 1.4).

An experiment revealing profound origins of the Faraday rotation was reported in a pioneering paper by Komarov et al. [2]. They showed that the Faraday rotation can

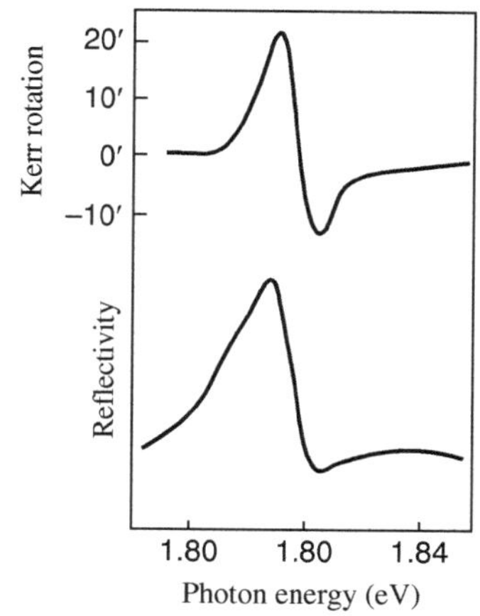

Fig. 1.4 Comparison between reflectivity and magnetooptical Kerr rotation spectra in the interband region of $Cd_{0.95}Mn_{0.05}Te$ shows common excitonic origin of the two effects. Reprinted with permission from [1]

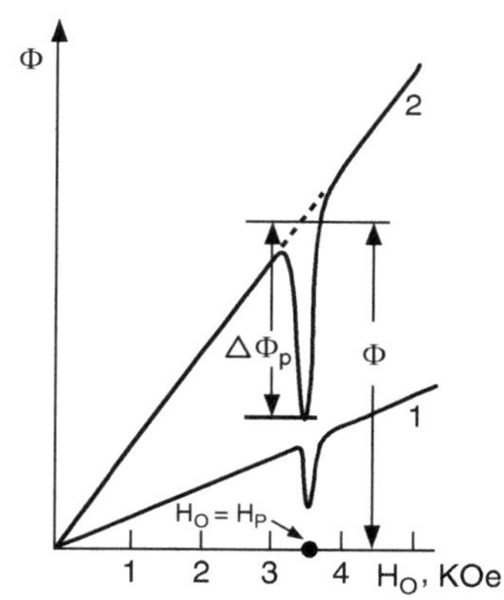

Fig. 1.5 Faraday rotation of CdTe:Mn under microwave illumination, measured at photon energies 1.5 eV (1) and 1.568 eV (2). A dip at 3.3 kOe corresponds to the spin resonance of Mn^{++} ions. The two curves correspond to two different wavelengths of light, whose polarization direction is monitored. Reprinted from [2] with kind permission of Springer Science and Business Media

be strongly attenuated by resonant heating of the Mn^{++} ion spins using microwaves. The Faraday rotation angle measured in CdTe:Mn sample under microwave illumination, plotted versus magnetic field, (Fig. 1.5) exhibits a strong dip at the magnetic field corresponding to the spin resonance of the Mn^{++} ions.

This result pinpoints the crucial role of the manganese ions in the physical mechanism of the giant Faraday rotation. The above qualitative results indicate the elements to be included in a model description of the giant Faraday rotation.

1.2.1.4 Dispersion Relations

Before undertaking the search for a quantitative model, it is useful to consider **dispersion relations** [3], which link the real and imaginary part of the complex refraction index (refraction index and extinction coefficient, respectively): $\tilde{n}(\omega) = n(\omega) + i\kappa(\omega)$

$$n(\omega) - 1 = \frac{1}{\pi}\mathcal{P}\int_{-\infty}^{\infty}\frac{\kappa(\omega')\mathrm{d}\omega'}{\omega' - \omega} \text{ and } \kappa(\omega) = -\frac{1}{\pi}\mathcal{P}\int_{-\infty}^{\infty}\frac{(n(\omega') - 1)\mathrm{d}\omega'}{\omega' - \omega}.$$

The causality principle imposes symmetry conditions which in the presence of a magnetic field take the form

$$n(-\omega, -B) = n(\omega, B) \text{ and } \kappa(-\omega, -B) = -\kappa(\omega, B)$$

At $B = 0$ n and κ are thus even and odd functions of ω, respectively.

This symmetry allows us to reduce the integrals in the dispersion relations to the physically meaningful range of positive frequencies, yielding the usual Kramers–Kronig form

$$n(\omega) - 1 = \frac{2}{\pi}\mathcal{P}\int_{0}^{\infty}\frac{\omega'\kappa(\omega')\mathrm{d}\omega'}{\omega'^2 - \omega^2} \text{ and } \kappa(\omega) = -\frac{2\omega}{\pi}\mathcal{P}\int_{0}^{\infty}\frac{(n(\omega') - 1)\mathrm{d}\omega'}{\omega'^2 - \omega^2},$$

$\mathcal{P}$ denoting the principal value of the integral.

If a nonzero magnetic field is present, the above simple scheme of reducing the integrals to the range of positive frequencies does not work any more, since passing from ω to $-\omega$ we have to change the sign of the field at the same time, resulting in n and κ being neither even nor odd functions of ω. However, for $\Delta\omega$ and $\Delta\kappa$ we can write $\Delta n(-\omega, B) = n_+(-\omega, B) - n_-(-\omega, B) = n_-(-\omega, -B) - n_+(-\omega, -B) = n_-(\omega, B) - n_+(\omega, B) = -\Delta n(\omega, B)$ and $\Delta\kappa(-\omega, B) = \Delta\kappa(\omega, B)$, where indices $+$ and $-$ represent two circular polarizations of the light.

Thus, Δn and $\Delta\kappa$ are odd and even functions of ω, respectively (in contrast to n and κ at $B = 0$, which are even and odd, respectively) and therefore the dispersion relations take the form

$$\Delta n(\omega) = \frac{2\omega}{\pi}\mathcal{P}\int_{0}^{\infty}\frac{\Delta\kappa(\omega')\mathrm{d}\omega'}{\omega'^2 - \omega^2} \text{ and } \Delta\kappa(\omega) = -\frac{2}{\pi}\mathcal{P}\int_{0}^{\infty}\frac{\omega'\Delta n(\omega')\mathrm{d}\omega'}{\omega'^2 - \omega^2}$$

The difference between this form of the dispersion relations and the standard Kramers–Kronig one is important when we are interested in a wide spectral range; in the vicinity of a narrow resonance it is negligible.

It must be stressed that the above expressions do not apply to naturally occuring optical activity of a medium with helical microscopic structure (e.g., sugar and its water solutions). Since no magnetic field is present, the usual form of Kramers–Kronig relations is valid also for the differences of real and imaginary part of the refractive index between the two circular polarizations. Another important difference between the Faraday rotation and natural optical activity concerns the sign of rotation when the propagation direction is reversed and the light passes across the sample a second time. In the case of the Faraday rotation, the polarization of the reflected light continues to turn in the same sense, producing a total rotation twice as big as in the case of a single passage, whereas for a helical medium the sense of the rotation is reversed when the light passes across the sample in the opposite direction. As a result, the total rotation vanishes (see Fig. 1.6) in the natural case.

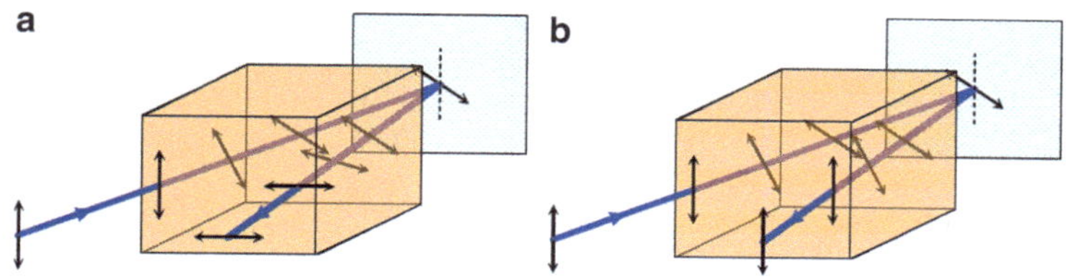

Fig. 1.6 Polarization rotation of light passing again the same sample after reflection is enhanced in case of the Faraday rotation (**a**) or compensated in case of a helical medium (**b**)

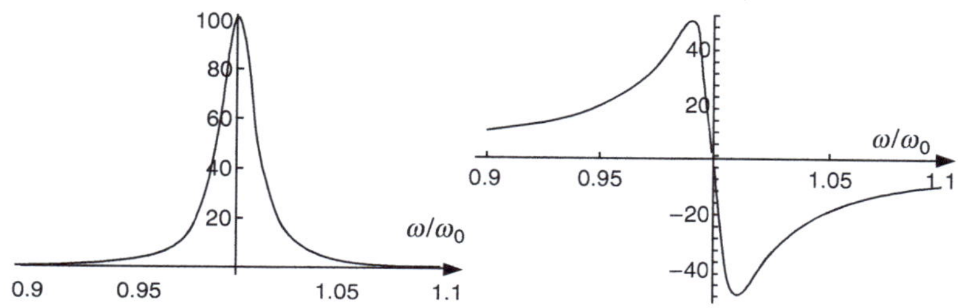

Fig. 1.7 Extinction coefficient (*left*) and refraction index (*right*) spectra for a Lorentzian line ($\omega_0 = 1$, $A = 1$, $\Gamma = 0.01$)

The difference between the two cases can be seen as a result of time reversal symmetry. Without magnetic field, the change of the light propagation direction to the opposite value corresponds to a time reversal and therefore it should bring the status of the light polarization back to the original value. In presence of a magnetic field, the time reversal symmetry would impose a sign change of the pseudovector of the magnetic field during the second passage of the light through the sample. If we keep the magnetic field unchanged, the second rotation enhances the first one instead of compensating it.

1.2.1.5 Three Types of the Faraday Rotation

As a simple example, let us consider a single Lorentzian optical transition line (Fig. 1.7). The extinction coefficient κ and refraction index n can be expressed, respectively, as imaginary and real parts of the complex refraction index[1] $\tilde{n}(\omega) = 1+\frac{A}{\omega_0-\omega-\mathrm{i}\Gamma}$, where ω_0 is the resonance frequency, 2Γ is full width at half maximum (FWHM) and A is an amplitude.

The Faraday rotation spectrum for such a line will differ depending on how the magnetic field influences the line parameters [4]. Usually, we think about the influence of the magnetic field on the resonance frequency (the line position), resulting in

[1] The chosen simplified Lorentzian form does not satisfy the symmetry conditions discussed above, therefore it can be used only close to the resonance energy.

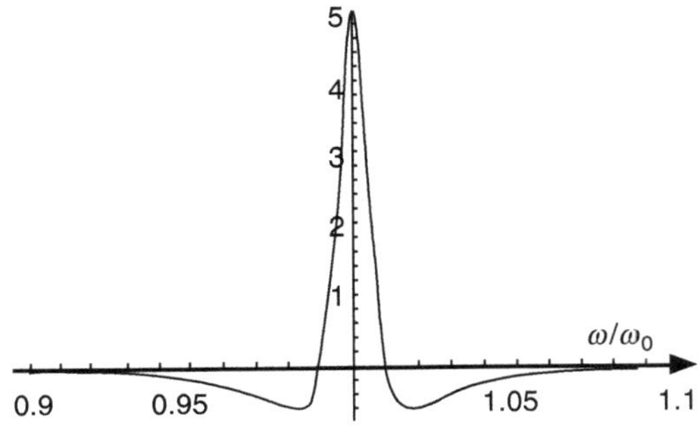

Fig. 1.8 Zeeman-type Faraday rotation spectrum for a Lorentzian line ($\omega_0 = 1$, $A = 1$, $\Gamma = 0.01$, $c = 1$, $\Delta\omega_0 = 0.001$)

the Faraday rotation spectrum originating from the Zeeman effect. In the low-field approximation, it reads

$$\vartheta(\omega)/d = \frac{\omega}{2c}\Delta n = \frac{\omega}{2c}\Delta\omega_0 \frac{\partial n(\omega)}{\partial\omega_0} = -\frac{\omega}{2c}\Delta\omega_0 A \frac{(\omega_0-\omega)^2 - \Gamma^2}{((\omega_0-\omega)^2+\Gamma^2)^2} \tag{1.2}$$

As shown in Fig. 1.8, the spectrum of the Faraday rotation is to a good approximation symmetrical related to the resonance frequency.

However, the magnetic field can also influence the line width and/or the amplitude of the line, resulting in two other different types of the Faraday rotation. The low-field analytical expressions read for the amplitude-type effect

$$\vartheta(\omega)/d = \frac{\omega}{2c}\Delta A \frac{\partial n(\omega)}{\partial A} = \frac{\omega}{2c}\Delta A \frac{\omega_0-\omega}{(\omega_0-\omega)^2+\Gamma^2} \tag{1.3}$$

and for the line width type

$$\vartheta(\omega)/d = \frac{\omega}{2c}\Delta\Gamma \frac{\partial n(\omega)}{\partial\Gamma} = -\frac{\omega}{c}\Delta\Gamma A\Gamma \frac{\omega_0-\omega}{((\omega_0-\omega)^2+\Gamma^2)^2} \tag{1.4}$$

As shown in Fig. 1.9, the amplitude- and linewidth-type Faraday rotation spectra have a spectral shape completely different from the Zeeman type spectrum, being approximately odd functions of detuning $\omega - \omega_0$.

1.2.1.6 Description of Examples of Experimental Faraday Rotation Spectra

The above examples are not as academic as they may seem to be: excitonic transitions either in bulk crystals or in quantum structures can be often described in such a simple approximation. In most cases, the simplest assumption of the Zeeman-type Faraday effect is sufficient but some interesting cases (e.g., charged excitons – see Chap. 9) may require a different description.

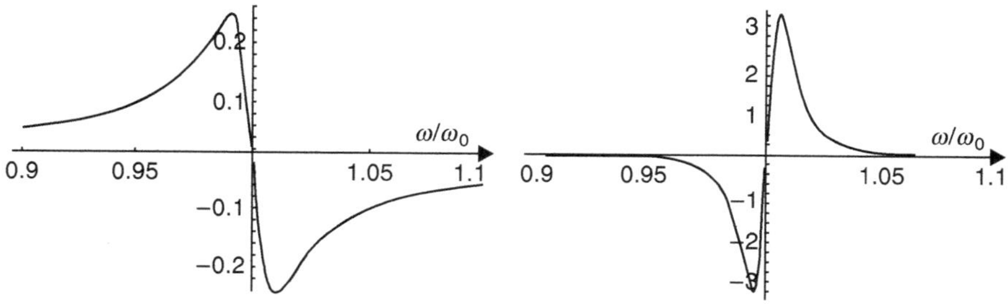

Fig. 1.9 Amplitude- and line width type Faraday rotation spectra for a Lorentzian line ($\omega_0 = 1$, $A = 1$, $\Gamma = 0.01$, $c = 1$, $\Delta A = 0.01$ and $\Delta\Gamma = 0.001$, respectively)

An important feature of the Faraday rotation is its long range in the spectrum: it may extend quite far from the energies where the actual optical transitions (and absorption due to them) occur. Therefore, e.g., in transmission experiments on bulk crystals of macroscopic thickness, where we do not access directly the excitonic region, the measured interband Faraday rotation will be influenced not only by the 1S excitonic transition but also by transitions from a relatively large spectral range above the band gap.

1.2.1.7 Model Description of Faraday Rotation in (Cd,Mn)Te

The interpretation of the first Faraday rotation spectra measured on (Cd,Mn)Te crystals was centered on a search for the simplest model of optical transitions responsible for the obtained spectra.

When we do not approach too closely the transition energies in the Faraday rotation measurements, we can neglect the broadening of the optical transitions. In such a situation, we can ask what is the type of singularity at the transition energy. For example, in the case of a single line, the extinction coefficient κ can be represented by a delta-function centered at the transition energy $\kappa(\omega) = A\delta(\omega - \omega_0)$, corresponding to a refraction index $n(\omega) - 1 = \frac{2}{\pi}\frac{A\omega_0}{\omega_0^2-\omega^2}$.

The Zeeman-like Faraday rotation in the linear approximation is then

$$\vartheta(\omega)/d = \frac{\omega}{2c}\Delta n = \frac{\omega}{2c}\Delta\omega_0\frac{\partial n(\omega)}{\partial\omega_0} = -\frac{\omega}{c}\frac{A}{\pi}\frac{\omega_0^2 + \omega^2}{(\omega_0^2 - \omega^2)^2}\Delta\omega_0. \tag{1.5}$$

The above expression possesses a second order singularity at $\omega \to \omega_0$. The single line can be associated with creation of the ground ($1s$) excitonic state or an atomic-like transition due to an impurity.

The Faraday rotation due to interband transitions (excitonic effects neglected) has been calculated, e.g., by Kołodziejczak et al. [5]. Their model assumes a square root shaped absorption edge, and as a consequence, a singularity of the type of $(\omega_0 - \omega)^{-1/2}$.

If we want to propose a more realistic description of the fundamental absorption edge, including excitonic effects, the first rough approximation may take the form of a step function $\kappa(\omega) = A\Theta(\omega_0 - \omega)$. For the Zeeman-like Faraday effect, we shall have in this case a δ-shaped circular dichroism $\Delta\kappa(\omega) = A\Delta\omega_0\delta(\omega_0 - \omega)$. Using the dispersion relations, we obtain then

$$\vartheta(\omega)/d = \frac{\omega}{2c}\Delta n = -\frac{\omega^2}{\pi c}\frac{A\Delta\omega_0}{\omega_0^2 - \omega^2}. \tag{1.6}$$

The above expression contains a first-order singularity at $\omega \to \omega_0$.

Gaj, Gałązka, and Nawrocki [1] compared the three model expressions with results of the Faraday rotation measurements in (Cd,Mn)Te. They found that the square root edge model is the most remote from the experimental spectra, which have an intermediate form between the single line model and step function model curves. It has been found that the singularity is located at the frequency of the excitonic structure occurring in reflectivity spectra. This fact led the authors to assign the observed Faraday rotation to interband transitions, which in medium- and wide-gap semiconductors are strongly modified by presence of excitons.

An improved model based on calculations of spectral moments was shown to describe the Faraday rotation of (Cd,Mn)Te with a fair accuracy [6] (Fig. 1.10).

In this model, the spectrum of the extinction coefficient is represented by a combination of a single line and a step function.

The best currently available description of the refractive index due to interband transitions is due to Tanguy [7, 8] for bulk crystals and low-dimensional structures. The model can certainly provide a precise description of the interband Faraday rotation.

Summarizing, we identify Zeeman split interband transitions as the source of the giant Faraday rotation observed in (Cd,Mn)Te. The size of the effect implies giant Zeeman splittings of these transitions. The next challenge is to confirm experimentally the existence of the giant Zeeman effect and to describe its relation to the spin state of Mn^{++} ions. This will be done in the following sections.

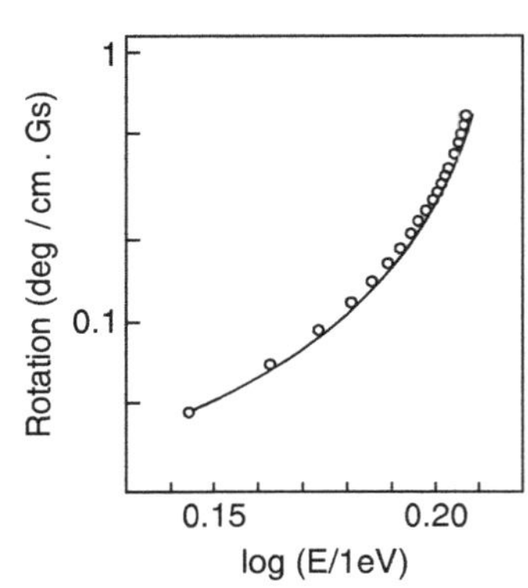

Fig. 1.10 Model description (*solid line*) of the Faraday rotation data of (Cd,Mn)Te (points). After [6]

1.2.2 Excitonic Zeeman Effect in (Cd,Mn)Te

1.2.2.1 Zeeman Effect Measurements

Reliable measurements of excitonic Zeeman splitting can be obtained from measurements performed directly in the spectral region of the excitonic absorption. While reflectivity measurements do not present any special experimental difficulty, absorption (which represents the most direct probe of optical transitions) requires thin samples (below 1 μm). Such samples have been first obtained from bulk DMS crystals by Twardowski [9], who was able to detect not only the excitonic ground $1s$ states but also a wealth of excited states in a magnetic field. Magnetoabsorption and magnetoreflectivity measurements on (Cd,Mn)Te at low temperatures [9, 10] revealed that the $1s$ excitonic ground state transition splits into six components: two of them visible in each circular polarization in the Faraday configuration (light propagating along magnetic field, Fig. 1.11a) and additional two visible in π polarization ($\boldsymbol{E} \parallel \boldsymbol{B}$) in Voigt configuration (light propagation perpendicular to the magnetic field, Fig. 1.11b).

Typical reflectivity spectra of (Cd,Mn)Te measured at 1 T and 1.8 K in various polarizations are shown in Fig. 1.12. Two strong and two weak components are visible in circular polarizations (Faraday configuration) and two additional ones in Voigt configuration. The observed six component splitting pattern can be expressed in terms of a band-splitting scheme shown in the same figure.

1.2.2.2 Energy Band Splitting Pattern

The scheme shown in Fig. 1.12 assumes symmetrical splitting of the conduction band in two components and of the valence band in four equidistant components. It imposes an interdependence between the energies of the observed components: each component observed in π polarization lies in energy in the middle between a strong and a weak components observed in the opposite circular polarizations. After experimental verification of this regularity, the Voigt configuration results were not used in further discussion, as their energies do not carry any new information. We

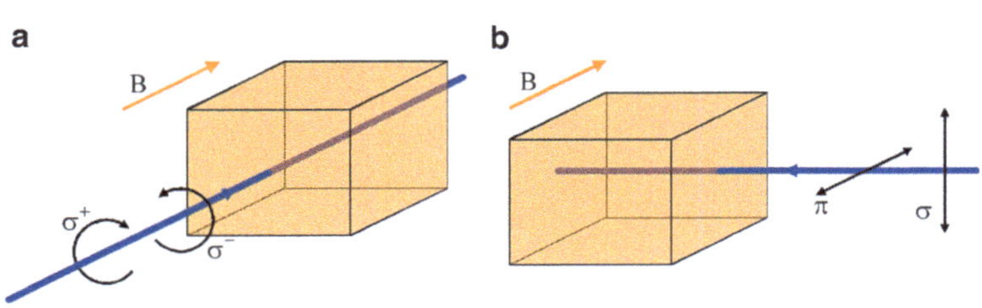

Fig. 1.11 Faraday configuration (**a**) and Voigt configuration (**b**) used in magnetooptical experiments. Linear (σ and π) and circular (σ^+ and σ^-) polarizations of respective eigenmodes indicated

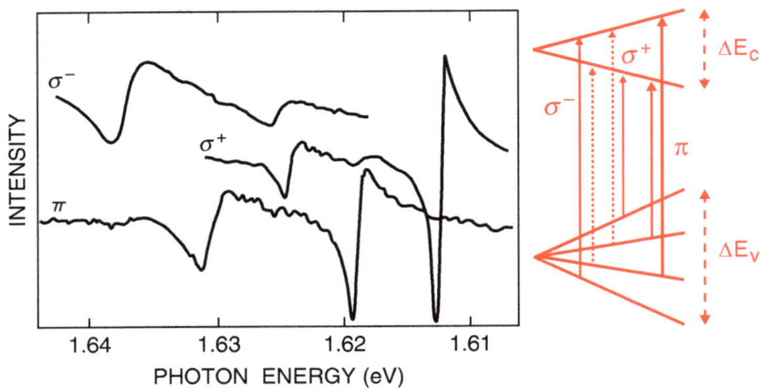

Fig. 1.12 Reflectivity spectra of (Cd,Mn)Te measured at 1.7 K in magnetic field 1.1 T in indicated polarizations. The scheme on the right depicts the giant Zeeman split conduction and valence band edges and optical transitions allowed in various polarizations. Relative transition intensities: *dotted/solid/thick solid lines* = 1/3/4. Reprinted with permission from [10]

will show in the following that this splitting scheme corresponds to J_z eigenstates of the conduction and valence electrons. This means that we can associate an exciton (keeping in mind the excitonic character of the optical transitions) with each pair of exchange-split components of energy bands. Such a result is obtained by neglecting electron–hole exchange interaction which would couple excitonic states associated with various pairs of the subbands. In fact, electron–hole exchange interaction energy in CdTe and in similar compounds is typically of the order of 1 meV, very small compared to giant Zeeman splittings, which at liquid He temperatures and in a field of a few Tesla are often of the order of 100 meV. Assuming that the excitonic effects are identical for all subbands in question, the relative transition intensities can be calculated for simple interband transitions, assuming conduction wavefunctions of the form [11]

$$|1/2\rangle = |S\rangle\uparrow, \quad |-1/2\rangle = |S\rangle\downarrow \tag{1.7}$$

and the valence band wavefunctions as

$$|3/2, 1/2\rangle = \sqrt{\tfrac{1}{6}}(|X + \mathrm{i}Y\rangle\downarrow -2|Z\rangle\uparrow),$$
$$|3/2, -1/2\rangle = \sqrt{\tfrac{1}{6}}(|X - \mathrm{i}Y\rangle\uparrow +2|Z\rangle\downarrow)$$

and

$$|3/2, 3/2\rangle = \sqrt{\tfrac{1}{2}}|X + \mathrm{i}Y\rangle\uparrow, \quad |3/2, -3/2\rangle = \sqrt{\tfrac{1}{2}}|X - \mathrm{i}Y\rangle\downarrow \tag{1.8}$$

for light- and heavy-hole bands, respectively. Six nonvanishing transition probabilities are then obtained as squares of absolute values of electric dipole matrix elements between the corresponding states. Denoting the electric dipole operator $\hat{\boldsymbol{D}} = (\hat{X}, \hat{Y}, \hat{Z})$, we use directly its components for linearly polarized light. In Voigt configuration (light propagation, say, along y direction), the only nonvanishing interband matrix elements for π polarization are

$$\langle 1/2|\hat{Z}|3/2, 1/2\rangle = -\sqrt{\tfrac{2}{3}}\langle S|\hat{Z}|Z\rangle$$

and

$$\langle -1/2|\hat{Z}|3/2, -1/2\rangle = \sqrt{\tfrac{2}{3}}\langle S|\hat{Z}|Z\rangle \tag{1.9}$$

whereas for σ polarization only

$$\langle 1/2|\hat{X}|3/2, 3/2\rangle = \sqrt{\tfrac{1}{2}}\langle S|\hat{X}|X\rangle = \langle -1/2|\hat{X}|3/2, -3/2\rangle$$

and

$$\langle 1/2|\hat{X}|3/2, -1/2\rangle = \sqrt{\tfrac{1}{6}}\langle S|\hat{X}|X\rangle = \langle -1/2|\hat{X}|3/2, 1/2\rangle \tag{1.10}$$

do not vanish.

In the Faraday configuration, when the light propagates along magnetic field (z direction), the electric dipole operator components corresponding respectively to σ^+ and σ^- circular polarizations

$$\hat{D}^+ = \sqrt{\tfrac{1}{2}}(\hat{X} + \mathrm{i}\hat{Y}) \text{ and } \hat{D}^- = \sqrt{\tfrac{1}{2}}(\hat{X} - \mathrm{i}\hat{Y}) \tag{1.11}$$

produce nonvanishing interband matrix elements

$$\langle 1/2|\hat{D}^+|3/2, -1/2\rangle = \sqrt{\tfrac{1}{3}}\langle S|\hat{X}|X\rangle = \langle -1/2|\hat{D}^-|3/2, 1/2\rangle$$

and

$$\langle 1/2|\hat{D}^-|3/2, 3/2\rangle = \langle S|\hat{X}|X\rangle = \langle -1/2|\hat{D}^+|3/2, -3/2\rangle \tag{1.12}$$

where we used equivalence of x and Y direction in a cubic crystal. Remembering that the Z direction is equivalent to each of them, we see that the relative intensities of all the interband optical transitions are given by the above results. We obtain thus a splitting pattern with four components in the Faraday configuration of relative intensities 6 (heavy holes) and 2 (light holes) in each circular polarization, whereas in the Voigt configuration we obtain two components of relative intensities 4 in p polarization and four components: two of relative intensity 3 (heavy holes) and two of relative intensity 1 (light holes). Experimental spectra confirm qualitatively the calculated selection rules (see Fig. 1.12). The splitting pattern described here was used in the previously described model description of the Faraday rotation spectra [6]: the assumed model shape of the absorption edge was assumed to split in four components (two weak and two strong ones) active in the Faraday configuration.

The intensity ratio 1:3 of weak to strong components was assumed according to the matrix element values calculated above.

1.2.2.3 Magnetization Measurements

A next step in the analysis of the magnetooptical results is to describe quantitatively the relation between the giant Zeeman effect and the spin state of the magnetic ions, proved qualitatively in [2]. It was done by comparing the giant Zeeman splitting with the magnetization of the crystal. As CdTe is known to be diamagnetic [12], the magnetization of a (Cd,Mn)Te crystal is at low temperature dominated by the alignment of the magnetic moments of the Mn^{++} ions. A series of magnetization measurements was performed on (Cd,Mn)Te samples with Mn mole fraction values x ranging from 0.005 to 0.3. Typical results are shown in Fig. 1.13.

At low field magnetization of all the samples increases linearly, showing at higher fields a tendency to saturation, most pronounced for the most dilute samples. For such samples, the magnetization can be fairly well described by a Brillouin function, characteristic for a system of noninteracting spins

$$B_S(\xi) = \frac{2S+1}{2S}\mathrm{cth}\left(\frac{2S+1}{2S}\xi\right) - \frac{1}{2S}\mathrm{cth}\left(\frac{1}{2S}\xi\right), \qquad (1.13)$$

where in our case $S = 5/2$ is Mn spin value. An empirical description valid for any Mn composition at pumped Helium temperatures up to about five Tesla can be obtained [13] by introducing two empirical parameters: effective spin saturation

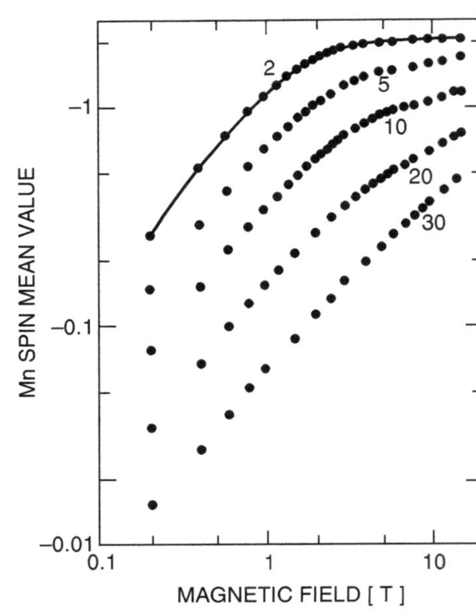

Fig. 1.13 Magnetization of (Cd,Mn)Te crystals of various Mn mole fraction x values (indicated in %). With increasing x a tendency to saturation at higher field weakens. Reprinted with permission from [13]

value S_0 and a temperature correction T_0. The magnetization, expressed as mean spin value per Mn ion, takes then the form

$$\langle S_z \rangle = S_0 B_S \left(\frac{g \mu_B S B}{k_B (T - T_0)} \right), \tag{1.14}$$

where $g = 2$ is Mn gyromagnetic factor, μ_B, k_B and T are Bohr magneton, Boltzmann constant, and temperature respectively. The parameters S_0 and T_0 depend on Mn mole fraction x of the alloy – this dependence was discussed in [14]. This magnetization behavior is related to ion–ion $d-d$ interaction, and will be discussed in Sect. 1.4.

1.2.2.4 Zeeman Splittings vs. Magnetization

The splitting of the strong components visible in reflectivity (such as the ones in Fig. 1.12) was plotted versus magnetization measured under approximately the same experimental conditions. The results are shown in Fig. 1.14.

Magnetization is presented in Fig. 1.14 as mean Mn spin value per unit cell

$$x \langle S_z \rangle = \frac{M}{N_0 g \mu_B}, \tag{1.15}$$

where N_0 is the number of unit cells per unit volume.

In this way, results obtained in samples with different x values can be directly compared. The points obtained at different magnetic fields for samples with seven

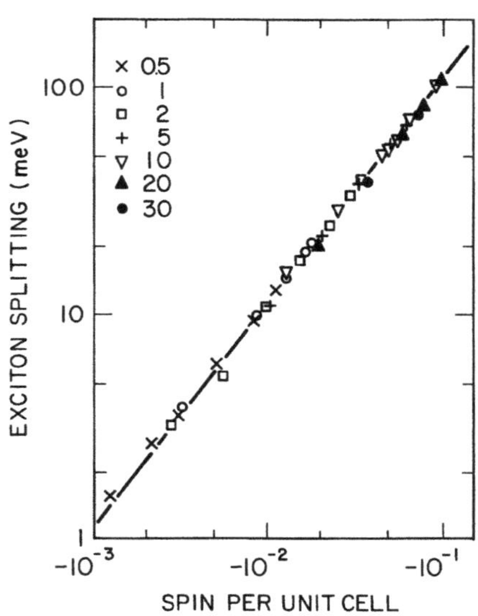

Fig. 1.14 Splitting of the "strong" (heavy hole) exciton component of (Cd,Mn)Te crystals of various Mn mole fraction values (indicated in %) vs. magnetization. Reprinted with permission from [13]

different x values form a unique linear dependence (represented by the straight line). This proportionality of the Zeeman splittings to the magnetization is a general property of a large part of DMS, with energy gap values around 1 or 2 eV. It will be interpreted in what follows in terms of ion-carrier $sp-d$ interaction described in the mean field approximation. Narrow gap DMS (e.g., (Hg,Mn)Te) and large gap ones (such as (Zn,Mn)O) do not obey this rule for different reasons. Their peculiarities will be discussed in Sect. 1.5 and Chap. 2, respectively.

1.2.3 Mean Field Approximation, Ion-carrier ($sp-d$) Exchange Integrals in (Cd,Mn)Te

We shall describe the influence of magnetic ions on the charge carriers in terms of $s-d$ (for the conduction band) or $p-d$ (for the valence band) exchange interaction. In fact its nature is purely electrostatic (see, e.g., Ashcroft & Mermin [15]), but it can be expressed by a scalar product of the interacting spins. In the form known as Heisenberg Hamiltonian [16]

$$\hat{H} = -\sum_i J(\boldsymbol{r} - \boldsymbol{R}_i)\boldsymbol{\sigma} \cdot \boldsymbol{S}_i, \tag{1.16}$$

where $\boldsymbol{\sigma}$ and $\boldsymbol{S}_i$ denote (vector) spin operators of the carrier and i-th magnetic ion, respectively, the magnetic ion being placed at the position characterized by the $\boldsymbol{R}_i$ lattice vector. J is an effective operator acting on the carrier's spatial coordinates. The physical origin of the $sp-d$ interaction in DMS is discussed in detail in Chap. 3.

While describing the influence of the carrier–ion interaction on the properties of a semimagnetic semiconductor with a moderate energy gap it is important to keep in mind the relative numbers: we have usually to do with a number of magnetic ions comparable to the total number of atoms in the crystal, while the number of carriers is smaller by orders of magnitude;[2] the material is often semi-insulating and the only carriers are those created by light. A delocalized, itinerant carrier interacts usually with a great number of magnetic ions. This disparity makes the influence of the carriers on the magnetic ions negligible, justifying the most commonly used approach: an external magnetic field aligns the magnetic ions, which in turn act on carriers via ion–carrier interaction. This action has been described successfully in the mean field approximation, where the spin operators of magnetic ions have been replaced by their thermal average. Furthermore, the mean field approximation is usually completed by the virtual crystal approximation, restoring crystal periodicity: the random distribution of magnetic ions and host cations over the cation sublattice is replaced by a periodical structure with artificial cations possessing properties of the host cation and the magnetic ion, averaged using occupation probabilities as

[2] Semiconductors or their structures containing large density of free carriers possess different properties, which are not discussed here.

weights. Within these approximations the carrier-ion Hamiltonian becomes

$$\hat{H} = -N_0 \alpha x \sigma_z \langle S_z \rangle \tag{1.17}$$

for the conduction electrons and

$$\hat{H} = -N_0 \beta x \sigma_z \langle S_z \rangle \tag{1.18}$$

for the valence band carriers, where N_0 denotes number of unit cells per unit volume, $\alpha = \langle S|J|S\rangle$ and $\beta = \langle X|J|X\rangle$ are exchange integrals for the conduction- and valence-electrons, respectively. This form of the Hamiltonian applied to the states at Γ-point of the Brillouin zone produces a splitting of the conduction band in two components of opposite spins separated by $\Delta E_c = |N_0 \alpha x \langle S_z \rangle|$ and a splitting of the valence band in four equidistant components with the external ones split by $\Delta E_v = |N_0 \beta x \langle S_z \rangle|$. Since in low x limit, the substances in question are paramagnetic, a non-vanishing magnetization can be obtained by applying the magnetic field. Acting on the conduction- and valence-band states, this field will modify the wavefunctions of the carriers, produce Landau quantization, usual spin splittings, etc. These effects are of primary importance in narrow gap materials, but in moderate- and wide-gap semiconductors often they may be neglected compared to the giant spin splitting due to ion–carrier interaction. In such an approximation, we can still apply the notion of the wavevector and consider unperturbed bands except for the exchange splittings (Fig. 1.15).

In the case of Mn^{++} ions possessing no orbital magnetic moment, the thermal average $g\mu_B N_0 \langle S_z \rangle$ represents the magnetization per magnetic ion and $g\mu_B N_0 x \langle S_z \rangle$ represents the magnetization per cation.

Considering nonvanishing wavevectors (kinetic energies non-negligible with respect to the exchange splittings) within the above approximations, Gaj, Ginter, and Gałązka [17] obtained a fairly complicated band structure as a result of combined carrier-ion exchange- and $k \cdot p$ interactions.

For the interpretation of the excitonic optical spectra, it is enough to limit oneself to the vicinity of the Γ-point. Using the approximations mentioned above (negligible e-h exchange interaction and constant exciton binding energy), we can express

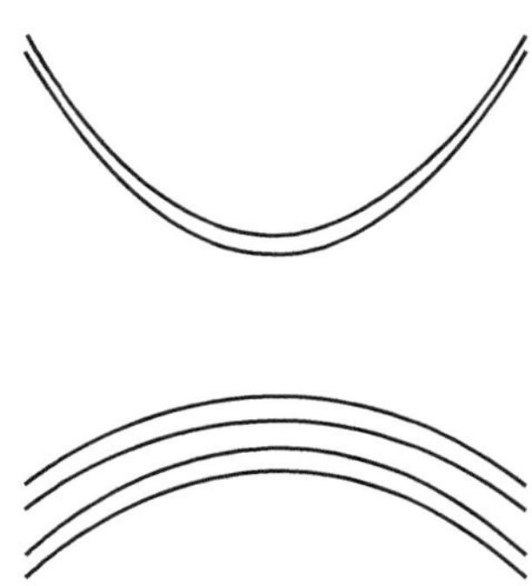

Fig. 1.15 Conduction- and valence bands of a typical moderate gap II–VI semimagnetic semiconductor in the vicinity of Γ point, split by exchange interaction (a schematical representation)

excitonic Zeeman splitting as differences of the corresponding band splittings. By plotting, thus obtained, excitonic Zeeman splittings versus the magnetization we obtain straight lines in excellent agreement with experimental data (see Fig. 1.14), with the slope determined by exchange integrals. Thus for the "strong" components observed in the Faraday configuration (heavy hole exciton), the Zeeman splitting in the mean field approximation (Fig. 1.12) is

$$\Delta E_e = E_- - E_+ = N_0(\beta - \alpha)x\langle S_z\rangle \tag{1.19}$$

where indices $-$ and $+$ denote circular polarizations (please note that $\langle S_z\rangle$ is negative). For the "weak" components (light hole exciton), the splitting is

$$\Delta E_e = E_- - E_+ = N_0(\beta/3 + \alpha)x\langle S_z\rangle. \tag{1.20}$$

Such plots have been shown to be common for various Mn mole fraction values. They supply reliable values of exchange integrals. In (CdMn)Te, commonly accepted values are $N_0\alpha = 0.22\,\text{eV}$ and $N_0\beta = -0.88\,\text{eV}$ for the conduction- and valence-band, respectively [13].

1.2.4 Giant Zeeman Effect in Narrow Gap Materials

In the beginning of 1970s, narrow gap semiconductors such as HgTe or HgSe and also ternary compounds with Cadmium, e.g., (Hg,Cd)Te were studied quite extensively because of their applications in infrared detectors. Therefore, narrow gap semiconductors (or even those with an inverted band gap) made of Mercury chalcogenides with Manganese were in fact among the first semimagnetic crystals studied in an extensive way. The interest in those compounds stemmed, at least partly, from the fact that Morrissy [18] in his Ph.D. work had observed anomalously high electron mobility in (Hg,Mn)Te. Narrow gap semiconductors represent a slightly more complicated case than (Cd,Mn)Te discussed above because proximity of the conduction and topmost valence bands precludes their separate consideration. In fact, the smallest set of states to be included in the calculation consists of four states: s-like conduction (light hole, in the case of inverted band structure), p-like valence band states (heavy and light holes degenerate at Γ point of the Brillouin zone, or heavy hole and conduction band states in the case of the inverted band structure) and p-like spin orbit split-off band. Away from the zone center, these states are considerably intermixed due to $k \cdot p$ interaction. Therefore, the description of the coupling of magnetic ions with, say, the conduction band will involve not only terms proportional to α but also due to β exchange integrals.

Since the basic method of investigation of these compounds involved either the interband magnetoabsorption [19, 20] or observation of Shubnikov de Haas oscillation of the magnetoconductivity [21, 22], therefore, all relevant calculations had to be done in the presence of an external magnetic field. It was realized fairly early that the modifications of the band structure of (Hg,Mn)Te upon doping with Mn were spin dependent. A very clear demonstration of this fact is shown in Fig. 1.16, which shows the spin and Landau splittings of nonmagnetic (Hg,Cd)Te and magnetic (Hg,Mn)Te having comparable energy gaps.

While the Landau splitting is essentially identical in both compounds (as it is ruled by the width of the gap, the momentum matrix element being very similar in most II–VI semiconductors), the spin splitting is very much different, being, moreover, temperature dependent in the case of Mn-containing samples. Because of narrow band gap, and therefore, small effective mass of conduction electrons ($\varepsilon_0 \sim 1/m^*$), the Landau quantization cannot be neglected. Technically, one has to diagonalize two 4×4 matrices each being a sum of "nonmagnetic" matrix due to $k \cdot p$ interaction and that due to $sp{-}d$ interaction: $h_L^a + h_S^a$ and $h_L^b + h_S^b$. The form of these matrices depends, of course, on the choice of the basis functions. Here, we have used the basis states given by $[|\frac{1}{2}\rangle, |\frac{3}{2}, \frac{3}{2}\rangle, |\frac{3}{2}, \frac{1}{2}\rangle, |\frac{1}{2}\rangle_{\rm so}]$ and $[|-\frac{1}{2}\rangle, |\frac{3}{2}, -\frac{3}{2}\rangle, |\frac{3}{2}, -\frac{1}{2}\rangle, |-\frac{1}{2}\rangle_{\rm so}]$ defined above in the case of the conduction and valence band states (heavy and light holes, (1.7)–(1.8)) together with

$$|-1/2\rangle_{\rm so} = -\sqrt{\tfrac{1}{3}}(|X - {\rm i}Y\rangle \uparrow -2|Z\rangle \downarrow),\ |1/2\rangle_{\rm so} = \sqrt{\tfrac{1}{3}}(|X + {\rm i}Y\rangle \downarrow +2|Z\rangle \uparrow)$$

to additionally describe the presence of the spin–orbit split-off valence band. This choice results in

$$H_L + H_S = \begin{pmatrix} h_L^a & 0 \\ 0 & h_L^b \end{pmatrix} + \begin{pmatrix} h_S^a & 0 \\ 0 & h_S^b \end{pmatrix},$$

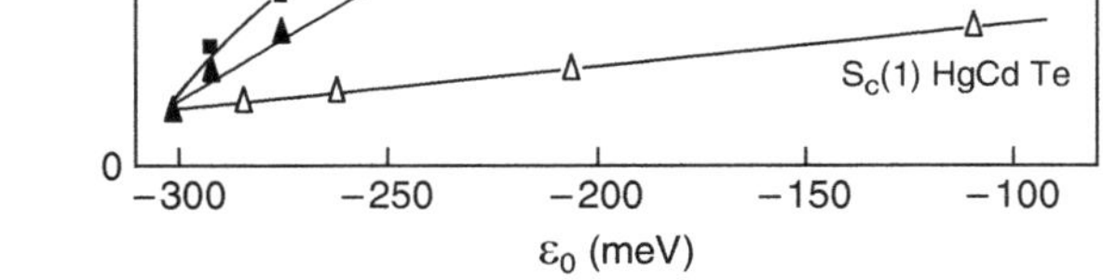

Fig. 1.16 The sum of cyclotron splitting of the first Landau levels in the conduction and heavy hole valence bands $\hbar\Omega$ and the sum spin splitting of the same subbands $s_c(1)$ measured directly in magnetoabsorption in (Hg,Mn)Te (*solid symbols*) and (Hg,Cd)Te (*empty symbols*) for various crystal compositions (and therefore various band gaps ε_0) at 2 T. Note that the cyclotron splitting follows the same trend in both materials, and is temperature independent between 4.4 K and 2 K, while the spin splitting is considerably larger in (Hg,Mn)Te, where it shows temperature dependence. Reprinted with permission from [19]

where $h_L^{a,b}$ represent four-dimensional Luttinger–Kohn (or Pidgeon and Brown) matrices [23,24] that describe the Landau levels within the general $k \cdot p$ approach in the presence of an external magnetic field in narrow gap semiconductors and whose terms depend on the energy gap, spin–orbit splitting energy, momentum matrix element and four parameters that describe interactions with bands more remote than the four (doubly degenerate) bands of Γ_6, Γ_8, and Γ_7 symmetries that are taken into account explicitly. The matrices $h_S^{a,b}$ (also 4×4 dimensional) represent additions due to $sp{-}d$ interaction and may be written explicitly as [25]

$$h_S^a = \begin{pmatrix} a & 0 & 0 & 0 \\ 0 & 3b & 0 & 0 \\ 0 & 0 & -b & -2\mathrm{i}\sqrt{2}b \\ 0 & 0 & \mathrm{i}2\sqrt{2}b & b \end{pmatrix} \quad \text{and} \quad h_S^b = \begin{pmatrix} -a & 0 & 0 & 0 \\ 0 & -3b & 0 & 0 \\ 0 & 0 & b & 2\mathrm{i}\sqrt{2}b \\ 0 & 0 & -\mathrm{i}2\sqrt{2}b & -b \end{pmatrix}$$

with $a = \frac{1}{2}N_0 x\alpha\langle S_z\rangle$ and $b = \frac{1}{6}N_0 x\beta\langle S_z\rangle$.

Let us notice that even in the simplest approximation (virtual crystal + mean field) the terms proportional to the magnetization appear not only on the diagonal of those matrices but also mix somewhat the light hole and spin–orbit split-off band. Fortunately, strong spin–orbit interaction in HgTe-based crystals and similar compounds makes this mixing a negligible effect. Small band gap combined with a strong spin–orbit interaction in mercury chalcogenides result in huge g-factors describing the spin splitting of the band states even in absence of magnetic ions. Nevertheless, inclusion of Mn in, say, HgTe leads to a truly huge g-factors that can sometimes exceed the orbital Landau splitting, which adds to the difficulty in proper identification of the optical transition lines. Numerous crossings of the various bands, occurring when the external magnetic field is varied, can be expected (and indeed are observed). One has to mention at this point that these are, in the present approximation, true crossings without any tendency of "quantum repulsion" or self-avoiding behavior. An illustration of the discussed behavior is shown in Fig. 1.17.

Variation of the spin splitting with the temperature leads to a spectacular phenomenon in the oscillatory magnetotransport experiment (Shubnikov de Haas oscillations). The oscillation, crudely speaking, arise because the Fermi level crosses the (oscillatory) density of states as the Landau levels are pushed up by the magnetic field in those samples. While the period of the oscillations of the magetoresistance is determined by the ratio of the Fermi energy and Landau splitting, the amplitude usually increases with the field. As a function of the temperature, the amplitude is usually quickly suppressed because of an increased role of various modes of scattering of the charge carriers. In fact, the rate of the amplitude decrease with the temperature can be used as a method of extraction of the value of the effective mass from such data. The method works, however, if both the mass and the spin splitting themselves are only weakly temperature dependent. Unfortunately, this is not the case of the g-factor now since it is a sensitive function of the temperature via the term containing the magnetization. Therefore, a temperature may exist, at which the spin splitting equal half of the cyclotron splitting. Under such conditions, one

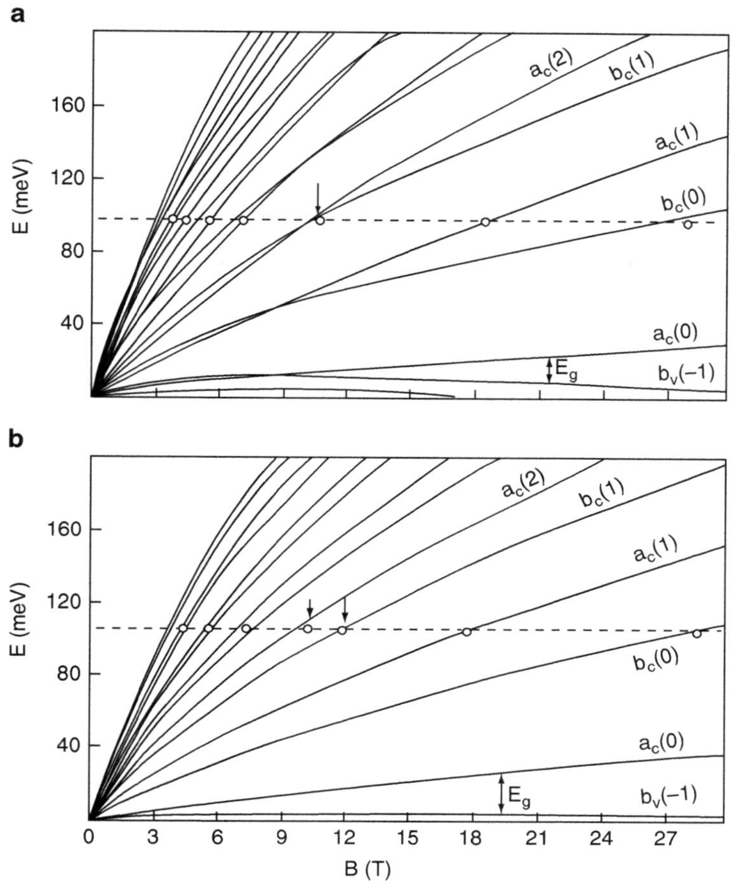

Fig. 1.17 Calculated spin split Landau levels in the conduction band of (Hg,Mn)Te (containing $x = 0.02$ Mn) and one, topmost Landau level belonging to the heavy hole set (denoted by $b_v(-1)$) at two temperatures: 4 K (upper panel) and 36 K (lower panel). Note that at 4 K the band gap opens only at elevated magnetic fields (above, approximately 8 T). Broken line shows the position of the Fermi level and the *open symbols* mark the observed maxima of the resistivity in the Shubnikov de Haas oscillation experiments. Note that at the higher temperature the peak indicated by the *arrow* splits in two components. The *lines* cross without showing an anticrossing behavior since the corresponding states do not mix. Reprinted with permission from [22]

expects the amplitude of the Shubnikov de Haas oscillations to vanish. In fact, theoretical expression for the amplitude of contains a factor $\cos(\pi g m^*/2m_0)$, which vanishes whenever the argument is equal to an odd multiple of $\pi/2$. With g being a strong function of the temperature this indeed may happen in an available temperatures range in diluted magnetic semiconductors. An example of such behavior is shown in Fig. 1.18. Sometimes several "spin-splitting zeros" of the amplitude are observed (c.f. [26]).

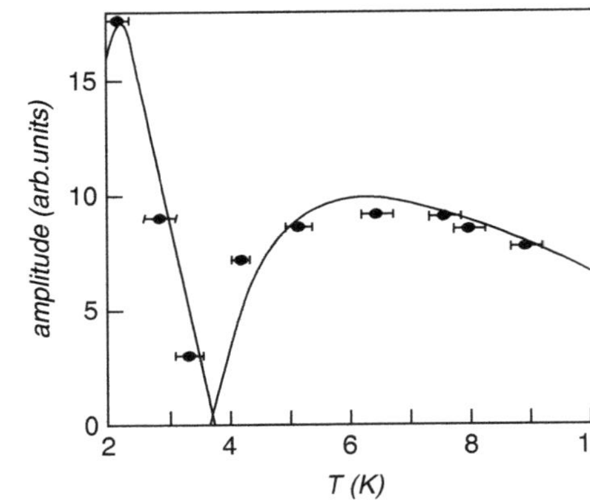

Fig. 1.18 The amplitude of the Shubnikov de Haas oscillations in (Hg,Mn)Te ($x = 0.02$) showing a clear spin-splitting zero. The *solid line* was calculated using the model described above and inserting the resulting values of the g-factor g and effective mass m^* to the expression $\cos(\pi g m^*/2m_0)$. Reprinted with permission from [21]

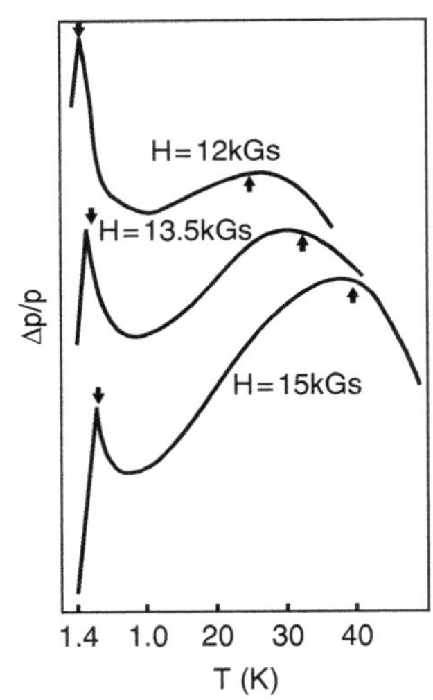

Fig. 1.19 Thermo-oscillation in magnetotransport of n-type doped (Hg,Mn)Te ($x = 0.009$) at several constant magnetic fields as a function of the sample temperature. The *arrows* show the calculated points of coincidence of the Fermi level and bottoms of the Landau levels in the conduction band. Reprinted with permission from [27]

In the case of an inverted band gap material (say, in (Hg,Mn)Te with less than about 7 molar percent of Mn, $x < 0.07$), the initial upward shift of the heavy hole subband can even lead to an effective closure of the gap making the compound a half-metal. Of course, when the magnetic field increases and the magnetization saturates, the "normal" huge g-factor overcomes the $sp{-}d$-induced modification and the band ordering returns to that observed in ordinary nonmagnetic narrow gap materials. Noticeably, the position of the band states becomes, via the magnetization, sensitive to the temperature at moderate fields. Thus, a phenomenon termed thermooscillations was predicted and recorded (see Fig. 1.19) in magnetotransport experiments [27].

Similar analysis can be done in the case of other narrow gap semimagnetic compounds such as PbTe:Mn [28,29] and Cd_3As_2:Mn [26] and related compounds.

Determination of the $sp-d$ exchange constants in narrow gap materials is much less reliable that that in wide band gap semiconductors. This is related to the fact that the analysis is done simultaneously for at least four bands together with determination of such parameters as momentum matrix element, the band gap (which is a strong function of x), and the higher band Luttinger parameters, the latter often not known with a very high precision even in starting nonmagnetic end-point materials, such as HgTe.

1.3 Values of $sp-d$ Exchange Integrals

1.3.1 Experimental Determination

As discussed above, the interband magnetoabsorption or magnetoreflectivity combined with magnetization measurements represent the basic experimental tool used for determination of $sp-d$ exchange integrals. However, other methods are also useful.

1.3.1.1 Photoluminescence in Magnetic Field

The optical transitions used in magnetoabsorption can be also exploited in photoluminescence (PL). However, several phenomena limit the applicability of the PL measurements, especially in bulk crystals. Photoluminescence often originates from recombination of localized excitons. In such cases, the approximation discussed in Sect. 1.4 of negligible influence of the carriers on the magnetic ions may no longer be valid, leading to creation of bound magnetic polarons (BMP), discussed in Chap. 8. Another limitation occurs in case of excitons bound to neutral centers, such as a neutral acceptor (A^0X state). These states contain a pair of identical carriers (holes for A^0X) in a singlet spin state, which cannot profit from the exchange with the magnetic ions within their range. As a result, destabilization of such bound excitons occurs at sufficient magnetization [30]. This is similar to the destabilization, which occurs in QWs for charged excitons (trions). The properties of doped DMS quantum wells are discussed in detail in Chap. 10. The effects mentioned above make the mean field expressions (1.19) and (1.20) inapplicable. However, if the BMP or 2D carrier gas effects are negligible (at magnetization values high enough) and at the same time, the magnetization is below destabilization value, (1.19) and/or (1.20) can be used. This possibility is particularly valuable in microphotoluminescence mapping, where the Zeeman shift measured in a moderate field range can be fitted with modified Brillouin function (1.14) and used to determine local magnetic ion concentration [31].

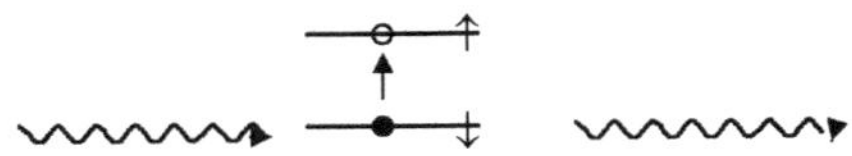

Fig. 1.20 Scheme of optical transitions in Spin Flip Raman Scattering

1.3.1.2 Spin Flip Raman Scattering

Excitonic magnetooptics provides information on combinations of exchange splitting of conduction- and valence-bands. These splittings can be measured separately in spin-flip Raman scattering (SFRS).

Figure 1.20 shows a scheme of a typical SFRS event. A photon gives a part of its energy to a carrier, inducing its transfer to another spin state. The spin flip energy, equal within the mean field approximation to the Zeeman splitting of the conduction band, can be determined from SFRS spectrum as Stokes shift, i.e. energy difference between the energies of the incident and scattered photons. In n-type DMS such as (Cd,Mn)Se, spin flip Raman scattering (SFRS) is useful for determination of $s-d$ exchange integral values, as demonstrated first by Nawrocki et al. [32]. Raman scattering selection rules favor Voigt configuration (light propagation perpendicular to the magnetic field). Typical SFRS experiments are performed in backscattering geometry with crossed linear polarizations of incident and scattered light. By tuning the incident light photon energy slightly below the energy gap, a resonant enhancement of the scattered light intensity can be obtained. SFRS spectra obtained in such experiments [32] are shown in Fig. 1.21.

The Stokes shift shown in inset exhibits to a first approximation a Brillouin-like behavior expected in the mean field approximation. However, a non-zero electron spin flip energy at zero field shows clearly that the mean field model is not fully applicable. This behavior originates from magnetic polaron effects, discussed in Chap. 8. The only case known so far of SFRS results describable within the mean field approximation is represented by (Cd,Fe)Se. These experiments were first performed by Heiman et al. [33]. Figure 1.22 shows clearly the proportionality of the Stokes shift to magnetization, in contrast with the results obtained on (Cd,Mn)Se.

Such experiments determine the conduction band exchange integral α from the coefficient of the proportionality. In case of materials where magnetic polaron effects are present, the slope should be measured at higher field values, where they are no longer significant.

1.3.1.3 Knight Shift

Another way of circumventing the difficulty in determination of $sp-d$ integrals created by high carrier concentration is related to the initial idea of Furdyna (see Introduction). Free carriers can influence electron paramagnetic resonance (EPR)

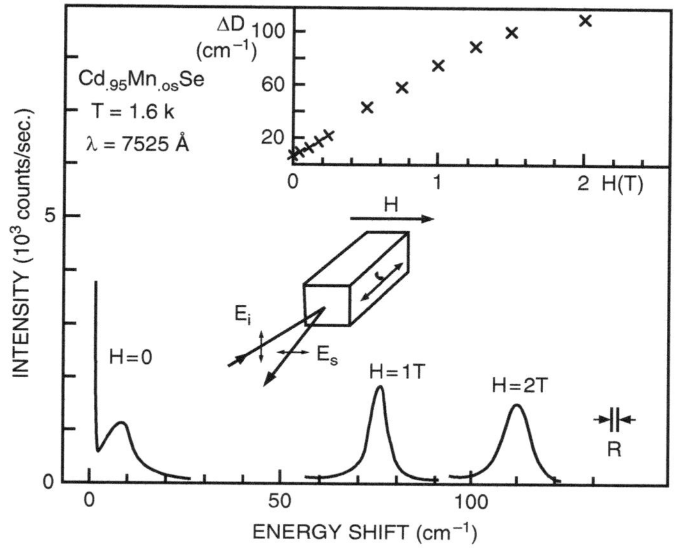

Fig. 1.21 Raman scattering spectra of CdMnSe is indicated configuration. Energy shift vs. magnetic field shown in inset. Reprinted with permission from [32]

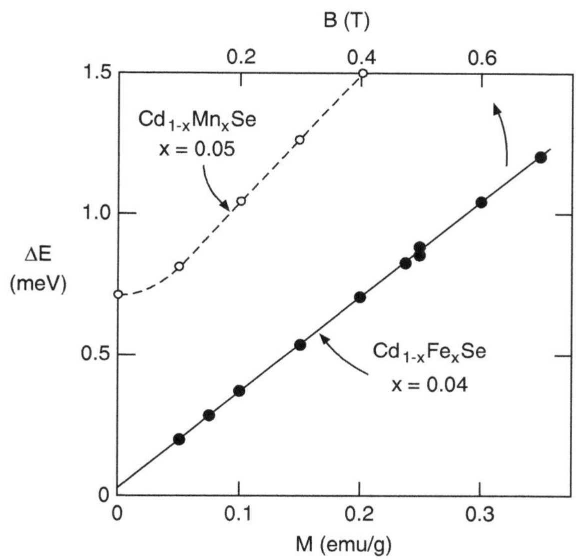

Fig. 1.22 Conduction electron spin flip energy for (Cd,Fe)Se represented vs. magnetization obeys well the mean field approximation, as opposed to (Cd,Mn)Se (bound magnetic polaron effects). Reprinted with permission from [33]

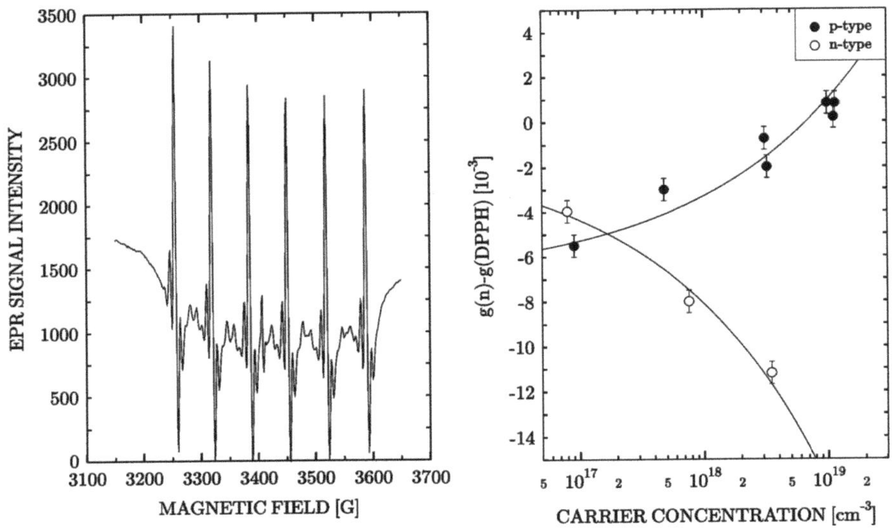

Fig. 1.23 Example of EPR spectra of (Pb,Mn)Te crystals (*left*) and the influence of concentration of electrons or holes on the measured g-factor value (*right*). Reprinted with permission from [34]

of magnetic ions. This modification can be expressed in the form of a variation of effective gyromagnetic factor with carrier concentration – see Fig. 1.23.

Story et al. [34] have shown that this modification, known under the name of Knight shift, can be used for determination of both sign and absolute value of carrier-ion exchange integrals in (Pb,Mn)Te and (Sn,Mn)Te. The modification of the effective g-factor can be explained by the presence of the effective magnetic (exchange) field B_{eff} of the carriers, acting on spins of magnetic ions. This field is proportional to the spin polarization and the density n of the carriers.

$$B_{\text{eff}} = \frac{J n \langle \sigma_z \rangle}{g \mu_{\text{B}}},$$

and adds to or subtracts from the external magnetic field, leading to an enhancement or decrease of the effective g factor, depending on the sign of carrier-ion coupling constant J.

1.3.1.4 Peculiarities of Large Gap Materials

Large gap semiconductors, such as those based on ZnO (3.4 eV) or GaN (3.5 eV), contain typically light anion elements. Therefore the spin–orbit splitting of the valence band is in such materials much smaller than, say, in CdTe. Furthermore, large effective masses lead to strong excitonic effects, in particular to large electron–hole exchange interaction. Both these effects modify the dependence of the excitonic energies on the magnetization, making it strongly nonlinear. In addition, small interatomic distances in these materials enhance the carrier–ion exchange interaction, which may lead to creation of electronic states bound to isoelectronic magnetic impurities. Because of all these effects, determination of $sp-d$ exchange integrals in large gap DMS is by far more complex than the simple procedures described here. The peculiarities of large gap DMS are discussed in detail in Chap. 2.

1.3.2 Numerical Values of sp−d Exchange Integrals, Chemical Trends

The procedure described in 1.2.3 has been applied to many other semimagnetic semiconductors resulting in determination of the $sp-d$ exchange constants. The obtained values are collected in Table 1.1. It is interesting to note positive values of valence band $p-d$ constant in Cr-containing alloys. The origin of this interesting result was discussed, e.g., by Blinowski and Kacman [50].

The mechanisms leading to those values are discussed in Chap. 3 by Merkulov and Rodina further in this volume. For more numerical values, the reader is refered to Landolt-Bornstein tables [51, 52]. Let us mark here only the trend that while α is relatively stable (with possible exception of narrow gap (Hg,Mn)Te and

Table 1.1 $sp-d$ exchange integrals for various diluted magnetic semiconductors

Compound	$N_0\alpha$ [eV]	$N_0\beta$ [eV]
(Cd,Mn)Te [13]	0.22	−0.88
(Cd,Mn)Se [35]	0.23	−1.27
(Cd,Mn)S [36]	0.22	−1.8[a]
(Zn,Mn)Te [37]	0.19	−1.09
(Zn,Mn)Se [38]	0.26	−1.31
(Be,Mn)Te [39]		−0.34/−0.4[c]
(Cd,Fe)Te [40]	0.30	−1.27
(Cd,Fe)Se [41]	0.26	−1.53
(Zn,Fe)Se [42]	0.22	−1.74
(Zn,Co)Te [43]	0.31	−3.03
(Cd,Co)Se [44]	0.28	−1.87
(Zn,Cr)Se [45]	0.2[b]	+0.95
(Zn,Cr)Te [45]	0.2[b]	+4.25
(Zn,Cr)S [45]	0.2[b]	+0.62
(Cd,Cr)S [46]	0.22	+0.48
(Hg,Mn)Te [47, 48]	0.4	−0.6[d]
(Hg,Mn)Se [49]	0.4	−0.7[d]

[a] High x value (see Sect. 1.4)
[b] Assumed typical value
[c] Light/heavy hole value
[d] We reversed the sign of the original determination to stay consistent with the convention adopted in this volume

(Hg,Mn)Se), β varies considerably: the smaller the ionic radii of the host compound the greater is the value of β. This trend (see Chap. 2) will break down in the case of semiconductor hosts with extremely large forbidden energy gap, i.e., those involving BeTe, ZnO, and GaN. Here, we quote only the value determined for (Be,Mn)Te.

1.3.3 Zeeman Effect for Excitons Above the Fundamental Energy Gap

Twardowski et al. [53] reported studies of excitons in (Cd,Mn)Te involving holes from the spin–orbit split-off valence band at energies about 2.5 eV ($E_0 + \Delta_0$ in the Fig. 1.24). The measured Zeeman splitting was described in terms of the same $sp-d$ interaction constants as for the transitions at the fundamental energy gap. These measurements allowed the authors to determine also a precise value of the spin–orbit splitting Δ_0 of the CdTe valence band, equal 0.948 eV.

Zeeman effect was also studied for transitions at the L-point of the Brillouin zone. These studies were initiated by Dudziak et al. [55] in 1982. The first quantitative results have been obtained on (Cd,Mn)Te by Ginter et al. [56]. Coquillat et al. [54] have performed measurements for (Zn,Mn)Te and (Hg,Mn)Te. In all cases, the measured splittings were much smaller than the width of the reflectivity structures. Therefore only some kind of averaged splitting values was possible to obtain by polarization modulation techniques (Fig. 1.25).

The splitting values of the E_1 transition were found to be proportional to those measured at Γ point (E_0 transition), and therefore to the magnetization (see Fig. 1.26), in agreement with the mean field approximation.

The splittings in question were found to be considerably smaller than those measured at the center of the Brillouin zone. An attempt to compute Zeeman splittings at the L point in terms of a tight binding model [56] suggested two reasons for the observed reduction: reduction of splittings of the energy bands and relaxation of selection rules, both due to mixing between the conduction- and valence-band wavefunctions. However, they calculated a reduction smaller than that measured

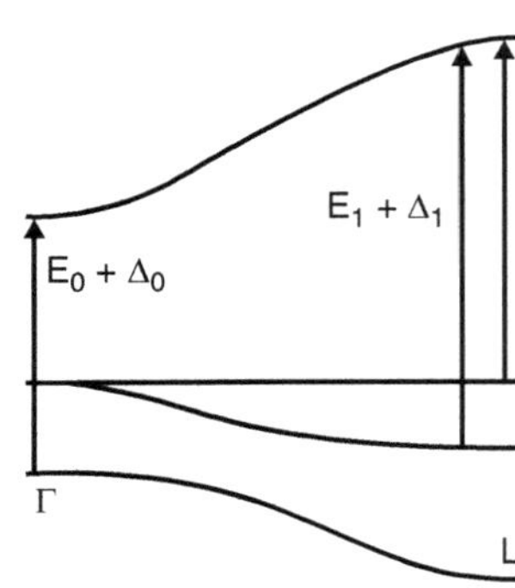

Fig. 1.24 Scheme of optical transitions in the Brillouin zone of (Cd,Mn)Te

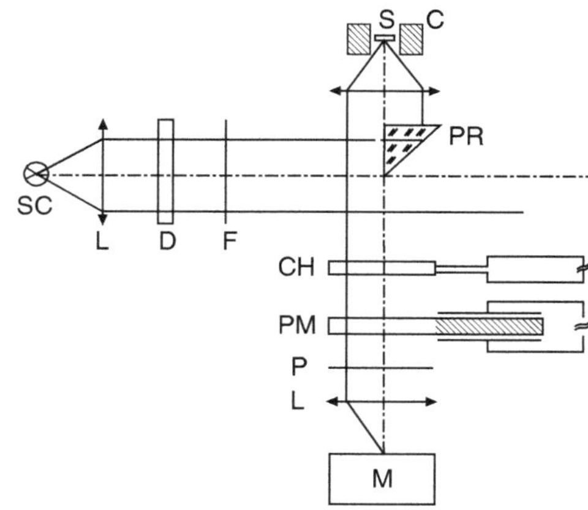

Fig. 1.25 Typical setup for modulation magnetoreflectivity measurements, where light reflected from the sample (S) placed in a superconducting coil (C) passes through a photoelastic modulator (PM) and is analyzed in a monochromator (M). Reprinted from [54]

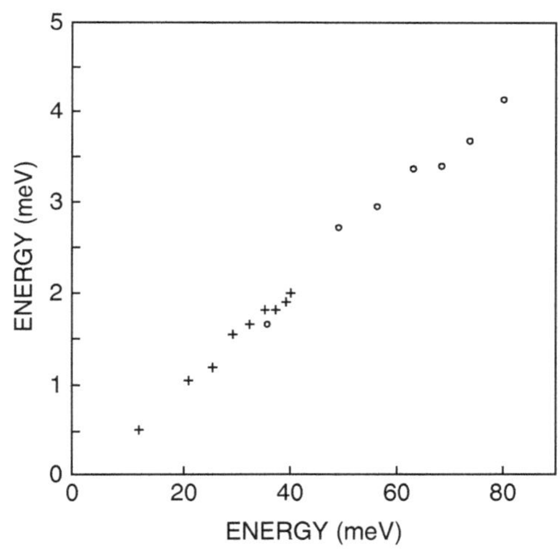

Fig. 1.26 Splitting of the E_1 reflectivity structure in $Zn_{1-x}Mn_xTe$ at 4.5 K vs. that of the two strong components of the E_0 structure, after Coquillat et al. [54]; $x = 0.02$ (*crosses*) and 0.17 (*circles*)

experimentally. To explain the difference, a k-dependence of the $p-d$ exchange integral was postulated. Besides, the calculations predicted for E_1 and $E_1 + \Delta_1$ transitions (involving two valence band components split by spin–orbit interaction) splittings identical in absolute value and opposite in sign (see Fig. 1.27). This result has been confirmed by experimental data [54] and later by a more realistic calculation performed by Bhattacharjee [57].

Dependence of $sp-d$ interaction on wave vector in quantum wells has been also reported by Mackh et al. [58]. It was deduced from Zeeman effect results in short period (Cd,Mn)Te/(Cd,Mg,Mn)Te superlattices.

The confinement-induced reduction of giant Zeeman splittings was analyzed theoretically by Bhattacharjee [59], who explained qualitatively the results of Mackh et al. and predicted much stronger effects in quantum dots.

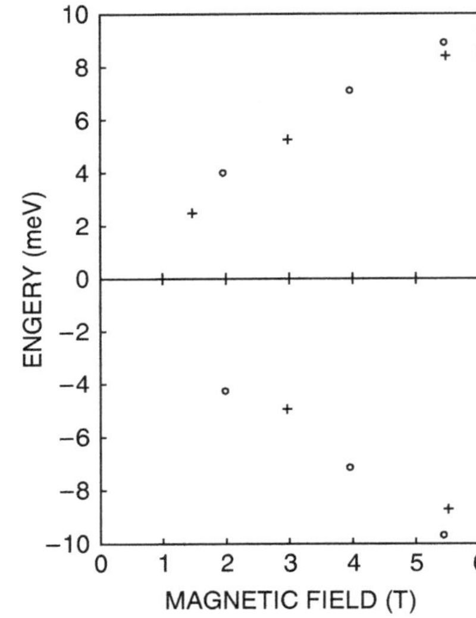

Fig. 1.27 Positive and negative values of Zeeman splitting measured in $Hg_{1-x}Mn_xTe$ for E_1 and $E_1 + \Delta_1$ transitions, respectively. Mn mole fraction $x = 0.05$ (*crosses*) and 0.2 (*circles*). After Coquillat et al. [54]

1.4 Beyond the Mean Field Approximation

That the mean field approximation is of limited applicability became evident fairly early in the course of investigation of diluted magnetic semiconductors. Namely, the attempts to determine the $sp-d$ exchange constants in CdS doped with minute amounts of Mn showed that these constants were (a) anomalously large and (b) strongly depended on the Mn molar fraction present in the crystal [60]. It was shown by Dietl [61] that this is a consequence of an increased probability of the presence of the s or p-like carriers in the vicinity of a magnetic ion.

Another example of a DMS object that could not be understood in terms of mean field approach was magnetic polaron [32,62]. Going further it is now very likely that quasi zero-dimensional quantum dots containing many Mn ions represent a polaron-like entity and thus are also not describable in terms of a simple mean field approach. In fact, Wojnar [63] used a very similar model to successfully describe photoluminescence spectra from Mn-containing CdTe quantum dots (muffin tin model) that previously was applied to the case of acceptor bound magnetic polarons [64, 65]. The most striking consequences of the usually neglected terms of the interaction Hamiltonian is that the energy of formation of the polaron is strongly dependent on (thermal) fluctuation of the magnetization in the region penetrated by the mobile carrier wave function.

Finally, anomalies in temperature dependence of the energy gap itself (observed in (Cd,Mn)Te [66] and ZnMnSe [67]) were interpreted due to higher order perturbation (and, thus, due to $\langle S_x \rangle$ and $\langle S_y \rangle$) by the $sp-d$ interaction. Again, the contribution may be expressed in terms of the magnetic susceptibility.

1.5 Ion–ion ($d-d$) Exchange Interaction

The magnetic moments localized on substitutitional transition metal ions in diluted magnetic semiconductors may intereact not only with delocalized valence and conduction band carriers that we have dealt with so far. They, of course, may also interact between each other either indirectly or directly. All the collective magnetic phenomena are, of course, caused by interactions between those microscopic magnetic moments. The most important of these interactions is the exchange interaction. It is also purely electrostatic by nature (see, e.g., Ashcroft and Mermin), and can have several underlying mechanisms. In the simplest case, it can be described by a spin Hamiltonian containing scalar products of the interacting spins. In this simple isotropic case in 3D space the corresponding Hamiltonian is called Heisenberg Hamiltonian and for two interacting spins $\boldsymbol{S}_1$ and $\boldsymbol{S}_2$ can be written as follows

$$\hat{H} = -J\boldsymbol{S}_1 \cdot \boldsymbol{S}_2,$$

where J denotes here the interaction strength; its positive sign corresponds to a tendency to parallel alignment (i.e., ferromagnetic) of the two spins. The exchange interaction can be either direct (then J is always positive) or mediated by some states residing between the two interacting magnetic moments: bond electrons, free carriers etc. Then, the sign of J can be such that an antiparallel (antiferromagnetic) alignment of neighboring spins may be more favorable. Usually, the exchange interaction is relatively short-ranged and it often suffices to limit oneself to spins occupying nearest neighboring sites neglecting interaction between those that are further apart. The case of antiferromagnetic exchange interactions may bring in a qualitatively new situation when not all spins can be aligned in a way that is energetically most favorable. The phenomenon is called frustration and the simplest example is that of three spins occupying a triangular lattice. While each two can be aligned antiferromagnetically, the third will always be ferromagnetic (i.e., configured energetically not favorably) to one of the first two. Formation of the spin glass is likely under such conditions. Indeed, a behavior characteristic for spin glasses is often observed when studying susceptibility or specific heat of diluted magnetic semiconductors [68]. The spin glasses in themselves are very interesting objects. For example, they exhibit intensive behavior of the dynamics of the localized spins [69]. For the sake of this discussion it will suffice, however, to envisage the collection of the localized spins as just giving rise to a static magnetization. This is, within the mean field/virtual crystal approximation, all that is needed to calculate the near the band gap modification of the band structure induced by the presence of the transition metal ions in our materials.

Another issue that is related to the mechanism of the interaction is the question how quickly it disappears with growing distance between the two interacting ions. The issue was studied for a variety of diluted magnetic semiconductors in [70], where it was found that the observed trends sometimes deviate from the predicted dependence $\exp(-r^2/a)$ found by Larson et al. [71] in the theory assuming superexchange as the main contributor to J. The deviations may indicate that other mechanism apart from the superexchange (e.g., Bloembergen–Rowland mechanism) are also present. In any case, the interaction strength drops down rather quickly and it is not a bad approximation to limit ourselves to interaction between nearest neighbors only. Experimentally, it is also observable that the smaller is the distance between cations occupying nearest neighbor sites in the crystalline lattice the greater is the value of J. So the interaction constant increases in the family of, say, Zn chalcogenides with Mn as one goes from tellurides down to sulfides. The superexchange theory provides a natural explanation of this trend: hybridization between p-states of the mediating anions is greater when the distance between atoms in the lattice is smaller. For the same reason, also the compressive strain which may arise in certain superlattices involving diluted magnetic semiconductors leads to a small increase of the strength of the exchange interaction.

As mentioned, very often to interpret various phenomena in DMSs, one needs only the values of magnetization that result from a parameterization of experimental values of the magnetization. One of the convenient parametrizations was already mentioned above in this chapter (see Sect. 1.2.2). While strictly speaking, the parametrization has no strict derivation it is intuitively understandable. T_0 is directly related to J (and can be approximately expressed with the help of high temperature series expansion [72]), while x_0 (smaller that actual x) defined as $x_0 = S_0 x/S$ reflects the fact that only a fraction of localized spins contributes to the magnetization; a part of localized spins is frozen by strong and predominantly antiferromagnetic interactions. In the crudest approximation (valid only in a limited range of small Mn molar fractions), $x_0 \approx x(1-x(1-x)^{12})$ provided that the crystal structure is that of fcc zinc blende type. An experimentally established values of x_0 and T_0 are collected in [73] for $Cd_{1-x}Mn_xTe$.

To determine the real values of the inter ion exchange interaction J, there are two methods that are especially accurate and therefore worth mentioning. The first makes use of magnetization steps observed in, e.g., optical measurements in strong magnetic fields, strong enough to disrupt the antiferromagnetic coupling between a pair of nearest neighboring Mn ions. The field when magnetic contribution to the energy becomes equal to that of the coupling is reflected by a relatively sudden increase of the magnetization and is observable as a step of, say, observed exciton transition line. The mechanism of formation of up to five steps is schematically depicted in Fig. 1.28.

Another accurate method of determination of J makes use of inelastic scattering of neutrons by pairs of interacting magnetic ions [75]. Of course, fitting of various models to the measured susceptibility and specific heat do provide also information concerning the value of J, however, this is (1) model dependent and (2) usually less reliable. For a tabulation of variously determined values of J the reader is referred to the existing compilations [51, 52]. Let it suffice to say here that J in bulk CdMnTe for nearest neighboring Mn ions amounts to about 6–10 K while for next nearest neighboring ions it is smaller, 1–2 K. Similarly, in ZnMnTe which has a smaller lattice parameter, those two exchange constants are, respectively 10 K and 0.6 K, as determined by the magnetization step method. Note that there is often a difference

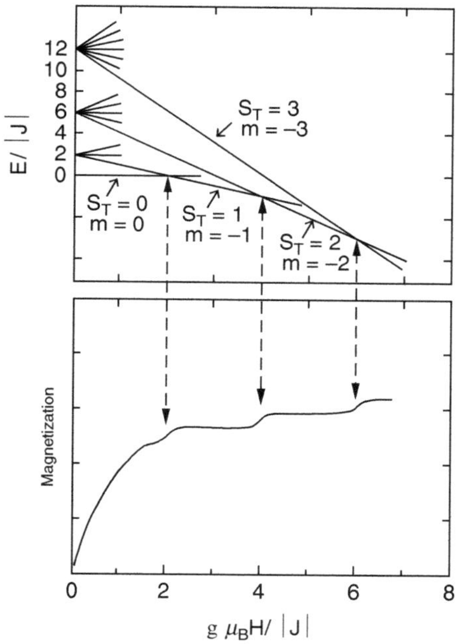

Fig. 1.28 *Upper panel*: energy spectrum of the triplet state of antiferromagnetically interacting (via the Heisenberg interaction) pair of localized magnetic moments as a function of an external magnetic field. The spin singlet with $m = 0$ does not contribute to the magnetization (*lower panel*) and at low field the latter is given only by those moments which do not have a nearest neighbor. With the field increasing the higher states with nonvanishing m become the ground state of the pair and their contribution to the magnetization manifests itself as a sudden-step like feature in the magnetization. Taken from [74]

by a factor of 2 in the definition of J, as is the case of the former determination, and therefore – differences in quoted values.

From the theoretical side the question of the inter-ion exchange interaction was addressed by Larson et al. [71]. The conclusion of that work was that in typical DMS materials the dominating channel of exchange coupling is that of superexchange, and proportional to hybridization of the valence band electrons with d-states localized on transition metal substituting the host cations. Thus, one expects that J is proportional to β^2. Indeed trends of this type are observed experimentally [76].

In a very concentrated case MnTe-rich materials seldom exists in zinc blende form, but tend to have nickel arsenide structure. However, by use of molecular beam epitaxy, one can grow cubic zinc blende MnTe as well. In such a form, it is also antiferromagnetic and the arrangement of localized spins in such a case is discussed in detail in Chap. 12 of this volume.

We have so far assumed that the localized spin–spin interaction is of Heisenberg form. That the Dzialoshinsky–Moriya interaction (involving a cross product of spins), although much weaker in strength, is present can be inferred from EPR (electron paramagnetic resonance) experiments reported by Samarth and Furdyna [77]. In most of the cases, this more complicated interaction can be safely neglected.

Finally, let us mention that when the band carriers are particularly abundant in the material an indirect coupling between two localized spins is possible as shown in lead salt-containing DMSs by Story et al. [78]. In fact, this specific interaction is of central interest in "newer" III–V-based diluted magnetic semiconductors and is a source of hopes of applications of these latter materials in semiconductor spintronics.

References

1. J.A. Gaj, R.R. Gałązka, M. Nawrocki, Solid State Commun. **25**, 195 (1978)
2. A.V. Komarov, S.M. Ryabchenko, O.V. Terletskii, I.I. Zheru, R.D. Ivanchuk, Zh. Eksp. & Teor. Fiz. **73**, 608 (1977)
3. for an introduction to dispersion relations see, e.g., F.W. Byron, R.W. Fuller, *Mathematics of Classical and Quantum Physics* (Dover, NY, USA, 1992)
4. W. Maślana, W. Mac, J.A. Gaj, P. Kossacki, A. Golnik, J. Cibert, S. Tatarenko, T. Wojtowicz, G. Karczewski, J. Kossut, Phys. Rev. B **63**, 165318 (2001)
5. J. Kołodziejczak, B. Lax, Y. Nishina, Phys. Rev. **128**, 2655, (1962)
6. J.A. Gaj, in *Diluted magnetic semiconductors* (Semiconductors and semimetals, vol 25), eds. by J.K. Furdyna, J. Kossut (Academic Press, San Diego, CA, USA, 1988), p. 296
7. C. Tanguy, Phys. Rev. Lett. **75**, 4091, (1995)
8. C. Tanguy, Sol. State Commun. **98**, 65 (1996)
9. A. Twardowski, M. Nawrocki, J. Ginter, Phys. Stat. Sol. **B96**, 497 (1979)
10. J. Gaj, P. Byszewski, M.Z. Cieplak, G. Fishman, R.R. Gałązka, J. Ginter, M. Nawrocki, N. The Khoi, R. Planel, R. Raymond, A. Twardowski, Phys. Conf. Ser. **23**, 113 (1979)
11. G.L. Bir, L.G. Pikus, *Symmetry and Strain Induced Effects in Semiconductors* (Willey, New York, 1974)
12. U. Sondermann, J. Magn. Magn. Matter **2**, 216 (1976)
13. J.A. Gaj, R. Planel, G. Fishman, Sol. State Commun. **29**, 435 (1979)
14. J.A. Gaj, W. Grieshaber, C. Bodin-Deshayes, J. Cibert, G. Feuillet, Y. Merle d'Aubigné, A. Wasiela, Phys Rev. **B50**, 5512 (1994)
15. N.W. Ashcroft, N.D. Mermin, *Solid State Physics* (Brooks Cole, Boston, 1976)
16. J. Kossut, Phys. Stat. Sol. (b) **72**, 359 (1975)
17. J.A. Gaj, J. Ginter, R.R. Gałązka, Phys. Stat. Sol. (b) **89**, 655 (1978)
18. C. Morrissy, Ph.D. Thesis, Oxford 1973, unpublished
19. G. Bastard, C. Rigaux, Y. Guldner, J. Mycielski, A. Mycielski, J. de Phys. **39**, 87 (1978)
20. M. Dobrowolska, W. Dobrowolski, M. Otto, T. Dietl, R.R. Gałązka, J. Phys. Soc. Jpn. **49**, 815 (1980)
21. M. Jaczyński, J. Kossut, R.R. Gałązka, Phys. Stat. Sol. (b) **88**, 73 (1978)
22. P. Byszewski, K. Szlenk, J. Kossut, R.R. Gałązka, Phys. Stat. Sol. (b) **95**, 359 (1979)
23. J.M. Luttinger, W. Kohn, Phys. Rev. **97**, 869 (1955)
24. C.R. Pidgeon, R.N. Brown, Phys. Rev. **146**, 575 (1966); see also W. Leung, L. Liu, Phys. Rev. B **8**, 3811 (1973)
25. R.R. Galazka, J. Kossut, in *Lecture Notes in Physics*, vol 133, ed. by W. Zawadzki. Narrow Gap Semiconductors Physics and Applications (Springer, Berlin, 1980), p. 245
26. J.J. Neve, J. Kossut, C.M. van Es, F.A.P. Blom, J. Phys. C **15**, 4795 (1982)
27. M. Dobrowolska, W. Dobrowolski, R.R. Gałązka, J. Kossut, Solid State Commun. **30**, 25 (1979)
28. J. Niewodniczańska-Zawadzka, T. Piotrowski, J. Kossut, J. Cryst. Growth **72**, 398 (1985)

29. J. Niewodniczańska-Zawadzka, G. Selinger, L. Palmetshofer, E.J. Fantner, G. Bauer, A. Lopez-Otero, W. Zawadzki, Physica B+C **117–118**, 458 (1983)
30. R. Planel, J. Gaj, C. Benoit à la Guillaume, J. Phys. C **41**, C5–C39 (1980)
31. W. Maślana, P. Kossacki, P. Płochocka, A. Golnik, J.A. Gaj, D. Ferrand, M. Bertolini, S. Tatarenko, J. Cibert, Appl. Phys. Lett. **89**, 052104 (2006)
32. M. Nawrocki, R. Planel, G. Fishman, R. Gałązka, Phys. Rev. Lett. **46**, 735 (1981)
33. D. Heiman, A. Petrou, S.H. Bloom, Y. Shapira, E.D. Isaacs, W. Giriat, Phys. Rev. Lett. **60**, 1876 (1988)
34. T. Story, C.H.W. Swüste, P.J.T. Eggenkamp, H.J.M. Swagten, W.J.M. de Jonge, Phys. Rev. Lett. **77**, 2802 (1996)
35. M. Arciszewska, M. Nawrocki, J. Phys. Chem. Sol. **47**, 309 (1986)
36. M. Nawrocki et al. Math. Res. Soc. Symp. Proc. **89**, 65 (1987)
37. A. Twardowski, P. Świderski, M. von Ortenberg, R. Pauthenet, Solid State Commun. **50**, 509 (1984)
38. A. Twardowski, M. von Ortenberg, M. Demianiuk, R. Pauthenet, Solid State Commun. **51**, 849 (1984)
39. D.R. Yakovlev, C. Dem, W. Ossau, L. Hansen, A. Waag, G. Landwehr, L.W. Molenkamp, Proc. 26th ICPS Edinbourgh 2002, CD edn., code D97, eds. by A.R. Long, J.H. Donids, A. Nakamura, N. Hori, N. Kato, K. Akita, R. Akomoto, ibid., code D99
40. C. Testelin, C. Rigaux, A. Mycielski, M. Menant, M. Guillot, Solid State Commun. **78**, 659 (1991)
41. D. Scalbert, M. Guillot, A. Mauger, J.A. Gaj, J. Cernogora, C. Benoit à la Guillaume, A. Mycielski, Solid State Commun. **76**, 977 (1990)
42. A. Twardowski, K. Pakuła, I. Perez, P. Wise, J.E. Crow, Phys. Rev. B **42**, 7567 (1990)
43. M. Zieliński, C. Rigaux, A. Lemaître, A. Mycielski, J. Deportes, Phys. Rev. B **53**, 674 (1996)
44. M. Nawrocki, F. Hamdani, J.P. Lascaray, Z. Gołacki, J. Deportes, Solid State Commun. **77**, 111 (1991)
45. W. Mac, A. Twardowski, M. Demianiuk, Phys. Rev. B **54**, 5528 (1996)
46. M. Herbich, W. Mac, A. Twardowski, K. Ando, Y. Shapira, M. Demianiuk, Phys. Rev. B **58**, 1912 (1998)
47. M. Dobrowolska, W. Dobrowolski, R.R. Gałązka, A. Mycielski, Phys. Stat. Sol. (b) **105**, 477 (1981)
48. G. Bauer, J. Kossut, R. Faymonville, R. Dornhaus, Phys. Rev. B **31**, 2040 (1985)
49. M. Dobrowolska, W. Dobrowolski, J. Phys. C **14**, 5689 (1981)
50. J. Blinowski, P. Kacman, Phys. Rev. **B46**, 12298 (1992)
51. R.R. Gałązka, J. Kossut, in *Landolt-Boernstein*, ed. by O. Madelung, vol 17b (Springer, Berlin, 1982)
52. R.R. Gałązka, J. Kossut, T. Story, in *Landolt-Boernstein*, vol 41 (Springer, Berlin, 1999), p. 650
53. A. Twardowski, E. Rokita, J.A. Gaj, Solid State Commun. **36**, 927 (1980)
54. D. Coquillat, J.P. Lascaray, J.A. Gaj, J. Deportes, J.K. Furdyna, Phys. Rev. **B39**, 10088 (1989)
55. E. Dudziak, J. Brzeziński, L. Jędral, in *Proceedings of the 11th Conference on Physic of Semiconducting Compounds*, Warsaw, 1982, ed. by J.M. Langer, p. 166
56. J. Ginter, J.A. Gaj, Le Si Dang, Solid State Commun. **48**, 849 (1983)
57. A.K. Bhattacharjee, in *Proceedings of the XX ICPS*, World Scientific, New York, 1990, p. 763
58. G. Mackh, W. Ossau, A. Waag, G. Landwehr, Phys. Rev. B **54**, R5227 (1996)
59. A. Bhattacharjee, Phys. Rev. B **58**, 15660 (1998)
60. S.I. Gubarev, M.G. Tyazhlov, Sov. Phys. – Solid State **36**, 373 (1989)
61. C. Benoit à la Guillaume, D. Scalbert, T. Dietl, Phys. Rev. B **46**, 9853 (1992)
62. P.A. Wolff, in *Semiconductors and Semimetals*, eds. by J.K. Furdyna, J. Kossut, vol 25 (Academic Press, Boston, 1988), p. 413
63. P. Wojnar, J. Suffczyński, K. Kowalik, A. Golnik, G. Karczewski, J. Kossut, Phys. Rev. B **75**, 155301 (2007)
64. A. Golnik, J. Ginter, J.A. Gaj, J. Phys. C **16**, 6073 (1983)
65. R. Beaulac, L. Schneider, P.I. Archer, G. Bacher, D.R. Gamelin, Science **325**, 973 (2009)

66. J.A. Gaj, A. Golnik, Acta Phys. Polon. A **71**, 197 (1987)
67. R.B. Bylsma, W.M. Becker, J. Kossut, Phys. Rev. B **33**, 8207 (1986)
68. R.R. Gałązka, S. Nagata, P.H. Keesom, Phys. Rev. B **22**, 3344 (1980)
69. Y. Zhu, C. Rigaux, A. Mycielski, M. Menant, W. Bontemps, Phys. Rev. B **40**, 8111 (1989)
70. J.M. Swagten, C.E.P. Gerrits, A. Twardowski, W.J.M. do Jonge, Phys. Rev. B **41**, 7330 (1990)
71. B.E. Larson, K.C. Hass, H. Ehrenreich, A.E. Carlsson, Phys. Rev. B **37**, 4137 (1988)
72. J. Spalek, A. Lewicki, Z. Tarnawski, J.K. Furdyna, R.R. Gałązka, Z. Obuszko, Phys. Rev. B **33**, 3407 (1986)
73. W. Grieshaber, A. Haury, J. Cibert, Y. Merle d'Aubigné, J.A. Gaj, Phys. Rev. B **53**, 4891 (1996)
74. J. Kossut, W. Dobrowolski, in *Handbook of Magnetic Materials*, ed. by K.H.J. Buschow, vol 7 (North Holland, Amsterdam, 1993), p. 231
75. T. Giebułtowicz, J.J. Rhyne, W.Y. Ching, D.L. Huber, J.K. Furdyna, B. Lebech, R.R. Gałązka, Phys. Rev. B **39**, 6857 (1989)
76. T.M. Giebułtowicz, J.J. Rhyne, J.K. Furdyna, P. Kłosowski, J. Appl. Phys. **67**, 5096 (1990)
77. N. Samarth, J.K. Furdyna, Solid State Commun. **65**, 801 (1988)
78. T. Story, R.R. Gałązka, R.B. Frankeland, P.A. Wolff, Phys. Rev. Lett. **56**, 777 (1986)

Chapter 2
Optical Spectroscopy of Wide-Gap Diluted Magnetic Semiconductors

Wojciech Pacuski

Abstract We focus here on striking differences between the physics of diluted magnetic semiconductors described in the preceding Chapter and those whose constituent semiconductor matrix is a wide gap material, as exemplified primarily by ZnO and GaN. We demonstrate the importance of two effects, usually negligible in moderate and narrow gap semiconductors: electron–hole exchange in excitons and small spin–orbit splitting of the valence band, both producing deviations from proportionality of giant Zeeman splitting to magnetization. We also show that the $sp-d$ exchange constants, expected to be large, do not result in very large exciton spin splitting. Therefore, a notion of apparent exchange constants is introduced and discussed.

2.1 Introduction

We decided to devote a separate Chapter to wide-gap diluted magnetic semiconductors such as DMSs based on ZnO and GaN, because many properties of these relatively new materials differ from those in other II–VI or III–V counterparts described in the previous Chapter.

2.1.1 Specific Properties of Wide Gap Diluted Magnetic Semiconductors

First, large band gap energy not only shifts the excitonic optical transitions to UV region, but it also, as it turns out, affects the $p-d$ exchange, thus changing interaction-induced Zeeman splitting or the charge state of substitutional magnetic

---

W. Pacuski
Institute of Experimental Physics, University of Warsaw, Hoża 69 00-681 Warsaw, Poland
e-mail: Wojciech.Pacuski@fuw.edu.pl

J. Kossut and J.A. Gaj (eds.), *Introduction to the Physics of Diluted Magnetic Semiconductors*, Springer Series in Materials Science 144,
DOI 10.1007/978-3-642-15856-8_2,

ions. It is possible to introduce transition metal ions into GaN as electrically neutral centers, although GaN is a III–V compounds. This makes (Ga,Mn)N similar to II–VI compounds, rather than to typical III–V DMSs such as (Ga,Mn)As. For example, a direct observation of the excitonic giant Zeeman splitting is possible in (Ga,Mn)N, but not in (Ga,Mn)As, where the high hole concentration practically excludes exciton spectroscopy. Next, small values of the lattice parameters of the wide-gap DMSs are expected to lead to strongly increased $p-d$ hybridization, and consequently, to strengthen the $p-d$ exchange interaction [1, 2]. This expected increase was one of the reasons why room temperature ferromagnetism in these materials was predicted [3]. The same increase was also shown in the case of (Cd,Mn)S to lead to new phenomena in the valence band, e.g., giant Zeeman splitting increasing nonlinearly with Mn content [4–6]. (Cd,Mn)S has a smaller lattice parameter compared to (Cd,Mn)Te (the archetypical II–VI DMS), but wide-gap DMSs have lattice parameter still much smaller then this material. In consequence, it is not evident if the model presented in the previous Chapter, developed for description of the giant Zeeman splitting of the valence band in typical II–VI DMS, can be directly applied to the valence band of wide-gap DMS. Moreover, when the spin–orbit coupling is weak, as it is the case of wide gap ZnO and GaN, strong electron–hole interaction leads to significant influence of excitonic effects, normally negligible in DMSs, such as (Cd,Mn)Te. Finally, wide-gap DMSs have wurtzite structure and exhibit important magnetic and optical anisotropy due to trigonal crystal field. All these reasons show that magnetooptical study of wide-gap DMS can make new contributions to DMS physics. On the other hand, making ZnO and GaN magnetic is interesting for spectroscopy of the wide gap semiconductors. Direct influence of magnetic fields on ZnO and GaN is only weak, but introducing magnetic ions increases excitonic g-factors by one or even two orders of magnitude. It means that the splittings observed so far only at very high magnetic field can be easily studied using standard magnetooptical methods. Thus, an opportunity arises for understanding fundamental properties of excitons in host ZnO itself, such as symmetry and excitonic mixing.

2.1.2 Quest for Room Temperature Ferromagnetism

Although interest in transition metals such as Co, Ni or Fe embedded in ZnO matrix has been exhibited since the sixties of the previous century [7–11], the real career of wide-gap DMS, such as DMS based on ZnO and GaN, started when theoretical predictions of room temperature ferromagnetism appeared [1, 12–14]. These predictions have been supported by first reports of room temperature ferromagnetism in (Ga,Mn)N [15, 16], (Zn,Mn)O [17], and (Zn,Co)O [17–23]. However, the origin of the observed ferromagnetic behavior remains up to now controversial. In many cases, the observed ferromagnetism can be explained by presence of small precipitates [24] of other phases or different chemical species. In this Chapter, we do not focus on the search for room temperature ferromagnetism in wide-gap DMS, but rather on spectroscopy of very homogeneous layers with high optical quality, and doped with low concentration of magnetic ions, so exhibiting only paramagnetic behavior.

2.2 Magnetooptical Spectroscopy of Excitons in Wide Gap DMS

The interaction between magnetic ions and excitons in diluted magnetic semiconductors leads to magnetooptical phenomena, such as giant Faraday rotation and giant Zeeman splitting. In the previous Chapter, these effects have been presented using, as examples, (Cd,Mn)Te and similar alloys. In wide-gap DMSs, such as ZnO and GaN with transition metals, analogical effects can be observed, but the magnitude of the observed effects is much weaker. Despite the small splitting, reaching only a few meV, we will still use the term giant Zeeman splitting, when we discuss splitting induced by $s, p{-}d$ exchanged interaction. First, let us describe the experimental findings.

2.2.1 *Reflectivity*

Reflectivity is a very efficient technique of studying free excitons in wide-gap DMSs. Using the reflectivity measurements, one can resolve three, typical for the wurtzite structure, excitons called A, B, and C. They are usually difficult to be resolved using other experimental techniques, such as transmission or photoluminescence. Figure 2.1 shows impact of concentration of magnetic ions (x) on

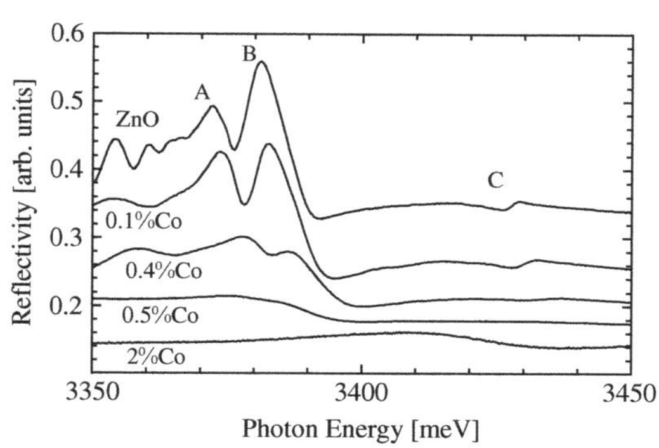

Fig. 2.1 Reflectivity spectra measured at $T = 1.6$ K with incidence angle of 45 degrees. Topmost spectrum is for ZnO, other spectra for $Zn_{1-x}Co_xO$ with increasing Co concentration 0.1%, 0.4%, 0.5%, 2%. Labels A, B, and C identify the three excitons, which are visible in ZnO and at low Co content; Fabry-Perot oscillations are also observed at low photon energy ($E < 3,370$ meV). After [25].

the reflectivity spectra of (Zn,Co)O. Excitonic lines shift to high energy when x increases, which indicates an increase of the energy gap due to doping with magnetic ions. Linewidth of A and B excitons increases with x due to chemical disorder introduced by magnetic ions. As a consequence, in sample with $x = 2\%$, A and B excitons cannot be resolved any more.

Broadening of the exciton lines limits observation of the excitonic giant Zeeman splitting to dilute samples, which partially explains why observation of truly huge splittings is usually not possible for wide-gap DMSs. But even for diluted wide-gap DMSs, as shown in Fig. 2.2c, the effect of the magnetic field on optical spectra is much stronger than in the case of undoped ZnO shown in Fig. 2.2a. The splitting induced by $s, p{-}d$ exchange interaction (Fig. 2.2d) is much larger than Zeeman

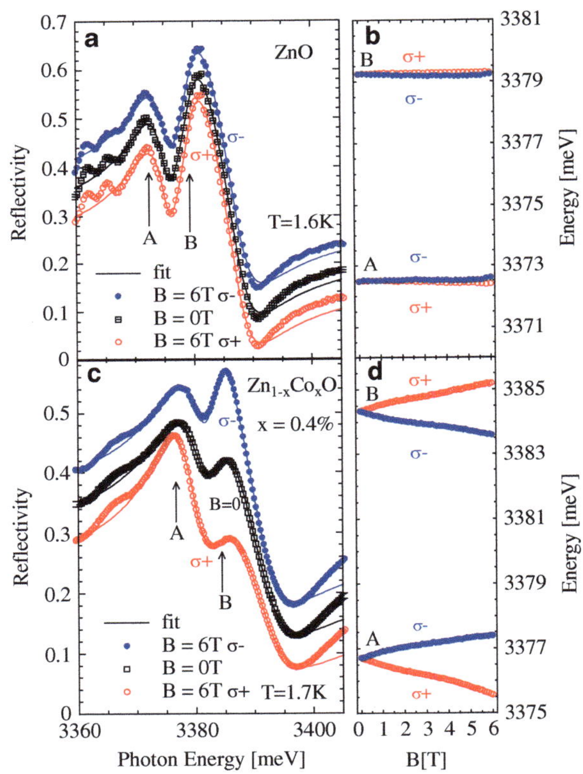

Fig. 2.2 (**a**) Reflectivity spectra of ZnO at $B = 0$ T and at $B = 6$ T in σ^+ and σ^- circular polarizations (symbols). The positions of excitons A and B, as determined from the fit of the $B = 0$ T spectrum (*solid line*) are marked by *arrows*. (**b**) The position of excitons A and B in ZnO, as determined from the fits of the reflectivity spectra shown in (**a**). (**c**) Reflectivity spectra of $Zn_{0.996}Co_{0.004}O$ at $B = 0$ T, and $B = 6$ T in σ^+ and σ^- polarizations. *Solid lines* are the fits to the spectra calculated with polariton model. (**d**) The position of excitons A and B, vs. magnetic field. At $B = 6$ T, the value of the splitting is -1.8 meV for exciton A and 1.6 meV for B. After [25]

splitting of free excitons in pure ZnO (Fig. 2.2b). Therefore, the adjective "giant" seems to be still justifyable for the exchange effects in wide-gap DMSs. Note that energy position of excitons in Fig. 2.2b, d is determined using a fit of a polariton model represented by the solid lines. Polariton effects can be observed in many semiconductors, but they are particularly important for strong optical transitions of ZnO [26, 27].

It is characteristic for wide-gap DMSs that the splitting of excitons A and B has an opposite sign (see Fig. 2.2d). It can be explained in a simple way: the exchange splitting depends on the spin momentum, and the circular polarization depends on orbital momentum of a hole. Excitons A and B are separated mainly because of the spin–orbit interaction, so in one band the exciton spin is parallel to the orbital momentum of the hole in the other the spin is antiparallel to the orbital momentum of the hole (see Fig. 2.14a). Therefore, excitons A and B shift to high (or low) energy in opposite circular polarization.

As a consequence of an opposite splitting of A and B excitons, for one circular polarization (in the case of (Zn,Co)O, for σ^-), the separation energy between those two excitons is reduced and the interaction between the excitons themselves starts to play a key role. Excitons interact due to electron–hole exchange which decreases the oscillator strength of the lower energy exciton and increases the oscillator strength of the higher energy exciton. Moreover, near the anticrossing the excitonic shifts are reduced compared to the shifts of the bands. Bands do not interact (being different eigenstates), so they cross freely (compare Fig. 2.15a, c). Electron hole exchange interaction within the exciton has been shown to be responsible for a twofold reduction of the excitonic giant Zeeman shift in the (Zn,Co)O epilayer [25]. Thanks to clear asymmetry of the excitonic shifts in σ^+ and σ^- polarizations (see Fig. 2.3a), the value of electron–hole exchange energy γ was found to be about 3.4 meV. This is a particularly large value compared to other semiconductors, but an even larger value, 5.6 meV, was reported for ZnO (see Table 2.1 and [28]). Note that 2γ is the separation energy exactly at the anticrossing point, which is almost reached at high field in σ^- polarization (see Fig. 2.3a). The fit shown in Fig. 2.3 calculated using the model including excitonic interactions (see Sect. 2.3.2) and the magnetization values corresponding to Co^{2+} (2.21). The fit leads to a value of $N_0|\beta - \alpha| = 0.8$ eV for Co^{2+} in ZnO. This is much smaller than the values expected for a wide-gap DMS, which suggests that β and α observed in magnetooptical experiments are not equal to real exchange integrals. In such a situation, it is reasonable to denote these effective exchange integrals as $\beta^{(\text{app})}$ and $\alpha^{(\text{app})}$ as the parameters describing giant Zeeman splittings (see also Sect. 2.2.5). Moreover, determination of both $\beta^{(\text{app})}$ and $\alpha^{(\text{app})}$ separately was not possible, because neither the exciton C nor the interaction of the exciton B with exciton C were observed in experimental configuration $\boldsymbol{B} \parallel c \parallel k$ (parallel magnetic field, c-axis, and the propagation direction of light). Therefore, only an effective exchange integrals difference $N_0|\beta^{(\text{app})} - \alpha^{(\text{app})}| = 0.8$ eV could be determined.

Determination of the sign of the apparent exchange integrals requires the valence band ordering to be established first. In ZnO, it is still controversial. Therefore, we need to consider two possible cases (see Fig. 2.13). Assuming that the valence band

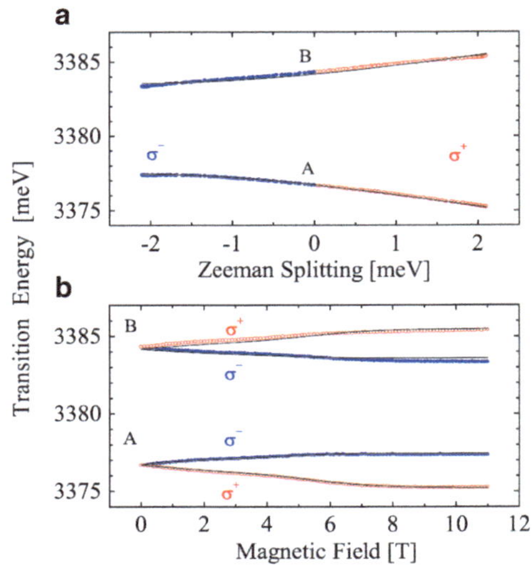

Fig. 2.3 (**a**) Position of excitons A and B in (Zn,Co)O epilayers (0.4% Co), the same sample as in Fig. 2.2c, measured at different values of the applied field. The horizontal axis is the splitting measured on exciton A, plotted as positive if the σ^+ line is at higher energy. Asymmetry between shift in two polarizations indicates anticrossing of excitons. (**b**) Position of the A and B excitons, as a function of the applied field for the same sample as in Fig. 2.2c (with 0.4% Co). *Symbols* represent experimental data: *full symbols* for σ^-, *empty symbols* for σ^+, lines are calculated. After [25]

Table 2.1 Band parameters and the energy of electron–hole exchange in selected wurtzite semiconductors and in zinc-blende CdTe

Semiconductor	Δ_1 (meV)	Δ_2 (meV)	Δ_3 (meV)	γ (meV)	References
CdSe	68.8	138	150.7	0.4	[28]
CdS	28.4	20.9	20.7	2.5	[28]
ZnO	36.3	1.9	7.4	5.6	[28]
GaN	10	6.2	5.5	0.6	[65, 72]
CdTe	0	316	316	0.045	[65]

For zinc-blende structure usually only one parameter is used, Δ_{SO}, because $\Delta_1 = 0$, $\Delta_2 = \Delta_3 = \Delta_{SO}/3$. We present three parameters for CdTe to compare them with parameters of ZnO and GaN. For values of parameters of wurtzite DMSs determined in the presence of magnetic ions, see [25, 67, 68, 70, 71]

ordering in ZnO is, from top to bottom on the energy scale, $\Gamma_{9(5)}$, $\Gamma_{7(5)}$, $\Gamma_{7(1)}$ (this corresponds to the positive sign of the spin–orbit coupling [28–31]) the shift of the exciton A to the lower energy observed in σ^+ polarization (shown in Fig. 2.2c) leads to a conclusion that $\beta^{(\mathrm{app})}$ is negative for ZnO doped with Co^{2+}. In contrast, assuming that the valence band ordering is, from highest to lowest in energy, $\Gamma_{7(5)}$, $\Gamma_{9(5)}$, $\Gamma_{7(1)}$ (this corresponds to the negative sign of the spin–orbit coupling [32–34]), the shift of the exciton A to lower energy observed for σ^+ polarization with the magnetic field leads to the conclusion that $\beta^{(\mathrm{app})}$ is positive for ZnO doped with Co^{2+}.

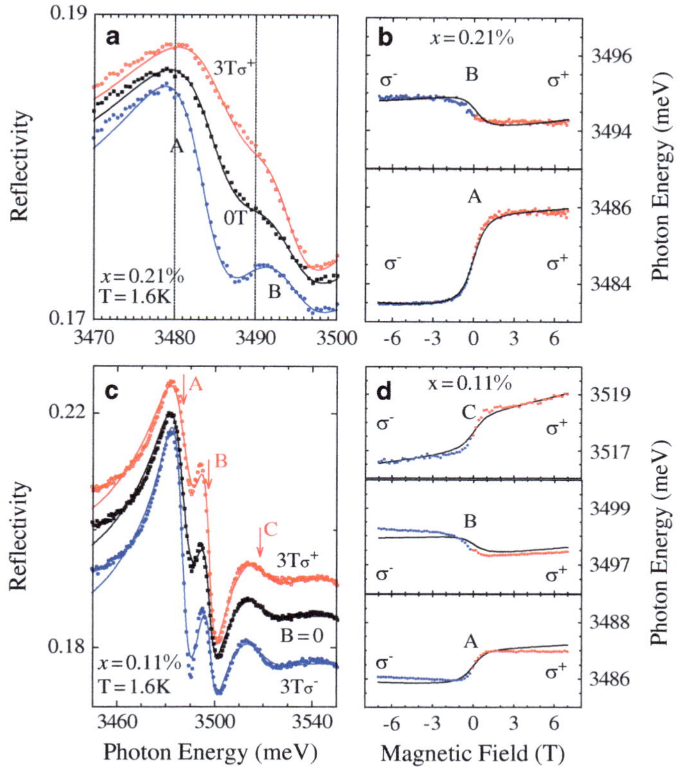

Fig. 2.4 Reflectivity of $Ga_{1-x}Fe_xN$ in Faraday configuration at 1.6 K for $x = 0.21\%$ (**a**) and 0.11% (**c**), where particularly well resolved excitons A, B, and C are visible in σ^- polarization. Panels (**a, c**) show experimental data (*symbols*) and their fitting with the polariton model (*solid lines*). Symbols in (**b, d**) present exciton energies obtained from such a fitting at various magnetic fields; *solid lines* are the expectation of the exciton model. After [36]

In the first case, the exciton A is interpreted as related to the $\Gamma_{9(5)}$ valence band, in the second case, the exciton A is interpreted as related to the $\Gamma_{7(5)}$ valence band.

In the case of GaN, there is no ambiguity concerning the valence band ordering. It is well established that the ordering is $\Gamma_{9(5)}$, $\Gamma_{7(5)}$, $\Gamma_{7(1)}$ [35], therefore the lowest energy exciton (exciton A) is related to the Γ_9 band. In such a case, the shift of the exciton A to lower energy observed in σ^- polarization (opposite to that in (Zn,Co)O) corresponds to positive $\beta^{(\mathrm{app})}$ for GaN doped with Fe^{3+} (shown in Fig. 2.4) and Mn^{3+} (shown in Fig. 2.5).

The example of (Ga,Fe)N is particularly interesting, because for this material it is possible to observe giant Zeeman splitting for all three excitons A, B, and C. Therefore, it was possible to determine independently $N_0\beta^{(\mathrm{app})} = +0.5 \pm 0.2$ eV and $N_0\alpha^{(\mathrm{app})} = +0.1 \pm 0.2$ eV [36]. The value of $N_0\alpha^{(\mathrm{app})}$ is within experimental error, very close to typical value $N_0\alpha = 0.2$ eV observed in many DMSs (except

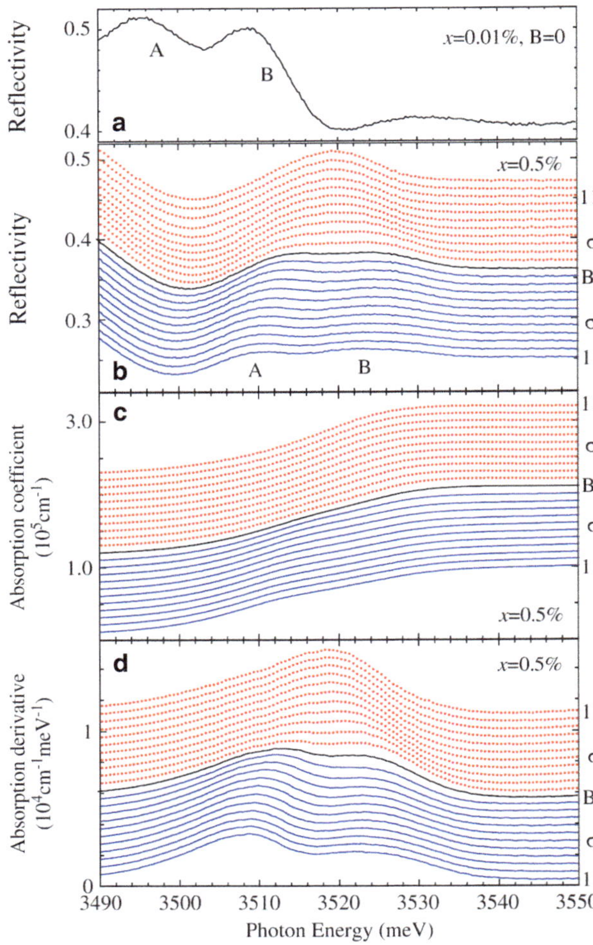

Fig. 2.5 (**a**) Reflectivity of A and B excitons measured in $Ga_{1-x}Mn_xN$ with $x = 0.01\%$. (**b**) Reflectivity, (**c**) absorption and (**d**) derivative of the absorption, for $x = 0.5\%$, in Faraday configuration, in σ^+ (*top, red*) and σ^- (*bottom, blue*) circular polarizations, at $T = 1.7$ K. After [37]

for mercury compounds [38]). However, the positive and relatively small $\beta^{(\mathrm{app})}$ is clearly a very different from β, which is usually large and negative (see the previous Chapter).

It is important that the giant Zeeman splitting in wide-gap DMSs is proportional to the magnetization. Figure 2.6 shows that the giant Zeeman splitting in $Ga_{1-x}Fe_xN$ indeed exhibits the same temperature and magnetic field dependence as the magnetization. Moreover, the giant Zeeman splitting increases linearly with magnetic ion concentration (at least for small concentrations, below $x =$

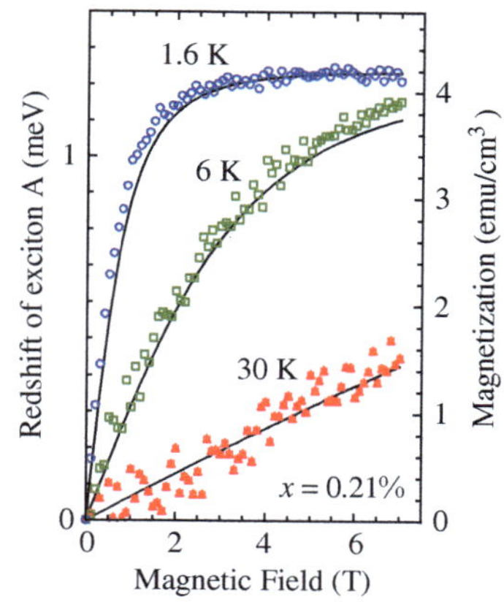

Fig. 2.6 Comparison between the computed magnetization (*solid lines, right axis*) and the redshift of exciton A in σ^- polarization (*symbols, left axis*) for $Ga_{1-x}Fe_xN$ with $x = 0.21\%$ at three temperatures. The reflectivity spectra are shown in Fig. 2.4a. After [36]

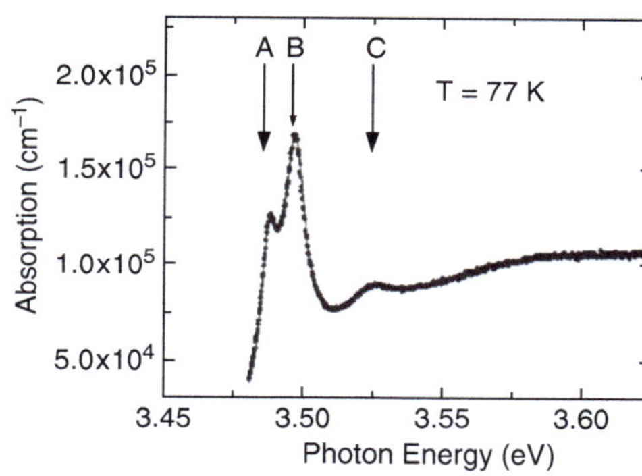

Fig. 2.7 Absorption spectrum of GaN after Muth et al. [40]. Three excitons A, B, and C are marked by *arrows*. Excitonic lines of pure GaN are significantly sharper than exctionic lines of (Ga,Mn)N shown in Fig. 2.5c

1%), as shown for $Ga_{1-x}Fe_xN$ [36], $Ga_{1-x}Mn_xN$ [37], $Zn_{1-x}Co_xO$ [25], and $Zn_{1-x}Mn_xO$ [39].

2.2.2 Absorption

The absorption spectrum of pure GaN is shown in Fig. 2.7. Three excitons A, B, and C are marked by arrows. Observation of such clearly resolved excitonic structures in absorption spectra of DMS based on GaN or ZnO requires very thin and homogeneous layers. Moreover, a usual growth technique makes use of nonmagnetic buffer below the magnetic layer. Since pure ZnO or GaN have smaller energy gap than ZnO or GaN doped with Mn or Co, excitons in the buffer layer affect strongly these in the magnetic layer. An efficient way to observe excitons in the absorption spectra is to grow corresponding layers on buffers with high energy gap. Figure 2.5c shows A and B excitonic structures of (Ga,Mn)N grown on AlN. Comparing to GaN shown in Fig. 2.7, the structures of (Ga,Mn)N shown in Fig. 2.5c are broad, due to strain and disorder, so a precise determination of their energy position is difficult. Some authors overcome this problem using a peak in the derivative spectra to determine the shift of the exciton positions under magnetic fields (Fig. 2.5d). As expected, structures observed in the absorption (Fig. 2.5c) and in the reflectivity (Fig. 2.5b) exhibit the same behavior under the magnetic field: excitons A and B separate out in σ^- circular polarization and get closer in σ^+ circular polarization. Excitonic shifts in (Ga,Mn)N can be described by effective exchange integrals $N_0(\beta^{(\mathrm{app})} - \alpha^{(\mathrm{app})}) = 1.2 \pm 0.2$ eV [37].

2.2.3 Magnetic Circular Dichroism

Observation of well resolved excitonic features either in the reflectivity or in absorption spectra is usually not possible in wide-gap DMS doped with magnetic ions to concentrations higher than $x > 2\%$. Nevertheless, one can observe magnetic circular dichroism (MCD) in the transmission or reflectivity spectra (Kerr effect). MCD is particularly strong near excitonic resonances. Since MCD spectra near the energy gap of ZnO or GaN are affected by overlapping and interacting excitons A, B, and C, the MCD signal usually gives no reliable information about the splitting, neither of the excitons nor the bands; but if it is measured at photon energy near the energy gap, it is approximately proportional to the magnetization (or mean spin) via the $s, p-d$ exchange interaction. Proof of such proportionality is shown in Fig. 2.8 for the reflectivity MCD measured in (Zn,Mn)O and in Fig. 2.9 for transmission MCD measured on (Zn,Co)O. Note that the magnetization curves of this two materials are significantly different. We will describe this difference in Sect. 2.4.

Figure 2.10 shows MCD spectra of (Zn,Co)O with various Co concentration. Significant MCD signal in our samples can be induced not only by the splitting of the

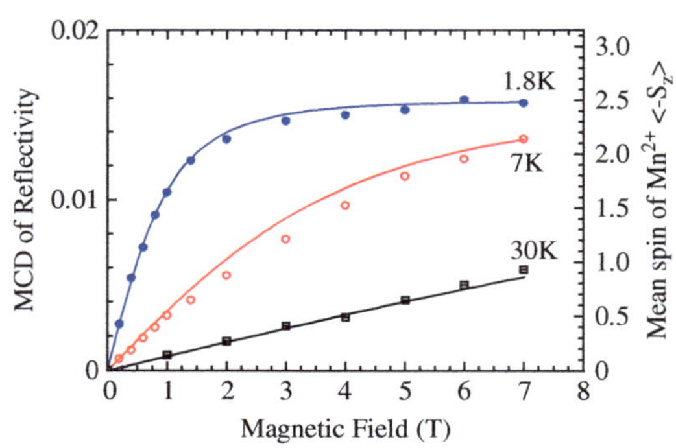

Fig. 2.8 Analysis of the near band gap reflectivity of diluted $Zn_{1-x}Mn_xO$ ($x < 1\%$). Amplitude of Magnetic Circular Dichroism in the reflectivity (*points, left axis*) is compared to the mean spin of Mn^{2+} ions (*solid lines, right axis*) calculated using the Brillouin function $B_{5/2}$. MCD is defined here as $(I_+ - I_-)/(I_+ + I_-)$, where $I_\pm$ is intensity of reflected light in $\sigma^\pm$ circular polarization. After [41]

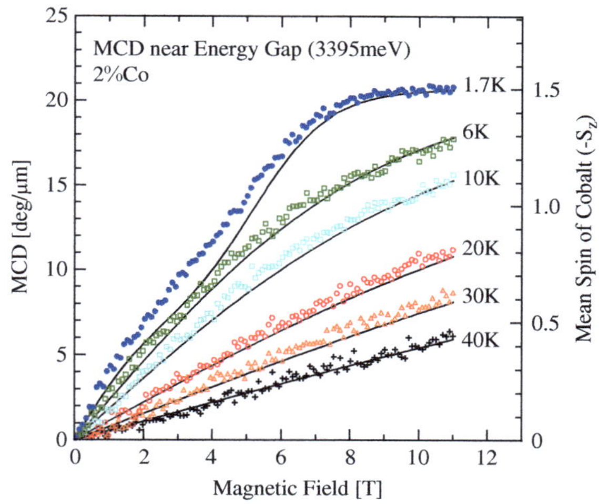

Fig. 2.9 *Symbols, left axis*: Magnetic circular dichroism of $Zn_{0.98}Co_{0.02}O$ in transmission, at photon energy 3,395 meV, close to the bandgap; lines (*right vertical axis*) mean spin of isolated Co ions, as calculated in Sect. 2.4. After [25]

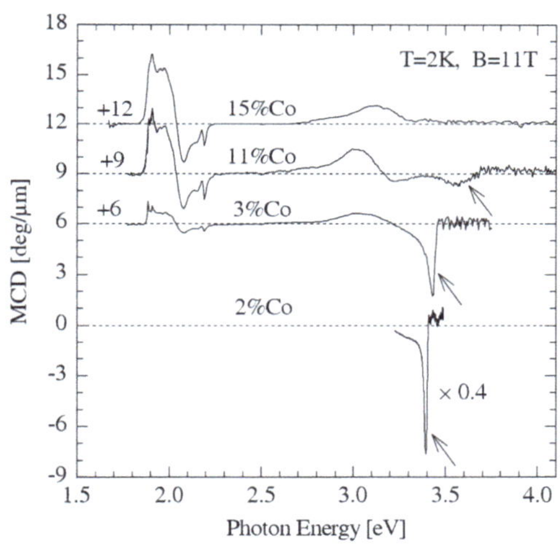

Fig. 2.10 Magnetic circular dichroism (MCD) in transmission in $Zn_{1-x}Co_xO$ for samples with various Co concentrations. The MCD signal contains three contributions: first (pointed by *arrows*) is most intense in the sample with the lowest Co concentration, 2% Co. It appears close to the bandgap at 3.4 eV. The second occurs below the gap, and the last one (most intense in the sample with the highest Co concentration, 15% Co) appears in coincidence with the internal transitions of Co (1.8–2.3 eV). Spectra are shifted for clarity. After [25]

excitons, but also by intraionic transitions or charge transfer transitions [17, 22, 42], or by a dispersion on precipitates, e.g., of Co [24, 43]. MCD related to intraionic transitions is observed near characteristic absorption bands and lines of magnetic ions, e.g., for Co^{2+} it is observed close to 2 eV. MCD due to charge transfer transitions is usually broader and consequently much more difficult to identify and

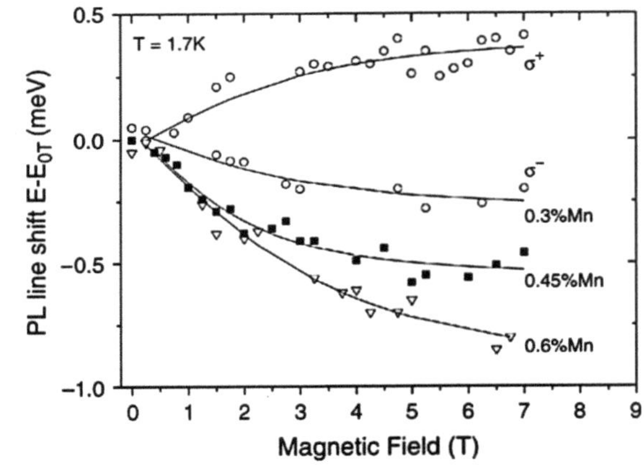

Fig. 2.11 The shift of the A^0X-bound exciton line vs. magnetic field measured at three $Zn_{1-x}Mn_xO$ samples ($x = 0.003$, 0045, 006) in two circular polarizations, at low temperature ($T = 1.7$ K). The exchange-induced shift is very weak, and it can be described by $N_0|\beta^{(app)} - \alpha^{(app)}| = 0.1$ eV After [39].

interpret. Typically, these transitions can be observed below the energy gap. MCD due to dispersion on precipitates exhibits very weak spectral dependence. Often, it is quite difficult to distinguish between the two latter MCD signals and therefore their interpretation is a matter of controvercy, as it is in the case of (Zn,Co)O. In [44], a broad MCD signal is interpreted as related to charge transfer transitions. In [24, 43] very similar signal is attributed to secondary ferromagnetic phases of Co, which coexist with paramagnetic Co in diluted ZnO.

2.2.4 Photoluminescence

Excitonic photoluminescence of wide-gap DMSs is usually very weak and difficult to observe. Diluted (Zn,Mn)O is an exception and it exhibits PL lines strong enough to perform a magnetooptical study of the giant Zeeman effect. Figure 2.11 shows the exchange–induced shifts of the PL line observed in three (Zn,Mn)O samples. Effective exchange integrals extracted from these data are very small, $N_0|\beta^{(app)} - \alpha^{(app)}| = 0.1$ eV. They are small even compared to (Zn,Co)O.

2.2.5 Effective Exchange Integrals

Let us focus for a while on the measured values of $\beta^{(app)}$. Material trends in DMSs are shown in Fig. 2.12, where p–d exchange energy is plotted vs. cation density. For semiconductors with small cation density (large lattice parameter) a negative p–d exchange constant increases with the cation density, as shown by the dashed line. This trend was observed using magnetooptical experiments such as reflectivity or PL, and confirmed using calculations based on Schrieffer-Wolff theory with parameters determined from theory or X-ray spectroscopy. From this material trend,

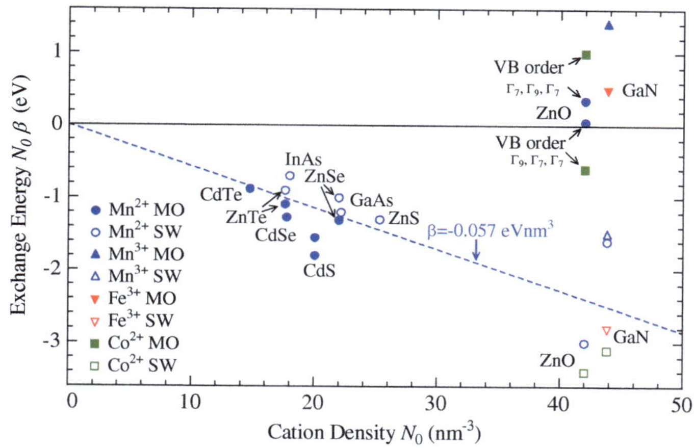

Fig. 2.12 Exchange energy vs. cation density N_0. *Full symbols*: experimental $N_0\beta^{(\mathrm{app})}$, magnetooptical determination (MO) [25,36,37,39,45–48]. *Empty symbols* (SW): $N_0\beta$, calculation based on formulae of Larson et al. [49] derived from Schrieffer-Wolff theory [50, 51], with parameters determined from theory [52] or X-ray spectroscopy [2, 53–57]. For wide-gap DMSs (right hand site of the plot), three electronic configuration are taken into account: d^5 (Mn^{2+} and Fe^{3+}), d^4 (Mn^{4+}), and d^7 (Co^{2+}). For other DMSs only d^5 electronic configuration is plotted for clarity. Compare with [45, 58].

for ZnO and GaN, so for semiconductors with high cation density (small lattice parameter), we expect large negative p–d exchange energy. Indeed, determination based on Schrieffer-Wolff theory shows a very large negative $N_0\beta$ for ZnO and GaN-based DMSs. However, the excitonic giant Zeeman splitting observed in magnetooptical measurements can be described only by relatively small p–d exchange energy, as has been reviewed in this Chapter. Moreover, at least in the case of GaN (the case of ZnO is not so clear, see Sect. 2.2.1), magnetooptical measurements lead to the conclusion that p–d exchange energy is positive. It indicates that giant Zeeman splitting in wide-gap DMSs is not directly governed by the parameter $N_0\beta$. In fact, Dietl remarked in [61, 62] that one could expect in oxides and nitrides that the transition metal ion would bind the hole. This trend was already suggested by the experimental evidence of strong deviations from VCA in (Cd,Mn)S [6]. It is observed also by the analysis of ab-initio calculations [59, 60]. A calculation by Dietl [61] shows that the spin splitting of extended states involved in the optical transitions remains proportional to the magnetization of the localized spins, but the apparent exchange energy $\beta^{(\mathrm{app})}$ becomes significantly renormalized. For the expected coupling strength, it is predicted that $-1 < \beta^{(\mathrm{app})}/\beta < 0$. This is the reason why we use $\beta^{(\mathrm{app})}$, not β in (2.2) and (2.3), which describe the giant Zeeman splitting in wide-gap DMSs. Independently, exchange interactions between electrons and holes bound to Mn acceptors leads to renormalization of effective s–d exchange integral [63, 64]. Therefore, we also use $N_0\alpha^{(\mathrm{app})}$ instead of $N_0\alpha$.

2.3 Description of the Giant Zeeman Effect in Wurtzite DMS with Large Energy Gap

ZnO and GaN-based diluted magnetic semiconductors naturally crystallize in the wurtzite structure, so they are member of the same family as CdS or CdSe-based DMSs. Although the band structures of all wurtzite semiconductors can be described using the same equations, the material parameters can vary significantly between different members of the family. Therefore, the importance of various terms included in the excitonic Hamiltonian for large gap materials differs from those for CdS and CdSe. For example, the role of electron–hole exchange interaction is not so important for description of the giant Zeeman effect in CdS or CdSe-based DMSs, but this effect is crucial for ZnO. This section contains analysis of various terms contributing to the excitonic Hamiltonian, as found to be appropriate for description of experimental values of the giant Zeeman effect in wide-gap diluted magnetic semiconductors. Step by step, several effects are introduced. First, we describe the valence bands of semiconductor with the wurtzite structure. We discuss the standard description of the giant Zeeman splitting of the conduction and valence bands induced by the coupling to magnetic ions. Next, we discuss selection rules of band-to-band optical transitions and the direct influence of the magnetic field on the bands as well as on the excitons. Finally, the electron–hole exchange interaction within the exciton is taken into account and a complete description of the excitonic giant Zeeman splitting is given.

In the absence of the magnetic field, the valence states are doubly degenerate, and the energy of the hole is given by [65]:

$$H_v = -\tilde{\Delta}_1(l_z^2\text{-}1) - 2\Delta_2 l_z s_z - 2\Delta_3(l_x s_x + l_y s_y), \tag{2.1}$$

where $\tilde{\Delta}_1$ describes the effect of the trigonal component of the crystal field; Δ_2 and Δ_3 are the parameters of the anisotropic spin–orbit interaction: $\Delta_2 = \Delta_{SO||}/3$, $\Delta_3 = \Delta_{SO\perp}/3$, where Δ_{SO} is a value of spin–orbit coupling in the cubic approximation when $\Delta_{SO\perp} = \Delta_{SO||}$ [32]; l_α and s_α are components of the orbital and spin momenta, respectively. The z direction is parallel to the crystal c-axis. The biaxial strain, which is typical for epitaxial layers, can be included in the energy of the band gap and in the value of $\tilde{\Delta}_1$ (Δ_1 is reserved for the crystal field parameter in the case of no strain present). For the case of the uniaxial strain, see [65, 66].

Further we will use the following basis consistent with the above Hamiltonian: $|p^+\uparrow\rangle$ and $|p^-\downarrow\rangle$ (which form $\Gamma_{9(5)}$ state), $|p^+\downarrow\rangle$ and $|p^-\uparrow\rangle$ (which contribute mainly to $\Gamma_{7(5)}$ state), $|p^z\downarrow\rangle$ and $|p^z\uparrow\rangle$ (which contribute mainly to $\Gamma_{7(1)}$ state). Here, the arrows denote sign of the spin projection ($s_z = \pm 1/2$). We denote by p^+, p^-, and p^z the hole states with orbital momenta $l_z = +1, -1$, and 0, respectively. Γ_9 and Γ_7 denote the symmetry of the valence states when the spin–orbit coupling is taken into account. The number in parenthesis (5 and 1 in $\Gamma_{9(5)}$, $\Gamma_{7(5)}$, $\Gamma_{7(1)}$) indicates the parent state without the spin–orbit coupling, as shown in Fig. 2.13. The state $\Gamma_{7(1)}$ consists mainly of functions with the orbital momentum $l = 0$. The state $\Gamma_{7(5)}$ is built mainly from the functions with the orbital momenta $l = \pm 1$

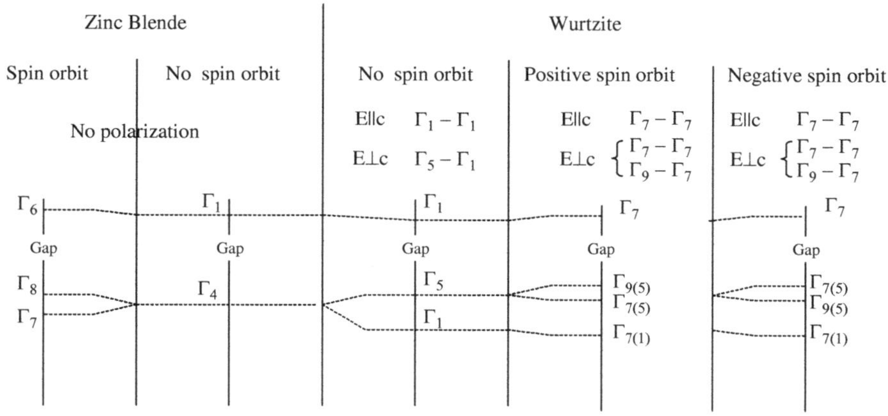

Fig. 2.13 Band structure of zinc blende and wurtzite semiconductors at $k = 0$. Impact of spin–orbit interaction on the symmetry and valence band ordering is shown. Polarization dependent selection rules are also given (compare [69])

with spins antiparallel to the orbital momentum. The state $\Gamma_{9(5)}$ consists mainly of the functions with orbital momenta $l = \pm 1$ with the spins parallel to the orbital momentum.

The states $\Gamma_{9(5)}$, $\Gamma_{7(5)}$, $\Gamma_{7(1)}$ described by the Hamiltonian (2.1) form three valence band edges. Their splitting is a result of the combined effect of the trigonal crystal field (described by $\tilde{\Delta}_1$) and the anisotropic spin–orbit coupling (determined by two parameters, Δ_2 and Δ_3). According to the values of the parameters given in Table 2.1, the spin–orbit coupling in ZnO is much smaller than the trigonal crystal field (more precisely, Δ_3 is much smaller than $\Delta_1 - \Delta_2$). Hence, the trigonal field splits the p-like states which form the bottom of the valence band into a doublet (Γ_5 in-plane p-states) and a singlet (Γ_1 out-of-plane p^z states). As a result of the spin–orbit coupling, the orbital doublet is split further into two doublets $\Gamma_{7(5)}$, $\Gamma_{9(5)}$, and the orbital singlet forms a $\Gamma_{7(1)}$ doublet with a small admixture of the $\Gamma_{7(5)}$ states. In bulk GaN, the mixing between valence states $\Gamma_{7(1)}$ and $\Gamma_{7(5)}$ is much stronger. However, for epitaxial layers an in-plane compressive, biaxial strain usually decreases this mixing by increasing an in-plane $\Gamma_{7(1)} - \Gamma_{7(5)}$ splitting. Figure 2.13 compares the band structure of zinc blende and wurtzite semiconductors. The most important difference is a strong optical anisotropy induced by the crystal field in the wurtzite materials. The sign of the spin–orbit coupling determines a relative energy position of $\Gamma_{9(5)}$ and $\Gamma_{7(5)}$ states. In bulk GaN, the valence band ordering is $\Gamma_{9(5)}$, $\Gamma_{7(5)}$, $\Gamma_{7(1)}$ [35], but the position and the symmetry of the valence edges are still a matter of controversy in bulk ZnO [32, 73]. If a positive spin–orbit coupling is assumed, the valence band ordering is like that in GaN; if a negative spin–orbit coupling is assumed, the ordering is different: $\Gamma_{7(5)}$, $\Gamma_{9(5)}$, $\Gamma_{7(1)}$. Both possibilities are shown in Fig. 2.13.

Table 2.1 shows that the spin–orbit coupling parameters Δ_2 and Δ_3 are strongly reduced in wide gap semiconductors, which contain relatively light anions. But the electron–hole exchange interaction γ (to be described in Sect. 2.3.2) is particularly strong in ZnO. We will show that this leads to significant consequences concerning exciton states.

2.3.1 Giant Zeeman Splitting of Bands

The effective Hamiltonian describing properly the giant Zeeman effect in many II–VI semiconductors has been introduced in Chap. 1. For DMSs with moderate energy gap, the giant Zeeman splitting is proportional to the magnetization and to exchange integrals $N_0\alpha$ and $N_0\beta$. When the cation density increases, such an approach fails in description of the giant Zeeman effect in such DMSs as (Cd,Mn)S, because the giant Zeeman splitting in different samples of such material does not scale with the magnetization with only one proportionality constant. In ZnO and GaN-based DMSs, the giant Zeeman splitting is approximately proportional to the magnetization, but the parameters of the proportionality, probably, are not $N_0\alpha$ and $N_0\beta$ as given by theoretical expectations any more (see Sect. 2.2.5). To take this fact into account, the effective exchange integrals $N_0\alpha^{(\mathrm{app})}$ and $N_0\beta^{(\mathrm{app})}$ have been introduced [36,63,74]. Determination of such effective parameters does not give direct information on $p{-}d$ or $s{-}d$ hybridization energy. Effective exchange integrals describe only the splitting of the bands. Therefore, we rewrite effective Hamiltonian for the conduction band electron

$$\hat{H}^{(\mathrm{app})}_{s-d} = -N_0\,\alpha^{(\mathrm{app})} x \langle \boldsymbol{S} \rangle \cdot \boldsymbol{s}_{\mathrm{e}}, \tag{2.2}$$

and for the valence band hole

$$\hat{H}^{(\mathrm{app})}_{p-d} = -N_0\,\beta^{(\mathrm{app})} x \langle \boldsymbol{S} \rangle \cdot \boldsymbol{s}_{\mathrm{h}}, \tag{2.3}$$

where $N_0\alpha^{(\mathrm{app})}$ and $N_0\beta^{(\mathrm{app})}$ are the effective exchange integrals, N_0, as usual, denotes the number of cations per unit volume, x is the mole fraction of the magnetic ions, $\langle \boldsymbol{S} \rangle$ is the mean spin of the magnetic ions, $\boldsymbol{s}_{\mathrm{e}}$ and $\boldsymbol{s}_{\mathrm{h}}$ are the spin operators of the electron and hole, respectively. To use the Hamiltonian (2.2) or (2.3), one has to know the magnetization or the mean spin of magnetic ion $\langle \boldsymbol{S} \rangle$. For d^5 electronic configuration (e.g., Mn^{2+} or Fe^{3+}), the fact that the total orbital momentum is zero reduces the effect of anisotropy on the magnetization, so it can be well approximated using the Brillouin function $B_{5/2}$ mentioned in the previous Chapter. For other electronic configurations, the effect of the spin–orbit coupling within the magnetic ion itself needs to be taken into account (see Sect. 2.4).

According to (2.2), the conduction band at the center of the Brillouin zone is split by the $s{-}d$ exchange interaction and its energy is given by:

$$E = E_0 \pm G_{\mathrm{e}}, \tag{2.4}$$

where E_0 denotes band gap energy, and $G_e = \frac{1}{2}N_0\alpha^{(\mathrm{app})}x\langle -S_i\rangle$. The sign of $\pm G_e$ corresponds to the electron spin projection $s_e = \pm 1/2$ on the magnetic field direction i, and to the conduction band wavefunction $|s\uparrow\rangle$ and $|s\downarrow\rangle$, respectively.

In the case of the effective $p-d$ Hamiltonian (2.3), we have to take into account the anisotropic structure of the valence band described by equation (2.1). Consequently, the exchange splitting is different for the magnetic field parallel and perpendicular to the c-axis. We can express explicitly the energy of the valence band for the magnetic field parallel to the c-axis. The energy of the hole containing contribution of $p^{\pm}$ the wavefunction is given by [67, 68, 70, 75–78]:

$$E^{p^{\pm}}_{\Gamma_{9(5)}} = \Delta_2 \pm G_h, \tag{2.5}$$

$$E^{p^{\pm}}_{\Gamma_{7(5)}} = \frac{\tilde{\Delta}_1 + \Delta_2}{2} - E_{\pm}, \tag{2.6}$$

$$E^{p^{\pm}}_{\Gamma_{7(1)}} = \frac{\tilde{\Delta}_1 + \Delta_2}{2} + E_{\pm}, \tag{2.7}$$

where

$$G_h = \frac{1}{2}N_0\beta^{(\mathrm{app})}x\langle -S_z\rangle \tag{2.8}$$

$$E_{\pm} = \sqrt{\left(\frac{\tilde{\Delta}_1 - \Delta_2}{2} \pm G_h\right)^2 + 2\Delta_3^2},\, x \tag{2.9}$$

The above equations are relatively simple because the valence state $\Gamma_{9(5)}$ does not mix with $\Gamma_{7(5)}$ and $\Gamma_{7(1)}$, when the magnetic field is parallel to c-axis.

In the case of the magnetic field perpendicular to the c-axis, all valence band states are mixed. A 6×6 matrix describing the giant Zeeman splitting of the valence band for $B\perp c$ is given in [75]. The direct influence of the magnetic field due to the usual Zeeman effect and the diamagnetic shift will be implemented in a model described in Sect. 2.3.2.3. Usually, these contributions are small enough to be safely neglected, as it was in the case of moderate gap DMSs.

2.3.2 *Giant Zeeman Splitting of Excitons*

2.3.2.1 Selection Rules in the Magnetic Field

The wavefunctions corresponding to the valence band states are strongly anisotropic. If $\Delta_3 = 0$, transitions involving $\Gamma_{9(5)}$ and $\Gamma_{7(5)}$ (i.e., optical transitions denoted previously by A and B) valence states are polarized perpendicular to c-axis ($E \perp c$), and transitions involving $\Gamma_{7(1)}$ (C) state are polarized parallel to c-axis ($E \parallel c$). This is shown in Fig. 2.14a. When the mixing between $\Gamma_{7(5)}$ and $\Gamma_{7(1)}$ increases, these

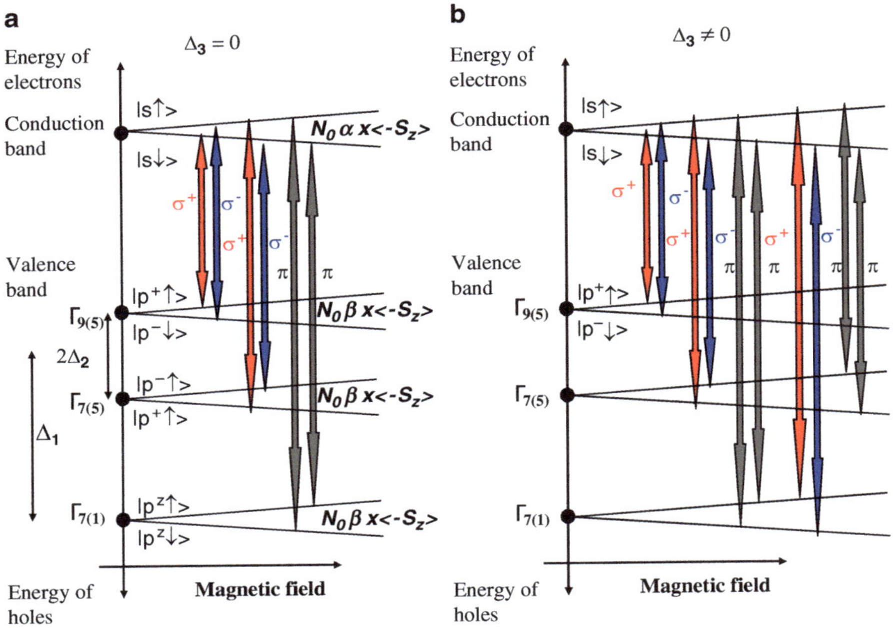

Fig. 2.14 Scheme of splittings, wavefunctions, and allowed transitions in a magnetic field parallel to the c-axis. Neglecting spin–orbit coupling parameter Δ_3 in (**a**) is a good approximation in ZnO

selection rules are relaxed, as shown in Fig. 2.14b. Distinguishing between strong and weak transitions is justified for some wide-gap semiconductors (ZnO and GaN), where the spin–orbit coupling (Δ_3) is relatively weak. In other wurtzite semiconductors such as CdS and CdSe, all transitions shown in Fig. 2.14b have comparable strength.

In the case of wide-gap DMSs, the giant Zeeman splitting of the bands cannot be directly translated into the giant Zeeman splitting of excitons, by calculating the difference between energy of initial and final state of transition, as usually done for DMS [68, 70, 75–78]. Exciton mixing related to electron–hole exchange interaction has to be taken additionally into account, especially in the case of ZnO (see next Section).

2.3.2.2 Electron–Hole Exchange

For the ground state of the exciton, the electron–hole interaction within the exciton itself results in

$$H_{\mathrm{e-h}} = -R^* + 2\gamma \boldsymbol{s}_{\mathrm{e}} \boldsymbol{s}_{\mathrm{h}}, \tag{2.10}$$

where R^* is the conventional electrostatic binding energy, $\boldsymbol{s}_{\mathrm{e}}$ is the electron spin and $\boldsymbol{s}_{\mathrm{h}}$ is the hole spin, and the electron–hole exchange [35, 65] is parameterized by the exchange integral γ. The binding energy and γ are different for the ground state and for excited states of the exciton.

2.3.2.3 Direct Influence of the Magnetic Field on Excitons

Following Luttinger [79] and Stępniewski et al. [35] we present here a standard description of effects linear in the magnetic field. The effective Hamiltonian describing the Zeeman splitting of the conduction band electron is

$$H_{\text{Zeeman}}^{\text{CB}} = g_{\text{e}}\mu_{\text{B}}\boldsymbol{B}\boldsymbol{s}_{\text{e}}, \tag{2.11}$$

and of the valence band hole it is

$$H_{\text{Zeeman}}^{\text{VB}} = 2\mu_{\text{B}}\boldsymbol{B}\boldsymbol{s}_{\text{h}} - \mu_{\text{B}}(3\tilde{\kappa} + 1)\boldsymbol{B}\boldsymbol{l}, \tag{2.12}$$

where g_{e} is an effective g factor of the electron, μ_{B} is the Bohr magneton, $\boldsymbol{B}$ is magnetic field, and $\boldsymbol{s}_{\text{e}}$ stands for spin of the electron, $\boldsymbol{s}_{\text{h}}$ and $\boldsymbol{l}$ are spin and orbital momentum of the hole, respectively. The effective parameter $\tilde{\kappa}$ describes directly effective Landé g factor of the hole in Γ_9 state of nonmagnetic material. In the magnetic field parallel to the c-axis $6\tilde{\kappa} = g_{\Gamma_9\parallel}$. Calculation of the splitting of Γ_7 bands is more complex. That splitting depends on a separation energy between $\Gamma_{7(5)}$ and $\Gamma_{7(1)}$, so it depends on all effects that affect this energy, e.g., strain [see Hamiltonian (2.1)]. Calculation of the splitting of optical excitonic transitions requires taking additionally into account the electron–hole exchange interaction.

The magnetic field induces also a diamagnetic shift of the excitonic transition energy, which can be approximated by a simple quadratic dependence on the magnetic field and by an effective Hamiltonian [35]:

$$H_{\text{diam}} = \text{d}B^2. \tag{2.13}$$

A value of the constant d depends not only on the material but also on the exciton properties: it increases with the exciton radius.

2.3.2.4 Excitonic Hamiltonian

Finally, putting all contributions together, the excitonic Hamiltonian takes the following form:

$$H = E_0 + H_{\text{v}} + H_{\text{e-h}} + H_{sp-d}^{(\text{app})} + H_{\text{Zeeman}} + H_{\text{diam}}, \tag{2.14}$$

where E_0 is the bandgap energy, $H_{sp-d}^{(\text{app})} = H_{s-d}^{(\text{app})} + H_{p-d}^{(\text{app})}$, $H_{\text{Zeeman}} = H_{\text{Zeeman}}^{\text{CB}} + H_{\text{Zeeman}}^{\text{VB}}$. In a general case, the basis of twelve excitons resulting from six holes states and two electron states is needed to calculate the eigenvalues of this Hamiltonian.

Excitons are given the same labels as the valence band of the hole forming an exciton, A, B, and C. Therefore, we shall call them in the order of increasing energy.

2.3.2.5 Giant Zeeman Effect in the Faraday Configuration, $B \parallel c \parallel k$

For the magnetic field parallel to the c-axis and light propagating along the same direction, we limit ourselves to six excitons, all of which have Γ_5 admixture if $\Delta_3 \neq 0$, therefore they are optically active in the σ polarization (see Fig. 2.14b). The Hamiltonian (2.14) separates into two operators acting in two subspaces corresponding, respectively, to excitons active in the σ^+ and σ^- circular polarizations. For σ^+ we use the following basis: $|s \downarrow\ p^+ \uparrow\rangle$ and $|s \uparrow\ p^+ \downarrow\rangle$, (which are active in σ^+ polarization and will give the main contribution to excitons A and B), and $|s \uparrow\ p^z \uparrow\rangle$ (which is optically inactive since it is spin-forbidden, but will give the main contribution to exciton C, which is observable in σ^+ polarization). The same matrices, with opposite giant Zeeman terms, apply in σ^- polarization with the basis $|s \uparrow\ p^- \downarrow\rangle$, $|s \downarrow\ p^- \uparrow\rangle$, and $|s \downarrow\ p^z \downarrow\rangle$. In this basis, the Hamiltonian (2.14) can be written [36, 37, 65, 80] as:

$$H_v = \begin{pmatrix} -\Delta_2 & 0 & 0 \\ 0 & \Delta_2 & -\sqrt{2}\Delta_3 \\ 0 & -\sqrt{2}\Delta_3 & \tilde{\Delta}_1 \end{pmatrix}, \tag{2.15}$$

$$H_{e-h} = -R^* + \frac{\gamma}{2}\begin{pmatrix} -1 & 2 & 0 \\ 2 & -1 & 0 \\ 0 & 0 & 1 \end{pmatrix}, \tag{2.16}$$

$$H^{\sigma\pm}_{sp-d} = \pm\frac{1}{2}N_0 x\langle -S_z\rangle \begin{pmatrix} \beta-\alpha & 0 & 0 \\ 0 & \alpha-\beta & 0 \\ 0 & 0 & \alpha+\beta \end{pmatrix}. \tag{2.17}$$

$$H^{\sigma\pm}_{\text{Zeeman}} = \pm\mu_B B \begin{pmatrix} -\frac{1}{2}g_e - 3\tilde{\kappa} & 0 & 0 \\ 0 & \frac{1}{2}g_e - 2 - 3\tilde{\kappa} & 0 \\ 0 & 0 & \frac{1}{2}g_e + 1 \end{pmatrix}. \tag{2.18}$$

Diagonalizing the Hamiltonian gives the energy of the three excitons A, B, and C in each circular polarization, as shown in Fig. 2.15. The corresponding oscillator strength marked by thickness of a given line is deduced from the projection of the corresponding eigenvector, $|\psi\rangle$, onto the relevant subspace active in circular polarization: it is proportional to $|\langle s \downarrow\ p^{\pm} \uparrow |\psi\rangle + \langle s \uparrow\ p^{\pm} \downarrow |\psi\rangle|^2$, for $\sigma^{\pm}$ circular polarization, respectively.

In zero field, the relative position of the excitons is determined mainly by the trigonal component of the crystal field including strain $\tilde{\Delta}_1$ and the parallel spin–orbit interaction Δ_2. The giant Zeeman shift of the exciton is induced by s, p–d interactions. The electron–hole exchange interaction γ governs the anticrossing and mixing of the A and B excitons. It strongly alters the excitonic oscillator strength.

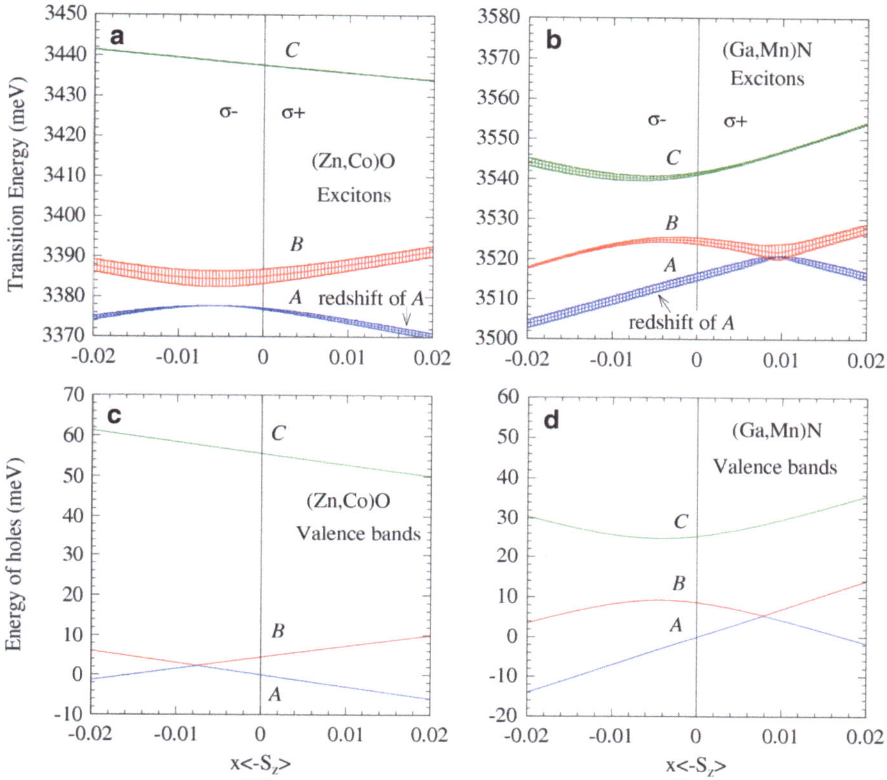

Fig. 2.15 Results of the model calculation: (**a, b**) Energies of the A, B, and C excitonic transitions vs. magnetization $x\langle -S_z\rangle$, for σ^- (left half) and σ^+ (right half) circular polarization. *Bars* show the oscillator strength. (**c, d**) Energies of the A, B, and C valence band holes. Note that A and B valence bands cross, while an anticrossing occurs for A and B excitons. We used a following set of parameters determined from preliminary analysis of experimental data: $\tilde{\Delta}_1 = 51$ meV, $\Delta_2 = 3$ meV, $\Delta_3 = 6.3$ meV, $\gamma = 3.4$ meV, and $N_0(\beta^{(\mathrm{app})} - \alpha^{(\mathrm{app})}) = -0.8$ eV for (Zn,Co)O and $\tilde{\Delta}_1 = 13$ meV, $\Delta_2 = 7$ meV, $\Delta_3 = 5.5$ meV, $\gamma = 0.6$ meV, and $N_0(\beta^{(\mathrm{app})} - \alpha^{(\mathrm{app})}) = 1.2$ eV for (Ga,Mn)N. We assumed a usual value of $N_0\alpha^{(\mathrm{app})} = N_0\alpha = 0.2$ eV for both materials. After [80]

The perpendicular spin–orbit interaction (Δ_3) is responsible for the anticrossing and mixing between the B and C excitons. As a result, the exciton C acquires a strong oscillator strength if the giant Zeeman splitting is large.

It is an important result that the giant Zeeman splitting of A and B excitons is not proportional to the magnetization due to magnetic ions [lines are curved in Fig. 2.15a,b]. The best way to probe the giant Zeeman energy [$N_0(\beta^{(\mathrm{app})} - \alpha^{(\mathrm{app})})x\langle -S_z\rangle$] is to measure the redshift of exciton A, which appears to be proportional to the magnetization as shown by the straight line in Fig. 2.15b calculated for GaN-based DMS. For ZnO-based DMS (Fig. 2.15a), electron–hole exchange interaction should be taken into account in any, qualitative or quantitative, interpretation of excitonic shifts.

2.3.2.6 Other Configurations of the Magnetic Field, Crystal c-axis, and Propagation Vector

There are many interesting configurations of the magnetic field applied externally and the electric field of incoming light, e.g., $\boldsymbol{B} \parallel c \perp k$ ($\boldsymbol{E} \perp c$ and $\boldsymbol{E} \parallel c$), $\boldsymbol{B} \perp c \perp k$ ($\boldsymbol{E} \perp c$ and $\boldsymbol{E} \parallel c$), $k \parallel c \perp \boldsymbol{B}$ ($\boldsymbol{E} \perp \boldsymbol{B}$ and $\boldsymbol{E} \parallel \boldsymbol{B}$), $\boldsymbol{B} \parallel k \perp c$ (with elliptical polarization).

In comparison with the usual configuration $\boldsymbol{B} \parallel k \parallel c$, we need larger matrices to describe the above configurations, which are in fact only rarely reported for wide-gap DMS. More common case of an experimental configuration is a measurement of polycrystals in the Faraday configuration. Calculation of an excitonic giant Zeeman effect of such system requires needs taking into account the fact that magnetic field forms a different angle with the c-axis in each microcrystal.

2.4 Magnetic Anisotropy

The Brillouin function given in Chap. 1 is appropriate for description of the magnetization in the case of DMS with a negligible anisotropy, i.e., for majority of DMS with the zinc blende structure. In wurtzite structure and, particularly, in wide-gap DMS, the trigonal crystal field combined with spin–orbit interaction induces strong magnetic anisotropy. Only magnetic ions with zero orbital momentum, such as Mn^{2+} or Fe^{3+} ($S = 5/2$), can be approximately described using the Brillouin function. For other localized spin configurations, effect of the spin–orbit coupling within the magnetic ion has to be taken in to account. The ground A_1^6 state of the isolated magnetic ion in the magnetic field can be described by spin Hamiltonian [81]:

$$H_s = \mu_B g_\parallel B_z S_z + \mu_B g_\perp (B_x S_x + B_y S_y) + D S_z^2, \tag{2.19}$$

where S is the spin and D describes the zero-field splitting; $g_\parallel$ and $g_\perp$ are the effective g-factors for directions of the magnetic filed parallel and perpendicular to the c-axis, respectively. This Hamiltonian contains only the terms, which are the most important for the calculation of the magnetization. A more advanced description links the phenomenological parameters of (2.19) with more general properties of the crystal and magnetic ions (see [9] and [82]). For instance, they include hyperfine interactions [83] or Jahn Teller effect [84–86].

Parameters of the Hamiltonian (2.19) are given in Table 2.2 for selected magnetic ions in wurtzite semiconductors. The Zeeman splitting calculated for various spin configurations is shown in Fig. 2.16 for $\boldsymbol{B} \parallel c$-axis. Magnetic ions with $S = 5/2$ have two orders of magnitude smaller zero-field splitting than other spin configurations. The g-factors for $S = 5/2$ spin configuration are also almost independent of the direction of the magnetic field, and the values of both $g_\parallel$ and $g_\perp$ are close to 2. Ions with $S = 5/2$ have five electrons on the d shell, so orbital momentum is equal to zero. Since the main contribution to the anisotropy results from spin–orbit interaction, ions with $S = 5/2$ remain almost isotropic even in a strongly anisotropic

Table 2.2 Parameters describing the anisotropy of selected magnetic ions in wurtzite semiconductors

Ion	Host	Spin	D (meV)	$g_{\parallel}$	$g_{\perp}$	Reference
Co^{2+}	ZnO	3/2	0.341	2.2384	2.2768	Jedrecy et al. [83]
Co^{2+}	ZnO	3/2	0.342	2.238	2.2755	Sati et al. [87]
Co^{2+}	ZnO	3/2	0.345	2.28		Pacuski et al. [25]
Co^{2+}	CdSe	3/2		2.295	2.303	Hoshina [88]
Co^{2+}	CdSe	3/2	0.062			Lewicki et al. [89]
Co^{2+}	CdS	3/2	0.084			Lewicki et al. [89]
Mn^{3+}	GaN	2	0.27	1.91	1.98	Marcet et al. [84]
Fe^{3+}	ZnO	5/2	−0.0074	2.0062	2.0062	Heitz et al. [11]
Mn^{2+}	ZnO	5/2	−0.0027	2.0016	2.0016	Chikoidze et al. [90]
Fe^{3+}	GaN	5/2	−0.0093	1.990	1.997	Heitz et al. [85]
Fe^{3+}	GaN	5/2	−0.0095	2.009	2.005	Malguth et al. [86]
Fe^{3+}	GaN	5/2		2.009	2.005	Bonanni et al. [91]

Note that particularly significant anisotropy can be observed for ions with spin other than 5/2 in wide-gap DMS

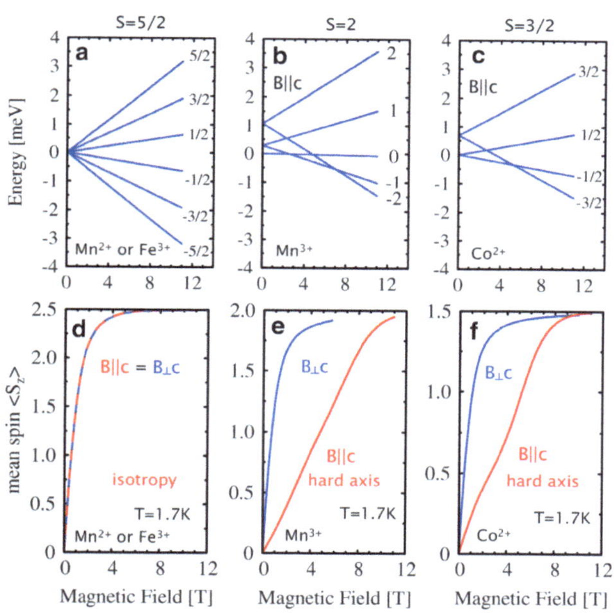

Fig. 2.16 **(a, b, c)** Zeeman splitting of the fundamental state of various magnetic ions in wide-gap DMS. Magnetic field is parallel to the c-axis. **(d, e, f)** Calculated mean spin of magnetic ions. It is calculated for two directions of magnetic field. Compare above curves to magnetic field dependence of the giant Zeeman splitting (Fig. 2.6) and MCD signal (Figs. 2.9 and 2.8) in Sect. 2.2

wurtzite structure. Consequently, the magnetization of an $S = 5/2$ can be described well by the Brillouin function. It is different in the case of $S = 3/2$ or $S = 2$, for which the zero-field splitting is comparable to $k_B T$ under typical experimental conditions. For example, the liquid helium temperature $T = 4.2$ K corresponds to

$k_B T = 0.36$ meV. The zero-field splitting for $S = 3/2$ is equal to $2D$, and it equals to 0.69 meV for Co^{2+} in ZnO. Hence, the zero-field splitting is twice larger than $k_B T$. Similar situation is for Mn^{3+} in GaN. Therefore, the anisotropy is crucial for the magnetization of $S = 3/2$ and $S = 2$ spin configurations in wide-gap DMS. In CdSe and CdS, the zero-field splitting is about four times smaller, and $2D$ values are roughly equal to $k_B T$ at superfluid helium temperatures.

Below we give analytic expressions describing the magnetization curves for the magnetic field parallel to the c-axis (B_z). We calculated them by combining the Hamiltonian (2.19) with the Maxwell-Boltzmann distribution. The projection of the mean spin for a magnetic ion with $S = 2$ (e.g., Mn^{3+}, Cr^{2+}, Fe^{2+}, Co^{3+}) is given by:

$$\langle -S_z \rangle = \frac{e^{-d} \sinh(\delta) + 2e^{-4d} \sinh(2\delta)}{\frac{1}{2} + e^{-d} \cosh(\delta) + e^{-4d} \cosh(2\delta)}, \tag{2.20}$$

while for ions with $S = 3/2$ (e.g., Cr^{3+}, Co^{2+}, V^{2+}):

$$\langle -S_z \rangle = \frac{\frac{1}{2} \sinh(\frac{1}{2}\delta) + \frac{3}{2} e^{-2d} \sinh(\frac{3}{2}\delta)}{\cosh(\frac{1}{2}\delta) + e^{-2d} \cosh(\frac{3}{2}\delta)}, \tag{2.21}$$

and for ions with $S = 1$ (e.g., Co^{+}):

$$\langle -S_z \rangle = \frac{e^{-d} \sinh(\delta)}{\frac{1}{2} + e^{-d} \cosh(\delta),} \tag{2.22}$$

where

$$d = \frac{D}{k_B T}, \tag{2.23}$$

$$\delta = \frac{g_{||} \mu_B B_z}{k_B T}. \tag{2.24}$$

Figure 2.16 shows that the anisotropy cannot be neglected for low temperature magnetization of Co^{2+} in ZnO and Mn^{3+} in GaN, when the magnetic field is along the hard magnetization axis (c-axis). For the easy magnetization plane (perpendicular to c-axis), the deviation from the Brillouin function is much weaker.

We discuss here only the magnetization of isolated magnetic ions, because observation of the giant Zeeman effect in wide-gap DMSs is usually limited to low concentrations of magnetic ions, where ion–ion interactions play only minor role.

2.5 Conclusions

Spectroscopy of ZnO and GaN-based DMS revealed several interesting properties of these materials:

- Excitons split under the magnetic field due to $s, p{-}d$ exchange interaction (giant Zeeman splitting is observed).

- Effect of s, $p-d$ interactions on excitons can be described using effective exchange integrals $N_0\alpha^{(\mathrm{app})}$ and $N_0\beta^{(\mathrm{app})}$. Observed $N_0\beta^{(\mathrm{app})}$ has a weak amplitude and at least in the case of GaN, positive sign. This shows that $N_0\beta^{(\mathrm{app})}$ is significantly different from expected large and negative $N_0\beta$. The difference can be explained by strong potential of magnetic ions, which bind holes and indirectly influence delocalized states involved in excitonic transitions.
- Electron hole exchange induces significant excitonic interactions. It decreases excitonic spin splitting induced by magnetic field. Neglecting electron–hole exchange interaction in interpretation of giant Zeeman splitting of free excitons in ZnO based DMS can lead to the error by factor of 2.
- Magnetic anisotropy is crucial for understanding of magnetic field and temperature dependence of excitonic splitting induced by s, $p-d$ interaction in wide-gap DMSs with configuration other than d^5.

Acknowledgements

Results reviewed in this Chapter originate from collaborations with A. Bonanni, E. Chikoidze, C. Deparis, T. Dietl, Y. Dumont, H. Mariette, J. Kossut, C. Morhain, A. Navarro-Quezada, E. Przeździecka, E. Sarigiannidou, P. Sati, A. Stepanov, M. Wegscheider, and many others. We acknowledge for their contribution. Special thanks should be given to J. Cibert, D. Ferrand, J.A. Gaj, and P. Kossacki, who were the supervisors of my thesis devoted to the spectroscopy of wide-gap DMSs [45].

We acknowledge financial support from the Polish Ministry of Science and Higher Education, the French Ministry of Foreign Affairs, the Marie Curie Actions (contract number MTKD-CT-2005-029671), the Foundation for Polish Science, the Deutscher Akademischer Austausch Dienst, and the Alexander von Humboldt Foundation.

References

1. T. Dietl, H. Ohno, F. Matsukura, Phys. Rev. B **63**, 195205 (2001)
2. J.I. Hwang et al., Phys. Rev. B **72**, 085216 (2005)
3. T. Dietl, H. Ohno, F. Matsukura, J. Cibert, D. Ferrand, Science **287**, 1019 (2000)
4. S.I. Gubarev, Zh. Eksp. Teor. Fiz **80**, 1174 (1983)
5. M. Nawrocki, J.P. Lascaray, D. Coquillat, M. Demianiuk, in *Diluted Magnetic (Semimagnetic) Semiconductors*, eds. by S. von Molnar, R. Aggarwal, J.K. Furdyna, Math. Res. Soc. Symp. Proc. **89**, 65 (1987)
6. C. Benoit à la Guillaume, D. Scalbert, T. Dietl, Phys. Rev. B **46**, 9853 (1992)
7. R. Pappalardo, D.L. Wood, R.C. Linares Jr, J. Chem. Phys. **35**, 2041 (1961)
8. R.S. Anderson, Phys. Rev. **164**, 398 (1967)
9. P. Koidl, Phys. Rev. B **15**, 2493 (1977)
10. H.-J. Schulz, M. Thiede, Phys. Rev. B **35**, 18 (1987)
11. R. Heitz, A. Hoffmann, I. Broser, Phys. Rev. B **45**, 8977 (1992)
12. K. Sato, H. Katayama-Yoshida, Jpn. J. Appl. Phys., Part 2, **39**, L555 (2000)

13. J. Blinowski, P. Kacman, T. Dietl, Math. Res. Soc. Symp. Proc. 690; arXiv: cond-mat/0201012 (2002)
14. N.A. Spaldin, Phys. Rev. B **69**, 125201 (2004)
15. H. Hori et al., Phys. B: Condens. Matter **324**, 142 (2002)
16. V.A. Chitta et al., Appl. Phys. Lett. **85**, 3777 (2004)
17. K.R. Kittilstved, N.S. Norberg, D.R. Gamelin, Phys. Rev. Lett. **94**, 147209 (2005)
18. K. Ueda, H. Tabata, T. Kawai, Appl. Phys. Lett. **79**, 988 (2001)
19. S.-W. Lim, D.-K. Hwang, J.-M. Myoung, Solid State Commun. **125**, 231 (2003)
20. K. Rode et al., J. Appl. Phys. **93**, 7676 (2003)
21. S. Ramachandran, A. Tiwari, J. Narayan, Appl. Phys. Lett. **84**, 5255 (2004)
22. A.C. Tuan et al., Phys. Rev. B **70**, 054424 (2004)
23. M. Venkatesan, C.B. Fitzgerald, J.G. Lunney, J.M.D. Coey, Phys. Rev. Lett. **93**, 177206 (2004)
24. K. Ando, H. Saito, V. Zayets, M.C. Debnath, J. Phys. Condens. Matter **16**, S5541 (2004)
25. W. Pacuski et al., Phys. Rev. B **73**, 035214 (2006)
26. J. Lagois, Phys. Rev. B **16**, 1699 (1977)
27. J. Lagois, Phys. Rev. B **23**, 5511 (1981)
28. D.W. Langer, R.N. Euwema, K. Era, T. Koda, Phys. Rev. B **2**, 4005 (1970)
29. D.C. Reynolds et al., Phys. Rev. B **60**, 2340 (1999)
30. B. Gil et al., Jpn. J. Appl. Phys. **40**, L 1089 (2001)
31. B. Gil, Phys. Rev. B **64**, 201310 (2001)
32. W.R.L. Lambrecht, A.V. Rodina, S. Limpijumnong, B. Segall, B.K. Meyer, Phys. Rev. B **65**, 075207 (2002)
33. T. Gruber et al., J. Appl. Phys. **96**, 289 (2004)
34. D.G. Thomas, J. Phys. Chem. Solids **15**, 86 (1960)
35. R. Stępniewski et al., Phys. Rev. B **60**, 4438 (1999)
36. W. Pacuski et al., Phys. Rev. Lett. **100**, 037204 (2008)
37. W. Pacuski et al., Phys. Rev. B **76**, 165304 (2007)
38. M. Dobrowolska, D. Dobrowolski, R.R. Galazka, A. Mycielski, Phys. Stat. Sol. B **105**, 477 (1981)
39. E. Przeździecka et al., Solid State Commun. **139**, 541 (2006)
40. J.F. Muth et al., Appl. Phys. Lett. **71**, 2572 (1997)
41. E. Chikoidze, Y. Dumont, H. von Bardeleben, W. Pacuski, O. Gorochova, Superlattice. Microstruct. **42**, 176 (2007)
42. D.A. Schwartz, N.S. Norberg, Q.P. Nguyen, J.M. Parker, D.R. Gamelin, J. Am. Chem. Soc. **125**, 13205 (2003)
43. K. Ando, *7th Symposium PASPS 2001*, condmat/0208010 (2002)
44. K.R. Kittilstved et al., Appl. Phys. Lett. **89**, 062510 (2006)
45. W. Pacuski, Ph. D. thesis, University of Warsaw and Université Joseph Fourier, Grenoble, http://tel.archives-ouvertes.fr, 2008
46. J. Gaj, R. Planel, G. Fishman, Solid State Commun. **29**, 435 (1979)
47. A. Twardowski, P. Swiderski, M. von Ortenberg, R. Pauthenet, Solid State Commun. **50**, 509 (1984)
48. A. Twardowski, P. Swiderski, M. von Ortenberg, R. Pauthenet, Solid State Commun. **51**, 849 (1984)
49. B.E. Larson, K.C. Hass, H. Ehrenreich, A.E. Carlsson, Phys. Rev. B **37**, 4137 (1988)
50. J.R. Schrieffer, P.A. Wolff, Phys. Rev. **149**, 491 (1966)
51. J.R. Schrieffer, J. Appl. Phys. **38**, 1143 (1967)
52. J. Blinowski, P. Kacman, Acta Phys. Pol. A **100**, 343 (2001)
53. J. Okabayashi et al., Phys. Rev. B **58**, R4211 (1998)
54. J. Okabayashi et al., Phys. Rev. B **59**, R2486 (1999)
55. T. Mizokawa, T. Nambu, A. Fujimori, T. Fukumura, M. Kawasaki, Phys. Rev. B **65**, 085209 (2002)
56. J. Okabayashi et al., Phys. Rev. B **65**, 161203 (2002)
57. J. Okabayashi et al., J. Appl. Phys. **95**, 3573 (2004)

58. T. Dietl, *Exchange Interactions and Nanoscale Phase Separations in Magnetically Doped Semiconductors*, in *Spintronics*, eds. by T. Dietl, D.D. Awschalom, M. Kaminska, H. Ohno. Semiconductors and Semimetals, vol 82 (Elsevier, San Diego, 2008)
59. T. Chanier, F. Virot, R. Hayn, Phys. Rev. B **79**, 205204 (2009)
60. R. Bouzerar, G. Bouzerar, T. Ziman, Europhys. Lett. (EPL) **78**, 67003 (5pp) (2007)
61. T. Dietl, Phys. Rev. B **77**, 085208 (2008)
62. W. Pacuski et al., Phys. Rev. Lett. **100**, 037204 (2008)
63. C. Śliwa, T. Dietl, Phys. Rev. B **78**, 165205 (2008)
64. R.C. Myers, M. Poggio, N.P. Stern, A.C. Gossard, D.D. Awschalom, Phys. Rev. Lett. **95**, 017204 (2005)
65. M. Julier, J. Campo, B. Gil, J.P. Lascaray, S. Nakamura, Phys. Rev. B **57**, R6791 (1998)
66. B. Gil, A. Alemu, Phys. Rev. B **56**, 12446 (1997)
67. M. Arciszewska, M. Nawrocki, J. Phys. Chem. Solids **47**, 309 (1986)
68. F. Hamdani et al., Phys. Rev. B **45**, 13298 (1992)
69. J.L. Birman, Phys. Rev. Lett. **2**, 157 (1959)
70. M. Herbich et al., Phys. Rev. B **58**, 1912 (1998)
71. Y.G. Semenov, V.G. Abramishvili, A.V. Komarov, S.M. Ryabchenko, Phys. Rev. B **56**, 1868 (1997)
72. B. Gil, O. Briot, R.-L. Aulombard, Phys. Rev. B **52**, R17028 (1995)
73. B. Gil, J. Appl. Phys. **98**, 086114 (2005)
74. T. Dietl, Phys. Rev. B **77**, 085208 (2008)
75. R.L. Aggarwal et al., Phys. Rev. B **28**, 6907 (1983)
76. M. Nawrocki, F. Hamdani, J.P. Lascaray, Z. Golacki, J. Deportes, Solid State Commun. **77**, 111 (1991)
77. A. Twardowski, K. Pakula, I. Perez, P. Wise, J.E. Crow, Phys. Rev. B **42**, 7567 (1990)
78. W.Y. Yu, A. Twardowski, L.P. Fu, A. Petrou, B.T. Jonker, Phys. Rev. B **51**, 9722 (1995)
79. J.M. Luttinger, Phys. Rev. **102**, 1030 (1956)
80. W. Pacuski et al., Acta Phys. Pol. A **110**, 303 (2006)
81. A. Abragam, M.H.L. Pryce, Proc. R. Soc. Lond. Ser. A, Math. Phys. Sci. **205**, 135 (1951)
82. P. Sati, C. Deparis, C. Morhain, S. Schafer, A. Stepanov, Phys. Rev. Lett. **98**, 137204 (2007)
83. N. Jedrecy, H.J.V. Bardeleben, Y. Zheng, J.L. Cantin, Phys. Rev. B **69**, R041308 (2004)
84. S. Marcet et al., Phys. Rev. B **74**, 125201 (2006)
85. R. Heitz et al., Appl. Phys. Lett. **67**, 2822 (1995)
86. E. Malguth et al., Phys. Rev. B **74**, 165202 (2006)
87. P. Sati et al., Phys. Rev. Lett. **96**, 017203 (2006)
88. T. Hoshina, J. Phys. Soc. Jpn. **21**, 1608 (1966)
89. A. Lewicki, A.I. Schindler, I. Miotkowski, B.C. Crooker, J.K. Furdyna, Phys. Rev. B **43**, 5713 (1991)
90. E. Chikoidze, H.J.V. Bardeleben, Y. Dumont, P. Galtier, J.L. Cantin, J. Appl. Phys. **97**, 10D316 (2005)
91. A. Bonanni et al., Phys. Rev. B **75**, 125210 (2007)

Chapter 3
Exchange Interaction Between Carriers and Magnetic Ions in Quantum Size Heterostructures

I.A. Merkulov and A.V. Rodina

Abstract A very general theoretical discussion of the values of $sp-d$ exchange constants is given starting from symmetry properties of diluted magnetic semiconductors. Complex structure of the valence band is fully taken into account with interacting heavy and light holes as well as spin–orbit split-off states kept in the consideration. The form of the Hamiltonian is derived and shown to be, in a general case, anisotropic. A series of approximations is made that reduces the Hamiltionian to a simple form introduced in Chap. 1. Specific cases of two-dimensional quantum wells and zero-dimensional spherical quantum dots are considered. Finally, the role of $sp-d$ interaction in spin relaxation processes is discussed.

3.1 Introduction

In semiconductor nano-heterostructures the potential energy, that a charge carrier experiences, changes on the length scale that is comparable to the characteristic wavelength of the carriers. In such a case, the quantum-size effects play a dominant role. The quantum confinement significantly affects the exchange interaction between carriers and localized spins of the magnetic ions. The effective exchange Hamiltonian becomes anisotropic because the presence of interfaces reduces the symmetry. In narrow structures with high potential barriers, the quantum confinement changes the ground state energy of the carriers to such an extent that the changes of the ground energy can be comparable to the band gap energy of a bulk

I.A. Merkulov (✉)
Ioffe Physico-Technical Institute, Russian Academy of Sciences, 194021 St. Petersburg, Russia
and
Condensed Matter Sciences Division, Oak Ridge National Laboratory, Oak Ridge, TN 37831-6016, USA
e-mail: imp@mail.ornl.gov

A.V. Rodina
Ioffe Physico-Technical Institute, Russian Academy of Sciences, 194021 St. Petersburg, Russia
e-mail: anna.rodina@mail.ioffe.ru

J. Kossut and J.A. Gaj (eds.), *Introduction to the Physics of Diluted Magnetic Semiconductors*, Springer Series in Materials Science 144,
DOI 10.1007/978-3-642-15856-8_3, © Springer-Verlag Berlin Heidelberg 2010

semiconductor. This results in significant modification of the exchange parameters. In this Chapter, we provide simple examples that illustrate these effects.

Section 3.2 contains a brief review of the multiband envelope wave functions and the energy spectrum of confined carriers in diamond-like semiconductors. We describe also the model heterostructures that we consider later in this Chapter. In Sect. 3.3, we present the exchange Hamiltonian in quantum-size heterostructures based on semimagnetic (diluted magnetic) semiconductors: the exchange Hamiltonian for the valence band holes is derived and analyzed to illustrate the general form of the exchange Hamiltonian. In Sect. 3.4, we study the renormalization of the exchange interaction constants that occurs with increase of the electron kinetic energy. The last Sect. 3.5 is devoted to the most interesting experiments that illustrate general theoretical conclusions. The material of this Chapter will be also used in Chaps. 7 and 8. In the Appendix, we review general properties of the eight-band Kane model and discuss general boundary conditions for the envelope wave function at heterointerfaces in narrow and deep quantum wells.

3.2 Energy Band Structure and Wave Functions of the Electrons and Holes

The detailed theoretical description of the energy band structure of semiconductor heterostructures can be found in the recently published monograph by E. Ivchenko [1] as well as in the previous book by Ivchenko and Pikus [2] (see also Chap. 4). Here, we provided only a short introduction summarizing the main results related to the description of the exchange interaction between carriers and magnetic ions in semimagnetic heterostructures.

3.2.1 *Electrons and Holes in 3D GaAs-like Crystals*

The bottom of the conduction band in bulk semiconductors with zinc-blende structure including also (Cd,Mn)Te semimagnetic compounds is in the center of the Brillouin zone (Γ-point). It is twice degenerate by spin ($s_{\mathrm{e}} = 1/2$) (Fig. 3.1). In the vicinity of the Γ-point, the electron kinetic energy is given by the conventional expression

$$\hat{K}_{\mathrm{e}} = \frac{1}{2m_{\mathrm{e}}^*} \sum_{\alpha=x,y,z} \hat{p}_{\alpha}^2, \tag{3.1}$$

where $\hat{p}_{x,y,z} = -\mathrm{i}\hbar\partial/\partial r_{x,y,z}$ is the operator of the quasimomentum projection on one of three crystallographic axes [100], [010], and [001], and m_{e}^* is an electron effective mass. The operator $\hat{\boldsymbol{p}}$ acts on the electron envelope wave function. In the ideal crystal, the envelope function is described by traveling plane wave with the wave vector $\boldsymbol{k}$. Thus, near the bottom of the conduction band the full electron wave function can be written as

$$\varphi_{\nu}(\boldsymbol{k},\boldsymbol{r}) \approx u_{\nu}^{\mathrm{c}}(\boldsymbol{r}) \exp\{\mathrm{i}\,(\boldsymbol{k}\cdot\boldsymbol{r})\}\,. \tag{3.2}$$

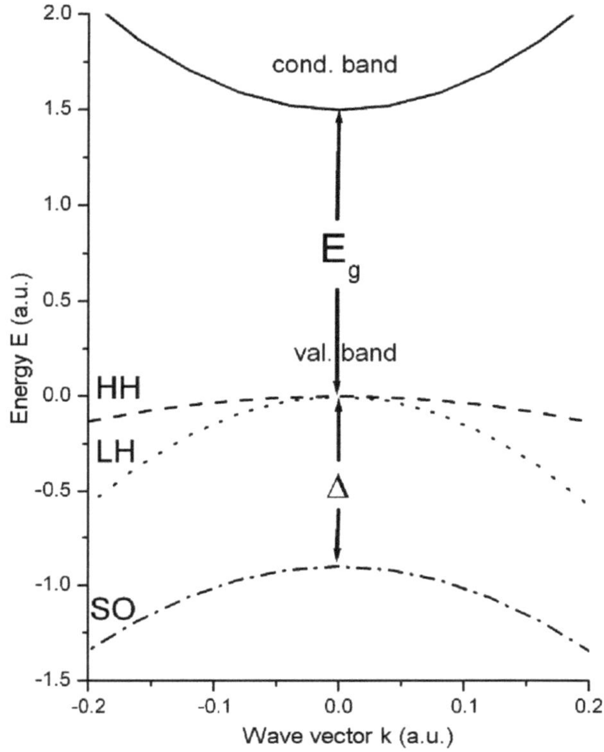

Fig. 3.1 Energy band structure near the Γ-point of the semiconductors with zinc-blende structure: the top of the valence band and the bottom of the conduction band are shown. The energy and the wave vector are given in the arbitrary units

Here, $u_\nu^c(\boldsymbol{r}) \equiv \varphi_\nu(0, \boldsymbol{r})$ is a Bloch amplitude oscillating with the period of the crystal lattice. In the spherical approximation, it has the symmetry of an s-type atomic orbital: $u_\nu^c = |S\rangle u_\nu$, where $|S\rangle$ is the eigenfunction of the internal[1] orbital momentum $l_c = 0$ and $\nu = \pm 1/2$ denotes the spin states so that $u_{1/2}$ and $u_{-1/2}$ are the eigenfunctions of the projection of the spin operator $\hat{s} = 1/2\hat{\boldsymbol{\sigma}}$ (where $\hat{\boldsymbol{\sigma}} = \{\hat{\sigma}_x, \hat{\sigma}_y, \hat{\sigma}_z\}$ are the Pauli matrices) on the arbitrarily chosen Z-axis. The normalization constant in (3.2) is omitted, which corresponds to the choice of the sample volume $V = (2\pi)^3$.

The valence band in these crystals has also an extremum at the Γ-point. It consists of three subbands that, in the absence of the spin–orbit interaction, have the same energy in the center of the Brillouin zone [3]. The respective Bloch amplitudes

[1] This momentum describes the orbital motion of the quasiparticle inside the central cell and is called internal in contrast to the external orbital momentum that describes the orbital motion of the quasiparticle over the crystal (see, Sect. 3.2.4).

have the symmetry of p-type atomic orbitals $|R_x\rangle = |X\rangle, |R_y\rangle = |Y\rangle, |R_z\rangle = |Z\rangle$ so that the combinations $(|R_x\rangle \pm i|R_y\rangle)/\sqrt{2}$ and $|R_z\rangle$ are the eigenfunctions of the Z-projection of the internal orbital momentum operator $\boldsymbol{l}_v$ ($l_v = 1$) with eigenvalues ± 1 and 0, respectively. Spin–orbital interaction removes partly this degeneracy and the states are described by the eigenfunctions of the Z-projection of the total internal momentum operator $\boldsymbol{I} = \boldsymbol{l}_v + \boldsymbol{s}$. This total internal momentum $I = |\mathbf{I}| = 3/2$ is commonly known as the spin of the hole. The states with the total internal momentum length $I = 3/2$ form the fourfold degenerate top of the valence band:

$$u^{\mathrm{v}}_{\pm 3/2} = \mp\frac{|X \pm iY\rangle}{\sqrt{2}}u_{\pm 1/2}, \quad u^{\mathrm{v}}_{\pm 1/2} = \sqrt{\frac{2}{3}}|Z\rangle u_{\pm 1/2} \mp \frac{|X \pm iY\rangle}{\sqrt{6}}u_{\mp 1/2}, \tag{3.3}$$

while the states with total momentum length $I = 1/2$

$$u^{\mathrm{so}}_{\pm 1/2} = \pm\frac{1}{\sqrt{3}}\left[|Z\rangle u_{\pm 1/2} \pm |X \pm iY\rangle u_{\mp 1/2}\right] \tag{3.4}$$

are shifted by the energy Δ and correspond to the so-called "spin–orbit split-off subband".[2]

The valence band is four-time degenerate only at the Brillouin zone center, at $k = 0$. The degeneracy is partly removed for nonzero wave vector k. There the valence band splits into two subbands, which correspond to the light (m_{lh}) and heavy (m_{hh}) effective masses of the holes. In the spherical approximation, the heavy hole subband is characterized by the maximal absolute value of the total momentum $\boldsymbol{I}$ projection on the hole wave vector ($I_{\boldsymbol{k},\mathrm{hh}} = \pm 3/2$), while for the light holes this projection has minimal absolute value ($I_{\boldsymbol{k},\mathrm{lh}} = \pm 1/2$). The Luttinger effective Hamiltonian [4] describes the hole kinetic energy in the general case with the account taken for the cubic anisotropy of the crystal lattice:

$$\hat{K}_h^{(L)} = \frac{1}{2m_0}\left[\left(\gamma_1^{\mathrm{L}} + \frac{5}{2}\gamma_2^{\mathrm{L}}\right)\hat{p}^2 - 2\gamma_2^{\mathrm{L}}\left(\hat{p}_x^2\hat{I}_x^2 + \hat{p}_y^2\hat{I}_y^2 + \hat{p}_z^2\hat{I}_z^2\right)\right.$$
$$\left. -2\gamma_3^{\mathrm{L}}\left(\hat{p}_x\hat{p}_y(\hat{I}_x\hat{I}_y + \hat{I}_y\hat{I}_x) + \hat{p}_y\hat{p}_z(\hat{I}_y\hat{I}_z + \hat{I}_z\hat{I}_y) + \hat{p}_z\hat{p}_x(\hat{I}_z\hat{I}_x + \hat{I}_x\hat{I}_z)\right)\right]. \tag{3.5}$$

Here, m_0 is the free electron mass, γ_1^{L}, γ_2^{L}, and γ_3^{L} are the Luttinger parameters. The hole effective masses can be expressed via γ_1^{L}, γ_2^{L}, and γ_3^{L} for different orientations of $\boldsymbol{I}$ and $\boldsymbol{p}$ with respect to three crystallographic axes [4].[3] The nonparabolicity of the energy dispersion of the electron and hole states increases with increase of the

---

[2] There is some arbitrariness in the choice of the Bloch amplitude phases. Here, we choose the phase factors as in [1]. Some alternative choices of phase factors are presented, for example, in [3,4]. Although any finally meaningful value is independent on the particular choice of the Bloch amplitude phases, the same choice of phases should be kept in all expressions during the derivation.

[3] In the spherical approximation $\gamma_3^{\mathrm{L}} = \gamma_2^{\mathrm{L}}$, and the heavy and light hole effective masses are given by $m_{\mathrm{hh}} = m_0/\left(\gamma_1^{\mathrm{L}} - 2\gamma_2^{\mathrm{L}}\right)$ and $m_{\mathrm{lh}} = m_0/\left(\gamma_1^{\mathrm{L}} + 2\gamma_2^{\mathrm{L}}\right)$, respectively.

wave vector k. The deviation of the Bloch amplitudes of such states from those at the Γ-point increases as well. However, for the relatively small values of the wave vector the wave functions and the energy spectra of the carriers can be found with the help of the $\boldsymbol{kp}$-perturbation theory [3, 5].

As it follows from expressions (3.1) and (3.5), the spin–orbit interaction is nearly negligible for the conduction band electrons but significant for the valence band holes. As a result, an electron keeps the memory about its initial spin direction a relatively long time, whereas scattering processes lead to very fast spin relaxation for holes.

3.2.2 Electrons in a Symmetrical 2D Quantum Well

In this subsection, we consider the quantum confinement of electrons in a symmetric $2D$ square quantum well as a model example. We put the well interfaces at the (001) planes $z = \pm L/2$ and assume infinite potential barriers for electrons. The translation symmetry in the quantum well plane allows one to characterize the electron states by the $2D$-wave vector $k_\perp \equiv (k_x, k_y)$ and to describe the motion along the quantum well plane by the wave function $\psi_\perp(\rho) = \exp\{\mathrm{i}(k_\perp \cdot \rho)\}$, where $\rho \equiv (x, y)$. The spin–orbital interaction for electrons is negligible. The motion along the quantum well plane and the motion in Z direction are independent of each other (3.1). Inside the well, the electron wave functions in Z direction are either symmetric or antisymmetric with respect to the well center ($z = 0$) and are described by the even or odd standing waves, respectively,

$$\phi(z) \propto \begin{cases} u^{\mathrm{c}}_{\pm 1/2} \cos(\mathrm{k}_n z) & n = 1, 3, 5, \ldots \\ u^{\mathrm{c}}_{\pm 1/2} \sin(\mathrm{k}_n z) & n = 2, 4, 6, \ldots \end{cases} \tag{3.6}$$

Here, n is a number of the quantum size subband, $u^{\mathrm{c}}_{\pm 1/2}$ are Bloch amplitudes of the conduction band states the same as in (3.2), $\mathrm{k}_n = \pi n/L$ is the respective wave vector along Z direction, and L is the quantum well width. In Fig. 3.2, we sketch the electron wave functions and edges of the energy levels for the two first subbands $n = 1$ and $n = 2$ in the quantum well with infinite potential barriers.

The problem becomes more complicated for very narrow and deep quantum wells. In this case, the electron quantum size energy, even at the subband bottom, is large and comparable with the band gap energy. Therefore, the interband mixing of the conduction band and valence band states caused by the $\boldsymbol{kp}$-perturbation is important. The energy level structure and electron wave functions for such states are considered in Appendix. In Sect. 3.4 of this Chapter, we will discuss corrections to the electron exchange interaction caused by the interband mixing.

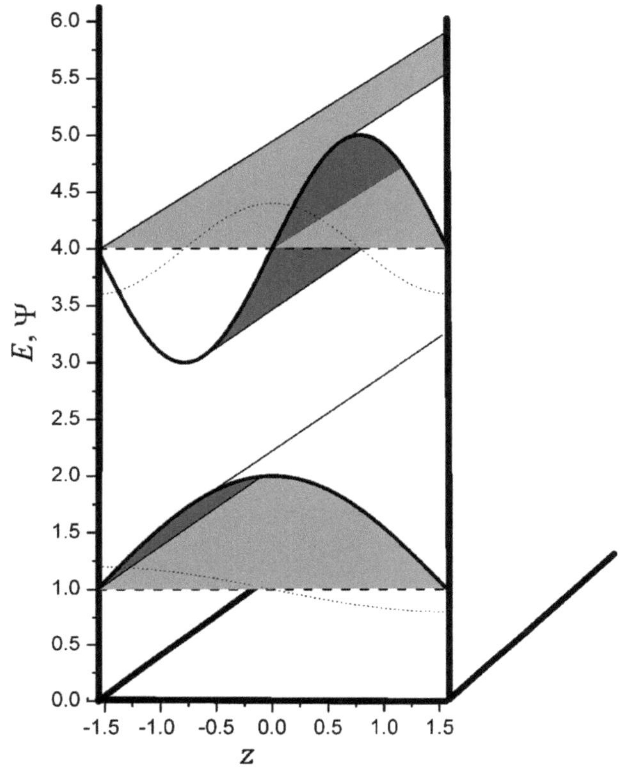

Fig. 3.2 Energy levels and electron wave functions at the bottom of first two $2D$ subbands for the electron confined in the quantum well with infinite potential barriers. For both $2D$ subbands, the conduction band component of the wave function vanishes at the interfaces. It has an antinode at the middle of the well for the first subband and a node for the second. The admixture of valence band component to the electron wave function in the framework of the Kane model is shown by the *dashed line*. Here, the limit case of strongly asymmetric boundary condition with $\left(\Psi_c/\Psi_z^{\mathrm{v}} = 0\right)$ is considered. The conduction band and valence band components have the same wavelength but have the shift of phases equal to the $\pi/2$, quarter of the wavelength. The length is measured in the inverse of the electron wave vector (k_1^{-1}) at the first quantum size level, and the energy of this level is taken as the energy unit

3.2.3 *Holes in a Symmetrical 2D Quantum Well*

The quantum confinement of the valence band holes in a square quantum well was first considered by Nedorezov [6]. The hole wave functions at the bottom of $2D$ subbands are similar to those given by (3.6). However, in this case a strong spin–orbital interaction in the valence band forms two sets of subbands that correspond to size quantization of light and heavy holes, respectively. At the bottom of the light hole subband, the holes have the internal momentum projection on the Z axis $I_z = \pm 1/2$ (Bloch amplitude $u^{\mathrm{v}}_{\pm 1/2}$), at the bottom of the heavy hole subband the internal momentum projection is equal to $I_z = \pm 3/2$ (Bloch amplitude $u^{v}_{\pm 3/2}$). In Sect. 3.3.2, we will show that this fact alone may cause a strong anisotropy of the exchange interaction at the bottom of $2D$ hole subband. Such anisotropy is especially large for the heavy hole subband.

Away from $k = 0$, the subbands mix. The mixing increases with an increase of the in-plane wave vector. As a result, for nonzero $k_\perp$ the hole wave functions cannot be characterized by the definite value of the momentum projection on the Z axis, and the general expressions for them are quite complicated (see, for example, [7]). Here, we restrict the analysis to the heavy holes subband that has the lowest energy in the square quantum well with infinite potential barriers. With an accuracy up to the linear in $k_\perp$ terms, we obtain the following expression for the wave function:

$$\Psi_{\mathrm{hh},\pm 3/2}(z, \boldsymbol{k}_\perp) = \sqrt{\frac{2}{L}}\left[u^{v}_{\pm 3/2}\cos(\mathrm{k}_1 z) \pm \sqrt{3}Lk_\pm u^{v}_{\pm 1/2}\psi_{\mathrm{lh}}(z)\right]\mathrm{e}^{\mathrm{i}(\boldsymbol{k}_\perp \boldsymbol{\rho})}, \quad (3.7)$$

$$\psi_{\mathrm{lh}} = \mathrm{i}\frac{8\gamma_3}{\pi L}\sum_{n=1}^{\infty}\frac{(-1)^{n-1}\sin(\mathrm{k}_{2n}z)\,\mathrm{k}_{2n}}{\left((2n)^2-1\right)\left((\gamma_1-2\gamma_2)\,\mathrm{k}_1^2-(\gamma_1+2\gamma_2)\,\mathrm{k}_{2n}^2\right)}$$
$$\approx -\mathrm{i}\frac{8\gamma_3}{3\pi^2}\frac{\sin(\mathrm{k}_2 z)}{((\gamma_1+2\gamma_2)\,4-(\gamma_1-2\gamma_2))}, \quad (3.8)$$

where $k_n = \pi n/L$ and $k_\pm = k_{\perp x} \pm i k_{\perp y}$. The $k_\perp^2$ terms lead to additional state mixing.

3.2.4 Holes in a Spherical Quantum Dot

Now let us consider the valence band hole states confined in a spherical quantum dot. Due to the spherical symmetry of the confining potential, the ground state is fourfold degenerate, similar to the valence band top in the bulk. However, the internal momentum of the hole ($I = 3/2$) is not a "good quantum number" any more. Instead, the ground state is characterized by the total momentum operator $\boldsymbol{F} = \boldsymbol{I} + \boldsymbol{L}$ ($F = 3/2$), which is the sum of $\boldsymbol{I}$ and external orbital momentum $\boldsymbol{L}$ ($L = 0, 2$) that describes the orbital motion of the hole between different cells inside the dot. The ground state wave function in the spherical quantum dot with infinite potential barrier has the form [1]:

$$\Psi_{F,F_z}(\boldsymbol{r}) = \frac{1}{\sqrt{4\pi}R^{3/2}}\sum_m u^{v}_m \times\langle 3/2, m|\left\{f_0\left(\frac{r}{R}\right) - f_2\left(\frac{r}{R}\right)\left[\left(\hat{\boldsymbol{F}}\cdot\frac{\boldsymbol{r}}{r}\right)^2 - \frac{5}{4}\right]\right\}|3/2, F_z\rangle, \quad (3.9)$$

where R is the dot radius, $r = |\boldsymbol{r}|$ is the radial coordinate of the hole in the spherical coordinate system, u^{v}_m are the Bloch amplitudes of the valence band top

($m = \pm 1/2, \pm 3/2$) given by (3.3). The matrix components of the vector operator $\hat{\boldsymbol{F}}$ are described by conventional expressions for spin $F = 3/2$. The radial functions f_0 and f_2 depend only on the ratio $x = r/R$ and can be written as

$$f_L(x) = C\left[j_L(\phi x) - (-1)^{L/2}\frac{j_0(\phi)}{j_0\left(\sqrt{\beta}\phi\right)}j_L\left(\sqrt{\beta}\phi x\right)\right], \tag{3.10}$$

where j_L are spherical Bessel functions, $\beta = m_{\text{lh}}/m_{\text{hh}}$ is the ratio of light to heavy hole mass, and C is the normalization constant determined by:

$$\int_0^1 \left(f_0(x)^2 + f_2(x)^2\right)x^2 \mathrm{d}x = 1. \tag{3.11}$$

The dimensionless parameter ϕ can be found from the condition $f_2(1) = 0$ and determines the energy of the state via

$$E^{(h)} = \frac{\hbar^2\phi^2}{2m_{\text{hh}}R^2}. \tag{3.12}$$

Figure 3.3a shows the radial functions f_0 and f_2 calculated for the mass ratio $\beta = 0.25$ that is close to the typical mass ratio values for widegap semimagnetic semiconductors. Using the explicit form of the wave functions, one can easily analyze the spatial distribution of the hole spin $\langle \boldsymbol{I}(\boldsymbol{r})\rangle$ quantum mechanically averaged

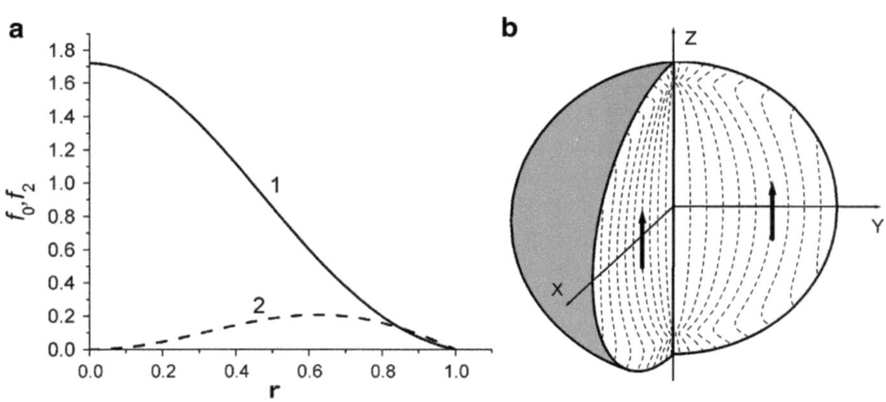

Fig. 3.3 The wave function of the hole confined in spherical quantum dot with infinite potential barrier: (**a**) – radial wave functions f_0 and f_2 for the components with orbital momentum $L = 0$ and $L = 2$, respectively. The probability to find a hole in the state with $L = 0$ is maximal in the center of the dot, while for $L = 2$ it has a maximum approximately at 2/3 from the center. The distance is measured in the units of the quantum dot radius. (**b**) – Spatial distribution of the average hole spin $\langle \boldsymbol{I}\rangle$ direction over the quantum dot volume in the state $F_z = 3/2$. The average spin $\langle \boldsymbol{I}\rangle$ is directed along the tangent to the dashed curve

over the ground state with definite projection F_z. For example, the spatial distribution of the $\langle \boldsymbol{I} \rangle$ directions for the state with $F_z = 3/2$ is shown in Fig. 3.3b. The average hole spin $\langle \boldsymbol{I}(\boldsymbol{r}) \rangle$ is tangent to the field lines shown in the Figure. It can be easily seen that even for the strongly polarized state with $F_z = 3/2$ the direction of $\langle \boldsymbol{I}(\boldsymbol{r}) \rangle$ deviates significantly from the Z direction near the dot surface. All components of $\langle \boldsymbol{I} \rangle$ are not conserved. Their spatial distributions over the quantum dot volume can serve as a visual characteristic of the steady hole state.

3.3 Anisotropic Exchange Interaction Between Carriers and Magnetic ions

3.3.1 *Carriers and Magnetic Ions Exchange Interaction in 3D Crystals*

In magnetic and semimagnetic semiconductors, the exchange scattering of carriers by magnetic ions is very strong. The exchange interaction between magnetic ions and carriers in the vicinity of the Brillouin zone center (electrons near the bottom of the conduction band or holes near the top of the valence band) is described by the Kondo Hamiltonian

$$\hat{H}^{\mathrm{ex}}_{\mathrm{e,h}} = -\alpha_{\mathrm{e,h}} \sum_m \left(\hat{\boldsymbol{s}}_{\mathrm{e,h}} \hat{\boldsymbol{S}}_m \right) \delta(\boldsymbol{r} - \boldsymbol{R}_m), \tag{3.13}$$

where $\hat{\boldsymbol{s}}, \boldsymbol{r}$ and $\hat{\boldsymbol{S}}_m, \boldsymbol{R}_m$ are the spin operator and position operator of the carrier and of the magnetic ion with number m, respectively, $\delta(\boldsymbol{r} - \boldsymbol{R}_m)$ is Dirac δ-function, and $\alpha_{\mathrm{e,h}}$ are the exchange interaction constants for electrons and holes, respectively.[4] For the electrons in (Cd,Mn)Te crystals studied in the experiments that we review below in Sect. 3.5, $\alpha_e N_0 \equiv \alpha N_0 \approx 220$ meV [8] (here $N_0 \approx 1.5 \times 10^{28}\ m^{-3}$ – is the lattice site concentration in the Cd sublattice), while for the holes $\alpha_h N_0 \equiv \beta N_0 \approx -4\alpha N_0 \approx -880$ meV.

The meaning of the Dirac δ-function in the exchange Hamiltonian (3.13) is the following. The scattering potential is formed by the d- or f-shell of the magnetic ion and acts on the length scale of the central cell size. The characteristic wave length of the carriers is much larger than the central cell size, and the envelope

---

[4] Conventionally, the exchange interaction constant for the electrons is denoted by α, and for the holes by β. Below we will use this notation except the case of the general expressions like (3.13), which are valid simultaneously for the electrons and holes. In the latter case, we denote the exchange interaction constant by α with indexes e or h for conduction band electrons or valence band holes, respectively. We note also that in this Chapter we use the hole representation for the valence band and the constant α_h is equal to the constant α_v introduced in Chap. 1 for the valence band electrons.

wave function of the initial $\Psi_i(\boldsymbol{r})$ and final $\Psi_f(\boldsymbol{r})$ states are nearly constants in this scale. Therefore, the matrix elements of the exchange interaction calculated with these functions are proportional to the product $\Psi_f^*(\boldsymbol{R}_m)\Psi_i(\boldsymbol{R}_m)$. This justifies the δ-function approximation of the potential.

The exchange constants α for electrons and β for holes have different signs. This is caused by predominance of one of the two qualitatively different mechanisms of the exchange interaction – potential or kinetic, respectively:

$$\alpha_{\mathrm{e,h}} = \alpha_{\mathrm{e,h}}^{\mathrm{pot}} + \alpha_{\mathrm{e,h}}^{\mathrm{kin}}, \quad \alpha_{\mathrm{e,h}}^{\mathrm{pot}} > 0,\ \alpha_{\mathrm{e,h}}^{\mathrm{kin}} < 0. \tag{3.14}$$

Both experimental data and theoretical calculations show that the potential exchange mechanism is dominant near the bottom of the conduction band while for the valence band holes the kinetic exchange prevails. (For details see [8,9] and Chap. 1).

Strong spin–orbit interaction in the valence band allows one to use the truncated Hamiltonian of the exchange interaction. It takes into account only the scattering between heavy hole and light hole subbands thus neglecting the scattering into the spin–orbital split-off subband

$$\hat{H}_{\mathrm{h}}^{\mathrm{ex}} = -\frac{\beta}{3}\sum_m \left(\hat{\boldsymbol{I}}\hat{\boldsymbol{S}}_m\right)\delta(\boldsymbol{r} - \boldsymbol{R}_m). \tag{3.15}$$

Here, $\hat{\boldsymbol{I}} = \hat{\boldsymbol{s}}_h + \hat{\boldsymbol{l}}_h$ is the operator of the internal hole momentum $I = 3/2$. As it has been discussed above, the quantum confinement effect partly removes the fourfold degeneracy of the valence band top. As a result, the exchange interaction of $2D$ holes with magnetic ions becomes anisotropic. It is more convenient to describe this anisotropy by considering the $2D$ holes as quasiparticles characterized by pseudospin $j = 1/2$ and anisotropic g-factor (see, for instance, [10]).

3.3.2 The General form of the Exchange Hamiltonian in 2D Heterostructures

The pseudospin description of the Kramers dublets has long been used in the theory of electron paramagnetic resonance. Theoretical details of such description can be found, for example, in [11]. Here, we will consider only the application of the pseudospin description for the Kramers dublets that originate from the fourfold degenerate valence band top due to the reduced symmetry.[5]

Let us consider the hole state in the $2D$ subband characterized by an arbitrary nonzero in-plane wave vector $\boldsymbol{k}_\perp$. Such state is doubly degenerate due to the time

[5] We will not take into account the effects of the bulk inversion asymmetry as well as the quantum well structure asymmetry on the wave functions of the Kramers dublet. These effects are usually taken into account in zero-th order of the perturbation theory by adding the additional operator of spin–orbit interaction in the $2D$ Hamiltonian (for details see [1]).

inversion symmetry. We demand that the wave functions of these degenerate states transform with the time inversion operation like the eigenfunctions of the spin 1/2 [12] and we obtain:

$$\Psi^{(n)}_{-1/2}(\boldsymbol{k}_\perp, \boldsymbol{r}) = \sum_{m=-3/2}^{3/2} \psi^{(n)}_{-1/2,m}(\boldsymbol{k}_\perp, \boldsymbol{r}) u_m = -i\left(\hat{\sigma}_y \Psi^{(n)*}_{+1/2}(\boldsymbol{k}_\perp, \boldsymbol{r})\right),$$

$$\Psi^{(n)}_{+1/2}(\boldsymbol{k}_\perp, \boldsymbol{r}) = \sum_{m=-3/2}^{3/2} \psi^{(n)}_{1/2,m}(\boldsymbol{k}_\perp, \boldsymbol{r}) u_m, \tag{3.16}$$

where n denotes the type (heavy or light) and the sequential number of the $2D$ hole subband, $\boldsymbol{k}_\perp$ is the in-plane wave vector in this subband, u_m are Bloch amplitudes at the valence band top and the explicit form of the functions $\psi^{(n)}_{\pm 1/2}(\boldsymbol{k}_\perp, \boldsymbol{r})$ depends on the problem under consideration. For $\boldsymbol{k}_\perp = 0$, the functions $\psi^{(n)}_{\pm 1/2}(0, \boldsymbol{r})$ are the envelope functions at the bottom of the $2D$ hole subband. Generally, the matrix elements of the exchange interaction between the magnetic ion and holes described by these functions depend on the position of the magnetic ion $\boldsymbol{R}_m$

$$\left\langle \eta_1 \left| \hat{H}_{j,S} \right| \eta_2 \right\rangle = \beta \sum_{m,\alpha,m_1,m_2} \psi^*_{\eta_1,m_1}(\boldsymbol{R}_m)\psi_{\eta_2,m_2}(\boldsymbol{R}_m)\left\langle m_1 \left| I_\alpha \right| m_2 \right\rangle \hat{S}_{m,\alpha}$$

$$= \beta \sum_{m,\alpha,\gamma} |\Psi(\boldsymbol{R}_m)|^2 \left\langle \eta_1 \left| j_\gamma \right| \eta_2 \right\rangle G_{\gamma,\alpha}(\boldsymbol{R}_m)\hat{S}_{m,\alpha}, \tag{3.17}$$

where $\eta_i = \pm 1/2$, $\hat{G}(\boldsymbol{r})$ is the second rank tensor that gives the relation between local values of the matrix elements of the hole internal spin momentum $\boldsymbol{I}$ and its pseudospin $\boldsymbol{j}$ ($\sum_\beta \left\langle \eta_1 \left| j_\beta \right| \eta_2 \right\rangle G_{\beta,\alpha} = \left\langle \eta_1 \left| I_\alpha \right| \eta_2 \right\rangle$), and $|\Psi(\boldsymbol{r})|^2$ gives the probability to find the hole at the point $\boldsymbol{r}$. It is easy to derive the following expressions for the components of the tensor $\hat{G}(\boldsymbol{r})$

$$G_{z,\alpha}(\boldsymbol{r}) = 2 \sum_{m_1,m_2} \psi^*_{m_1}(\boldsymbol{r})\psi_{m_2}(\boldsymbol{r})\langle m_1|J_\alpha|m_2\rangle \Big/ |\Psi(\boldsymbol{r})|^2,$$

$$G_{x,\alpha}(\boldsymbol{r}) = 2\Re \sum_{m_1,m_2} \left((-1)^{1/2-m_1}\psi_{-m_1}(\boldsymbol{r})\psi_{m_2}(\boldsymbol{r})\langle m_1|J_\alpha|m_2\rangle\right) \Big/ |\Psi(\boldsymbol{r})|^2,$$

$$G_{y,\alpha}(\boldsymbol{r}) = 2\Im \sum_{m_1,m_2} \left((-1)^{1/2-m_1}\psi_{-m_1}(\boldsymbol{r})\psi_{m_2}(\boldsymbol{r})\langle m_1|J_\alpha|m_2\rangle\right) \Big/ |\Psi(\boldsymbol{r})|^2. \tag{3.18}$$

It follows from (3.18) that in general case the tensor $\hat{G}$ is not symmetric: $\hat{G}_{\alpha\beta} \neq \hat{G}_{\beta\alpha}$. Let us introduce the g-factor tensor g and the pseudovector $\boldsymbol{D}$ by

$$g_{\alpha\beta} = \left(\hat{G}_{\alpha\beta} + \hat{G}_{\beta\alpha}\right)/2, \quad D_\gamma = \epsilon_{\alpha\beta\gamma}\hat{G}_{\alpha\beta}/2. \tag{3.19}$$

Here, $\epsilon_{\alpha\beta\gamma}$ is Levi-Civita anti-symmetric tensor that relates the nonzero components of the antisymmetric tensor of the second range via the three components of the pseudovector.

Substituting (3.19) into (3.17), we obtain the final expression for the truncated Hamiltonian of the exchange interaction between holes and magnetic ions:

$$\hat{H}_{jd} = -\frac{\beta}{3}\sum_m \left[\left(\hat{j}\,\hat{g}^j(\boldsymbol{k}_\perp,\boldsymbol{r})\hat{\boldsymbol{S}}_m\right) + \left(\boldsymbol{D}(\boldsymbol{k}_\perp,\boldsymbol{r})[\hat{j}\times\hat{\boldsymbol{S}}_m]\right)\right]\delta(\boldsymbol{r}-\boldsymbol{R}_m). \tag{3.20}$$

The first term in this Hamiltonian describes the scalar exchange interaction whose anisotropy is given by the anisotropic tensor $\hat{g}$. The second term in the Hamiltonian (3.20) is the so-called Dzyaloshinskii–Moriya interaction. Note that there is some freedom in the choice of the anisotropic g-factor and the Dzyaloshinskii–Moriya vector. This freedom is caused by the arbitrary choice of the basis functions corresponding to the pseudospin projection on the Z-axis $\eta = \pm 1/2$. One can choose the basis function for each particular point $\boldsymbol{r}$ in such a way that the Dzyaloshinskii–Moriya interaction vanishes [11]. However, in general case it is not possible to do so for all positions simultaneously.

Formula (3.20) gives a general expression for the scalar operator $\hat{H}$ that depends linearly on two vector operators $\boldsymbol{j}$ and $\boldsymbol{S}$. It is valid when the energy of the exchange interaction depends for some reason only on the first power of the spin projections $S_{m,\alpha}$ of the magnetic ions. Note that by definition $j = 1/2$ and any matrix j_α^n of the α projection of the n-th power of the pseudospin operator have only four independent elements. Therefore, it can be written with the help of the unit matrix and three matrices $\hat{j}_\alpha \equiv \hat{\sigma}_\alpha/2$ so that the energy of the exchange interaction depends always only on the first power of the pseudospin operator components j_α.

3.3.3 Exchange Hamiltonian for Heavy and Light Holes at the Bottom of 2D Subband

As the first example let us consider the exchange interaction Hamiltonian for the holes at the bottom of the $2D$ subbands of the square well. Choosing the Z-axis to be normal to the quantum well plane, we obtain:

$$\Psi(k_\perp = 0, z)_{\mathrm{hh},1/2} = u_{3/2}\psi_{\mathrm{hh}}(z), \quad \Psi(k_\perp = 0, z)_{\mathrm{hh},-1/2} = -u_{-3/2}\psi_{\mathrm{hh}}(z), \tag{3.21}$$

$$\Psi(k_\perp = 0, z)_{\mathrm{lh},1/2} = u_{1/2}\psi_{\mathrm{lh}}(z), \quad \Psi(k_\perp = 0, z)_{\mathrm{lh},-1/2} = u_{-1/2}\psi_{\mathrm{lh}}(z). \tag{3.22}$$

In this case, the anisotropic g-factor has the same values for all points in the position space. Moreover, the choice of the pseudospin eigenfunctions (3.16) leads to $\boldsymbol{D}_{\mathrm{hh}}(\boldsymbol{r}) = \boldsymbol{D}_{\mathrm{lh}}(\boldsymbol{r}) = 0$. The anisotropic g-tensors for the heavy $g^{(\mathrm{hh})}$ and light $g^{(\mathrm{lh})}$ holes subbands are given by:

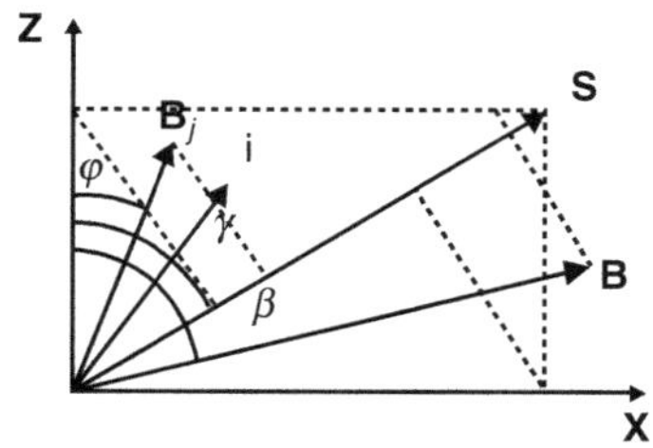

Fig. 3.4 Directions of the pseudospin $\boldsymbol{j}$ for a heavy hole with strongly anisotropic g-factor ($g_{zz} \gg g_{xx}, g_{yy}$) in a spin complex formed by the hole and magnetic ions. The direction of the average spin $\langle \boldsymbol{S} \rangle$ of the magnetic ions is determined by the sum of the external field $\boldsymbol{B}$ and exchange field $\bar{B}_j \propto \hat{g}\,\langle \boldsymbol{j} \rangle$ induced by the hole, while the hole pseudospin is directed along the exchange field $\boldsymbol{B}_{\mathrm{Mn}} \propto \hat{g}\,\langle \boldsymbol{S} \rangle$ induced by magnetic ions (usually $B_{\mathrm{Mn}} \gg B$). The strong anisotropy of the hole g-factor leads to the significant deviation of the directions of $\langle \boldsymbol{S} \rangle$ and $\langle \boldsymbol{j} \rangle$ from the direction of the external field $\boldsymbol{B}$

$$g^{(\mathrm{hh})} = \begin{pmatrix} 0\,0\,0 \\ 0\,0\,0 \\ 0\,0\,3 \end{pmatrix}, \tag{3.23}$$

$$g^{(\mathrm{lh})} = \begin{pmatrix} 2\,0\,0 \\ 0\,2\,0 \\ 0\,0\,1 \end{pmatrix}. \tag{3.24}$$

The heavy hole g-factor tensor has only one nonzero component $g_{zz}^{(\mathrm{hh})} = 3$. Therefore, the scalar product in (3.11) (the Heisenberg Hamiltonian) transforms into the Ising Hamiltonian $\propto (S_z j_z)$.

The introduction of the anisotropic g-factor is very helpful for the description of the exchange interaction between holes and magnetic ions in external magnetic field (see Fig. 3.4) It allows one to describe the anisotropic spin dynamics of the photoexcited holes [13–21] (see Chap. 10), the anisotropic properties of magnetic polarons [17–19], (see Chap. 8), and the magnetic anisotropy of the ferromagnetic ordered spin system of the Mn ions in the p-type doped heterostructures GaAsMn [22, 23].

3.3.4 Exchange Hamiltonian for Heavy and Light Holes near the Bottom of 2D Subband

The situation changes for the states with a nonzero in-plane wave vector $k_\perp$ away from the subband bottom. The hole wave functions are not characterized by the definite projections of internal angular momentum $\boldsymbol{I}$ on the Z-axis. For example, in the linear in $k_\perp$ approximation the even wave function of the first heavy hole subband has some admixture of the odd light hole wave functions. (see (3.7), (3.8) in Sect. 3.2). The matrix elements of the scattering Hamiltonian in this case depends on the wave vector values of the initial $\left(\boldsymbol{k}_{\perp}^{(\mathrm{i})}\right)$ and final $\left(\boldsymbol{k}_{\perp}^{(\mathrm{f})}\right)$ states.

Using the expression (3.7), (3.8) for the wave functions in the quantum well with infinite potential barriers we obtain in the linear in $k_{\perp}$ approximation

$$\left\langle \mathrm{hh}, \pm\frac{1}{2} \middle| J_x \middle| \mathrm{hh}, \pm\frac{1}{2} \right\rangle \approx \mp 3C_\gamma \sin\left(\pi\frac{z}{L}\right)\left[i\left(k_{\perp,x}^{(\mathrm{i})} - k_{\perp,x}^{(\mathrm{f})}\right) + \left(k_{\perp,y}^{(\mathrm{i})} + k_{\perp,y}^{(\mathrm{f})}\right)\right] L = \pm\frac{1}{2} G_{z,x}, \tag{3.25}$$

$$\left\langle \mathrm{hh}, \pm\frac{1}{2} \middle| J_y \middle| \mathrm{hh}, \pm\frac{1}{2} \right\rangle \approx \pm 3C_\gamma \sin\left(\pi\frac{z}{L}\right)\left[i\left(k_{\perp,y}^{(\mathrm{i})} - k_{\perp,y}^{(\mathrm{f})}\right) + \left(k_{\perp,x}^{(\mathrm{i})} + k_{\perp,x}^{(\mathrm{f})}\right)\right] L = \pm\frac{1}{2} G_{z,y}, \tag{3.26}$$

$$\left\langle \mathrm{hh}, \pm\frac{1}{2} \middle| J_z \middle| \mathrm{hh}, \mp\frac{1}{2} \right\rangle = 0 = \frac{1}{2}\left(G_{x,z} \mp i G_{y,z}\right), \tag{3.27}$$

$$\left\langle \mathrm{hh}, \pm\frac{1}{2} \middle| J_{(x,y)} \middle| \mathrm{hh}, \mp\frac{1}{2} \right\rangle = 0 = \frac{1}{2}\left(G_{x,(x,y)} \mp i G_{y,(x,y)}\right), \tag{3.28}$$

$$C_\gamma = \frac{8\gamma_3^{\mathrm{L}}}{3\pi^2\left(3\gamma_1^{\mathrm{L}} + 10\gamma_2^{\mathrm{L}}\right)}. \tag{3.29}$$

Thus, the only nonzero components of the tensor G are $G_{z,(x,y)}$ (3.25), (3.26). Therefore, the exchange Hamiltonian contains the terms $S_x j_z$, $S_y j_z$, while the inverse terms $S_z j_y$, $S_z j_x$ are absent. The Z- component of the hole pseudospin induces the exchange field acting on the magnetic ion in all three directions X,Y, and Z. At the same time, the magnetic ion spin $\boldsymbol{S}$ of arbitrary direction, induces the exchange field acting on the hole pseudospin only in Z direction: $\boldsymbol{H} \propto \left(G_{z,x}\langle S_x\rangle + G_{z,y}\langle S_y\rangle + G_{z,z}\langle S_z\rangle\right)\boldsymbol{e}_z$. Due to the precession of spins around the field, all three components of the magnetic ion spin are changed by the exchange interaction. For the hole, however, only two components $\left(j_x, j_y\right)$ are changing. The projection of the hole spin on the Z-axis denotes the spin subband and is conserved. Therefore, the exchange interaction does not induce the transition between different spin subbands. This asymmetry of the exchange interaction is reflected both by the Dzyaloshinskii–Moriya vector

$$\boldsymbol{D} = -3C_\gamma \sin(\pi z/L)\, L\left(\left(\boldsymbol{k}_{\perp}^{(\mathrm{i})} + \boldsymbol{k}_{\perp}^{(\mathrm{f})}\right) + i\left[\boldsymbol{e}_z \times \left(\boldsymbol{k}_{\perp}^{(\mathrm{i})} - \boldsymbol{k}_{\perp}^{(\mathrm{f})}\right)\right]\right) \tag{3.30}$$

and by the Heisenberg Hamiltonian with anisotropic g-factor

$$g_{z,x} = g_{x,z} = D_y, \quad g_{z,y} = g_{y,z} = -D_x \tag{3.31}$$

Here, e_z is the unit vector director normal to the quantum well plane.

In the case of a symmetric quantum well, the nondiagonal components g_{xz} and g_{yz} of the hole g-factor and the $\boldsymbol{D}$ vector vanish upon the averaging over $\boldsymbol{k}_{\perp}$. Therefore, upon the average, the spatially inhomogeneous magnetizations in X and Y-directions do not induce the exchange fields acting on the holes. This is in spite of the fact that the exchange field induced by the hole acts on all three spin components of each magnetic ion. This makes possible the direct energy transfer between the spin system of the magnetic ions and the moving $2D$ heavy holes. In a strong external magnetic field normal to the quantum well plane, such direct energy transfer mechanism dominates over the conventionally considered flip-flop processes. The latter are forbidden by the energy conservation law because of the giant spin splitting in the heavy hole subband [24] (see Chap. 9).

3.3.5 Exchange Hamiltonian for the Hole Scattering

It is necessary to note that in the problem of hole scattering, the initial and final values of the wave vector may be different and the resulting components of the tensor G are complex (3.25), (3.26). This indicates the elliptic polarization of the scattering field created by the hole on the magnetic ion. The scattering of polarized holes in the XY plane depends in this case on the polarization of magnetic ions along Z-axis. This effect is known as the Mott scattering [25]. It has long been used in nuclear physics in the investigations of scattering on polarized targets [26].

3.3.6 Exchange Hamiltonian for Holes in a Spherical Quantum Dot

Quantum confinement in two directions (realized in quantum wires) does not lead to any new feature in the exchange interaction Hamiltonian. It still has a form of the expression (3.20). However, the g-factor and the Dzyaloshinskii–Moriya vector in $1D$ quantum wires depend on the spatial position of the magnetic ion even for the states at the bottom of the $1D$ subband. The main problems by the studies of the exchange interaction in 1D systems are usually connected with the numerical calculations of the hole wave functions.

For the holes confined in the quantum dot of spherical symmetry, the Hamiltonian (3.20) does not represent the general form of the exchange interaction operator. As the total hole momentum $F = 3/2 > 1/2$, the general exchange operator may depend on (1) the vector $\hat{\boldsymbol{F}}$, (2) operator of the quadrupole momentum with components expressed via products of the total momentum projection $\hat{\boldsymbol{Q}}_{\alpha,\beta} \propto \left\{\hat{F}_\alpha \hat{F}_\beta\right\}$, and (3) operator of the octupole momentum $\hat{\boldsymbol{T}}_{\alpha,\beta,\gamma} \propto \left\{\hat{F}_\alpha \left\{\hat{F}_\beta \hat{F}_\gamma\right\}\right\}$. Here, the symbol { } denotes the anticommutator: $\left\{\hat{A}\hat{B}\right\} = \hat{A}\hat{B} + \hat{B}\hat{A}$. The products of the higher degrees of the spin projections for $F = 3/2$ can be always expressed via the unit matrix $\hat{\boldsymbol{E}}$ and the matrices $\hat{\boldsymbol{F}}$, $\hat{\boldsymbol{Q}}$ and $\hat{\boldsymbol{T}}$.

To construct the general exchange Hamiltonian, it is necessary to take into account its symmetry with respect to the operation of the time inversion. Only the operator $\hat{\boldsymbol{F}}$ and the magnetic ion spin $\hat{\boldsymbol{S}}$ depend on the time explicitly. They both change the sign with time inversion. The Hamiltonian depends only on the magnetic ion spin $\hat{\boldsymbol{S}}$ linearly. Therefore, it cannot contain the even powers of $\hat{\boldsymbol{F}}$ and can depend only on the first power of the spin $\hat{\boldsymbol{F}}$ and of the octuple momentum operator $\hat{\boldsymbol{T}}$. The combinations of $\hat{\boldsymbol{F}}$ and $\hat{\boldsymbol{T}}$ together with the components of the radius $\boldsymbol{r}$ and magnetic ion spin $\hat{\boldsymbol{S}}$ shall form a scalar. It can be easily seen that for the first power of $\hat{\boldsymbol{F}}$ only two combinations that meet these conditions are possible. These are the scalar product $\left(\hat{\boldsymbol{F}}\hat{\boldsymbol{S}}\right)$ and the operator

$$\mathrm{H}_{pd} \propto \left[\frac{1}{3}\left(\hat{\boldsymbol{F}}\hat{\boldsymbol{S}}\right) - \frac{1}{r^2}\left(\hat{\boldsymbol{F}}\boldsymbol{r}\right)\left(\hat{\boldsymbol{S}}\boldsymbol{r}\right)\right]. \tag{3.32}$$

The interaction described by the operator (3.32) is often called the pseudodipole interaction. Note that it can be also described with the help of the anisotropic local g-factor of the holes $\Delta g_{\alpha,\beta}(\boldsymbol{r}) \propto r_\alpha r_\beta / r^2$ in the scalar Hamiltonian of the exchange interaction.

There also exist two possible invariants with the octupole momentum operator $\hat{\boldsymbol{T}}$:

$$\mathrm{H}_{T1} \propto \frac{\left\{\left(\hat{\boldsymbol{F}}\boldsymbol{r}\right)^2\left(\hat{\boldsymbol{F}}\hat{\boldsymbol{S}}\right)\right\}}{r^2}, \quad \mathrm{H}_{T2} \propto \frac{\left(\hat{\boldsymbol{F}}\boldsymbol{r}\right)^3\left(\hat{\boldsymbol{S}}\boldsymbol{r}\right)}{r^4}. \tag{3.33}$$

Thus, the Hamiltonian of the exchange interaction between the hole and the surrounding magnetic ions in a spherical quantum dot has the following general form

$$\begin{aligned}\mathrm{H}_{\mathrm{QD}} = \Big(& a(r)\left(\hat{\boldsymbol{F}}\cdot\hat{\boldsymbol{S}}\right) + b(r)r^{-2}\left\{\left(\hat{\boldsymbol{F}}\cdot\boldsymbol{r}\right)^2\left(\hat{\boldsymbol{F}}\cdot\hat{\boldsymbol{S}}\right)\right\} \\ &+ a_1(r)r^{-2}\left(\hat{\boldsymbol{F}}\cdot\boldsymbol{r}\right)\left(\boldsymbol{r}\cdot\hat{\boldsymbol{S}}\right) + b_1(r)r^{-4}\left(\hat{\boldsymbol{F}}\cdot\boldsymbol{r}\right)^3\left(\boldsymbol{r}\cdot\hat{\boldsymbol{S}}\right)\Big).\end{aligned} \tag{3.34}$$

Using explicit expressions (3.9) for the hole wave functions, we obtain the following formulas for the radius-dependent coefficients:

$$a(r) = -\frac{\beta}{24\pi}\left(2f_0^2(r) + 5f_0(r)f_2(r) + 3f_2^2(r)\right), \tag{3.35}$$

$$b(r) = \frac{\beta}{12\pi r^2} f_2(r)\left(f_0(r) + f_2(r)\right), \tag{3.36}$$

$$a_1(r) = \frac{\beta}{24\pi r^2} f_2^2(r), \tag{3.37}$$

$$b_1(r) = -\frac{\beta}{6\pi r^4} f_2^2(r). \tag{3.38}$$

For the first time the expression (3.34) was obtained in [27], where the exchange interaction between magnetic ions and the hole bound at the acceptor was studied. For the radial functions of the hole confined in the spherical quantum dot, one can use the explicit expression (3.10).

To gain more insight into the influence of the spin–orbital effects on the giant spin splitting of the hole energy levels, it is instructive to compare the values of the exchange energy shift of the localized hole spin levels in the homogeneous field $H_z = He_z$ with maximum exchange energy in the inhomogeneous classical field $\boldsymbol{H}_{\text{pol}} = H\langle \boldsymbol{I}(\boldsymbol{r})\rangle / |\langle \boldsymbol{I}(\boldsymbol{r})\rangle|$. At each spatial position, the $\boldsymbol{H}_{\text{pol}}$ field is directed along the vector of the local quantum mechanical averaged spin value of the hole

$$\langle \boldsymbol{I}(\boldsymbol{r})\rangle_{\text{QM}} = \frac{\left(\Psi^*(\boldsymbol{r})\,\hat{I}_x\Psi(\boldsymbol{r})\right)e_x + \left(\Psi^*(\boldsymbol{r})\,\hat{I}_y\Psi(\boldsymbol{r})\right)e_z + \left(\Psi^*(\boldsymbol{r})\,\hat{I}_z\Psi(\boldsymbol{r})\right)e_z}{|\Psi(\boldsymbol{r})|^2} \tag{3.39}$$

In the first case, the homogeneous exchange field leads to the giant Zeeman energy spitting $E_m - E_{m'}$ of the hole states with $F_z = m, m'$.

$$\begin{aligned} E_m - E_{m'} &= \tilde{\beta} F_z \langle S_z(\boldsymbol{B})\rangle_T = \beta \langle F_z|I_z|F_z\rangle \langle S_z(\boldsymbol{B})\rangle_T \\ &= \beta \overline{\langle I(\boldsymbol{r})\rangle_{\text{QM}} \cos\vartheta(\boldsymbol{r})} \langle S_z(\boldsymbol{B})\rangle_T . \end{aligned} \tag{3.40}$$

Here, the bar denotes averaging over the quantum dot volume with the weight $|\Psi(\boldsymbol{r})|^2$, $\vartheta(\boldsymbol{r})$ is the angle between the vector $\langle \boldsymbol{I}(\boldsymbol{r})\rangle_{\text{QM}}$ and Z-axis, and $\langle S_z(\boldsymbol{B})\rangle_T$ is the thermodynamically averaged value of magnetic ion spins.

In the second case, ($\boldsymbol{H}_{\text{pol}} = H\langle \boldsymbol{I}(\boldsymbol{r})\rangle / |\langle \boldsymbol{I}(\boldsymbol{r})\rangle|$) represents the zero approximation for calculation of the energy levels of the magnetic polaron. In the latter, the polarization of magnetic ions is caused by their interaction with the hole and follows the direction of $\langle \boldsymbol{I}(\boldsymbol{r})\rangle_{\text{QM}}$ (see Chap. 8).

Direct calculation of the averaged spin value for the state shown in the Fig. 3.3b gives $\langle F_z = 3/2|\,\hat{I}_z\,|F_z = 3/2\rangle \approx 1.42$. The resulting renormalization of the exchange interaction constant is then $\tilde{\beta} = 0.94 \cdot \beta$. For the values averaged over the quantum dot volume, we obtain $\overline{\langle I_z(\boldsymbol{r})\rangle} \approx 0.95 \cdot F_z$ and $\overline{\cos\vartheta(\boldsymbol{r})} \approx 0.99$, respectively. Thus, the main contribution to renormalization of the exchange interaction constant comes from deviation of the quantum mechanically averaged spin value from its maximal possible value 3/2. On the other hand, the deviation of the local averaged $\left\langle \hat{\boldsymbol{I}}(\boldsymbol{r})\right\rangle$ from the Z-axis near the quantum dot surface is not very important.

The above analysis has been carried out for the spherical quantum dots such as investigated, for example, in [28]. Depending on the preparation method, the quantum dots may have different shape (see Chap. 5). The particular shape of the dot mostly affects the details of the exchange interaction description. The special features of the quantum dots with large number of carriers are analyzed in Chap. 6.

To conclude this Section, let us note that all the above considerations are based on the description of the exchange interaction as a small perturbation. However, the giant spin splitting of the hole levels is different for the different points in the quantum well. This may lead to significant changes of the potential profile in the well so that the simple expressions of the zero-order perturbation theory will not be valid any more. The calculation of the energy spectrum and carrier wave functions in such a case demands the whole zone engineering tool. It is necessary to consider the complete quantum confinement problem together with the strong exchange interaction using the general form of the Hamiltonian (3.15) [or even (3.13)], Luttinger operator of the kinetic energy and the realistic potential profile of the heterostructure. Examples of such calculations can be found in [29–31].

3.4 Exchange Interaction Between Electrons and Magnetic Ions in Narrow and Deep Quantum-confined Structures

As we have already discussed in the beginning of the Sect. 3.3, the exchange interaction between magnetic ion and conduction band electron in the vicinity of the Brillouin zone center is described by the direct Hamiltonian (3.13). Such exchange has a potential character [8]. The exchange constant is positive and nearly independent of the electron wave vector (see Chap. 1). On the other hand, this simple picture is not valid far away from the Γ-point. As the electron kinetic energy increases, the electron wave function gets the admixture of the states with the valence band Bloch amplitudes. For the valence band states, however, the kinetic exchange mechanism plays the leading role in the exchange interaction with magnetic ions [8,9]. For the kinetic exchange, the exchange constant is negative. As the result, the net value of the exchange interaction constant decreases and may even change sign [32]. Moreover, the traveling waves corresponding to the contribution of the conduction band states and the waves corresponding to the contribution of the valence band states into resulting wave function are shifted by a quarter of the wavelength with respect to each other. For the standing waves in the quantum-confined structures, this shift results in a giant spatial inhomogeneity of the exchange interaction.

3.4.1 Renormalization of the Exchange Interaction for 3D Electrons with High Kinetic Energy

First, let us consider how the electron–ion exchange interaction changes with increase of the free electron kinetic energy in the bulk semiconductor [32].

The admixture of the p-like wave functions to the conduction band makes the hybridization of the conduction band states with d orbitals of magnetic ions possible. Accordingly, the virtual transitions with capture of electrons from the conduction band by the half-filled d shell (see Fig. 3.5) become allowed. This means that the mechanism of kinetic exchange also has to be considered for the conduction band states. The exchange Hamiltonian acting at the eight component envelope

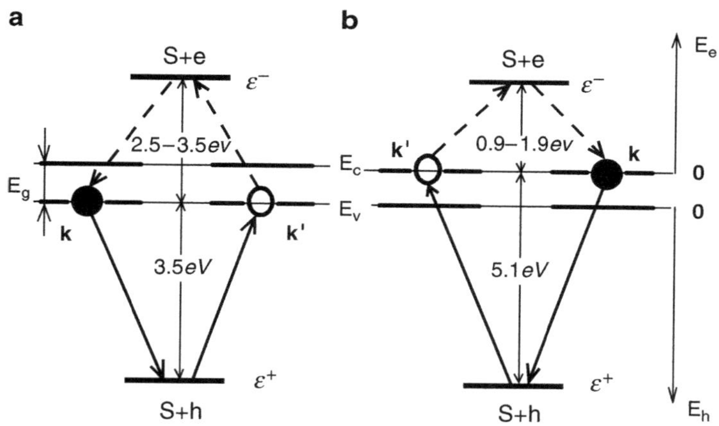

Fig. 3.5 Virtual transitions responsible for the kinetic exchange of carriers with magnetic ions: panel (**a**) sketches the transitions of holes (*black circle*) in the vicinity of the valence band top E_v and panel (**b**) sketches the transition of electrons (*white circle*) in the vicinity of the conduction band edge $E_c = E_v + E_g$. The axes E_e and E_h correspond to the electron and hole energy, respectively. Energy levels are shown for the magnetic ion $Mn^{(2+)}$ in $Cd_{0.98}Mn_{0.02}Te$ alloy. The hole is scattered from the state labeled $\boldsymbol{k}$ to the state $\boldsymbol{k}'$ or – what is exactly equivalent – the electron is scattered from the state $\boldsymbol{k}'$ to $\boldsymbol{k}$. In the intermediate state, the number of electrons in the d shell differs from usual number, which corresponds to the half-filled shell. The transitions during which the number of electrons in the d shell decreases ($d^5 \longrightarrow d^4$) and increases ($d^5 \longrightarrow d^6$) are depicted by *solid* and *dash arrows*, respectively. The half-filled d^5 state is spherically symmetric (S). Therefore, it is convenient to treat the ($d^5 \longrightarrow d^4$) as a capture of a hole by the d shell of Mn ion ($S + h$), while the ($d^5 \longrightarrow d^6$) transition a capture of a electron ($S + e$)

function $\begin{pmatrix} \Psi^c \\ \Psi^v \end{pmatrix}$ in the eight band Kane model (see Appendix) can be written as

$$\hat{H}^{\text{ex}}_{\text{Kane}} = \sum_n \begin{pmatrix} -\alpha\delta(\boldsymbol{r} - \boldsymbol{R}_n) & 0 \\ 0 & -\beta\delta(\boldsymbol{r} - \boldsymbol{R}_n) \end{pmatrix} (\boldsymbol{s}\boldsymbol{S}_n), \tag{3.41}$$

where $\boldsymbol{S}_n$ is the ion spin, $\boldsymbol{s} = 1/2\boldsymbol{\sigma}$, $\alpha = \alpha_{\text{pot}}$, and $\beta = \beta_{\text{pot}} + \gamma(E)\beta_{\text{kin}}$. (Here, α_{pot}, β_{pot}, and β_{kin} are the potential and kinetic exchange constants at the conduction band and the valence band edge, i.e., at $k = 0$). The coefficient $\gamma(E)$ accounts for possible dependence of the kinetic exchange on the electron kinetic energy. The dependence $\gamma(E)$ appears because the kinetic exchange is normally calculated in the second-order perturbation theory. As a result, $\gamma(E)$ depends drastically on the energy difference $E_i - E_n$ between the initial state and a virtual state by which the electron (or the hole) is captured on the d shell of the magnetic ion [8,9]: $\beta_{\text{kin}}(E_i) \propto |V_{pd}|^2 \sum_n (E_i - E_n)^{-1}$. The matrix element V_{pd} of the operator that describes the $p - d$ hybridization is assumed to be constant. The virtual states facilitating the kinetic exchange for the conduction electrons are the same as for the valence band holes. Their respective energies measured from the top of the valence band are denoted ε^- and ε^+ in Fig. 3.5. The energies of the initial states for kinetic exchange in the top of the valence and the bottom of the conduction bands differ by E_g. As a result

$$\gamma(E) = \frac{-\varepsilon^- \varepsilon^+}{(\tilde{E} - \varepsilon^+)(\varepsilon^- - \tilde{E})}, \tag{3.42}$$

where $\tilde{E} = E + E_g$ is the electron energy calculated from the valence band top and E_g is a band gap energy.

In the framework of the eight-band Kane model (see Appendix), the electron wave function with the nonzero wave vector directed along Z-axis $\boldsymbol{k} = k\boldsymbol{e}_z$ has the contributions of the conduction band and valence band amplitudes in the following form:

$$\Psi(r)^e_{\pm 1/2} = A_s(k)\mathrm{e}^{ikz} \times \left[|S > u_{\pm 1/2} + iA_1(k)|X \pm iY > u_{\mp 1/2} - iA_0(k)|Z > u_{\pm 1/2}\right]. \tag{3.43}$$

Explicit expressions for the coefficients A_s, A_1, A_0 are given by (3.70) (see Appendix). Hence, we obtain the following expression for the exchange energy operator of the electron with $\boldsymbol{k} = k\boldsymbol{e}_z$:

$$\left\langle \boldsymbol{k} | \hat{H}^{\mathrm{ex}}_{\mathrm{Kane}} | \boldsymbol{k} \right\rangle = -\frac{1}{V} \sum_n \left(\alpha_{\|}(E)\,(s_z S_{nz}) + \alpha_{\perp}(E)\left(s_x S_{nx} + s_y S_{ny}\right) \right). \tag{3.44}$$

Here, the effective exchange constants $\alpha_{\|}$ and $\alpha_{\perp}$ for the ions with spin $\boldsymbol{I}$ parallel to the wave vector $\boldsymbol{k}$ and the ions with spin perpendicular to the wave vector, respectively, are given by

$$\begin{aligned} \alpha_{\|}(E) &= \alpha_0(E) + \beta_0(E) + \beta_1(E), \\ \alpha_{\perp}(E) &= \alpha_0(E) + \beta_0(E). \\ \alpha_0(E) &= \alpha A_s^2(k), \\ \beta_0(E) &= \beta A_s^2(k) A_0^2(k), \\ \beta_1(E) &= -2\beta A_s^2(k) A_1^2(k) \approx 0. \end{aligned} \tag{3.45}$$

We see that the constant $\beta_1(E)$ characterizes the anisotropy of the exchange interaction. In the case of CdTe, this anisotropy is about 5%. Neglecting the anisotropy (with $\beta_1(E) \approx 0$), we obtain

$$\alpha_{\|}(E) \approx \alpha_{\perp}(E) \approx \alpha(E) \approx \alpha_0(E) + \beta_0(E) \approx \alpha_c \cos^2(\varsigma) + \beta_v \sin^2(\varsigma), \tag{3.46}$$

where

$$\cos^2(\varsigma) = A_s^2(k) = \frac{(\tilde{E}(\tilde{E} + \Delta))^2}{(\tilde{E}(\tilde{E} + \Delta))^2 + (Pk\Delta/3)^2 + (Pk(\tilde{E} + 2\Delta/3)^2}. \tag{3.47}$$

Here, P is the Kane matrix element describing the coupling of the conduction and valence bands (see Appendix), and Δ is the spin–orbit splitting of the valence band; $\tilde{E}$ is the electron energy calculated from the top of the valence band (c.f. (3.70) in the Appendix).

Thus, for the plane travelling waves, the admixture of the valence band Bloch amplitudes to the electron wave function leads to renormalization of the exchange interaction constant. The latter may even change its sign from positive to negative.

For small confinement energies ($E \ll E_g$), the linear expansion with respect to E is valid. Then we find from (3.46) and (3.47):

$$\frac{d\alpha}{dE} \approx -\frac{9E_g^2 + 12E_g\Delta + 2\Delta^2}{3E_g(E_g + \Delta)(3E_g + 2\Delta)} \left(\alpha_{pot} - \left[\beta_{pot} + \beta_{kin}\gamma(E)\right]\right). \quad (3.48)$$

For (Cd,Mn)Te, the value of $\varepsilon^+ \approx 3.5$ eV is established fairly accurately from photoemission experiments. For ε^-, no experimental data are available and only estimations in the range from 2.5 to 3.5 eV can be found in literature [33]. Thus, the value ε^- is the fitting parameter of the discussed theoretical model, which can be adjusted within certain limits. The dependences $\alpha(E)$ calculated after (3.45)–(3.47) for $Cd_{0.98}Mn_{0.02}Te$ for two values of $\varepsilon^- = 2.5$ eV and 3.5 eV are plotted in Fig. 3.6. We assume that $\beta_{pot} \ll \beta_{kin}$ and $-4\alpha_{pot} \approx (\beta_{pot} + \beta_{kin}) \approx \beta_{kin}$ [34]. The results of the linear approximation by (3.48) are shown by dash-dotted lines. Negative values of $E = \tilde{E} - E_g$ correspond to states below the band gap, i.e., the tunneling regime (as appropriate, for example, for the tails of electron wave functions penetrating into the potential barrier) or surface localized Tamm states [35,36]. One can see that the electron exchange constant decreases with increasing absolute value of E. For the state with E larger than 210 meV and smaller than −260 meV ($\varepsilon^- = 2.5$ eV), the exchange constant α became negative.

In the paper [37], the renormalization of the exchange interaction constant for the electrons with large kinetic energies has been calculated in the framework of

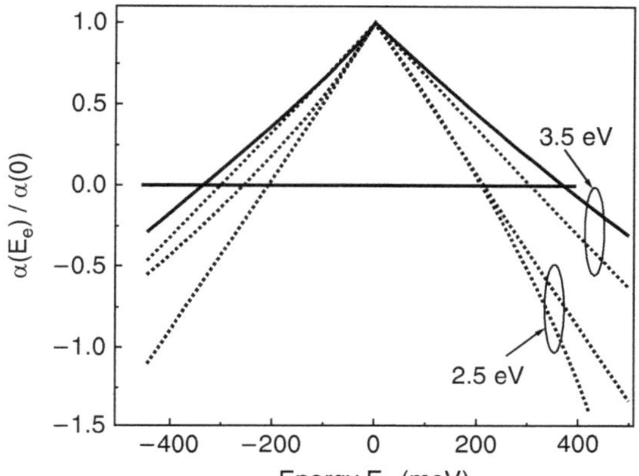

Fig. 3.6 Dependence of the electron exchange parameter α on the electron kinetic energy in $Cd_{0.98}Mn_{0.02}Te$ [32]. Calculations performed according to (3.46) and for a linear approximation of (3.48) (*dash-dotted lines*) are shown for two values of $\varepsilon^- - E_v$ of 2.5 and 3.5 eV

the $sp^3 - s^*$ tight-binding model. Qualitatively, the predictions of the eight-band $\boldsymbol{k}\,\boldsymbol{p}$-theory and of the tight-binding model are similar. In both models, the decrease of the exchange interaction constants and possible change of its sign is predicted. However, for the large energies ($E > 100$ meV), the $\boldsymbol{kp}$-model predicts that the effect is much stronger than in the tight-binding model. Such discrepancy for large energies is not surprising. Indeed, for large values of the wave vector more accurate calculations in the framework of the $\boldsymbol{kp}$ model are needed. One has to take into account the admixture of more basis bands, for example, the contribution from the higher conduction band states. This will correct the results of the $\boldsymbol{kp}$ model for the large energies. Still, the Figure presented in [37] shows large difference in the curvature of the exchange constant energy dependence in the region of small electron energies. In this energy region, the applicability of the eight-band $\boldsymbol{kp}$-model for the calculation of the electron energy states and wave functions is well justified and supported by comparison with numerous experimental data. All main parameters of the $\boldsymbol{kp}$-model allow direct experimental determination and verification. Therefore, the observed discrepancy between the $\boldsymbol{kp}$-model and the tight-binding model results cannot be related to the restrictions of the Kane model. The origin of this discrepancy was not analyzed in the paper [37].

Up to now we have considered free electrons moving in the bulk semiconductor. However, the concentration of the hot electrons with energies 100 meV is very small in the bulk and their effect on the properties of the diluted semimagnetic semiconductor is negligible. A different situation can be observed in quantum confinement structures with narrow and deep quantum wells. Due to the effect of size quantatization, the ground state electron energy may be large enough to cause considerable renormalization of the exchange interaction constant. We will discuss the experiments, where such effect was observed in Sect. 3.5. Here, we will focus on the new special features of the exchange interaction between magnetic ions and electrons with large confinement energy.

3.4.2 The Exchange Interaction for 2D Electrons in a Narrow Quantum Well

Let us consider the renormalization of the electron–ion exchange interaction in a narrow quantum well. For the waves traveling in a bulk semiconductor, all position points are equivalent. This is not the case for the standing waves in the quantum well. As a result, the exchange interaction becomes dependent on the position of the magnetic ion.

Let us consider electrons at the bottom of the $2D$ subband ($k_{\perp} = 0$) in a square quantum well with two interfaces at $z = \pm L/2$. For simplicity, we consider here only symmetric quantum well with symmetric interfaces modeled by infinite potential barriers (infinitely large band offsets both in the conduction band $\Delta V_{\mathrm{c}} \to \infty$ and in the valence band $\Delta V_{\mathrm{v}} \to \infty$). Then, independently on the particular boundary conditions at the interfaces, the conduction band component of the electron wave function in the well can be written as:

$$\Psi_c = A_s(k_n)\left(C^+\exp(ik_nz) + C^-\exp(-ik_nz)\right), \tag{3.49}$$

where $C^+ = C^- = C$ for the even states and $C^+ = -C^- = -iC$ for the odd states and C is determined by the normalization condition. The value k_n of the quantatized wave vector in z-direction is related to the energy E_n (calculated from the bottom of the conduction band) of the quantum confined state via

$$E_n = \frac{\hbar^2 k_n^2}{2m_c(E_n)}, \tag{3.50}$$

where $m_c(E_n)$ is electron effective mass given by (3.65) in Appendix. Then the total wave function for the even states is given by

$$\Psi(r)^{\mathrm{even}}_{\pm 1/2} = 2A_s(k_n)C\,\big[\cos(k_nz)|S> u_{\pm 1/2}$$
$$-\sin(k_nz)\left(A_1(k_n)|X \pm iY> u_{\mp 1/2} - A_0(k_n)|Z> u_{\pm 1/2}\right)\big], \tag{3.51}$$

and for the odd states by

$$\Psi(r)^{\mathrm{odd}}_{\pm 1/2} = 2A_s(k_n)C\,\big[\sin(k_nz)|S> u_{\pm 1/2}$$
$$-\cos(k_nz)\left(A_1(k_n)|X \pm iY> u_{\mp 1/2} - A_0(k_n)|Z> u_{\pm 1/2}\right)\big], \tag{3.52}$$

To find the energies E_n of these states, one needs to specify the boundary conditions for envelope functions at the interfaces. For the multiband envelope theory, taking into account the admixture of different band states, the problem of the boundary conditions is not simple [35, 36, 38–45]. Generally, determination of the boundary condition parameters requires the detailed analysis of the experimental data or first principle calculations with account for the microscopic properties of the interfaces. However, the problem gets simplified in the case of the infinite potential barriers, where the wave function vanishes outside the well. In this case, the boundary conditions are described by only one parameter $\tilde{a}$ which determines the ratio of Ψ^c/Ψ^v_z at the interface (see Appendix). To describe the exchange interaction of electrons with the energy E_n it is more convenient to use this energy as the parameter. Figure 3.2 shows the conduction band (solid lines) and valence band (dot lines) components for first two subbands in the extremely asymmetrical case $\left(\Psi_c/\Psi^v_z = 0\right)$. More generally, the conduction band component does not vanish at the interfaces, as it was found, for example, in [38, 46] by fitting the experimental data. Independently on the boundary conditions, the nodes and antinodes of the conduction and valence band components are shifted with respect to each other by the quarter of the wavelength similar to the situation in Fig. 3.2.

Diagonal matrix elements of the exchange Hamiltonian (3.41) calculated with the functions (3.51) and (3.52) have the form

$$\left\langle k_n|\hat{H}^{\mathrm{ex}}_{\mathrm{Kane}}|k_n\right\rangle = -\frac{1}{V}\sum_m \Big[\left(\alpha_\perp(E_n) \pm \alpha^{af}(E_n)\cos(2k_nz)\right)(s\,\mathbf{S}_m)$$
$$+\ \beta_1(E)(1 \mp \cos(2k_nz))\,(s_z S_{mz})\Big], \tag{3.53}$$

where upper (lower) sign is for the even (odd) states, $\alpha_\perp$ and β_1 are given by (3.45) with E_n substituted for E and k_n substituted for k, and $\alpha^{af}(E_n) = \alpha_0(E_n) - \beta_0(E_n)$ is the parameter of the position-dependent exchange interaction. Neglecting the anisotropy parameter $\beta_1 \approx 0$ so that $\alpha_\perp(E_n) \approx \alpha(E_n)$, we obtain

$$\begin{aligned} \alpha(E_n) &\approx \alpha\cos^2(\varsigma) + \beta\sin^2(\varsigma), \\ \alpha^{af}(E_n) &\approx \alpha\cos^2(\varsigma) - \beta\sin^2(\varsigma), \\ \alpha^{\pm} &= \alpha(E_n) \pm \alpha^{af}(E_n)\cos(2k_n z), \end{aligned} \tag{3.54}$$

and

$$\left\langle k_n \middle| \hat{H}^{\text{ex}}_{\text{Kane}} \middle| k_n \right\rangle = -\frac{1}{V}\sum_m \alpha^{\pm}(k_n z, \varsigma)\,(\boldsymbol{s}\boldsymbol{S}_m)\,. \tag{3.55}$$

Thus, the value of the exchange interaction in the case of the standing wave depends on the position of the magnetic ion. According to the probability distribution, this value has maximum in the antinodes of the standing wave and vanishes in its nodes. However, the electron wave function in the Kane model is given by the superposition of two waves with Bloch amplitudes of s- and p-types, respectively. As it was discussed above, these waves have the same wavelength but their phases are shifted by the quarter of the period with respect to each other. Nodes of the p-wave coincide with antinodes of the s-wave and vice versa. Correspondingly, the potential exchange mechanism dominates in the space points corresponding to the antinodes of the s-wave and the nodes of the p-wave. Conversely, the kinetic exchange caused by the admixture of the valence band states, dominates in the space points corresponding to the antinodes of the p-wave and the nodes of the s-wave. Thus, the spatial dependence of the exchange interaction constant inside the well in (3.54) is caused by two reasons: first, the probability distribution in the standing wave and second, the different contribution of the potential and kinetic exchange mechanisms for the magnetic ions situated in different position points in the quantum well.

To analyze the contribution of these two factors in the spatial dependence of the exchange interaction constant, it is convenient to rewrite the matrix element of (3.54) extracting the square of the wave function $\left|\Psi^{\pm}(kz,\varsigma)\right|^2$ in the space point z. Here, upper (lower) sign is for the even (odd) states. Assuming $A_1 \approx 0$ and $\beta_1 \approx 0$, we obtain:

$$\left\langle k_n \middle| \hat{H}^{\text{ex}}_{\text{Kane}} \middle| k_n \right\rangle = -\frac{\left|\Psi^{\pm}(kz,\varsigma)\right|^2}{V}\sum_m \tilde{\alpha}^{\pm}(k_n z, \varsigma)\,(\boldsymbol{s}\boldsymbol{S}_m)\,, \tag{3.56}$$

$$\tilde{\alpha}^{\pm}(kz,\varsigma) = ((\alpha+\beta) \pm (\alpha-\beta)\,f^{\pm}(kz,\varsigma))/2C^2, \tag{3.57}$$

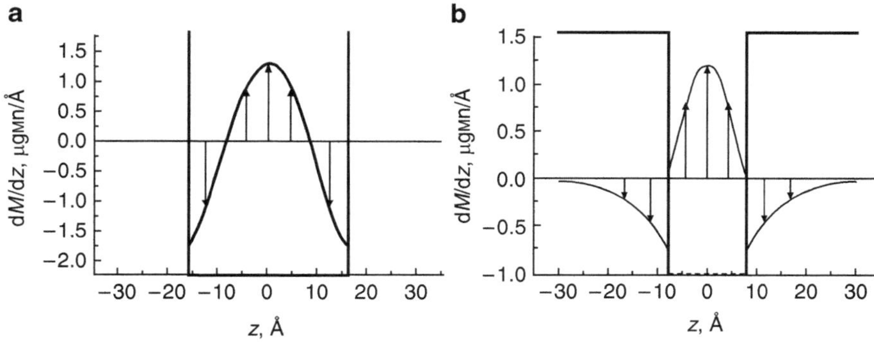

Fig. 3.7 Examples of the spatial distribution of the polarization of magnetic ions in the case of the antiferromagnetic magnetic polaron: (**a**) – in the well with infinitely large potential barriers in the conduction and valence bands (the asymmetric boundary condition $\left(\Psi_c/\Psi_z^{\mathrm{v}} = 0\right)$ at the interfaces is assumed) and (**b**) – in the well with relatively small potential barrier $\Delta V_c = 555$ meV) [47]

$$f^{\pm}(kz,\varsigma) = \frac{\cos(2\varsigma) + \cos\left(2\left(kz\right)\right)}{1 \pm \cos(2\varsigma)\cos\left(2\left(kz\right)\right)}, \tag{3.58}$$

$$\left|\Psi^{\pm}\left(kz,\varsigma\right)\right|^2 = 2C^2(1 \pm \cos(2\varsigma)\cos\left(2kz\right)). \tag{3.59}$$

As it was shown in [47], the spatial distribution of the exchange constant with alternating signs may lead to the formation of a so-called "nonmagnetic" (antiferromagnetic) polaron. This complex is formed by the localized electron bound to surrounding magnetic ions by the strong exchange interaction. The total magnetic momentum of such a complex is nearly zero. The spatial distribution of the total magnetic momentum of a nonmagnetic polaron is shown in Fig. 3.7 for two quantum wells. Figure 3.7a corresponds to the quantum well with infinite potential barriers (the limit case with $\Psi_c/\Psi_z^{\mathrm{v}} = 0$ at the interfaces), and Fig. 3.7b shows the complex formed in the quantum well with rather small finite potential barriers.

The results obtained above have some interesting consequences for symmetrical quantum wells with a monatomic layer of magnetic ions in the middle (δ -doping). The character of the exchange interaction between these ions and confined electrons will be qualitatively different for the electrons in the even [the uppers index "+" in (3.53)–(3.59)] and in the odd (the lower index "−" in (3.53)–(3.59) quantum size subbands. In the even subbands, the valence band component has a node in the middle of the well (Fig. 3.2). As a result, the exchange interaction has solely potential character and the exchange constant is positive. In the odd states, the s-wave has a node in the middle of the wave, the exchange interaction has a kinetic character and the constant is negative. Consequently, the sign of the giant spin splitting of quantum confinement electrons must be different for the even and odd subbands.

In the next Section, we turn to the brief description of the experiments supporting the theoretical conclusions about the carrier-magnetic ion exchange interaction in semimagnetic heterostructures discussed above.

3.5 Comparison with Experiment

In the Sections above, we have given rather general theoretical consideration of the carriers-magnetic ions exchange interaction in quantum confinement structures. The general tensor description of the anisotropic exchange interaction (see Sect. 3.2), for example, allows one to give simple interpretation of many experiments. At the same time, direct theoretical calculations (not involving the anisotropic g-factor formalism) give, of course, similar results but are more cumbersome. The advantage of the anisotropic g-factor formalism over the direct calculations appears especially noticeable in the analysis of the nonlinear processes, where the cumbersome direct calculations often do not allow one to see the physical meaning of the results. Expressions obtained in this Chapter will be used for the description of the effects of the energy transfer between the spin system of carriers and magnetic ions (Chap. 7) as well as for the description of the magnetopolaron states (Chap. 6). Here, we briefly describe only some key experiments that have stimulated the development of the present theoretical analysis and illustrate its results.

3.5.1 Anisotropy of the Giant Spitting of the Exciton States in Quantum Wells

The anisotropy of the giant spin splitting was detected and investigated for exciton state in quantum wells. Giant spin splitting of the electron and hole energy levels in the exchange field induced by polarized magnetic ions is usually detected optically. For this purpose, the magnetic ion spins are oriented by an external magnetic field. From (3.13), (3.20), (3.23), and (3.24), it follows that the average value of the giant spin splitting for the spin dublet states of the confined electrons and holes at the bottom of the $2D$ subbands is given by

$$\Delta E_{\mathrm{e,h}} = -\alpha_{\mathrm{e,h}} \left| \hat{g}_{\mathrm{e,h}} \left(N \langle \boldsymbol{S} \rangle \right) \right| . \tag{3.60}$$

Here, N is the concentration of magnetic ions, $\langle \boldsymbol{S} \rangle$ – is the averaged value of their spin, and the product $N \langle \boldsymbol{S} \rangle$ determines the spin density, which can be measured experimentally. The electron effective g-factor is $g_{\mathrm{e}} = 1$, and the hole effective g-factor $\hat{g}_{\mathrm{h}}$ is given by (3.23) for the heavy hole subband and by (3.24) for the light hole subband. Phenomenological dependence of the spin density on the temperature and the external magnetic field is conventionally described with the help of the modified Brillouin function (for details, see Chap. 1) [61].

Strong anisotropy of the giant spin splitting have been reported for the states at the bottom of the first heavy hole subband in a lot of papers. In [48], the theoretical calculations were compared with the experimental data for the heterostructure formed by the CdTe quantum well with semimagnetic barriers. Conversely, the $Cd_{0.9}Mn_{0.1}Te$ quantum well surrounded by nonmagnetic $Cd_{0.62}Mg_{0.38}Te$ barriers

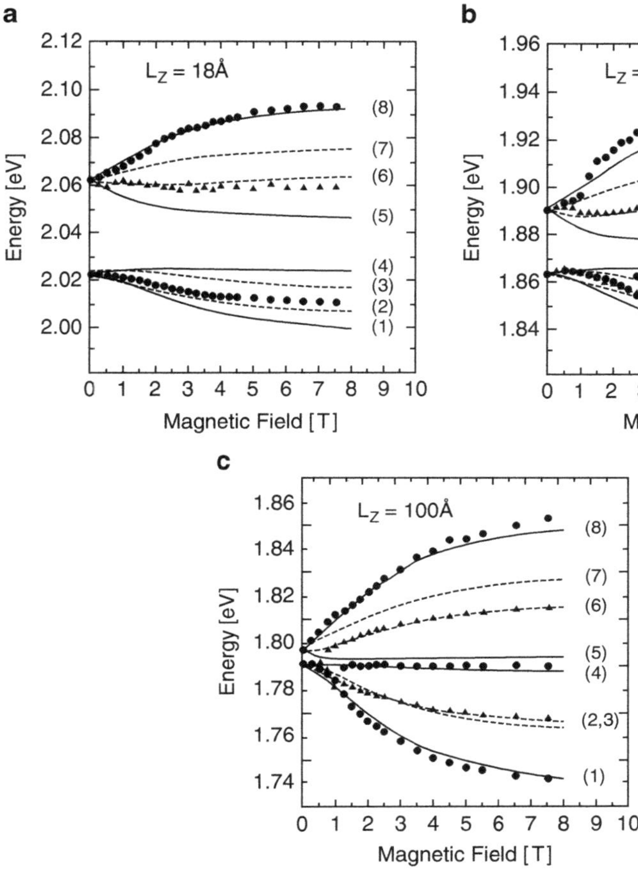

Fig. 3.8 Zeeman pattern of heavy- and light hole transitions in a (**a**) 18 Å, (**b**) 45 Å, and (**c**) 100 Å wide $Cd_{0.9}Mn_{0.1}Te/Cd_{0.6}Mg_{0.4}Te$ quantum well in an in-plane magnetic field [50]

was investigated in [49,50]. Qualitatively, these so different structures demonstrated very similar spectra, although there were naturally some noticeable quantitative differences. Figure 3.8 shows the energies of the exciton resonances determined from the photoluminescence excitation (PLE) spectra as well as the values of the spin splitting energies of the excitonic states from [49, 50]. Experimental studies and theoretical calculations were performed for three quantum wells with the width of 18, 45, and 100 Å, respectively.

In absence of an external magnetic field, there are only two optically active states corresponding to the light and heavy excitons. The external magnetic field in the Faraday geometry splits both these states into two states that are active with right- or left circular polarized light, respectively. The projections of the total spins of the electron and hole in these states on the magnetic field are ± 1. In the case of the heavy exciton, these projection values are equal to the difference between the projections of the electron and hole spins on the z-axis: $F_z^{\mathrm{hh}} = \pm 1 = \pm (3/2 - 1/2)$. The spin splitting of the heavy exciton line is then proportional to $\left(\tilde{\alpha} - \tilde{\beta}\right)$. Here, "tilde" is used to denote the renormalized values of the constants for the exchange interaction between the nonuniform Mn distribution in the heterostructure and the carriers bound in the exciton state. For the holes, the renormalization is caused by the spatial confinement of the hole movement in the quantum well plane and is similar to the renormalization of the exchange constant for the holes in the spherical quantum dot (see Sect. 3.2). Since the conduction band and valence band exchange constants have different signs the value of the exchange splitting is maximal. In the case of the light exciton, the situation is reversed. The electron and hole spins in the optically active state are directed in the same direction so that $\left(F_z^{\mathrm{lh}} = \pm 1 = \pm (1/2 + 1/2)\right)$, and the giant spin splitting is proportional to $\left(\tilde{\alpha} + \tilde{\beta}/3\right)$. Exchange splittings of the electron and hole states in great part compensate each other.

In the Voigt geometry much more complicated spectra are possible. If the quantum well is large enough the five excitonic lines corresponding to the σ^+, σ^- and π polarizations can be observed. In weak magnetic fields, the splitting of the heavy exciton states is practically absent because of the zero transverse g-factor of the heavy hole. In this region, one can clearly see that the hole energy shift is proportional to the square of the magnetic field. This shift describes the van Vleck paramagnetism of heavy holes. The spin splitting of the heavy hole levels is proportional to the third power of the exchange field. This splitting, as well as the spin splitting of the light hole levels, saturates together with the saturation of the magnetic ion polarization.

It is interesting to compare these results with those of [51] where the exciton spectra of the uniaxially deformed ZnMnSe were measured and calculated (see Fig. 3.9). All qualitative features of the magnetooptical anisotropy can be observed also in such a structure. This is because they are caused by practically the same geometrical reason – the symmetry reduction in the material under investigation. This symmetry reduction can be quantitatively characterized by the value of the quadruple splitting of the exciton states (the energy difference between the light and heavy exciton levels) $\Delta_{\mathrm{lh-hh}}$. The formalism of the anisotropic g-factor calculated with the nonperturbed wave function represent zero approximation on the exchange field and can be applied as far as the value of the giant spin splitting of the hole levels is much less than $\Delta_{\mathrm{lh-hh}}$.

3.5.2 Spin Dynamics for Carriers with Anisotropic g-factor

The anisotropic g-factor formalism is even more helpful for the description of the spin dynamics in a system formed by photoexcited hole and magnetic ions. These systems have demonstrated the identical special feature in a whole number of experiments. We consider this feature in detail for the Raman scattering experiments, as an example [13].

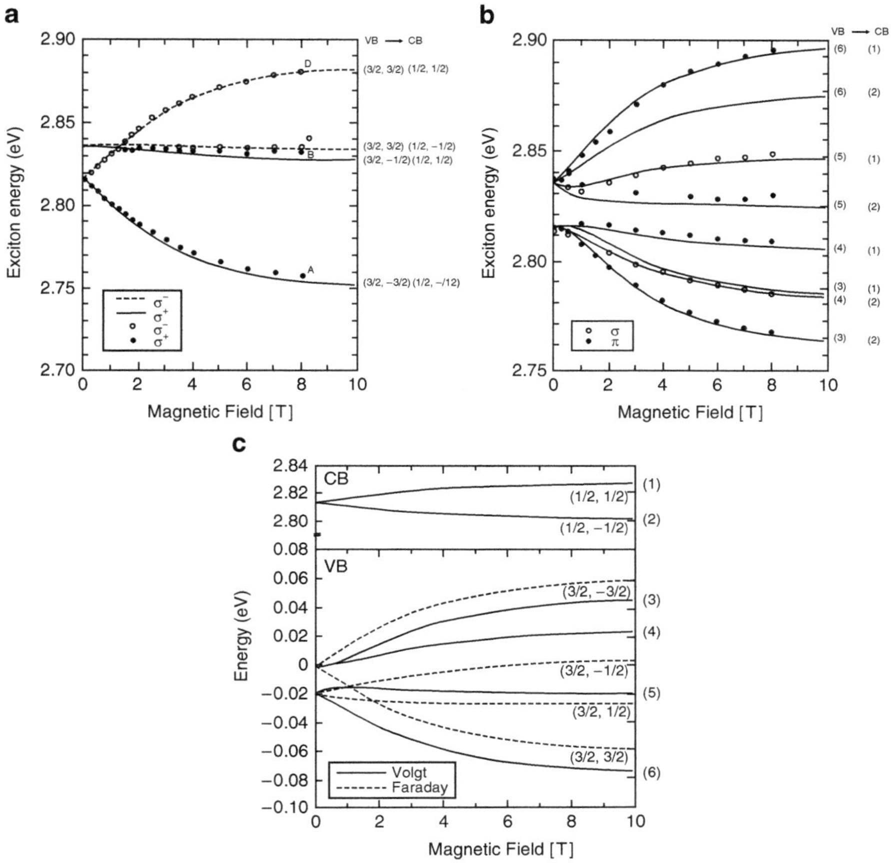

Fig. 3.9 Exciton energies as a function of the magnetic field for the $Zn_{0.917}Mn_{0.083}Se$ epilayer at $T = 4.2$ K [51]: (**a**) Faraday configuration ($\vartheta = 0^o$) (**b**) Voigt configuration ($\vartheta = 90^o$) (**c**) calculated energies of the electrons, heavy holes and light holes; *dashes lines*: Faraday geometry, *solid lines*: Voight geometry. The lines in (**a**) and (**b**) are the calculated interband transition energies

The Raman scattering spectra measured in the $Cd_{0.98}Mn_{0.02}Te/Cd_{0.76}Mg_{0.24}Te$ structures revealed a set of lines that were determined by the paramagnetic resonance involving collective spin flip processes for a large number of localized spins. Up to 15 resonance lines were observed in the quantum well with 18 Å width (Fig. 3.10). Maximal Raman signal was detected for the resonance excitation of the exciton ground state. As the external magnetic field applied in the Voigt geometry increases, the number of the Stokes peaks increases as well as the energy difference between the positions of those picks and of the excitation energy. The position of the peak with number n is given by a simple formula $E_n = ng\mu_B B$, where $g = 1.997 \cong g_{Mn}$ coincided with the g-factor of electron in the d-shell of the Mn ion with a good accuracy.

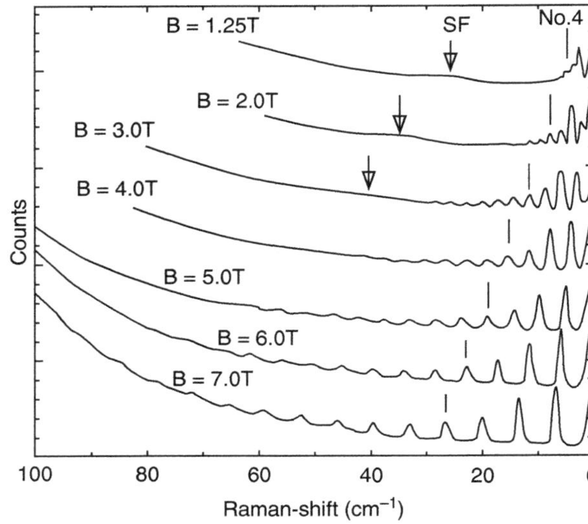

Fig. 3.10 Raman spectra of a single quantum well (18 Å) of $Cd_{0.76}Mn_{0.24}Te/Cd_{0.98}Mn_{0.02}Te$ for various magnetic fields B [13]. $B \parallel [110]$ (Voigt geometry) taken close to resonance with the heavy-hole (HH) dipole transition. The Paramagnetic resonance lines No. 4 have been marked as a guide. For $B_{ext} \leq 3.0\ T$ the lines No. 1 are partially suppressed by band pass function of the spectrometer. The increasing background is caused by the luminescence of the 18 Å well. SF represents electronic spin flip

Many of the observed Stokes peaks appear to be in contradiction with the selection rules for the optical transitions known for the bulk semiconductors of the cubic symmetry. However, this observation can be readily explained if one takes into account the strong anisotropy of the g-factor of carriers in the first $2D$ heavy hole subband in the quantum well (3.23). As a result of this anisotropy, the magnetic ($\boldsymbol{B}$) and exchange ($\boldsymbol{B}_{ex}$) fields directed in the quantum well plane (the XY plane) practically do not act on the spin of the photoexcited hole. For the situation when the spins of the photoexcited holes are oriented along the normal to the quantum well plane, and the magnetic field is in the plane, the hole spins do not change with time. At the same time, the total field is given by the sum of the external field and of the exchange field created by z component of the hole spin. This total field has a nonzero z component. Acting on the magnetic ion it causes multiple spin flip processes for the magnetic ion spins. Different modifications of this experiment as well as the theoretical calculations were presented in later papers [15–21]. The detailed experimental studies and the theoretical analysis with account taken for the small but nonzero components of the hole g-factor in the quantum well plane were performed in [16]. The analysis of the complex spin dynamics of the hole spin was based solely on the anisotropic g-factor formalism.

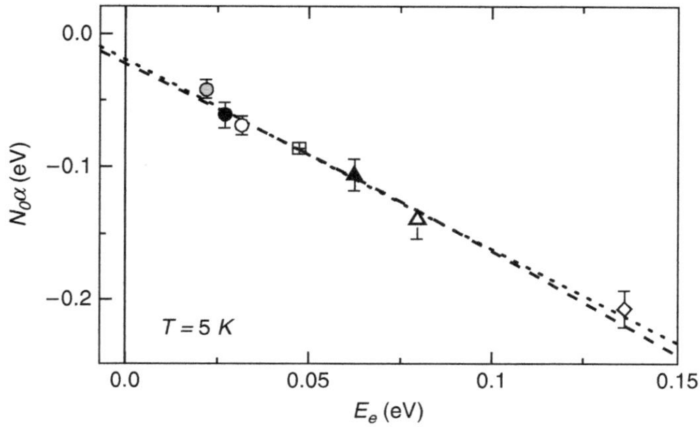

Fig. 3.11 Renormalization of $s - d$ exchange interactions in GaMnAs quantum wells [55]. $N_0\alpha$ plotted as a function of electron kinetic energy for quantum wells with $L = 10$ nm (*circles*), and $L = 7.5$ nm (*squares*), $L = 5$ nm (*triangles*), and $L = 3$ nm (*diamonds*). The *dotted line* is the linear approximation and the *dashed line* is an envelope function calculation based on [32]

3.5.3 Renormalization of the Exchange Interaction Between Magnetic Ions and Electrons Confined in Narrow Quantum Well

Renormalization of the exchange interaction constants for the electrons with large confinement energy was observed in the papers [32, 52–56]. In [32], the structure based on the widegap semiconductor (CdMnTe) was investigated, while in [52, 53] the narrow gap semiconductor structure (HgCdMnSe) was studied. In the papers [54, 55], this effect was investigated in GaMnAs structures. All the experiments clearly demonstrated that the effect increases with natural increase of the electron energy. Since the ratio of the energy E to the band gap energy serves as the main parameter of the theory, the effect in the narrow gap semiconductor is noticeable even for the small values of E.

In Fig. 3.11 taken from the paper [53], the experimental dependence of the exchange interaction parameter on the electron energy for the GaMnAs structure is compared with the results of the theoretical model [32]. This comparison demonstrates good agreement between the theory and experiment.

3.6 Conclusions

We have shown that the exchange interaction between magnetic ions and carriers in semiconductor heterostructures may be significantly different from the one in bulk material. For example, the values of the exchange interaction constants are renormalized. The symmetry of the Kondo Hamiltonian is reduced. The Hamiltonian contains the anisotropic g-factor and additional Dzyaloshinskii–Moriya term. Theoretical analysis is confirmed by numerous experiments.

Results reviewed in this Chapter originate from collaboration with Yakovlev, Kavokin, Keller, Ossau, and Landwehr. We appreciate greatly fruitful collaboration with these colleagues.

A portion of this research was conducted at the Center for Nanophase Materials Sciences, which is sponsored at Oak Ridge National Laboratory by the Division of Scientific User Facilities, U.S. Department of Energy. A.V. Rodina acknowledges the support from the Swiss National Science Foundation. The work was partly conducted within the project of the presidium of Russian Academy of Sciences "Spin dynamics and magnetic interactions in semiconductor nanostructures", partly within the project of CRDF number RUP1-2890-ST-07, partly within the project of RFBR number 09-0201296, and partly supported by the grant number 2396.2008.2 of the "Leading Scientific Schools", Russia.

Appendix: The Eight-band Kane Model

The energy band structure of cubic semiconductors for the wave vectors close enough to the center of the first Brillouin zone can be well described within the eight–band $\boldsymbol{kp}$ model [2, 3, 60, 62]. In homogeneous semiconductor, the full wave function can be expanded as:

$$\Psi(\boldsymbol{r}) = \sum_{\mu=\pm 1/2} \Psi^{\mathrm{c}}_{\mu}(\boldsymbol{r})|S\rangle u_{\mu} + \sum_{\mu=\pm 1/2}\sum_{\alpha=x,z,z} \Psi^{\mathrm{v}}_{\alpha\,\mu}(\boldsymbol{r})|R_{\alpha}\rangle u_{\mu}, \tag{3.61}$$

where $u_{1/2}$ and $u_{-1/2}$ are the eigenfunctions of the spin operator, $|S\rangle$ is the Bloch function of the conduction band edge and $|R_x\rangle = |X\rangle$, $|R_y\rangle = |Y\rangle$, $|R_z\rangle = |Z\rangle$ are the Bloch functions of the valence band edge at the Γ –point of Brillouin zone introduced in Sect. 3.2. The smooth functions $\Psi^{\mathrm{c}}_{\pm 1/2}(r)$ are the components of the conduction band spinor envelope function

$$\Psi^{\mathrm{c}} = \begin{pmatrix} \Psi^{\mathrm{c}}_{1/2} \\ \Psi^{\mathrm{c}}_{-1/2} \end{pmatrix},$$

and $\Psi^{\mathrm{v}}_{x\pm 1/2}(r)$, $\Psi^{\mathrm{v}}_{y\pm 1/2}(r)$, $\Psi^{\mathrm{v}}_{z\pm 1/2}(r)$ are the x, y, z components of the valence band spinor envelope vector

$$\boldsymbol{\Psi}^{\mathrm{v}} = \begin{pmatrix} \boldsymbol{\Psi}^{\mathrm{v}}_{1/2} \\ \boldsymbol{\Psi}^{\mathrm{v}}_{-1/2} \end{pmatrix} = \left\{ \begin{pmatrix} \Psi^{\mathrm{v}}_{x\,1/2} \\ \Psi^{\mathrm{v}}_{x\,-1/2} \end{pmatrix}, \begin{pmatrix} \Psi^{\mathrm{v}}_{y\,1/2} \\ \Psi^{\mathrm{v}}_{y\,-1/2} \end{pmatrix}, \begin{pmatrix} \Psi^{\mathrm{v}}_{z\,1/2} \\ \Psi^{\mathrm{c}}_{z\,-1/2} \end{pmatrix} \right\}$$

.

In bulk homogeneous semiconductor, the eight-component envelope function $\Psi(\boldsymbol{r}) \equiv \{\Psi^{\mathrm{c}}(\boldsymbol{r}), \boldsymbol{\Psi}^{\mathrm{v}}(\boldsymbol{r})\}$ is the solution of the Schrödinger equation

$$\begin{pmatrix} \hat{H}^{\mathrm{c}}\Psi^{\mathrm{c}} & \mathrm{i}P(\hat{\boldsymbol{k}}\boldsymbol{\Psi}^{\mathrm{v}}) \\ -\mathrm{i}P\hat{\boldsymbol{k}}\Psi^{\mathrm{c}} & \hat{H}^{\mathrm{v}}\boldsymbol{\Psi}^{\mathrm{v}} \end{pmatrix} = E \begin{pmatrix} \Psi^{\mathrm{c}} \\ \boldsymbol{\Psi}^{\mathrm{v}} \end{pmatrix}. \tag{3.62}$$

Here, the energy E is measured from the bottom of the conduction band, $\hat{\boldsymbol{k}} = -i\nabla$ is the wave vector operator, and $P = -\mathrm{i}\hbar\langle S|\hat{p}_z|Z\rangle/m_0$ is the Kane matrix element describing the coupling of the conduction and valence bands. The conduction band part of the Hamiltonian, $\hat{H}^{\mathrm{c}}$, acting on the spinor function Ψ^{c} has the form:

$$\hat{H}^{\mathrm{c}}(\hat{\boldsymbol{k}}) = \frac{\gamma_c \hbar^2}{2m_0}\hat{k}^2, \tag{3.63}$$

where γ_{c} takes into account the contribution of remote bands to the electron effective mass m_{e}^* given by [57]:

$$\frac{1}{m_{\mathrm{e}}^*} = \frac{1}{m_0}\left(\gamma_c + \frac{E_p}{3}\left[\frac{2}{E_{\mathrm{g}}} + \frac{1}{E_{\mathrm{g}} + \Delta}\right]\right). \tag{3.64}$$

Here, E_p is the Kane energy parameter $E_p = 2m_0 P^2/\hbar^2$, and Δ is the spin–orbit splitting of the valence band top. The valence band Hamiltonian $\hat{H}^{\mathrm{v}}$ in spherical approximation has a form [3,4,57]:

$$\hat{H}^{\mathrm{v}}(\boldsymbol{k}) = -E_{\mathrm{g}} - \frac{\hbar^2}{2m_0}\left[(\gamma_1 + 4\gamma)k^2 - 6\gamma\,(\boldsymbol{k}\boldsymbol{I})^2\right] + \frac{1}{3}\Delta\,[(\boldsymbol{I}\boldsymbol{\sigma}) - 1], \tag{3.65}$$

where E_{g} is the band gap energy, and γ_1 and γ take into account the contribution of remote bands to the Lutinger parameters $\gamma_1^{\mathrm{L}} = \gamma_1 + E_p/E_{\mathrm{g}}$ and $\gamma^{\mathrm{L}} = \gamma_2^{\mathrm{L}} = \gamma + E_p/E_{\mathrm{g}}$ [57]. The Pauli matrices, σ_x, σ_y, and σ_z, are acting on the spinor components of the wave functions ($\mu = \pm 1/2$). The Hamiltonian $\hat{H}^{\mathrm{v}}$ should be considered as the 2×2 matrix acting on the spinor vector $\boldsymbol{\Psi}^{\mathrm{v}}$ rather than the 6×6 matrix Hamiltonian acting on the six-component wave function as in [4]. Correspondingly, $\hat{\boldsymbol{I}} = \{\hat{I}_x, \hat{I}_y, \hat{I}_z\}$ is the vector operator of the internal orbital momentum. It is easy to show that in this representation $(\hat{\boldsymbol{I}}\boldsymbol{T})\boldsymbol{\Psi}^{\mathrm{v}} = i[\boldsymbol{T} \times \boldsymbol{\Psi}^{\mathrm{v}}]$, where $\boldsymbol{T}$ is an arbitrary vector.

Neglecting the remote band contributions one gets: $\gamma_{\mathrm{c}} = 1$, and $\gamma_1 = -1, \gamma = 0$. However, such neglecting leads to the incorrect energy dispersion for both electrons and hole states. To describe the electron dispersion properly, we will use the eight-band Kane model with dispersion for electrons only. This model uses the proper value of α and assumes $\gamma_1 = \gamma = 0$ [38]. The valence band contribution to the wave $\boldsymbol{\Psi}^{\mathrm{v}}$ can be expressed through $\boldsymbol{k}\Psi^{\mathrm{c}}$ as:

$$\boldsymbol{\Psi}^{\mathrm{v}} = \frac{-iP}{E_p}\left(\frac{m_0}{m_c(E)} - \gamma_{\mathrm{c}}\right)\boldsymbol{k}\Psi^{\mathrm{c}} - \frac{P}{2E_p}\left(g_c(E) - g_0\right)[\hat{\boldsymbol{\sigma}} \times \boldsymbol{k}]\Psi^{\mathrm{c}}, \tag{3.66}$$

where the energy-dependent electron effective mass $m_c(E)$ is given by (3.64) with the energy $E_g + E$ substituted for the band gap energy E_g (so that $m_c(0) = m_e^*$), and the energy-dependent electron g-factor $g_c(E)$ is given by:

$$g_c(E) = g_0 \left(1 - \frac{E_p}{3} \frac{\Delta}{(E_g + E)\left(E_g + E + \Delta\right)}\right), \tag{3.67}$$

where $g_0 = 2$ is the free electron g factor, and the remote band contribution is neglected.

The conduction band spinor Ψ^c in the homogeneous semiconductor satisfies the equation

$$\left(\frac{\hbar^2 \hat{k}^2}{2m_c(E)} - E\right)\Psi^c = 0. \tag{3.68}$$

For the free electron states one can write the conduction band spinor as $\Psi^c(\boldsymbol{r}) = A_s e^{i\boldsymbol{kr}}$, where the constant A_s is determined by the normalization condition $\int (|\Psi^c|^2 + (|\Psi^v|^2)\mathbf{d}^3\boldsymbol{r} = 1$ (the integration is carried out over whole sample volume $V = (2\pi)^3$). Directing the wave vector $\boldsymbol{k} = k\boldsymbol{e}_z$ in z direction one can write the total electron wave function $\Psi^e_{\pm 1/2}(\boldsymbol{r})$ [given by (3.61)] for the states with spin projection $\pm 1/2$, respectively, as:

$$\Psi^e_{\pm 1/2} = A_s(k)e^{ikz} \times \left[|S > u_{\pm 1/2} + iA_1(k)|X \pm iY > u_{\mp 1/2} - iA_0(k)|Z > u_{\pm 1/2}\right]. \tag{3.69}$$

Here,

$$\begin{aligned} A_s(k) &= \frac{1}{\sqrt{1 + A_0^2 + 2A_1^2}}, \\ A_1(k) &= \frac{Pk\Delta}{3\tilde{E}(k)(\tilde{E}(k) + \Delta)}, \\ A_0(k) &= \frac{Pk(3\tilde{E}(k) + 2\Delta)}{3\tilde{E}(k)(\tilde{E}(k) + \Delta)}, \end{aligned} \tag{3.70}$$

where $\tilde{E}(k) = E_g + E$ is the energy of electron calculated from the top of the valence band, and relation between E and k follows from (3.68).

Let us consider the application of the Kane model to the electron states in the narrow quantum well with impenetrable potential barriers (infinitely large band offsets both in the conduction and valence bands $\Delta V_c \to \infty$ and $\Delta V_v \to \infty$). We consider the square quantum well with two interfaces at $z = \pm L/2$. Generally, these interfaces may be microscopically asymmetric. This may occur even when two opposite interfaces are formed by contacts with the same materials [58, 59]. For the Kane model, the standard boundary conditions for the infinite potential barrier $\Psi^{e}_{\pm 1/2}(\pm L/2) = 0$ have no sense because the components Ψ^{c} and Ψ^{v} can not vanish simultaneously [35].

It has been shown in [35, 36, 38] that in the case of impenetrable barrier the general boundary conditions guarantee that the normal-to-surface components of the envelope flux density matrix vanish at the interface. For the model under consideration, this condition reads

$$J_z^{\lambda\eta}(\pm L/2) = \frac{i\hbar}{2m_0}\left[\gamma_{\mathrm{c}}\left(\Psi^{\mathrm{c}\eta}\nabla_z\Psi^{\mathrm{c}\lambda *} - \Psi^{\mathrm{c}\lambda *}\nabla_z\Psi^{\mathrm{c}\eta}\right)\right.$$
$$\left.\left. + \frac{2m_0 P}{\hbar}\left(\Psi_z^{\mathrm{v}\eta}\Psi^{\mathrm{c}\lambda *} - \Psi_z^{\mathrm{v}\lambda *}\Psi^{\mathrm{c}\eta}\right)\right]\right|_{z=\pm L/2} = 0, \quad (3.71)$$

where λ and η denote the eigenstates of the Hamiltonian with energies E_λ and E_η, respectively. To ensure the self-adjoinment of the Hamiltonian, this general requirement of (3.71) should be satisfied for any two arbitrarily chosen eigenfunctions $\Psi^\lambda = (\Psi^{\mathrm{c}\lambda}, \boldsymbol{\Psi}^{\mathrm{v}\lambda})$ and $\Psi^\eta = (\Psi^{\mathrm{c}\eta}, \boldsymbol{\Psi}^{\mathrm{v}\eta})$ with energies E_λ and E_η, respectively, if and only if they are the subject of the same boundary conditions (BCs). Generally, the requirement of (3.71) leads to the BC in the form:

$$\Psi^{\mathrm{c}}(\pm L/2) = \pm a^{\pm}\left(\gamma_{\mathrm{c}}\Psi^{\mathrm{c}'}(\pm L/2) - \frac{E_P}{P}\Psi_z^{\mathrm{v}}(\pm L/2)\right), \quad (3.72)$$

where $\Psi^{\mathrm{c}'}(\pm L/2) = \nabla_z\Psi^{\mathrm{c}}$ at $z = \pm L/2$, and $a^{\pm}$ are some surface parameters having the dimension of length. It is convenient to choose the value $L_K = \sqrt{\hbar^2/2E_P m_0}$ as a length unit [35] and introduce dimensionless parameters $\tilde{a}^{\pm} = a^{\pm}/L_K$. The condition (3.72) can be also written as

$$\Psi^{\mathrm{c}}(\pm L/2) = \pm\tilde{a}^{\pm}L_K\left.\left(\frac{m_0}{m_c(E)}\frac{\mathrm{d}\Psi^{\mathrm{c}}}{\mathrm{d}z} - \frac{g_0 - g_c(E)}{2}[\sigma \times \boldsymbol{k}_\perp]_z\Psi^{\mathrm{c}}\right)\right|_{z=\pm L/2}, \quad (3.73)$$

where $\boldsymbol{k}_\perp = (k_x, k_y)$ is the in-plane wave vector, and the relation given by (3.66) is used. One can see that in general case the boundary conditions are spin-dependent [36, 39]. This may result in the additional spin-splitting of the $2D$ energy subbands in asymmetric structures [36, 39].

The parameter $\tilde{a}$ characterizes the asymmetry of the interface with respect to the contributions from the conduction and valence band states $\Psi^{\mathrm{c}}/\Psi_z^{\mathrm{v}}$. In the limit case $\tilde{a} = 0$, the component of the conduction band Ψ^{c} vanishes at the interface. In the opposite limit case $\tilde{a} = \infty$, the valence band component Ψ_z^{v} vanishes at the interface. In real heterostructures the case $|\tilde{a}| \approx 1$ was observed [38, 46]. In the case of the infinite potential barriers, the value of the parameter $\tilde{a}$ depends generally on the ratio $\Delta V_{\mathrm{c}}/\Delta V_{\mathrm{v}}$ of the band off-sets in the conduction and valence bands. This ratio is finite even when both barrier heights ΔV_{c} and ΔV_{v} are infinitely large. In the limit cases, $\Delta V_{\mathrm{c}}/\Delta V_{\mathrm{v}} = \infty$ and $\Delta V_{\mathrm{c}}/\Delta V_{\mathrm{v}} = 0$, the extremely asymmetric conditions with $\tilde{a} = 0$ and $\tilde{a} = \infty$, respectively, can be considered. The spatial distribution of the conduction band and valence band components over the quantum well for the limit case $\tilde{a}^{\pm} = \tilde{a} = 0$ is shown in Fig. 3.2.

The review of the general boundary conditions for the interface with finite potential barriers can be found in [2, 35, 36, 39–45]. The parameters of the boundary condition can be determined from the comparison with experiment or taken from the corresponding first principle calculations [40, 41].

References

1. E.L. Ivchenko, *Optical Spectroscopy of Semiconductor Nanostructures* (Alpha Science International Ltd., Harrow U.K., 2005)
2. E.L. Ivchenko, G.E. Pikus, *Superlatices and Other Heterostructures, Symmetry and Optical Phenomena*, 2nd edn. (Springer, Berlin Heidelberg New York, 1995, 1997)
3. G.L. Bir, G.E. Pikus, *Symmetry and Strain-induced Effects in Semiconductors* (Wiley, New York, 1974)
4. J.M. Luttinger, Phys. Rev. **102**, 1030 (1956)
5. C. Kittel, *Quantum Theory of Solids* (Wiley, 1987)
6. S.S. Nedorezov, Fiz. Tverdogo Tela **12**, 2269 (1970) [Sov. Phys. Solid State **12**, 1814 (1970)]
7. I.A. Merkulov, V.I. Perel, M.E. Portnoi, ZhETF **99**, 1202 (1991)
8. A.K. Bhattacharjee, G. Fishman, B. Coqblin, Physica **117&118**, 449 (1983)
9. B.E. Larson, K.C.H.E. Ehrenreich, A.E. Carlsson, Phys. Rev. B **37**, 4137 (1988)
10. I.A. Merkulov, K.V. Kavokin, Phys. Rev. B **52**, 1751 (1995)
11. B. Bleaney, A. Abragam, *Electron Paramagnetic Resonance of Transition Ions*, Chap. 15 (Oxford Clarendon press, Oxford, 1970)
12. L.D. Landau, E.M. Lifshitz, *Quantum Mechanics*, 4th edn. (Pergamon Press, New York, 1989)
13. J. Stuhler et al., Phys. Rev. Lett. **74**, 2567 (1995)
14. J. Stuhler et al., J. Cryst. Growth **159**, 1001 (1996)
15. K.V. Kavokin, I.A. Merkulov, Phys. Rev. B **55**, R7371 (1997)
16. A.V. Kudinov et al., Phys. Solid State **45**, 1360 (2003)
17. I.A. Merkulov et al., Phys. Rev. B **55**, 5727 (1996)
18. I.A. Merkulov et al., Phys. Solid State **39**, 1859 (1997)
19. D.R. Yakovlev et al., Phys. Rev. B **56**, 9782 (1997)
20. S.A. Crooker, D.D. Awshalom, J.J. Baumbers, F. Flack, N. Samarth, Phys. Rev. B **56**, 7574 (1997)
21. R. Akimoto, K. Ando, F. Sasaki, S. Kobayashi, T. Tani, Phys. Rev. B **57**, 7208 (1998)
22. D. Kechrakos, N. Papanikolaou, K.N. Trohidou, T. Dietl, Phys. Rev. Lett. **94**, 127201 (2005)
23. Roger G. Melko, Randy S. Fishman, Fernando A. Reboredo, Phys. Rev. B **75**, 115316 (2007)
24. A.V. Akimov et al., Phys. Rev. B **73**, (2006)
25. N.F. Mott, Proc. R. Soc. A **124**, 425 (1929)
26. C.G. Shull, C.T. Chase, F.E. Myers, Phys. Rev. B **63**, 29 (1943)
27. Y.F. Berkovskaya et al., Zh.ETF **94**, 183 (1988)
28. D.M. Hoffman et al., Solid State Commun. **114**, 547 (2000)
29. C.E. Silva et al., Solid State Commun. **58**, 247 (1986)
30. T.Z. Kachishvili, Solid State Commun. **80**, 283 (1991)
31. T.Z. Kachishvili, Solid State Commun. **83**, 509 (1992)
32. I.A. Merkulov, D.R. Yakovlev, A. Keller, W. Ossau, J. Geurts, G. Landwehr, G. Karczewski, T. Wojtowicz, J. Kossut, Phys. Rev. Lett. **83**, 1431 (1999)
33. T. Dietl, in *Handbook of Semiconductors*, ed. by S. Mahajan (North-Holland, Amsterdam, 1994), p. 1254
34. A.K. Bhattacharjee, Phys. Rev. B **58**, 15660 (1998)
35. A.V. Rodina, A.Yu. Alekseev, Al.L. Efros, M. Rosen, B.K. Meyer, Phys. Rev. B. **65**, 125302 (2002)

36. A.V. Rodina, A.Yu. Alekseev, Phys. Rev. B. **73**, 115312 (2006)
37. A.K. Bhattacharjee, J. Perez-Conde, in *Proceedings of the 25th International Conference on the Physics of Semiconductors*, eds. by N. Miura, T. Ando (Springer Verlag, Berlin, 2000), p. 242
38. A.V. Rodina, Al.L. Efros, A.Yu. Alekseev, Phys. Rev. B **67**, 15530 (2003)
39. F.T. Vasko, A.V. Kuznetsov, *Electronic States and Optical Transitions in Semiconductor Heterostructures* (Springer, New York, 1999)
40. B.A. Foreman, Phys. Rev. B **72**, 165345 (2005)
41. M.G. Burt, J. Phys. Condens. Matter **4**, 6651 (1992)
42. I.V. Tokatly, A.G. Tsibizov, A.A. Gorbatsevich, Phys. Rev. B **65**, 165328 (2002)
43. M.V. Kisin, B.L. Gelmont, S. Luryi, Phys. Rev. B **58**, 4605 (1998)
44. Smadar de-Leon, B. Laikhtman, L.D. Shvartsman, J. Phys. Condens. Matter **10**, 8715 (1998)
45. S.R. White, G.E. Marques, L.S. Sham, Vac. Sci. Technol. **21**, 544 (1982)
46. J.A. Gupta, D.D. Awshalom, Al.L. Efros, A.V. Rodina, Phys. Rev. B **66**, 125307 (2002)
47. I.A. Merkulov, Phys. Solid State **42**, 132 (2000)
48. P.A.W. Peyla, Y. Merle d'Aubigne, D.E. Ashenford, B. Lunn, Phys. Rev. B **47**, 3783 (1993)
49. B. Kuhn-Heinrich, W. Ossau, E. Bangert, A. Waag, G. Landwehr, Solid State Commun. **91**, 413 (1994)
50. B. Kuhn-Heinrich, W. Ossau, A. Waag, G. Landwehr, in *Proceedings o the 22nd International Conference on the Physics of Semiconductors*, ed. by D.J. Lockwood (World Scientific, Singapore, 1994), p. 2529
51. W.Y. Yu, A. Twardowski, L.P. Fu, A. Petrou, B.T. Jonker, Phys. Rev. B **51**, 9722 (1995)
52. W. Hoerstel et al., Semicond Sci. Technol. **14**, 820 (1999)
53. G.G. Tarasov et al., J. Cryst. Growth **214/215**, 436 (2000)
54. R.C. Myers, M. Poggio, S.P. Stern, A.C. Gossard, D.D. Awshalom, Phys. Rev. Lett. **95**, 017204 (2005)
55. N.P. Stern, R.C. Myers, M. Poggio, A.C. Gossard, D.D. Awschalom, Phys. Rev. B **75**, 045329 (2007)
56. D.A. Bussian, S.A. Crooker, M. Yin, M. Brynda, Al.L. Efros, V.I. Klimov, Nat. Mater. **8**, 35 (2008)
57. C.R. Pidgeon, R.N. Brown, Phys. Rev. **146**, 575 (1966)
58. T. Reker, H. Im, L.E. Bremme, H. Choi, Y. Chung, P.C. Klipstein, H. Shtrikman, Phys. Rev. Lett. **88**, 056403 (2002)
59. A.V. Platonov, V.P. Kochereshko, E.L. Ivchenko, G.V. Mikhailov, D.R. Yakovlev, M. Keim, W. Ossau, A. Waag, G. Landwehr, Phys. Rev. Lett. **83**, 3546 (1999)
60. J.M. Luttinger, K.W. Kohn, Phys. Rev. **97**, 869 (1955)
61. J.A. Gaj, R. Planel, G. Fishman, Solid State Commun. **29**, 435 (1979)
62. E.O. Kane, J. Phys. Chem. Solids **1**, 249 (1957)

Chapter 4
Band-Offset Engineering in Magnetic/Non-Magnetic Semiconductor Quantum Structures

J.K. Furdyna, S. Lee, M. Dobrowolska, T. Wojtowicz, and X. Liu

Abstract Low-dimensional structures made of components fabricated from diluted magnetic semiconductors are studied in this Chapter. Starting with either DMS quantum wells of non-magnetic quantum wells contained in the DMS barriers is discussed first. Second, the notion of spin superlattices is introduced and discussed. In particular, it is shown how by spectroscopic methods one is able to study localization of carriers in various layers constituting the structures in question. A few remarks are made concerning interacting quantum dots made of DMSs.

4.1 Introduction

Diluted magnetic semiconductors (DMSs) are semiconducting alloys in which a part of the semiconductor crystal lattice is substitutionally replaced by magnetic transition metal ions. II–VI semiconductors in which a fraction of the group-II atoms is replaced by Mn^{++} are the best-known examples of such alloys (e.g., $Zn_{1-x}Mn_xSe$, $Zn_{1-x}Mn_xTe$, $Cd_{1-x}Mn_xSe$, etc.). The random distribution of magnetic ions over the cation sublattice in DMSs leads to novel important magnetic effects, e.g., the formation of spin-glass-like phase at low temperature, extremely large Zeeman splittings of electronic levels, giant Faraday rotation, magnetic-field-induced metal–insulator transition, and formation of bound magnetic polarons. These enhanced spin-dependent properties occurring in DMS systems were widely studied in the literature and are already summarized in many excellent review articles [1–7] (see also Chap. 1).

J.K. Furdyna (✉), M. Dobrowolska, and X. Liu
Department of Physics, University of Notre Dame, Notre Dame, IN 46556, USA
e-mail: Jacek.K.Furdyna.1@nd.edu

S. Lee
Department of Physics, Korea University, Seoul, 136-701, Korea
e-mail: slee3@nd.edu

T. Wojtowicz
Institute of Physics, Polish Academy of Sciences, Al. Lotników 32/46, 02-668, Warsaw, Poland
e-mail: Tomasz.Wojtowicz@ifpan.edu.pl

J. Kossut and J.A. Gaj (eds.), *Introduction to the Physics of Diluted Magnetic Semiconductors*, Springer Series in Materials Science 144,
DOI 10.1007/978-3-642-15856-8_4, © Springer-Verlag Berlin Heidelberg 2010

The ternary nature of DMSs provides the possibility of tuning the lattice constants and band parameters by varying the composition. Owing to such tunability, the $II_{1-x}Mn_xVI$ alloys are excellent candidates for the preparation of quantum wells, superlattices, and other heterostructures that involve band gap engineering. Advances in molecular beam epitaxy (MBE) growth have already made it possible to prepare a wide range of such DMS-based low-dimensional structures [8]. The presence of magnetic ions in such heterostructures additionally allows one to explore new spin-dependent phenomena at lower dimensionalities.

The presence of localized magnetic ions in DMSs leads to an exchange interaction between the *sp* band electrons and the *d* electrons associated with Mn^{++}, resulting in extremely large Zeeman splittings of electronic levels. The theoretical formulation of this process – which underlies most phenomena discussed below – is presented in detail in Chaps. 1 through 3 of this volume.

The consequence of the *sp–d* exchange interaction mentioned above that will be of key interest in this chapter is the dramatic enhancement of the Zeeman splitting of band edges in zinc-blende $II_{1-x}Mn_xVI$ alloys. This band edge splitting, shown schematically in Fig. 1.12 of Chap. 1 of this volume, can be conveniently monitored by exciton transitions, since exciton levels closely follow the band edges.[1] Thus, the Zeeman splitting of free exciton ground state provides one of the most direct and convenient situations which one can use for measuring exchange effects in typical DMSs [6,9–19]. This is illustrated in Fig. 3.9 of Chap. 3 of this volume by a typical magnetic field dependence of the exciton transition energy observed experimentally for $Zn_{0.917}Mn_{0.083}Se$ [20]. Here, the most dramatic effect is the very large energy difference between the two transitions observed with the σ^+ and σ^- polarizations. The splittings are of the order of 100 meV – a magnitude which in a nonmagnetic semiconductor would require fields of the order of a megagauss. Owing to such large energy splitting, this feature provides an exceptionally convenient experimental tool for 'tuning' the energy gap of a semiconductor system by using an external magnetic field. This ability to tune the bandgap has, in turn, important implications for studying quantum effects in DMS heterostructures, providing the experimenter with a powerful tool for varying the relative band alignment in a controlled and continuous manner simply by varying the applied field. This phenomenon will be illustrated in this chapter using a variety of quantum structures, especially those combining DMS and non-DMS constituents.

4.2 Single Quantum Well

4.2.1 Determination of Band Offset Using DMS Properties in Rectangular QWs

In semiconductor heterostructures, the band offset between layers comprising the structure is the most important parameter in determining the quantum effects

---

[1] See Chap. 2 of this volume for exceptions from this rule.

exhibited by the system. Many electrical, optical, and x-ray techniques have been applied to measure the band offset in various heterostructures based on diverse materials systems [21–25]. Since optical experiments – e.g., absorption, reflectance, and photoluminescence – are especially useful for directly determining the band gap of semiconductor systems, they are also frequently used for investigating band offsets in heterostructures.

The usefulness of optical studies of semiconductor heterostructures lies in their ability to precisely determine transition energies of the system, enabling one to calculate the energy levels of the structure. However, since interband transitions observed in optical experiments involve energy levels in both the conduction and the valence bands, this automatically requires knowledge of band offset in each band – a quantity that is usually rather difficult to determine. A significant advantage in determining band offsets in heterostructures can be achieved, however, by involving DMSs, in which a significant fraction of the band gap can be tuned by an external magnetic field. The band gap tunability of DMSs in heterostructures not only provides a variation of the transition energy with magnetic field, but also enables direct measurement of band offsets at the DMS/non-DMS interface, which is made possible by observing the intensity variation of transitions as a function of the applied magnetic field.

Such ability of DMS systems has been utilized in various hetrostructures consisting of DMS/non-DMS combinations, such as CdTe/CdMnTe [26–30] or ZnSe/ZnMnSe [31, 32]. Figure 4.1 illustrates the general idea of the band alignment of a single QW involving a DMS, and the variation of its band alignment in a magnetic field. Panel (b) shows the band diagram in the absence of an external magnetic field of a single DMS QW (e.g., ZnMnSe) between non-DMS barriers (e.g., ZnBeSe). In the presence of magnetic field [panels (a) and (c)], the band edges of the barrier material show negligible Zeeman splitting, while the band edges of the DMS well show giant shifts. The shift of the conduction-to-valence band separation

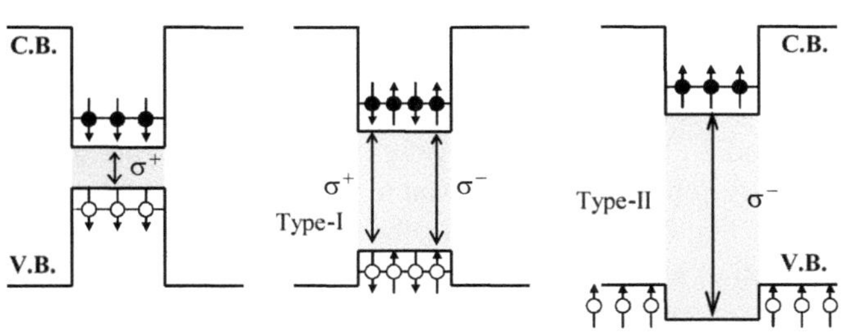

Fig. 4.1 Schematic band diagram of DMS QW between non-DMS barriers. At zero-magnetic field (*central panel*), the spin-up and spin-down conduction and valence band states are degenerate for both bands. In the presence of a magnetic field, the spin states separate. As a result, the well depth increases for spin-down orientation (*left panel*); and decreases for the spin-up states (*right panel*). As this continues, at some field value there occurs a transition from type-I to type-II band alignment

relative to its value at zero field is given by [5] (see Chap. 1 in this volume)

$$\Delta E = (\alpha - \beta) N_0 x \langle S_z \rangle, \tag{4.1}$$

$$\langle S_z \rangle = S_0 B_{5/2} \left(\frac{5 \mu_B g_{Mn} B}{2 k_B (T + T_0)} \right), \tag{4.2}$$

where αN_0 and βN_0 are the well-known exchange integrals for electrons and holes; x is the Mn concentration; and $B_{5/2}$ is the Brillouin function for spin $S = 5/2$. S_0 is the effective spin and T_0 is the difference between the effective and real temperatures. S_0 and T_0 are treated as phenomenological fitting parameters.

One should recall that α and β in (4.1) – quantities which determine the Zeeman splittings of the conduction and valence bands – are of opposite sign. Thus, for a given spin orientation the band edges will move in opposite directions. For example, as shown in Fig. 4.1, for the spin-down orientation the wells in both bands will become deeper [panel (a)], while for spin-up the wells in both bands become more shallow [panel (c)]. Since βN_0 is typically much larger than αN_0 in most $II_{1-x}Mn_xVI$ alloys (e.g., in ZnMnSe $\alpha N_0 = 0.27\,eV$, $\beta N_0 = -0.9\,eV$) [33], this effect is much stronger in the valence band than in the conduction band. The arrows in Fig. 4.1 show the allowed spin-conserving transitions. Thus, in the presence of an applied field B transitions observed in the σ^+ circular polarization (which connect the spin-down states) will be strongly red-shifted relative to their zero-field energy, and transition in the σ^- polarization will be strongly blue shifted. If one chooses the composition of the constituent alloys (i.e., their energy gaps) in such a way that the resulting band offset in one of the bands at $B = 0$ is smaller than the maximum value of the Zeeman shift, one can expect that at a certain magnetic field B the initial offset will be overridden for one of the spin orientations. This will lead to a change-over from the type-I to the type-II band alignment for that spin orientation, which in turn will strongly affect the optical properties of the system.

This phenomenon is clearly observed in the two ZnMnSe single quantum well (SQW) structures having different Be concentration in the ZnBeSe barrier layer, shown in Fig. 4.2. As an example, in the upper and lower panel of Fig. 4.2 we show the field dependence of the exciton transition energies for two SQW structures with 1% and 2.3% Be in the ZnBeSe barrier layers. Note that the excitonic transition between spin-up states observed in the σ^- circular polarization (the blue shifted branch of Fig. 4.1) disappears for the low Be-concentration sample (shown in the upper panel of Fig. 4.2) at $B \simeq 0.5\,T$, and in high Be-concentration sample (shown in the lower panel) at $B \simeq 1.75\,T$. Above those particular magnetic fields, the Zeeman shift of the heavy hole band exceeds the original band offset at zero-magnetic field. When this happens, the wave functions in the heavy hole band spread out, and the heavy-hole spin-up eigenstates cease to be localized. This indicates that the quantum well disappears in the valence band (changing over to a type-II band alignment, a situation illustrated on the right of Fig. 4.1), and thus the ground state transition can no longer be observed.

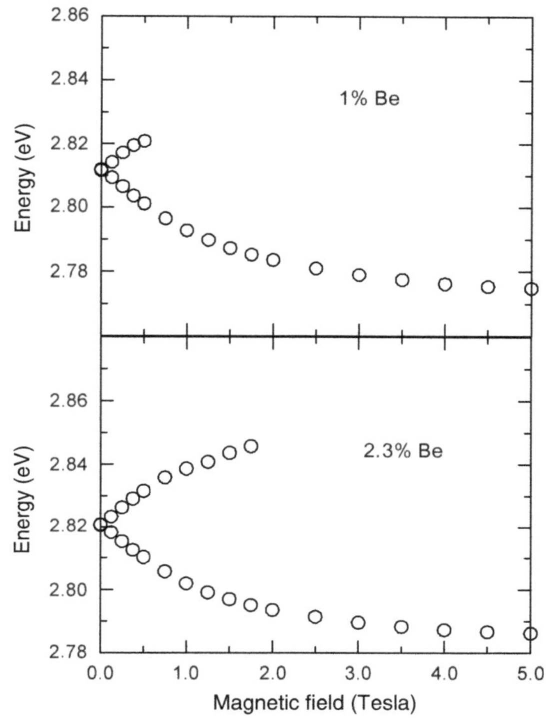

Fig. 4.2 Experimental data (*open circles*) for samples with different Be content in the barriers. Note that transitions observed in the σ^- circular polarization (*upper branch*) disappear at a certain magnetic field for both samples, as discussed in the text. Reprinted from [32], © (2000), with permission from Elsevier

The valence band offset can then be directly determined by comparing this result to the amount of the Zeeman shift at that same field in a control ZnMnSe epilayer grown with the same Mn concentration. In the specific example used here, it turns out that the offset in the valence band is about 40% of the total energy gap difference in the ZnMnSe/ZnBeSe QW structures. For more precise values of the band offset in such structures, strain and the effect of type-I/type-II transition on excitons must be included in the band structure model [34, 35], as has been done, e.g., in [32] and [36].

4.2.2 Graded Potential Quantum Wells

The precision in determining the value of the valence band offset with the use of optical methods and magnetic field band gap tuning in DMS structures can be further increased by the application of quantum structures having precisely controlled, semicontinuous spatial distribution of mixed crystal components in the growth direction. Such a grading in the composition of a ternary compound, achievable for instance with the use of molecular beam epitaxy [37–39], leads to the structures in which a bandgap does not change abruptly at the interfaces, as in most typical heterostructures, but changes also in semicontinuous, graded fashion. The pattern of the quantized energy levels in such structures depends on the particular shape of the confining potential produced by the band gap grading. Using appropriate composition grading, one can produce quantum wells with, e.g., triangular [39], parabolic [38,40,41] or half-parabolic [42,43] confining potential. The characteristic of graded potential quantum wells is that the distances between energetic levels of confined carriers, contrary to the case of rectangular QWs, depend sensitively and explicitly on the actual value of the band gap discontinuity between the materials used to produce these graded QWs [44]. This makes graded QWs very attractive as a tool for precise determination of band offsets.

For instance, quantized energy levels of a carrier with the mass m_i^* belonging to the i-th band ($i = e$, lh, hh) bound in the infinitely high parabolic potential $U_i = (K_i/2)z^2$ are given by:

$$E_{ni} = \left(n - \frac{1}{2}\right)\hbar\sqrt{\frac{K_i}{m_i^*}}, \quad K_i = Q_i K, \quad K = \frac{8\Delta E_{\mathrm{g}}}{L_z^2}, \tag{4.3}$$

where $Q_i = \Delta V_i/\Delta E_{\mathrm{g}}$, with ΔV_i being the discontinuity of the band edges, and ΔE_{g} being the energy gap discontinuity between the materials at the center and at the distance $\pm L_z/2$ from the center of the QW, and $n = 1, 2, 3, \ldots$ being the harmonic oscillator quantum number. Therefore, by plotting experimental values of the energy distance between two successive diagonal transitions in the heavy hole series (e.g., between levels with the same quantum number for holes and electrons: $HH_n \rightarrow E_m$ with $n = m$): $\Delta E = E_{(n+1)\mathrm{hh}} + E_{(n+1)\mathrm{e}} - E_{(n)\mathrm{hh}} - E_{(n)\mathrm{e}}$ versus square root of the total curvature K at the well center one should obtain a straight line. The slope of this line is determined by the value of the valence band offset Q_{hh} ($Q_{\mathrm{hh}} + Q_{\mathrm{e}} = 1$). In Fig. 4.3, the example of the application of such a procedure is given for the case of parabolic QWs (PQW) made of $Cd_{1-x}Mn_xTe$ DMS [40]. This simplest method does posses, however, some drawbacks, including the need of knowing the value of effective masses of carriers. These drawbacks can partially be overcome by observing and analyzing the energies of nondiagonal transitions (e.g., $HH_n \rightarrow E_m$ with $n \neq m$). Especially handy are series of transitions originating from the same level in the valence band, e.g., $HH_n \rightarrow E_n$, $HH_n \rightarrow E_{n+1}$, $HH_n \rightarrow E_{n+2}$ because energetic differences in such series do not depend on the hole effective mass, which typically is not well known, especially in the case of mixed crystals. For the observation of non-diagonal transitions especially useful are half-parabolic QWs (HPQW), in which all nondiagonal transitions are parity-allowed due to undetermined parity of the states (see further).

The methods of band-offset determination based on the studies of optical transitions in graded QWs relies obviously on the precise knowledge of the actual shape of the confining potential and on the correct identification of various optical transition, including the nondiagonal ones and those in which light hole levels are involved. The great advantage of using DMS graded QW structures or nonmagnetic graded QW structures with, however, additional magnetic component evenly distributed

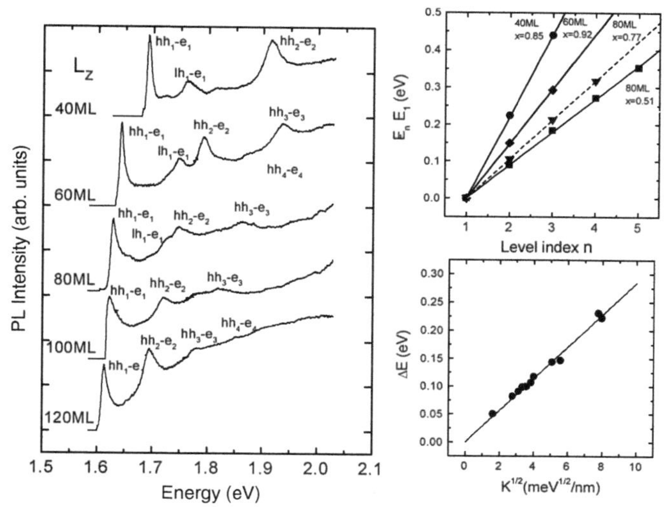

Fig. 4.3 Photoluminescence excitation spectra of five $Cd_{1-x}Mn_xTe$ PQW structures with $x \simeq 0.8$ and well width from 40 to 120 ML collected at $T = 1.8\,K$ (*left panel*). The identification of the excitonic lines is given by the labels $\mathrm{hh}_n - e_m$ and $\mathrm{lh}_n - e_m$, where n is the heavy and light hole level index and m is the level index of electrons. The top right panel shows the scaling of the main $\mathrm{hh}_n - e_n$ transition series with the level index n for a number of PQWs. The bottom right panel shows the average energy distance between the peaks in the hh diagonal ($n = m$) series plotted vs. square root of the curvature (at the well center) of the confining potential for all studied $Cd_{1-x}Mn_xTe$ PQWs. The *straight line* is the least square fit to the data and yields the valence band offset value $Q_{\mathrm{hh}} = 0.44 \pm 0.1$ in CdTe/MnTe system. Reprinted from [40], © (1996), with permission from Acta Physica Polonica A, Institute of Physics, Polish Academy of Sciences

throughout the structure [41], is again based on the possibility of large magnetic field tuning of the band gap, and is twofold. First, the unambiguous identification of various transitions is possible by taking advantage of their distinctive spin splitting. For instance excitonic transitions involving various electron confined levels originating from the same heavy hole level should all have similar spin splitting. That is because the splitting of heavy hole excitonic transitions is dominated by the splitting of the hh levels, since βN_0 is about four times larger than αN_0. The second advantage of tailoring the confining potential by grading of the magnetic component of the mixed crystal is that this provides a very sensitive method of precise determination of the actual shape of the potential via studying the field dependence of the spin splitting of various excitonic states. The method is based on the fact that the spin-splitting in DMSs is a nonlinear function of Mn concentration and, therefore, the application of an external field translates into a strong perturbation of the potential profile from its zero-field shape. These deviations are most pronounced in the valence band where, in the case of a parabolic potential shape in the absence of the field, for one of the spin species (namely, for the $|\,3/2, -3/2\rangle$ state) one can expect a 'camel-back' shape of the confining potential profile after application of a magnetic field, as seen in Fig. 4.4 [39]. The field-induced change of the potential shape for

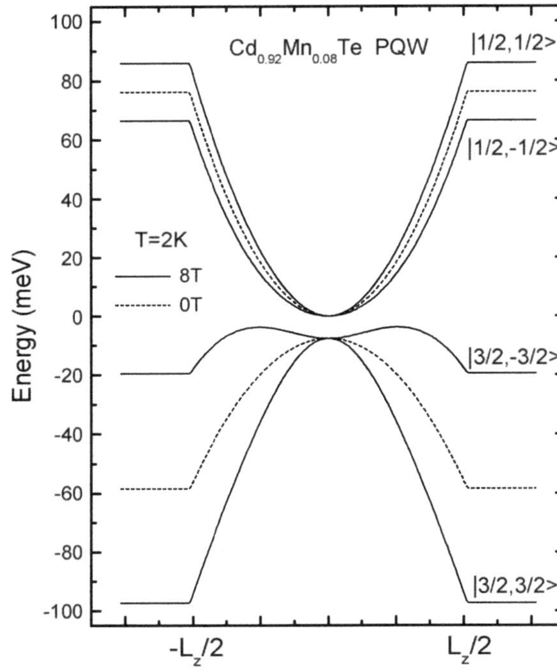

Fig. 4.4 Effective potential in the conduction and heavy hole valence bands of a parabolic quantum well made of $Cd_{0.92}Mn_{0.08}Te$ at 2 K in the absence of magnetic field (*dotted line*) and at 8 T. The spin indices are indicated in the figure. A pronounced 'camel-back' shape of the heavy hole $|\,3/2,-3/2\rangle$ spin subband is visible. Reprinted from [39], © (1997), with permission from Elsevier

different spin components, which determines to a great degree the field-dependence of confined energy levels and hence the spin splitting of different excitonic transitions, depends sensitively on the actual distribution of Mn ions (i.e., composition profile).

The value of the valence band offsets Q_{hh} was determined with the use of graded QWs for two materials systems: CdTe/MnTe and CdTe/MgTe to be 0.4 ± 0.05 [42, 43] and 0.45 ± 0.1 [41], respectively. Both parabolic and half-parabolic $Cd_{1-x}Mn_xTe$ and $Cd_{1-y}Mg_yTe$ QWs, having a wide range of curvatures K, which was varied both by the width of a graded region (from 40 to 100 monolayers, $1ML \simeq 3.24$ Å) and by the maximum composition of the mixed crystal (x, y from 0.14 to $\simeq 1$), were produced by the digital growth technique [37, 39]. The optical transitions were studied with the use of magneto-absorption [41, 43] and magneto-photoluminescence excitation spectroscopy [38–42, 45]. Up to five equally spaced diagonal transitions in heavy hole series were observed and additionally a number of nondiagonal transitions were unambiguously identified. In Fig. 4.5, an example of the spectra and the analysis of the data along the method described previously are presented for $Cd_{1-x}Mn_xTe$ parabolic QWs [40]. An example of studies of optical transitions in half-parabolic QW is presented in Fig. 4.5 for the 194 Å (60 ML) wide $Cd_{0.75}Mn_{0.25}Te$ well [43]. Top right panel presents the scheme of the levels and the observed transitions. Left panel shows the energy of 1s excitonic features observed in magnetotransmission experiment as a function of magnetic field. The lines represent a model calculation based on numerical solution of the Luttinger Hamiltonian

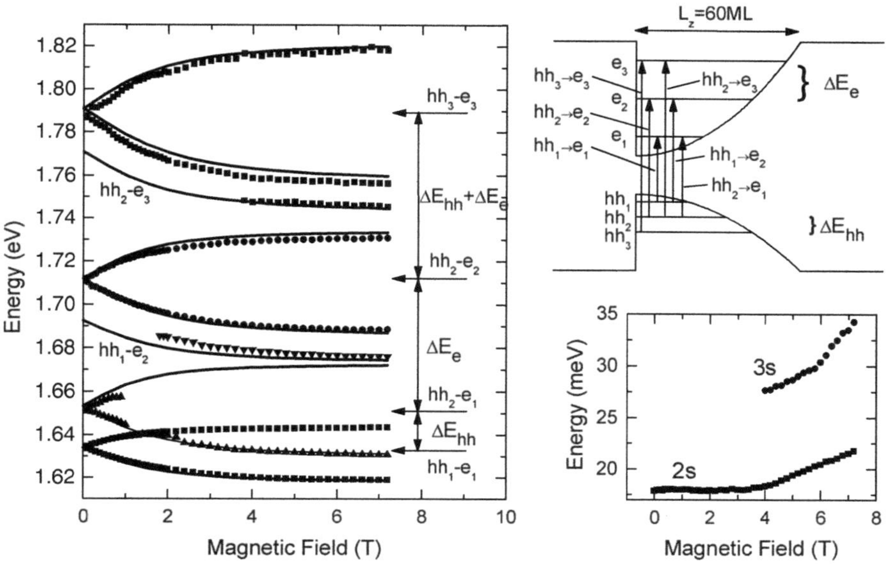

Fig. 4.5 *Left panel*: the energy of 1s excitonic features observed in optical magnetotransmission in the structure containing 10 identical 194 Å wide $Cd_{0.75}Mn_{0.25}Te$ HPQW as a function of magnetic field at $T = 1.8$ K. The *lines* represent the calculated values of the transition energy involving various spin split confined states. The *right panel* presents the scheme of the levels and the observed transitions (*arrows*). In the lower part of the right panel, we show the magnetic field dependence of the observed $2s - 1s$ and $3s - 1s$ energy differences of the $hh_1 - e_1$ exciton. Reprinted from [43], © (1998), with permission from Elsevier

describing the valence band, and of the conduction band with a quadratic dispersion, with an additional half-parabolic potential shape [42]. The calculations were performed with $Q_{hh} = 0.4$ and taking into account small deviations of the real potential shape from the ideal half-parabolic one caused by the finite reaction time of the Mn shutter during the MBE growth. Good agreement of calculated and experimental spin splitting of all 1s excitonic transitions, including those of nondiagonal type, is clearly visible. A similarly good agreement between experimentally observed spin splitting and that modeled theoretically was observed also for other half-parabolic [42] and parabolic [45] QWs, providing convincing confirmation of the precision in both potential profiling and determination of the band offset. It is worth mentioning that graded potential QWs made of DMS, apart from being attractive for band offset determination posses other interesting properties that can also serve as a basis for optoelectronic device applications. First of them is a strong – stronger than in the case of rectangular QWs with the same width – enhancement of the exciton binding energy in parabolic QWs. This can be easily understood by noting that the parabolic potential introduces a new length scale to the exciton problem [39, 43, 46, 47]. Using this characteristic length, one can estimate the binding energy of excitons within the fractional dimension approach [48]. The exciton binding energy enhancement can also be easily calculated using the variational method with strictly two-dimensional exciton wave function and one variational parameter [43]. It is interesting to note that such simple approach gives in the case of parabolic QWs, in contrast to rectangular QWs, quite accurate results, as was shown by comparison with corresponding results obtained with more sophisticated theoretical models [46–49]. First experimental confirmation of the strong enhancement of exciton binding energies in DMS QWS with parabolic confining potential comes from magnetooptical studies of half-parabolic QW [43]. In the magnetoabsorption spectra, apart from the 1s heavy-hole exciton feature, there were also observed 2s and 3s excited exciton states, which allows the exciton-binding energy of 22 meV in a 60 ML wide QW to be compared with 11 meV in the bulk (see Fig. 4.5).

Another interesting property of DMS-graded quantum structures is the possibility of a unique "spin splitting engineering" in such structures made of quaternary $Cd_{1-x-y}Mn_xMg_yTe$ [39]. Taking advantage of a very similar composition dependence of the band gap and nearly the same band offsets of $Cd_{1-x}Mn_xTe$ and $Cd_{1-y}Mg_yTe$, one can replace for a given position in the z direction a part of Mg atoms by the same amount of Mn atoms practically without affecting the potential profile in the absence of a magnetic field. By properly modeling the spatial distribution of the magnetic component, one can obtain a required change of the potential profile in the presence of the magnetic field. This 'spin engineering' should allow producing, for instance, a linear dependence of spin splitting as a function of a magnetic field. This method of modifying spin splitting of carriers in low-dimensional structures made of DMSs is additional to the other two methods, which are available also in rectangular QWs. One of these methods is based on the reduction of the *s–d* exchange constant with increasing electron confinement [50] and other on the modification of Mn spin temperature by electron gas heated by photo-excited carriers [51].

Parabolic quantum wells, including those made of DMSs, reveal also greater sensitivity to an external electric field [52]. In such structures, in contrast to rectangular QWs, the electric filed does not change the shape of the potential, but shifts the minima of potential parabolas of electrons and holes in opposite directions along the applied field. This property has already been used for optoelectronic control of spin dynamics at near terahertz frequencies in $Zn_{0.85}Cd_{0.15}Se$ parabolic quantum wells containing Mn [53].

4.3 The Double Quantum Well

4.3.1 *Control of Coupling Between Wells*

One of the most interesting semiconductor heterostructures is the double quantum well (DQW) geometry, comprised of two quantum wells separated by a thin barrier. In this structure, inter-well interaction (i.e., coupling) is an extremely sensitive function of the barrier separating the wells [54]. Clearly, the thinner or the lower the barrier, the greater will be the interaction between the wells. As a result of the interaction, each state occurring in an isolated single QW will split into a symmetric and an antisymmetric state, the two states having different energies. The splitting of the energy levels due to the coupling between the two wells is, of course, mainly determined by the barrier [55].

In a DQW structure in which the barrier is made from a DMS alloy, the giant Zeeman splitting characteristic of DMSs allows us to tune the barrier height continuously by varying an external magnetic field during the experiment. The DQW structure then provides a unique laboratory for investigating the effect of the barrier potential on inter-well coupling, and it is especially important that this can be done continuously in a single sample. Figure 4.6 shows such a DMS DQW structure, in which the barrier height can be tuned by an external magnetic field. In the specific case illustrated, the quantum wells consist of 45 Å nonmagnetic $Zn_{1-y}Cd_ySe$ layers with $y \simeq 0.2$; each DQW pair is bordered on the left and on the right by thick ZnSe layers; and the barriers separating the wells within each DQW pair consist of magnetic $Zn_{1-x}Mn_xSe$ layers with $x \simeq 0.2$. Investigations of such structures carried out by magneto-photoluminescence (magneto-PL) [56] are not suitable for determining the degree of coupling between constituent parts of the overall system, since these experiments only provide information on the spin-down component of the symmetric ground state of the DQW, to which all other states relax before recombination. When one is interested in inter-well coupling in a DQW structure, and thus

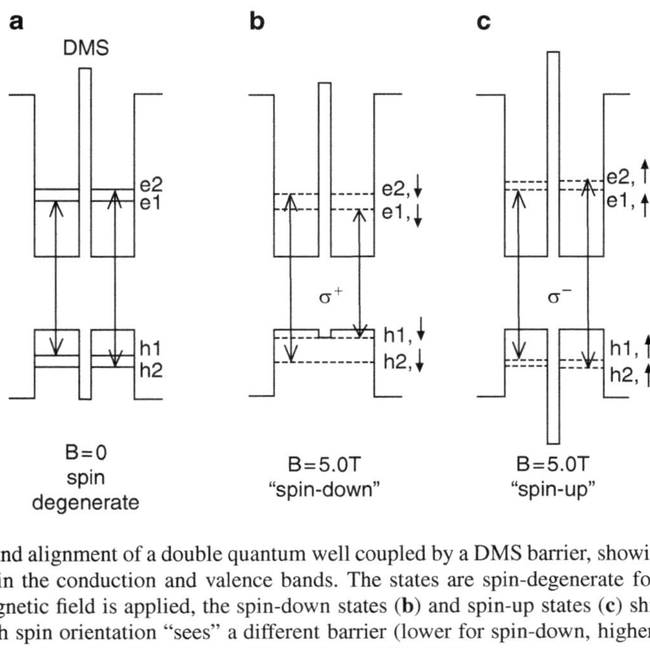

Fig. 4.6 Band alignment of a double quantum well coupled by a DMS barrier, showing two lowest eigenstates in the conduction and valence bands. The states are spin-degenerate for $B = 0$ (**a**). When a magnetic field is applied, the spin-down states (**b**) and spin-up states (**c**) shift differently, because each spin orientation “sees” a different barrier (lower for spin-down, higher for spin-up). *Vertical arrows* show the allowed optical transitions. After [19]

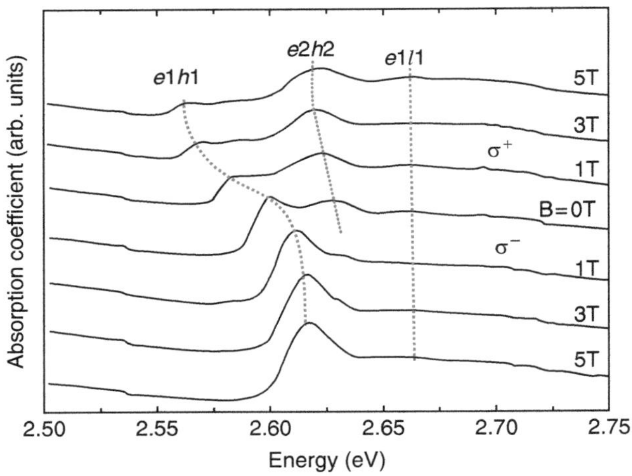

Fig. 4.7 Absorption spectra for a DQW with a 10 Å barrier at different magnetic fields. σ^+ and σ^- refer to spin-down and spin-up transitions, respectively. For structural parameters of this DQW system see [19]

in the splitting between the symmetric and the (higher-lying) antisymmetric states and its relation to barrier parameters, magneto-*absorption* is ideally suited for that purpose, since it provides information on transitions involving both symmetric and antisymmetric states and on their magnetic field dependence [19].

Figure 4.7 shows magneto-absorption spectra for a ZnCdSe/ZnMnSe DQW with a 10 Å magnetic barrier for several magnetic fields, for both σ^+ and σ^- circular polarizations. The two stronger absorption peaks, at 2.600 eV and at 2.629 eV, are identified as the heavy hole exciton transitions from the lowest symmetric (e_1h_1) and antisymmetric (e_2h_2) states, respectively. The magnetic field dependence exhibited by other DQWs having different barrier thicknesses is qualitatively similar to that shown in Fig. 4.7. Figure 4.8 is a summary plot, showing the observed transition energies for each polarization as a function of magnetic field for three DQW structures. The transitions related to the symmetric and the antisymmetric states (i.e., e_1h_1 and e_2h_2) in the sample with the thinnest barrier (10 Å) are clearly seen in zero-magnetic field, separated by over 29 meV. Although e_2h_2 is too weak to be observed at $B = 0$ in the DQW with a 35 Å barrier, extrapolation from high-field σ^+ transitions indicates the zero-field splitting to be only 16 meV for that DQW system, indicating much weaker coupling, as would be expected for a thicker barrier. In the sample with 100 Å barrier, the two peaks are not even resolved. Thus, as expected, the thinner the barrier, the stronger is the inter-well coupling, and the bigger is the splitting between the e_1h_1 and e_2h_2 transition energies.

The energy levels of each state of the DQW are spin-degenerate at zero-magnetic field. When the magnetic field is applied, however, barrier heights in both the conduction and the valence bands decrease for spin-down states, and increase for spin-up states. As a result, transitions associated with the spin-down states (those

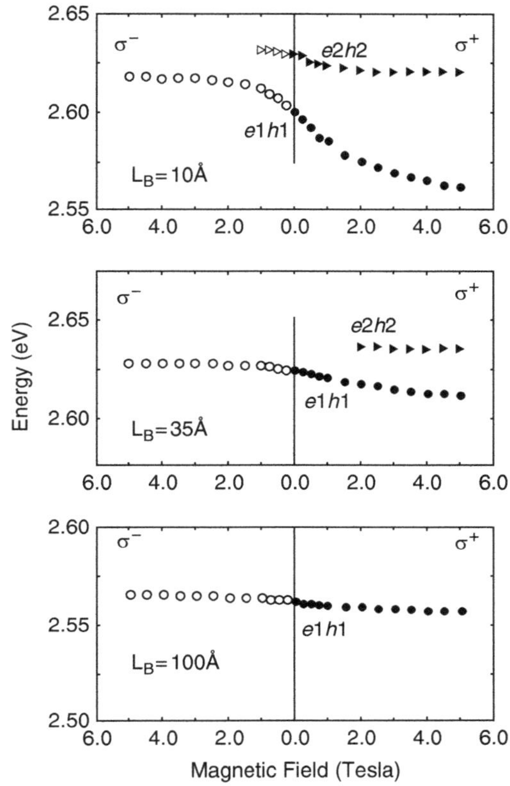

Fig. 4.8 Transition energies for the e_1h_1 and e_2h_2 transitions observed in σ^+ and σ^- circular polarizations, plotted vs. magnetic field for DQWs with 10, 35, and 100 Å DMS barriers. The σ^+ data are designated by *solid symbols*; σ^- by *open symbols*. For structural parameters of DQW systems used in this figure, see [19]

seen with σ^+ polarization of light) move further apart (stronger coupling), and those involving spin-up states (seen in σ^- polarization) move closer together. This is seen in Figs. 4.7 and 4.8: the antisymmetric state transitions (e_2h_2) show a significantly smaller shift compared to that of the symmetric state transitions as the barrier height changes with magnetic field. This is because the probability density $|\psi|^2$ (where ψ is the wave function) for the antisymmetric states has a node at the center of the DMS barrier, making this state much less sensitive to the barrier height than is the case for the symmetric state.

A particularly important feature of this experiment is the ability to continuously change the inter-well coupling with the magnetic field. For example, in the sample with the thinnest barriers (10 Å), which shows excitonic transitions from both the symmetric and the antisymmetric states at $B = 0$, we can clearly follow the change in the inter-well coupling as the barrier height of the DQW is varied by the applied magnetic field. Since the energy separation between the two transitions is a direct indication of the coupling strength, it is clear that the coupling between the wells is increased for the spin-down (σ^+) transitions, and is decreased for the spin-up (σ^-) transitions. This continuous change of inter-well coupling as a function of an applied magnetic field is quite obvious in Fig. 4.9. As the thickness of the barrier

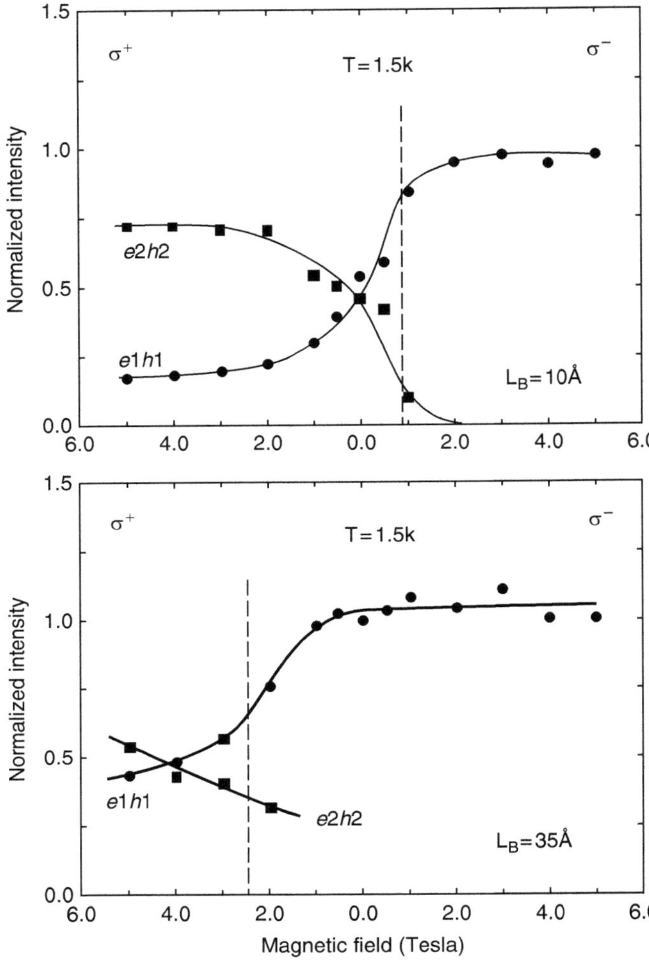

Fig. 4.9 Intensities for transitions e_1h_1 (*circles*) and e_2h_2 (*squares*) plotted as a function of magnetic field for the σ^+ and the σ^- circular polarizations for DQWs with barrier thicknesses of 10 Å and 35 Å. For structural parameters of DQW systems used in this figure, see [19]. The inter-well barrier increases continuously as one progresses from left to right. The continuous lines are guides for the eye. The *dashed vertical line* shows the field at which the coupling-induced splitting between e_1h_1 and e_2h_2 equals 20 meV, taken as the exciton binding energy. This point provides a (very rough) boundary between the mixed-exciton and the single-particle regimes

increases, interactions between the wells become weaker, and the observable variations of inter-well coupling become correspondingly smaller. In the example used the sample with the 35 Å barrier still shows a clear splitting between the two lowest spin-down transitions as the magnetic field is applied. However, the splitting of these transitions in the sample with the thickest (100 Å) barriers was not observed.

This is not surprising, since with such a large barrier thickness this DQW sample behaves in effect like two uncoupled single QWs, and the magnetic field does not alter this situation.

4.3.2 *Intra-Well and Inter-Well Excitons*

The interest in the optical behavior of double quantum wells (DQWs) is strongly related to their excitonic properties [54, 57–60]. As already noted, the important parameter that determines many of the properties of excitons in DQWs is the width and the height of the barrier between the wells, which regulates the inter-well coupling. Excitons in DQWs are easiest to understand when the wells are strongly coupled, i.e., when inter-well coupling is much stronger than the effect of Coulomb correlation [61]. In this limit, each pair of the single-well states in the conduction and in the valence band (which are exactly degenerate in a symmetric DQW) splits into a set of symmetric and antisymmetric states:

$$f_{\mathrm{s}}^{\mathrm{e}}(z_{\mathrm{e}}) = f_{\mathrm{s}}^{\mathrm{e}}(-z_{\mathrm{e}}), \quad f_{\mathrm{a}}^{\mathrm{e}}(z_{\mathrm{e}}) = -f_{\mathrm{a}}^{\mathrm{e}}(-z_{\mathrm{e}}) \tag{4.4}$$

and

$$f_{\mathrm{s}}^{\mathrm{h}}(z_{\mathrm{e}}) = f_{\mathrm{s}}^{\mathrm{h}}(-z_{\mathrm{e}}), \quad f_{\mathrm{a}}^{\mathrm{h}}(z_{\mathrm{e}}) = -f_{\mathrm{a}}^{\mathrm{h}}(-z_{\mathrm{e}}), \tag{4.5}$$

where ‘a’ and ‘s’ designate ‘antisymmetric’ and ‘symmetric’, and ‘e’ and ‘h’ stand for ‘electron’ and ‘hole’. Here, the z-axis is taken along the growth direction, and the positions of the electron and the hole are specified by the z coordinates z_{e} and z_{h}, respectively. The origin of the z-coordinates is taken to be at the center of the barrier of the DQW.

The exciton states in the DQW are just the products of these single-particle electron and hole states. The exciton ground state is the symmetric exciton state obtained from the combination of the symmetric electron and hole states given above,

$$f_{\mathrm{s}}^{\mathrm{ex}}(z_{\mathrm{e}}, z_{\mathrm{h}}) = f_{\mathrm{s}}^{e}(z_{\mathrm{e}}) f_{\mathrm{s}}^{h}(z_{\mathrm{h}}), \tag{4.6}$$

with a finite oscillator strength. The antisymmetric exciton, made up from the antisymmetric electron and hole states

$$f_{\mathrm{a}}^{\mathrm{ex}}(z_{\mathrm{e}}, z_{\mathrm{h}}) = f_{\mathrm{a}}^{e}(z_{\mathrm{e}}) f_{\mathrm{a}}^{h}(z_{\mathrm{h}}), \tag{4.7}$$

has a higher energy. Both $f_{\mathrm{s}}^{\mathrm{ex}}$ and $f_{\mathrm{a}}^{\mathrm{ex}}$ have finite – and comparable – oscillator strengths. Forbidden (‘dark’) excitons , made from the products of symmetric and antisymmetric electron and hole states but having a zero oscillator strength, occur at energies between the symmetric and the antisymmetric excitons [62]. This description of excitons in a DQW is referred to as a single-particle picture, and it is valid only when the inter-well coupling is much stronger than the Coulomb correlation of the electron and the hole forming the exciton.

The splitting between symmetric and antisymmetric single particle states of a DQW is roughly proportional to the coupling between the wells, and decreases exponentially with increasing barrier width or height. Thus for a sufficiently high barrier, the splitting can easily be smaller than the exciton binding energy (which is a measure of the Coulomb interaction). It has been shown that in this latter case the Coulomb interaction causes strong mixing of the symmetric and the antisymmetric single-particle states, f_s^{ex} and f_a^{ex} [63]. Such mixing in turn generates two different excitonic states: a ground state that is predominantly intra-well in nature (direct exciton); and a higher-energy state, which is predominantly inter-well in character (indirect exciton) [64,65]. For intra-well excitons, the electron and the hole *are both confined in the same well* at any one instant, so that the oscillator strength for the corresponding excitonic transition is high. For the inter-well exciton, however, the electron and the hole find themselves in different wells at any one moment, and consequently the oscillator strength for the excitonic transition is in that case very small. This feature can be identified by intensities of the excitonic transitions in absorption experiments.

The variation of absorption intensities of the excitonic transition for e_1h_1 and e_2h_2 states are clearly seen in Fig. 4.7 for the DQW structure with a 10 Å barrier. The highest field for the σ^- polarization (bottom curve in Fig. 4.7) corresponds to the highest barrier between the two wells; and the highest field for the σ^+ polarization (uppermost curve in Fig. 4.7) corresponds to the lowest value of the barrier height. It can be seen from Fig. 4.7 that, as the barrier height decreases, the *intensity* of the e_1h_1 transition decreases, while that of e_2h_2 increases. The dependence of absorption intensities of the excitonic transitions for both e_1h_1 and e_2h_2 states are plotted in Fig. 4.9 for two DQW structures, with 10 and 35 Å barriers. This interesting dependence of intensity on the coupling strength cannot be understood by the wave function overlap in the single particle picture. In particular, intensity variation of the transitions is not expected on the basis of wave function overlap in the region where the coupling between the wells is weak (i.e., for transitions involving the σ^- polarization in this case) [66].

The electron–hole Coulomb interaction, i.e., the 'exciton effect', provides the key for the understanding of this rather puzzling behavior of the transition intensities in the presence of varying degree of coupling. As has already been remarked, the Coulomb interaction tends to mix symmetric and antisymmetric single-particle states [63], the strength of such mixing depending on the coupling between the wells. Thus, on the one hand, it depends on the splitting of the single particle symmetric and antisymmetric states; and, on the other hand, it is determined by the exciton binding energies. In the case of a single quantum well, where the same mixing process is in principle present, subband mixing by Coulomb interaction is usually negligible, since the subband spacing is typically much larger than the exciton binding energy (except in very wide wells). In the case of coupled DQW structures, however, the subband separation can easily be much less than the exciton binding energy, and so the mixing via electron–hole Coulomb interaction can become very important. For the structures presented in Figs. 4.7–4.9, the splitting of the single particle states varies from zero to about 50 meV. The exciton binding energy for ZnSe bulk material is 20 meV, and this value can be further enhanced in quantum-confined systems. Therefore, in the case of transitions involving the σ^- polarization, we indeed have a situation in which the Coulomb interaction is comparable with, or greater than, the inter-well coupling.

Dignam and Sipe [63] used what they called 'two-well basis functions' to carry out the analysis of exciton states in coupled DQW structures. Such a two-well basis set is appropriate for determining the ground state of the exciton as well as its excited states. These authors calculated the dependence of the exciton energies on the barrier-thickness, as well as the oscillator strength of the ground and the first excited states of the excitons. In the specific case which Dignam and Sipe have treated, the variation of the coupling was introduced by varying the barrier thickness while keeping barrier height constant. In the current DQW structures, the variation of the coupling was achieved by changing barrier height while keeping barrier width unchanged. However, the strength of the coupling between the wells as a whole is determined by the amount of tunneling through the barrier, independent of physical details of the tunneling. The tunneling coefficient trough the barrier is given by [67]

$$| T |^2 \simeq \exp\left(-2\sqrt{(2m^* L_\mathrm{b}/\hbar^2)[V - E]}\right), \tag{4.8}$$

where m^* is the effective mass of the carrier (i.e., electron or hole), L_b is the barrier width, and V is the barrier potential. The above formula indicates that when we reduce the barrier height but keep the barrier thickness fixed, the results are essentially the same as those obtained upon keeping the barrier height constant and changing its thickness. Thus, as the barrier height is reduced, we would expect to obtain the same result as that obtained by reducing barrier thickness: a decrease in the oscillator strength for the ground state of the exciton, and an increase in the oscillator strength for the first excited state. This is just what the intensity variation of the two transitions exhibits in absorption experiments shown in Fig. 4.7.

Even though the physics is the same in the two cases (increasing the width or increasing the height of the barrier), exact calculation of the oscillator strength for our DQW structures using the approach of Digman and Sipe would be rather difficult. In the case when the variation of the coupling is achieved by changing the barrier *height*, the basis functions also change as the height of the barrier changes. Thus, one cannot use the four fixed basis functions, as was done in [63]. That is, the action quantity determining the tunneling probability through the middle barrier (i.e., the coupling between the two wells) will essentially vary as $\sqrt{VL_\mathrm{b}^2}$ for either case. Thus the barrier-thickness dependence of the oscillator strength calculated in [63] can be 'mapped' into barrier-height dependence by using appropriate scaling, which can be the exciton binding energy as the energy scale, and the Bohr radius as the length scale. The corresponding oscillator strengths obtained by such mapping for the DQW with a 10 Å barrier are shown in Fig. 4.10, where the oscillator strengths for the two excitonic transitions are plotted as a function of the

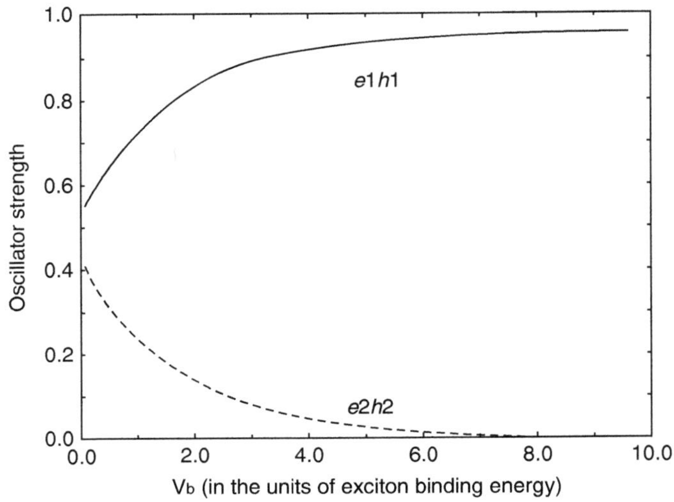

Fig. 4.10 Oscillator strength as a function of barrier height for the two optically active exciton states for a DQW with 10 Å barrier thickness, 'mapped' from Fig. 5 in [63], as discussed in the text. Reprinted with permission from [63], © (1991) by the American Physical Society

barrier height. The qualitative behavior of the two excitons explains very nicely the dependence of the intensities of the e_1h_1 and e_2h_2 transitions in the region of weak coupling observed in the absorption experiments (i.e., in the region of σ^- polarization in Fig. 4.9).

The origin of these results can be qualitatively understood as follows. For e_1h_1 exciton states, both the electron and the hole are very highly localized *in the same well* (intra-well exciton), while for the e_2h_2 exciton state the two carriers have a very high probability to be *in different wells* (inter-well exciton). This suggests that the exciton binding energy is enhanced for intra-well excitons, and reduced for inter-well excitons, because the Coulomb interaction is effectively much larger when the electron and the hole are in the same layer.

4.4 Multiple Quantum Wells

4.4.1 Wave Function Mapping

In quantum mechanical description, a particle in any multi-component structure is characterized by a wave function which is, in principle, distributed throughout the entire structure. The spatial distribution of the wave function amplitude then provides a measure of the localization of the particle. In multiple QWs coupled by reasonably narrow barriers, wave functions of the eigenstates in each well interact through the barrier with those of neighboring wells, and the energy levels are consequently split into multiplets. An interesting and fundamental feature of multiple QWs is the wave function distribution of the coupled states. For example, some wells can even act as *barriers* for particular eigenstates whose wave function vanishes in those wells, the state itself becoming insensitive to the properties of that well. This situation is quite different from the case of double QWs or superlattices, where the eigenstates of the system are always localized in *all* wells [68, 69].

Here, we select the specific case of three coupled QWs to illustrate these new features in wave function distribution of coupled states, bearing in mind that the manner in which the wave function is distributed over the components of the system reflects the spatial consequences of interactions between these components. The wave functions of the three coupled QWs are calculated using the model and the 'finite element method' [70]. Figure 4.11 represents the wave functions of the lowest coupled states for the conduction band in symmetric triple QWs, the picture being essentially the same for the valence band. The calculations show that the first and the third of the lowest triplet of states have some fraction of their wave functions in each of the wells. In the case of the second state, however, the wave function vanishes totally in the central well, representing in effect an 'electronic blind spot'. This interesting wave function distribution in a multiple quantum well systems can be 'mapped out' using as our tool the strong splitting of the band edges which occurs in DMS layers when a magnetic field is applied.

The central idea is that in structures consisting of alternating DMS and non-DMS layers, the Zeeman splitting of a given state will reflect its weighted probability distribution over the two media. In other words, the Zeeman splitting will be determined by how many localized magnetic moments does the electron 'see' when it is in a particular state. If the wave function of the electron is strongly localized in the non-DMS layers, the splitting will be negligible; if, on the other hand, it is strongly localized in the DMS layers, the splitting will be very large. This technique has

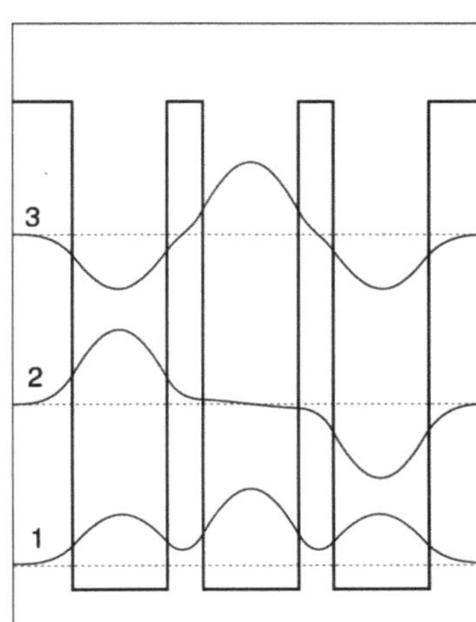

Fig. 4.11 Wave functions for the three lowest states in a symmetric triple QW. After [68]

been also used to pinpoint the above barrier states in single-barrier and superlattice systems [71,72]. Inspection of the wave function distribution shown in Fig. 4.11 for the triple QW suggests that the study of symmetric triple QW structures, in which DMS layers are used either for the center well (we will refer to this as the TQW1 geometry), or for the two side wells (TQW2), would be particularly informative. It is easy to see from Fig. 4.11 that when only the center well is made of DMS material, the first and the third state – which are partially localized in that layer – will be strongly influenced by the magnetic field, while the second state – which is localized only in the side wells – will show negligible dependence on the field. On the other hand, in a system where the two side wells consist of DMS layers, the second state will be affected by the magnetic field – in fact more so than the other states. By observing the Zeeman splitting of optical transitions between specific states in systems fabricated in this way, one can determine in which layer (i.e., where in physical space) a given transition actually occurs. Such structures are schematically shown in Fig. 4.12, where the DMS and non-DMS wells are indicated as darkly and lightly shaded regions, respectively.

As an example, absorption spectra in various magnetic fields observed on TQW1 at 30 K are shown in Fig. 4.13. These represent excellent illustrations of the splitting of the degenerate single well ground state into three states as the three wells become coupled. The transition from the m-th heavy-hole subband (h_m) to the n-th conduction band (e_n) is designated $e_n h_m$. Even though there actually is a total of six coupled states in TQW1 (three in the conduction band wells and three in the heavy hole band wells, as shown in Fig. 4.11), only three strong transitions are expected due to the $n = m$ selection rule (see [68]). Guided by this rule, the three left-most peaks in Fig. 4.13 can be attributed to transitions e_1h_1, e_2h_2, and e_3h_3, respectively, as indicated in the figure.

The energy shifts of these transitions with various magnetic fields are summarized in Fig. 4.14 [73]. For the TQW1 structure (DMS center well) it is clear that the strongest magnetic field dependence is seen for the e_1h_1 transition, and no noticeable energy variation with field is observed for e_2h_2, as would be expected from the wave function distributions shown in Fig. 4.11. This clearly indicates that the e_2 and h_2 states are indeed localized only in the two side wells, which in this case consists

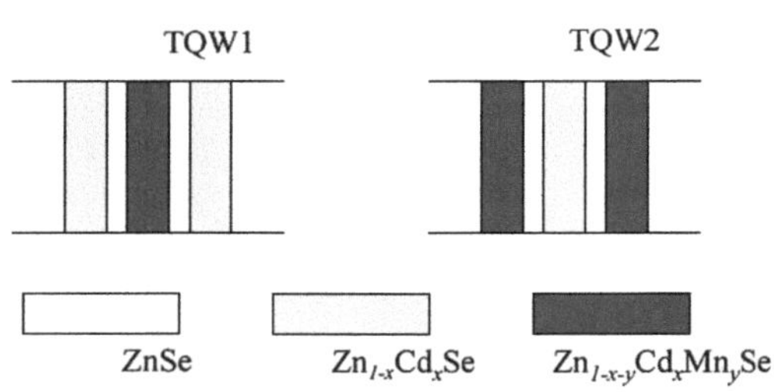

Fig. 4.12 Schematic diagrams for the symmetric multiple QWs (TQW1 and TQW2) consisting of $Zn_{1-y-x}Cd_xMn_ySe$ and of $Zn_{1-x}Cd_xSe$ wells, with ZnSe barriers. *Shaded* regions indicate wells (*dark shading* for $Zn_{1-y-x}Cd_xMn_ySe$ and *light shading* for $Zn_{1-x}Cd_xSe$ wells). *Unshaded* regions are ZnSe barriers

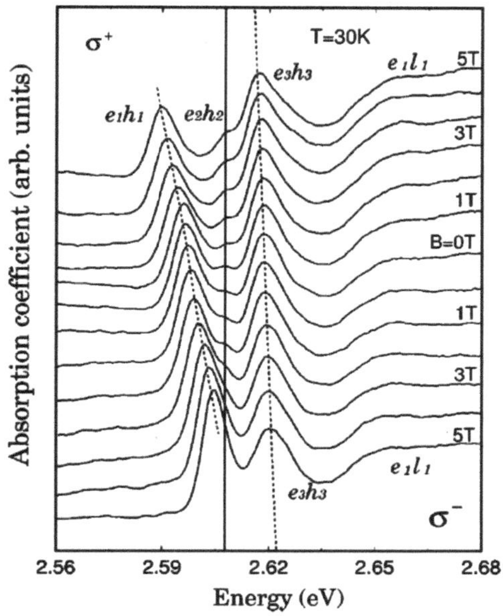

Fig. 4.13 Absorption spectra taken at 30 K for TQW1 at different magnetic fields. σ^+ and σ^- refer to spin-down and spin-up transitions, respectively. Magnetic field dependence of the observed transitions from the triplet of states in the conduction band to the corresponding triplet in the heavy hole band are clearly shown. The *dotted lines* through e_1h_1 and e_3h_3 are guides for the eyes. The thin solid line is drawn vertically at constant energy, showing that the position of e_2h_2 through which the line passes is independent of the field. For structural parameters of this TQW system, see [68]

of non-DMS layers. This is a beautiful example of Zeeman mapping of the wave function distribution in a system comprised of multiple quantum wells, since – as seen in Fig. 4.11 – the states involved in these transitions have a 'blind spot' in the central (DMS) well.

Note, however, that Fig. 4.11 was generated with the assumption of equal potential depth in all three wells at zero field. However, in our TQW Mn ions are added in the central well (about 4%), so that in practice the potential depth of the center well may be *slightly* different, which may make the wavefunction distribution slightly different from that shown in Fig. 4.11. More realistic wave function distributions for our TQWs are shown in Fig. 4.15, which provides a quantitative basis for observed larger Zeeman shift of the e_1h_1 transition than of e_3h_3 in this specific TQW system. The physical explanation of the results observed on TQWs is given in the next two paragraphs.

The Zeeman shift of each eigenstate from its zero-field position is determined by the overlap of the probability $|\ \psi\ |^2$ of that state with the DMS layer. In the first-order approximation, the shift of the n-th eigenstate can be written

$$\Delta E_n^{\mathrm{c,v}} = P_n \Delta E^{\mathrm{c,v}}, \tag{4.9}$$

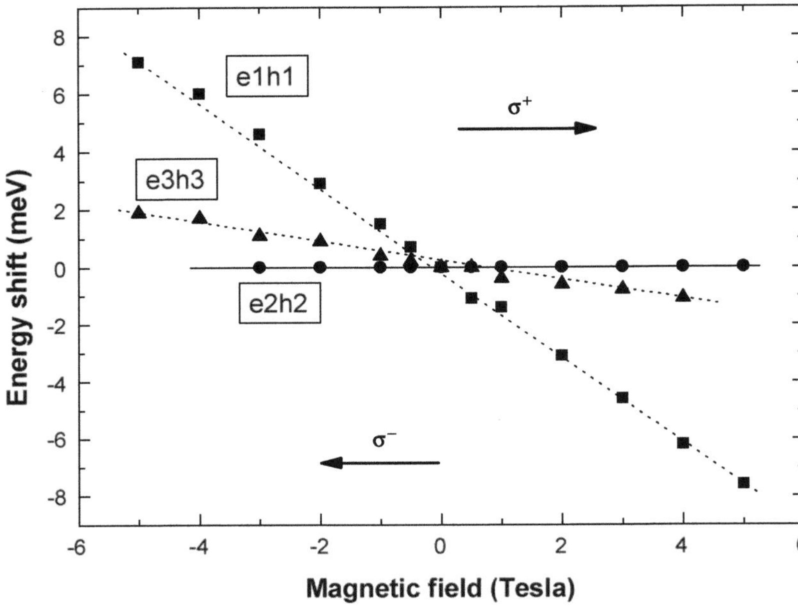

Fig. 4.14 Energy shift of the $\Delta n = 0$ transitions at 30 K for TQW1 at different magnetic fields. σ^+ and σ^- refer to spin-down and spin-up transitions, respectively. The amount of energy shift indicates the degree of localization of states in the DMS well. The *dotted and solid lines* are guides for the eye. Reprinted from [73], © (2003), with permission from Elsevier

where P_n is the probability density of carriers in the n-th eigenstate integrated over the DMS layer, and $\Delta E^{c,v}$ is the Zeeman shift of the conduction (E^c) or valence (E^v) band edge of the DMS layer, depending on whether the n-th state is in the conduction or the valence band. Thus, the observed Zeeman shift of a given transition will indicate the physical location of wave functions of the states involved in that transition.

Note that the Zeeman shift of the heavy-hole band edge is significantly larger (about four times) than that of the conduction band edge. In practice, the observed Zeeman shift of the transitions ΔE_n will thus be dominated by the contribution from the heavy-hole band. The Zeeman splitting between σ^+ and σ^- transition energies observed at 5 T, and the probability densities of the heavy hole states integrated over the DMS wells (i.e., over the central well for TQW1) are compared in Table 4.1. There is indeed a rather good correlation between the ratios of the Zeeman splittings of the various transitions and the ratios of the calculated probability densities of the corresponding heavy hole states in the DMS well.

Similar idea was applied to probe the wave function in a single CdTe QW. Prechtl et al. [74, 75] used magneto-photoluminescence experiments to map out the probability density for the ground state by placing a single magnetic probe of 1/4 ML of MnTe at various positions along the QW growth axis. To increase the sensitivity and to map out the probability density not only of a ground state but also of the first two

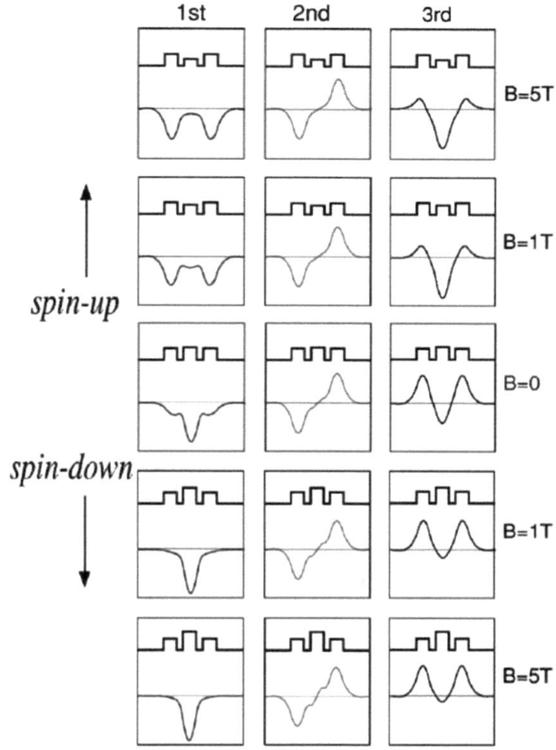

Fig. 4.15 Progression of potential profile with magnetic field, and corresponding wave functions for the heavy-hole band, for spin-up and spin-down states in the TQW1 structure. The first, second, and third columns correspond to the first, second, and third states of the ground-state triplet, respectively. Noticeable qualitative changes in heavy-hole localization occur in the first and the third states. After [69]

Table 4.1 Comparison of Zeeman splittings of successive transitions observed at 5T, and probabilities of the heavy hole states in the center (DMS) well for TQW1

TQW1	ΔE_n (meV)	14.5	0	3.3	$\mid \psi_n^v \mid^2$	0.73	0.01	0.17
	$\Delta E_n / \Delta E_1$	1	0	0.228	$\mid \psi_n^v \mid^2 / \mid \psi_1^v \mid^2$	1	0	0.233

excited states, Kłopotowski et al. [76] used nominally 2 ML wide $Cd_{0.78}Mn_{0.12}Te$ probe, placed at various positions inside deep CdTe QW with $Cd_{0.66}Mg_{0.33}Te$ barrier. The locations of the probe corresponded to the minima and maxima of the wave-functions of the first three confined states. The spin splitting of the heavy hole diagonal excitons ($h_n - e_n$, with $n = 1, 2, 3$) observed in magnetotransmission experiments correlated very well with the $P_{ne} \Delta E^c + P_{nh} \Delta E^v$.

4.4.2 *Wave Function Transfer in QWs at Off-Resonance Conditions*

In symmetric multiple quantum wells (MQW's), wave functions of the lowest multiplet of states are distributed within the system in a rather surprising manner, some of the states being localized only in certain wells and almost *entirely absent* in others. The mapping of such distribution of wave functions has already been discussed in the previous section. This interesting wave-function distribution, however, occurs only in the case of strong coupling between the wells (i.e., under resonant condition). To maintain this condition, in the previous section the depth variation of the DMS wells in an applied field was restricted to a relatively narrow range of energies (less than 10 meV), to assure that the system remains in *nearly resonant* condition during the entire experiment.

It is, however, also interesting to observe how the wavefunctions redistribute themselves when the system changes to off-resonant conditions (i.e., in the case of weak coupling between the wells), and how the coupling varies when the variation of the well depth is large. The TQW1 structure is the simplest and most informative system for illustrating this behavior. The band-edge profiles for the heavy-hole band at $B = 0$ (including the effect of strain) are shown for the TQW1 system in the third row of panels in Fig. 4.15. The corresponding wave functions for the three lowest heavy-hole states are also shown in that row. At $B = 0$ the lowest state (h_1) is clearly distributed over the three wells, while the second state (h_2) is localized only in the two side wells. The third state (h_3) is again distributed over the three wells. The evolution of the potential profile for spin-up and spin-down heavy-hole states with increasing field is shown for TQW1 in the upper (first and second) and the lower (fourth and fifth) rows of Fig. 4.15, respectively. For the spin-down states at high fields, the central valence-band well becomes significantly deeper than the side wells, resulting in a symmetric triple QW system with very unequal well depths. In this configuration, the states in the central well and in the side wells are no longer resonant with each other, and the coupling between them is dramatically reduced. The system is now, effectively, a combination of a single (central) QW and a double resonant QW comprised of the two side wells. The wave function of the h_1 state then approaches the ground-state wave function of a single quantum well, localized only in the central DMS layer due to the significantly lower potential of that layer compared to the side wells. The h_2 and h_3 wave functions, on the other hand, are mostly localized in the two side wells, and are thus determined primarily by those wells. This is not surprising for the h_2 state, since that state is derived from a wave function that is originally only localized in the two side wells of the triple QW with equal well depths [68]. However, the wave function of h_3, which was initially nearly equally distributed over the three wells, has now – somewhat unexpectedly – also redistributed itself primarily into the two side wells.

In contrast, for the spin-up states the depth of the DMS well in TQW1 becomes significantly shallower at high magnetic fields than that of the side wells. The whole system then begins to separate into a deep resonant double quantum well comprised of the side wells, for which the central well – no longer in resonance – becomes part of the barrier; and a single shallow central QW. Then the lowest state h_1 and the first excited state h_2 become those of a double QW, localized equally in the two side wells (which in the case of TQW1 correspond to the non-DMS wells), as shown in the first and second columns of the top panels in Fig. 4.15. Surprisingly, most of the wave function for the h_3 state is seen to localize in the central well, even though that

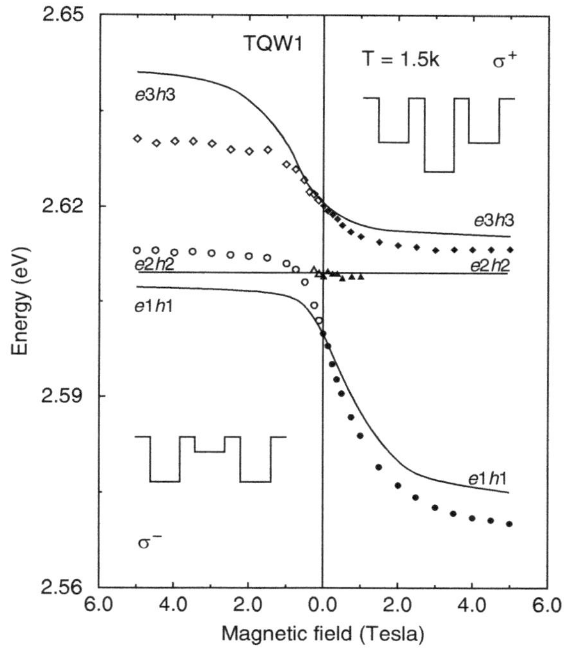

Fig. 4.16 Calculated energies for the $\Delta n = 0$ transitions for TQW1, plotted together with experimental results. The *solid lines* are calculated for actual structures used in the experiments, taking into account their detailed physical properties (strain, exact band alignment at $B = 0$, etc.), as discussed in the text. After [69]

well is at the highest potential. This process of wave function redistribution and its evolution can be readily mapped by the magnetooptical experiments, such as those described in this section.

Figure 4.16 shows transition energies for the TQW1 structure as a function of applied magnetic field for the two circular polarizations σ^+ and σ^-. The solid lines show transition energies calculated for TQW1, obtained by taking the Zeeman splitting measured on a 'companion' DMS epilayer as the potential variation of the DMS wells. The calculations in Fig. 4.16 were carried out using the $\boldsymbol{k} \cdot \boldsymbol{p}$ model and the 'finite element method' [70]. The parameters used have been discussed in [68] and [69], and the reader is referred to those references for additional details. Calculated energies for optical transitions involving the ground-state triplet in the TQW1 configuration (i.e., DMS layer in the center well) exhibit several interesting phenomena that were not present in the small perturbation regime discussed in the preceding section. One of the most striking features is the asymmetry of the Zeeman shift between the σ^+ and σ^- polarizations (i.e., the spin-down and spin-up transitions, respectively) exhibited by both e_1h_1 and e_3h_3 transitions. For the σ^+ polarization, the e_1h_1 transitions in the TQW1 configuration show the largest Zeeman shift, indicating that the states involved in these transitions are much more localized in the DMS wells than other states. For the σ^- polarization, however, the e_1h_1 transition shows considerably smaller Zeeman splittings and quickly saturates at magnetic fields higher than 0.75 T, indicating that the states involved in this transition reside primarily in the non-DMS layers. A similar asymmetry of the Zeeman shift, but in the opposite direction, is also observed for the e_3h_3 transition.

This asymmetric behavior of the e_1h_1 and e_3h_3 transitions – which differ from their behavior in the small perturbation regime, where the Zeeman shift is approximately symmetric – results from the difference in the distribution of the wave functions for the spin-up and spin-down states (see Fig. 4.15). This feature is characteristic of coupled MQW systems made up of DMS and non-DMS combinations, including superlattices [77, 78]. In a single QW consisting of DMS material, the ground state is always localized in the well, even though the well depth changes with magnetic field for different spin states. In multi-well systems, however, the lowest-lying state seeks out wells of the lowest potential, to which it tunnels when the well depths change relative to one another. Such wave-function transfer taking place in a triple QW structure is clearly seen for the h_1 and h_3 states in the first and the third columns of Fig. 4.15.

A fundamental feature of the TQW1 configuration can be appreciated by examining the three transition energies at 5 T for the σ^+ and σ^- polarizations. It is clear in that limit that two of the transitions are always close to each other, while a third one is farther removed. The former resemble the two lowest-energy transitions of a double QW, and the latter behaves like the ground-state transition of a single QW. This *subgroup* behavior results from the significant reduction in the coupling between the central and the side wells. In triple QWs under resonant conditions, all wells are strongly coupled to one another, and the states of the lowest multiplet are almost equally separated, as is the case for zero magnetic field in Fig. 4.16. However, as the wells move away from the resonance condition, their behavior becomes (nearly) independent from each other. Then each subgroup of QWs exhibits their own characteristics, without 'feeling' the existence of the other QWs. In the case of the TQW1 configuration, the system divides itself – due to the large changes of the DMS band edges induced by the applied magnetic field – into a single QW made of the DMS layer, and a double quantum well consisting of non-DMS materials.

Since the 'single QW subgroup' corresponds to the DMS well, the state representing the single quantum well behavior will always follow the magnetic shift of the DMS material. This behavior is illustrated by Fig. 4.16, where the e_1h_1 transition for σ^+ and the e_3h_3 transition for σ^- clearly follow the Zeeman shift characteristic of the DMS layer, showing a behavior that separates them from the other two transition lines. The transitions corresponding to the 'double QW subgroup' show a different characteristic behavior. The flatness of the energy-versus-field behavior of the two closely spaced transitions reflects the fact that states participating in these transactions reside in the non-DMS side wells. The difference in the transition doublets for the two spin orientations (e_1h_1 and e_2h_2 for σ^-; e_2h_2 and e_3h_3 for σ^+) arises from the fact that – even though the wells are identical – the degree of the inter-well coupling in this double-QW subgroup is different for the two spin orientations, and depends on magnetic field. In particular, it is clear from the larger separation exhibited by the transition energies in the σ^+ polarization that the coupling through a barrier containing a deep well is stronger than that through the more 'solid' barrier, in contrast to σ^-.

Experimental data observed for TQW1 in magneto-transmission are shown by points in Fig. 4.16. The observed magnetic-field dependences of the transitions agree qualitatively with the calculations. First, the calculated asymmetry of the Zeeman splitting between transitions for the σ^+ and σ^- polarizations is clearly observed for the e_1h_1 transition. Furthermore, by careful inspection of the experimental data one can also see (especially for the low-field region in the σ^- polarization) that transitions for this triple QW divide into subgroups in a continuous manner as the magnetic field is increased, essentially corroborating the wave-function picture of this system under large perturbation, as discussed above. Figure 4.16 is intended to illustrate the trends which govern the behaviors of TQWs. For additional details including quantitative discrepancies between data and simulation, see [69].

4.5 Superlattices

4.5.1 Spin Superlattice

A 'spin superlattice' (SSL) is a superlattice in which carriers with opposite spin states are confined in different layers. To achieve such spin modulation, one needs to start with a structure in which the energy gaps of the constituent layers are initially (i.e., in the absence of magnetic field) equal, and the band offsets at the interfaces are initially zero, as shown at the top of Fig. 4.17. When a magnetic field is applied, the large Zeeman splitting of the band edges in the DMS layers results in induced band offsets and, consequently, in a spatial separation of the spin-up and spin-down states, as shown in Fig. 4.17 (bottom). The SSL phenomenon

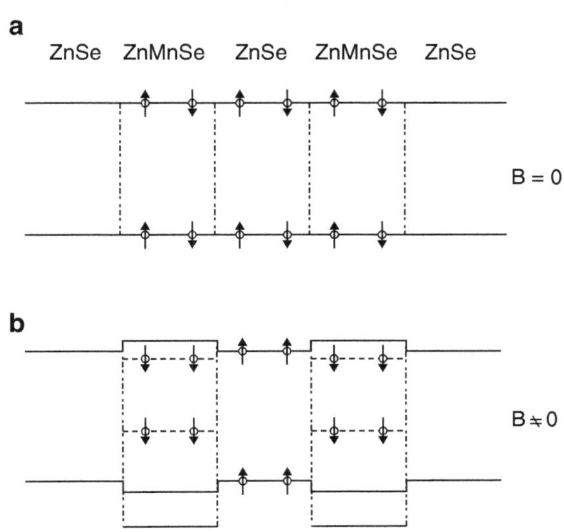

Fig. 4.17 A schematic diagram of band structure of a magnetic-field-induced spin superlattice. In the ZnMnSe layers in the lower picture ($B \neq 0$), the *dotted* and the *solid lines* show the spin-down and spin-up states of electrons and heavy holes, respectively. The Zeeman splitting in the ZnSe layers is negligible. The *arrows* show the orientation of spins localized in the different layers of the superlattice. After [78]

has already been observed experimentally in several DMS/non-DMS multiquantum well systems [78, 79]. One particularly attractive DMS system for achieving the structure described above – which will be used as an example in this section – is the ZnSe/$Zn_{1-x}Mn_xSe$ superlattice. The energy gap of $Zn_{1-x}Mn_xSe$ exhibits a rather striking bowing with Mn concentration x at low temperatures, first decreasing with x, and then rapidly increasing [14, 80]. It is thus possible to find a value of x at which $Zn_{1-x}Mn_xSe$ has the same energy gap as ZnSe (this occurs for $x \sim 0.04$ at 1.5 K). Furthermore, since the valence band offsets at ZnSe/$Zn_{1-x}Mn_xSe$ interfaces are close to zero by the common anion rule (which works reasonably well in typical $II_{1-x}Mn_xVI$ DMSs such as ZnMnSe [81]), the value of x which gives equal energy gaps in ZnSe and $Zn_{1-x}Mn_xSe$ automatically leads to zero (or, in practice, to very small) band offsets in both the conduction and the valence bands, thus satisfying the properties stipulated above.

When a magnetic field is applied, the band edge of $Zn_{1-x}Mn_xSe$ for spin-down electrons will move down, and it will move up in energy for spin-down holes. The $Zn_{1-x}Mn_xSe$ layers thus become wells for both carriers with a spin-down orientation. Similarly, for spin-up states the conduction band moves up in energy, and the heavy-hole band moves down, so that for this spin orientation the nonmagnetic ZnSe layers become the wells. This magnetic-field-induced band offset has therefore the consequence of *spatially separating* electron and hole states of different spin orientation. It is particularly convenient to observe the formation of such spin separation in magnetoabsorption experiments with the magnetic field applied perpendicular to the SSL layers (the Faraday geometry), in which only spin conserving transitions are allowed. Using circular polarizations σ^+ and σ^-, one can then observe separately transitions between spin-up valence and conduction band states (corresponding to σ^-), and those between spin-down states (σ^+).

Figure 4.18 shows the magnetic field dependence of excitonic transitions observed with σ^+ and $\sigma-$ polarizations, illustrating very nicely the spatial separation of spins occurring in the SSL. Considering first the spin-down (σ^+) transitions, we see the rapid red shift of the σ^+ absorption line, since the initial and final states of the transition are both in the DMS layers at finite magnetic fields, and follow the

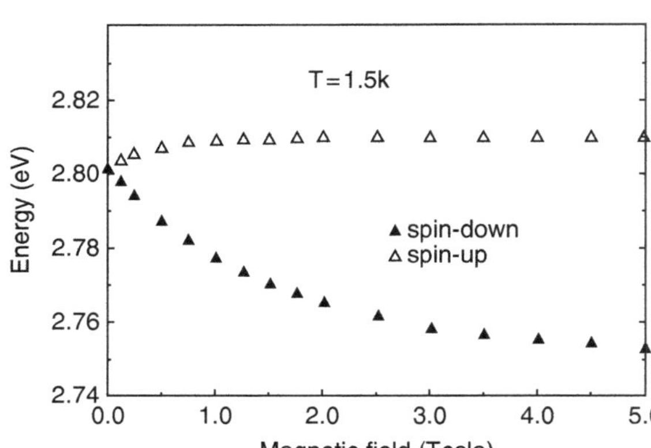

Fig. 4.18 Magnetic field dependence for excitonic ground state transitions observed in a ZnSe/$Zn_{0.96}Mn_{0.04}Se$ (10 nm/10 nm) SSL for the two-spin orientations. Note the striking asymmetry in the Zeeman shift observed in the SSL for the spin-up and spin-down transitions relative to $B = 0$. After [78]

field dependence of the valence and conduction band edges of the DMS material. In contrast, the spin-up transitions take place between states localized in ZnSe layers, which become the wells for this spin orientation when the field is applied. This is responsible for the flatness of the transition energy at higher fields, i.e., once the confinement energy becomes stable. The small initial rise in energy seen in Fig. 4.18 at low fields (i.e., when the offset is just beginning to form) arises from the fact that in that region of extremely shallow wells the state in the ZnSe well is much more sensitive to the height of the (DMS) barrier, which increases with the field. As the barrier continues to increase, the energy of the state confined in the ZnSe well becomes less and less sensitive to the barrier height and eventually flattens out, approaching the behavior of the nonmagnetic quantum well in which that state is confined [78, 79].

4.5.2 Magnetic-Field-Induced Type-I to Type-II Transition

As already noted, in DMS materials the Zeeman splitting for heavy holes is typically much larger than for electrons. At the same time, it has been shown that in a common-anion system (e.g., in ZnSe/$Zn_{1-x}Mn_xSe$), the valence band offset is smaller than the conduction band offset [82]. This implies that, as the field is applied, the heavy-hole Zeeman shift will overcome the band offset before this happens for electrons. Since the spin-up holes will become localized in ZnSe layers before the electrons as the field is raised, it is then expected that in a certain magnetic field range the spin-up holes will be localized in the ZnSe layers, while the spin-up electrons remain in the $Zn_{1-x}Mn_xSe$ layers. Such conversion from a type-I to a type-II superlattice can be manifested by a dramatic change in optical transition intensities, because in a type-II structure the carriers initially involved in the strongest optical transitions are now physically separated into different layers. This is indeed evidenced in the intensity of spin-up transition as a function of the magnetic field, as can be seen from the data shown in Fig. 4.19. Here, the triangles represent the integrated absorption for spin-up transitions in the ZnSe/$Zn_{1-x}Mn_xSe$ system. It is clear that, as the magnetic field increases from zero, the intensity of spin-up transitions first decreases, indicating the initial decrease of barrier height in the type-I case, and then the type-I to type-II conversion. When the Zeeman shift for the spin-up electrons overcomes the conduction band offset, however, spin-up electrons and spin-up holes both become localized in the ZnSe quantum wells. The band alignment for spin-up electrons and holes is thus once again type-I, with an accompanying increase in the absorption intensity. Beyond this magnetic field, the spin superlattice comes into existence.

The phenomenon of such band-alignment conversion has already been experimentally observed in several systems, e.g., in $Zn_{1-x}Fe_xSe$/ZnSe [83] and in $Cd_{1-x}Mn_xTe$/CdTe [84]. It is important to note that this phenomenon is particularly important in II–VI-based DMS superlattices involving a common-anion systems (e.g., ZnMnSe/ZnSe and the two materials combinations mentioned earlier), since these are typically characterized by small valence-band offsets. Such small offsets

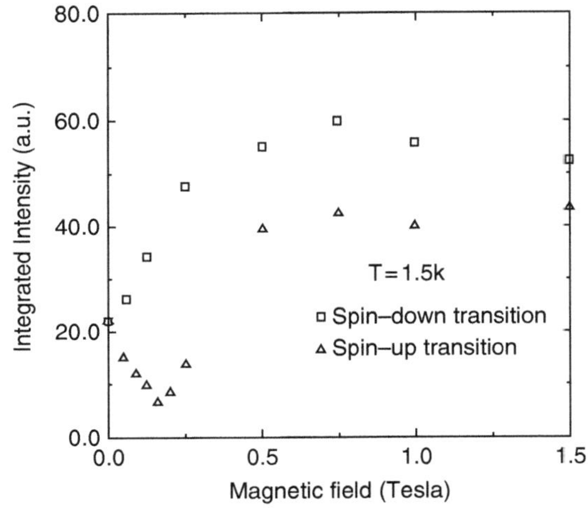

Fig. 4.19 Integrated intensities of the Zeeman-split excitonic transitions as a function of magnetic field in $ZnSe/Zn_{1-x}Mn_xSe$. The *squares* and *rectangles* represent the spin-down and spin-up transitions, respectively. After [78]

can be easily overcome by the large Zeeman shift of the valence band, thus changing the type of alignment even at moderate magnetic fields. When this is achieved, we automatically obtain spatial separation of spin states similar to the spin superlattice discussed earlier, but for one band only. As was the case with the SSL, such spin separation also leads to the characteristic transition asymmetry already pointed out in connection with Fig. 4.18.

4.6 Above-Barrier States

Most optical studies of heterostructures focus on states localized in quantum wells [85–87]. Although transitions between the ground states in single quantum wells and in superlattices are well understood, the interpretation of transitions involving excited states is less well established; and transitions involving states higher than the barriers are even less studied. In this section, we describe investigations of above-barrier states in semiconductor heterostructures containing DMS layers, which allow the band offsets in the multilayer to be tuned via the Zeeman splitting of the band edges, thus making it possible to identify the location of the states participating in specific transitions. This feature allows an unambiguous interpretation of the role of states above the barrier.

4.6.1 Single Barrier

Single quantum barriers have not been a common choice for optical studies because there exist no quantized states in the system. The energy spectrum consists of a continuum, starting from the top of the barrier. Contrary to common impression, however, the energy continuum of above-barrier states is not featureless. This can be seen from a simple calculation of tunneling that shows resonances when the wavelength λ of an electron in the barrier region is equal to $\lambda = NL_b/2$, where L_b is the width of the barrier and N is an integer.

In optical studies, there are two important factors which must be taken into account: the shape of the wave function, which determines the oscillator strength; and the density of states. The wave function of a single quantum barrier can be easily derived analytically. While for most of energy continuum above a single barrier the wave function is delocalized, it becomes quasi-localized in the barrier for energies corresponding to $\lambda = NL_b/2$ in the barrier region. In other words, the wave function will cover entire space, but its amplitude in the barrier region is larger than in the rest of the structure. This is shown in Fig. 4.20 (solid cure) by the quantity $\rho = \left(| A |^2 + | B |^2\right) / | I |^2$, where I is the coefficient of incoming electron wave outside the barrier region and A and B are the amplitude coefficients of electron waves traveling in opposite directions inside the barrier region. For situations shown in Fig. 4.20, the calculation was done for a barrier width of 20 nm and a height of 0.64 eV. The ratio ρ defined above provides a measure of quasi-localization in the single barrier, i.e., a large value of this quantity represents a state with most of its wave function in the barrier region. As can be seen from Fig. 4.20, the degree of localization is quite striking.

Since the structure of a single barrier does not have translational symmetry, calculation of the density of states is not straightforward. It can, however, be obtained in the manner described in [72], and the resulting 1-dimensional (growth direction) density of states is indicated in Fig. 4.20 by the dashed curve. It is evident from Fig. 4.20 that the density of states for a single barrier system is quite different from that for the bulk. In particular, there are peaks in the density of states at energies at which states are quasi-localized.

With the quasi-localization and the density-of-states characteristics discussed above, it was found that a single quantum barrier is in fact very similar to an *infinite single quantum well*. The only difference between the two is that for the infinite single quantum well we replace the two curves shown in Fig. 4.20 by delta functions

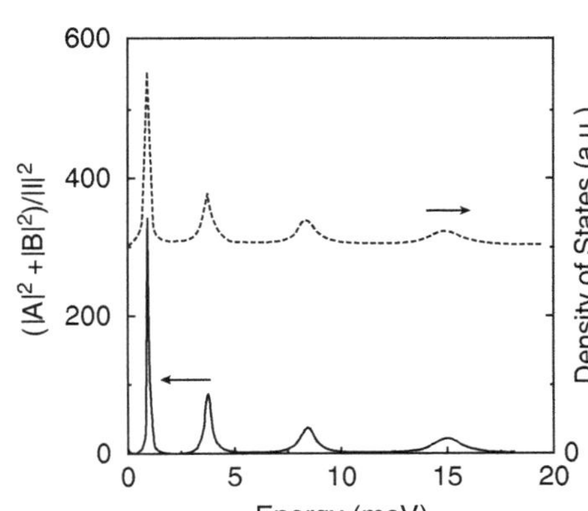

Fig. 4.20 The ratio ρ which provides a measure of localization of the wave function, as defined in the text (*solid line*). The *dashed line* represents the density of states, which peaks at the same energies as the localization. The energy is measured from the top of the barrier. After [72]

positioned at exactly the same energies as the peak positions for the two curves. This similarity should therefore lead to similar optical properties.

The ZnSe/ZnMnSe single barrier (in which ZnMnSe layer is the barrier for both the conduction band and the valence band) is one of the best examples for investigating above-barrier states, since the tuning ability of the DMS layer enables one to identify the above-barrier absorption peaks above the band edge of ZnSe. Absorption peaks for three different thicknesses of the ZnMnSe single barrier are shown in Fig. 4.21, where all peaks appear at higher energies than the ZnSe band gap (i.e, above 2.8 eV). Here, the above barrier transition is identified by the fact that the peaks show a blue shift with decreasing barrier width. As noted above, the results are clearly very similar to what is expected for a single quantum well.

The fact that the absorption peaks in Fig. 4.21 correspond to transitions between states quasi-localized in the barrier region can be confirmed by studying the magnetic field dependence of the spectrum. If the states involved are delocalized, there should be very small Zeeman splitting, since the structure is mostly made up of ZnSe. However, if the transition occurs between states that are quasi-localized in the barrier region (i.e., in the ZnMnSe layer), a large Zeeman splitting is expected. The experimental results are shown in Fig. 4.22. Also shown in the figure is the Zeeman splitting observed on a control ZnMnSe epilayer with the same Mn concentration as that of the single barrier. As one can see, the Zeeman splitting in the single quantum barrier is the same as in the control ZnMnSe epilayer. This unambiguously concludes that the transition peak observed in the absorption spectrum above the ZnSe band edge occurs between states quasi-localized in the ZnMnSe barrier layer.

4.6.2 *Above-Barrier States in Type-I Superlattice*

The observations of transitions involving above barrier states in superlattices were reported for several material systems [88, 89]. These observations made it clear that

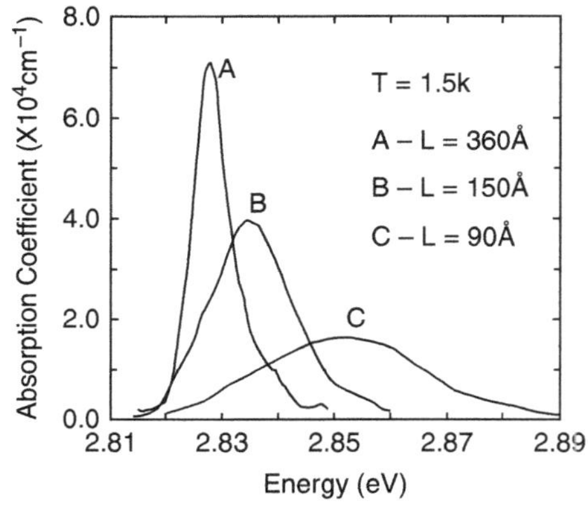

Fig. 4.21 Absorption coefficients of three single-barrier ZnMnSe/ZnSe systems. The *clear blue shift* of the absorption peak position with decreasing barrier width indicates that the states involved are localized in the barrier. From [72]

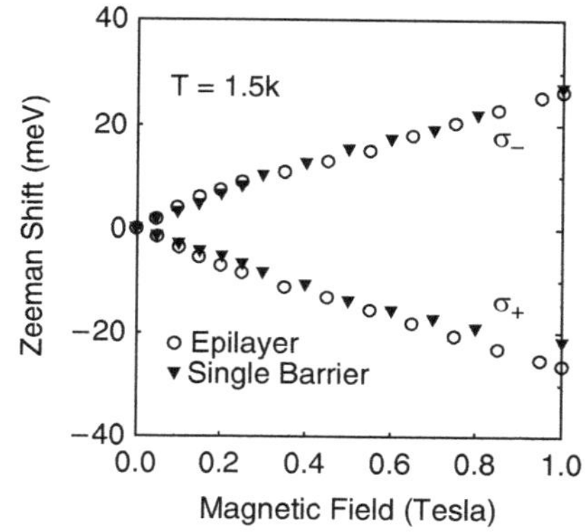

Fig. 4.22 Zeeman splitting observed for a 15 nm single barrier (*triangles*) and for a ZnMnSe control epilayer (*circles*) plotted as a function of magnetic field. After [72]

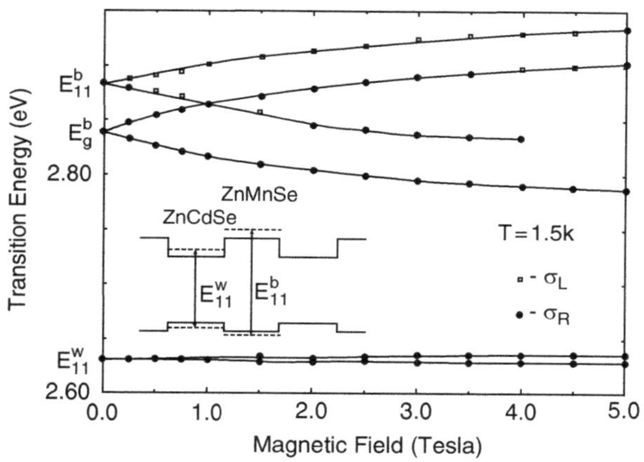

Fig. 4.23 Spin splittings of the excitonic transitions $E^{w}_{11}$and E^{b}_{11} observed in a $Zn_{0.88}Cd_{0.12}Se/Zn_{0.9}Mn_{0.1}Se$ superlattice, and of the E^{b}_{g} transition observed for a $Zn_{0.9}Mn_{0.1}Se$ control epilayer, as a function of magnetic field. The *solid lines* are guides for the eye. Band alignment for the ZnCdSe/ZnMnSe superlattice is shown in the inset. Reprinted from [90], © (1994), with permission from Elsevier

such high-lying subbands participate strongly in optical transitions due to the formation of *quasi-localized* states above the barriers. Here, we will discuss the case of type-I superlattices [71]. As an example of this we have chosen $Zn_{0.88}Cd_{0.12}Se/Zn_{0.9}Mn_{0.1}Se$ SL with well and barrier widths both equal to 75 Å. The band structure of this system is depicted in the inset of Fig. 4.23. Since for the Cd and Mn concentrations used in this case the energy gap of ZnCdSe is considerably smaller than that of ZnMnSe, the wells correspond to the nonmagnetic layers, while the barriers are magnetic. Although the structure exhibits several strong excitonic peaks, here we will concentrate only on the lowest energy peak labeled E^{w}_{11} in Fig. 4.23, which is identified as the free exciton transition between the first heavy-hole ground state and the conduction electron ground state of the superlattice; and on the peak marked E^b_{11}, which is identified as the transition between the first above-barrier heavy-hole state and the first above-barrier electron state (the lowest-energy above barrier transition). Note that E^b_{11} occurs at a higher energy than the energy gap of the barrier material. (To determine the band gap of the barrier, we use the exciton peak labeled E^b_g observed on a control epilayer of $Zn_{0.90}Mn_{0.10}Se$ grown under identical conditions as the superlattice). When an external magnetic field is applied, the band edges of the ZnMnSe barriers will be Zeeman split, leading in turn to the splitting of optical transitions in the superlattice.

Figure 4.23 shows the energies of the E^w_{11} and E^b_{11} transitions as a function of the applied field. Also shown in the figure is the Zeeman splitting of the exciton line E^w_g observed on the control epilayer. It is significant that the spin splitting of the above-barrier transition E^b_{11} is clearly much larger than that of the below-barrier transition E^w_{11}, and almost (*but not quite*) as large as that observed on the ZnMnSe control epilayer. We recall that the structures investigated are type-I superlattices, consisting of nonmagnetic wells and magnetic barriers. The relatively small (but observable) Zeeman splitting of the ground-state exciton transition (which originates and terminates in the nonmagnetic wells) thus arises from the partial penetration of the wave functions of the initial and final states into the magnetic barriers. By contrast, the much larger Zeeman splitting of E^b_{11} (almost the same as that for the bulk $Zn_{0.90}Mn_{0.10}Se$ material) indicates that the E^b_{11} transition originates and terminates on states localized predominantly in the DMS (i.e., the barrier) regions. The fact that the splitting of E^b_{11} is slightly (about 15%) below that observed for the $Zn_{0.90}Mn_{0.10}Se$ epilayer indicates, however, that a part of the above-barrier wave functions, which determine the transition extends into the non-DMS layers. The data shown in Fig. 4.23 thus provide direct evidence that above-barrier excitons in a type-I superlattice are localized in the barrier layers.

4.6.3 Above-Barrier States in Type-II Superlattices

Using the same techniques, it was also shown that in type-II superlattices there exist spatially direct type-I-like excitons including above barrier states [91,92]. The inset in Fig. 4.24 shows a schematic diagram of the type-II band alignment for a CdMnSe/ZnTe superlattice. Here, the quantum wells for electrons and for holes exist in different (i.e., adjacent) layers. Specifically, the conduction electrons at energies below the barriers are localized in the CdMnSe layer, while the below-barrier heavy holes are in the ZnTe layers. Excitonic transitions are weak in such a system because wave functions of the electrons and the holes are spatially separated, resulting in weak Coulomb attraction.

As seen in the inset, two types of excitonic transitions are possible in such type-II superlattices. One takes place between electron states localized in CdMnSe wells and hole states in ZnTe wells, as shown by the arrow marked ‘II’ in Fig. 4.24. This is the well-studied type-II excitonic transition. The other occurs between electron (or hole) subbands confined in the wells, and hole (or electron) subbands at

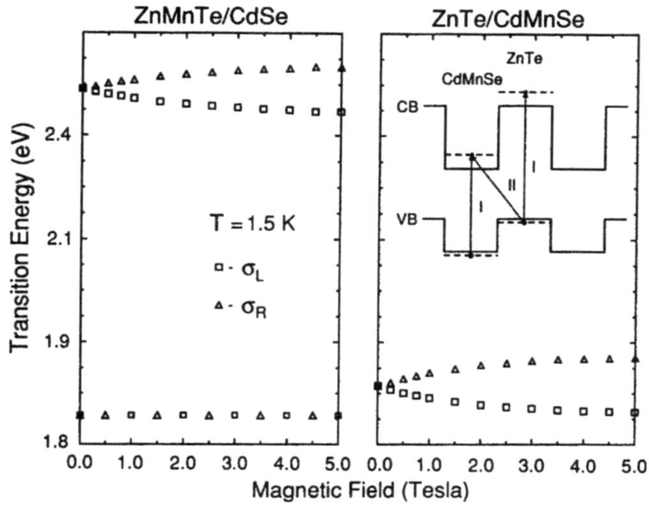

Fig. 4.24 Magnetic-field-dependence of spatially direct transitions occurring in a type-II CdSe/ZnMnTe superlattice (*left*); and in a type-II CdMnSe/ZnTe superlattice (*right*). The band alignment and possible transitions in a CdMnSe/ZnTe type-II structure are shown in the right-hand panel. Reprinted from [90], © (1994), with permission from Elsevier

above-barrier energies confined in the barriers. In that case, both the initial and the final state participating in a given transition are localized *in the same layer*, as shown by the arrows marked 'I' in Fig. 4.24. We shall refer to these spatially direct processes as type-I excitonic transitions. The existence of type-I excitons in type-II superlattices has been shown in two systems: CdSe/$Zn_{0.94}Mn_{0.06}Te$, where conduction electrons are localized in nonmagnetic wells; and $Cd_{0.92}Mn_{0.08}Se$/ZnTe, where electrons are localized in magnetic wells. In these two cases, the type-I excitons confined in the nonmagnetic layers will exhibit very different magnetooptical properties from those confined in the DMS layers.

The magnetoabsorption spectrum measured on CdSe/ZnMnTe superlattices revealed two clear excitonic absorption peaks, whose energies are shown on the left in Fig. 4.24. Here, the lower energy peak is identified as the type-I (spatially direct) excitonic transition between the lowest electron state confined in the CdSe conduction-band wells and the first above-barrier hole state confined in the CdSe valence-band barrier layers. The higher-energy peak is attributed to a similar type-I transition, in this case between the above-barrier electron states confined in ZnMnTe barriers and the hole states confined in ZnMnTe wells. The difference in the magnetooptical behavior of the two transitions is shown in Fig. 4.24, where energies observed for the σ^+ and σ^- circular polarizations are plotted as a function of magnetic field. It is quite striking that there is no observable Zeeman splitting for the lower-energy transition, confirming that this transition indeed takes place between states strongly localized in the nonmagnetic CdSe layers. In contrast, there is a large

Zeeman splitting for the higher-energy transition, which clearly indicates that it occurs between states localized in the DMS layer.

On the right of Fig. 4.24, we also show the Zeeman splitting of the lowest-energy absorption peak observed in the ZnTe/CdMnSe superlattice. This absorption line now exhibits a large Zeeman splitting, indicating that the transition occurs between electrons localized in CdMnSe wells and above-barrier holes localized in the same layers. (The higher-energy line, which now would correspond to states localized in the ZnTe layers, was obscured by the absorption due to the ZnTe buffer on which this structure was grown.) Both panels of Fig. 4.24 thus clearly establish the existence of type-I excitons in type-II superlattices, and identify their localization. The unambiguous experimental demonstration of how above-barrier states are distributed in space was accomplished by exploiting the difference in Zeeman splitting of states localized in DMS and in non-DMS regions comprising the superlattices. However, it should be emphasized that the effects discussed here are, of course, not limited to systems based on the DMSs, but represent *general properties* of type-I and type-II superlattices. The role of the DMS constituent is only used as a 'marker', in order to bring out these general properties of wave function distribution.

4.7 DMS-Based Quantum Dots: Inter-dot Spin–spin Interactions

DMS-based quantum dot (QD) systems are of great interest for several reasons. First, they provide a laboratory wherein one can investigate the DMS characteristics discussed in the preceding part of this chapter (giant Zeeman splittings, band offset tuning, etc.) in zero-dimensionality and under extreme localization. They also offer the possibility of investigating individual QDs, whose extremely narrow spectral lines offer unprecedented resolution in which to study spin-based effects by optical methods; and similarly, whose ultra-small size offer the possibility of 'zooming-in' on spin phenomena on the nanometer scale. Furthermore, the physics of inter-quantum-dot correlations is of great interest, in that it provides the opportunity for studying spin–spin interactions under conditions when carriers are strongly localized (as distinguished from quantum wells, in which carriers are unconstrained and can move in the plane of the well). Finally, in longer term, one needs to bear in mind that spin states in quantum dots (QDs) have been proposed as promising candidates for quantum bits in quantum computing. In this latter context, DMS-based QDs are especially interesting because of the strong spin-based effects which DMS materials bring into the picture, thus allowing highly promising options for spin manipulation in the QD geometry [93]. The properties of individual DMS-based QDs have been discussed in detail in Chap. 6 of this Volume. Using the properties of QDs presented in Chap. 6, here we will focus on the interactions between such dots, and especially on the role which spin plays in such interactions.

4.7.1 Inter-dot Interactions in DMS/Non-DMS Double Layer QD Systems

Spin states in quantum dots (QDs) have been proposed as viable candidates for quantum bits in quantum computing [93]. To obtain multi-bit gate functionality using this spin property, one must have a good understanding of inter-dot spin exchange interaction in coupled QD structures. Here, double-layer QD structures are especially important, since they constitute the simplest and at the same time the most informative QD system for investigating inter-QD spin interactions in a highly localized geometry [94–96]. In this section, we describe the behavior of spin polarization observed in especially designed *asymmetric* double-layer QD structures (DLQDs) involving DMS materials. The giant Zeeman splittings of the band edges which is characteristic of DMSs when a magnetic field is applied [5] leads to strong spin polarization of carriers in such DMS-based quantum structures [97], thus providing a powerful tool both for studying and for manipulation of spin phenomena in zero-dimensionality.

Polarization-selective photoluminescence (PL) is an extremely useful tool for investigating spin polarization of carriers in quantum structures, since transitions between spin-up states of the conduction and valence bands correspond to the σ^- circular polarization of the PL, while transitions between the spin-down states correspond to the σ^+ polarization. The intensities I^+ and I^- of the emitted σ^+ and σ^- polarizations then directly reflect the spin-down and spin-up populations of the carriers at the instant of recombination. One can thus obtain the net spin polarization of a given quantum structure from the measured degree of circular polarization, $P = (I^+ - I^-)/(I^+ + I^-)$, of the PL peak emitted by that specific quantum structure.

In the non-DMS system (for example, CdSe or CdZnSe in ZnSe matrix), the spin-up states are shifted down in energy in the conduction band and up in the valence band, thus becoming the ground states in both bands [98]. In DMS systems, on the other hand (for example, in CdMnSe or CdZnMnSe QDs in ZnSe matrix), the sequence of spin states is reversed relative to the CdZnSe QDs due to the *sp–d* interaction between spins of the extended band states and of the localized Mn ions, so that the *spin-down states* in both bands become the ground states of the system [99]. Although both spin orientations are excited with approximately equal probabilities, in each QD system a fraction of the higher-energy spin states will thermalize to the ground state via spin-flip transitions before recombining. Thus, the emission from CdZnSe QDs is expected to be increasingly dominated by the σ^- polarization, and the CdMnSe QD emission by the σ^+ polarization as the applied field is increased.

Figure 4.25 shows the magnetic field dependence of P observed for a CdMnSe/CdZnSe double layer quantum dot system (referred to as DLQD1), along with data observed on a single layer of CdZnSe QDs (referred to as SLQD1), which serves as a reference [100]. The open triangle and solid square symbols, respectively, represent P observed on CdZnSe QDs in the SLQD1 and DLQD1 geometries. The sign of P is positive for DMS QDs due to the sign of the g-factor, and the value of P

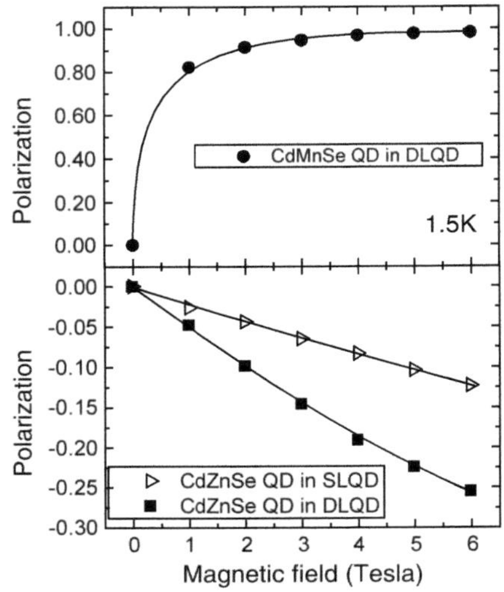

Fig. 4.25 Magnetic field dependence of the degree of polarization P observed for CdMnSe QDs (*upper panel*) and CdZnSe QDs (*lower panel*). *Open triangles* and *solid squares* in the *bottom panel* represent results obtained for CdZnSe QDs in SLQD1 and DLQD1 structures, respectively. The significant enhancement of P observed for CdZnSe QDs in DLQD1 indicates antiferromagnetic interaction between the carrier spins localized in coupled pairs of QDs in the two adjacent layers. Note that the direction of polarization in CdMnSe QDs (i.e., the positive P value) is opposite to that in CdZnSe QDs (i.e., negative P). The *lines* are a guide for the eye. Reproduced with permission from [100], © (2006) Wiley – VCH Verlag GmbH & Co. KGaA

reaches almost 100%, reflecting the large Zeeman splitting characteristic of DMS systems. The value of P is negative for CdSe QDs, indicating that the component of the σ^- polarization is dominant in the PL emitted by non-DMS QDs. It is evident that the degree of polarization of CdZnSe QDs is significantly enhanced in the DLQD1 structure compared to the SLQD1. Since the degree of polarization directly reflects the spin polarization of carriers in the dots, this result indicates that the spin polarization of CdZnSe QDs was significantly affected by neighboring CdMnSe QDs in the DLQD1 geometry [101].

In Fig. 4.26, we show the magnetic field dependence of P obtained on another double-layer QD system, consisting of CdSe and CdZnMnSe QD layers (which we refer to as DLQD2), along with a single-layer CdSe QD control structure (referred as SLQD2) [102, 103]. Comparison of the data from CdSe QDs in the SLQD2 and DLQD2 reveals that the degree of polarization in the DLQD2 system is much larger than that observed in SLQD2.

Since the PL energy of CdSe QDs is lower than that of CdZnMnSe QDs DLQD2 due to the incorporation of Zn, one can selectively excite the carriers in the CdSe layer of QDs by choosing a laser excitation energy above the PL emission of the CdSe QDs, but *below* that of the CdZnMnSe QD system. With this energy-selective experiment, one can study the spin polarization behavior of only the CdSe QDs, without the influence of the neighboring higher-energy CdZnMnSe QDs (since

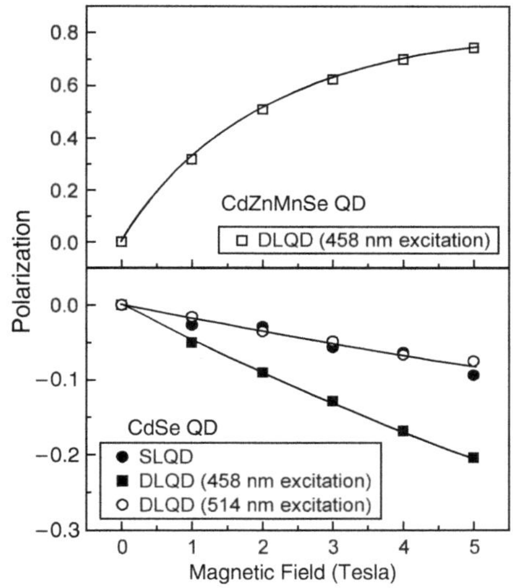

Fig. 4.26 Magnetic field dependence of the degree of polarization P for the CdSe and CdZnMnSe peaks of the SLQD2 and DLQD2 structures. The *open and solid squares* represent the results obtained from excitation of both CdSe and CdZnMnSe QDs in DLQD2. The *solid circles* represent the degree of polarization P for CdSe QDs in SLQD2. The *open circles* indicate the values of P for CdSe QDs when there are no carriers in the CdZnMnSe QDs. The enhancement of P observed for the CdSe QDs when both QD layers of the DLQD2 structure are excited (i.e., when carriers are present in *both* QD layers) indicates antiferromagnetic interaction between the carrier spins localized in coupled pairs of QDs from the two layers. The *lines* are guides for the eye. Reproduced with permission from [100], © (2006) Wiley – VCH Verlag GmbH & Co. KGaA

under this excitation no carriers are present in the DMS dots) [102]. The degree of PL polarization P observed for the CdSe QDs under such energy-selective excitation is plotted as open circles in Fig. 4.26. It is clear from Fig. 4.26 that the degree of polarization of the PL emitted by the CdSe QDs in the DLQD2 structure under energy-selective excitation (i.e., when only the CdSe QDs are excited) is much smaller than that obtained under simultaneous excitation of both CdSe and CdZnMnSe QDs. In fact, the value of P and its dependence on magnetic field is now very similar to that observed on the SLQD2 system. The distinct difference of the polarization behavior between the case *when only one QD layer* is selectively excited in the DLQD2 structure, and when both layers are excited, provides compelling evidence that the net spin polarization of carriers within the CdSe QDs is different in the two cases. Here, the contribution of direct spin injection from DMS dots or overlap with Mn ions, which could decrease the degree of σ^- polarization, is ruled out due to the observed enhancement of σ^- polarization in the DLQD2 system. Such difference in spin population of the CdSe QDs due to the excitation of CdZnMnSe QDs can therefore only arise as a consequence of *spin-dependent interaction between carriers* localized in the neighboring dot layers (i.e., carriers must be present in the CdZnMnSe layer for the inter-QD spin interaction between CdZnMnSe and CdSe QDs to occur).

4.7.2 Interaction Between Non-DMS Quantum Dots and a DMS Quantum Well

Experimental demonstration of spin–spin exchange interaction between pairs of similar QDs is quite difficult due to the fact that the PL emitted from the individual dots comprising the interacting system is indistinguishable. In the preceding section, we have used *asymmetric* coupled quantum dot structures, in which interacting QDs emit at different energies, to circumvent this difficulty [104]. However, in the case of QD–QD coupled structures the QDs in one layer need to be spatially paired with QDs in the neighboring layer for the interaction to be significant, imposing the additional requirement of aligning the dots from the two layers of the DLQD structure. This difficulty can be bypassed by designing a system comprised of a layer of QDs adjacent to a *quantum well* (QW), each having a different energy gap, so that one can distinguish the PL signal from the two quantum structures. The advantage of this combination is that the structure does not require vertical stacking of dots, as in the case with DLQD systems, because every QD will automatically be coupled to the adjacent QW. The QD-QW structure used here consists of a DMS (ZnCdMnSe) QW and a layer of nonmagnetic (CdSe) with ZnSe used as the barrier QDs. In such hybrid double-layer structure, the spin states of the carriers in the nonmagnetic QDs are expected to be affected by the carrier spins from the magnetic QW via exchange interaction between the QW and the QDs, the degree of interaction depending on the barrier separating the two quantum systems.

The field dependence of the polarization P for the PL emission in QD-QW double layer systems is plotted in Fig. 4.27 [105]. The solid squares, circles, and triangles in the lower panel represent the results for CdSe QDs with 40, 20, and 12 nm barriers, respectively; and the open circles in the upper panel represent the degree of polarization P from the ZnCdMnSe QW observed on the sample with a 20 nm barrier.

It is quite striking that the polarization P observed on the CdSe QDs in the double-layer geometry shows a strong dependence on the thicknesses of the ZnSe spacers. To compare the effect of the neighboring magnetic QW on the spin polarization in the non-magnetic CdSe QDs, in Fig. 4.27 the polarization for the single-layer CdSe QD sample is also shown (as open squares). The behavior of P observed on the single layer of CdSe QDs is very similar to that exhibited by CdSe QDs in the hybrid QD-QW layer structure with a 40 nm spacer. This similarity of P implies that the interaction between the QDs and the QW in the case of the widest ZnSe barrier is negligible, so that the QDs in that double-layer geometry are essentially decoupled from the QW when the barrier thickness is 40 nm or larger.

In sharp contrast, the observed degree of σ^- polarization in the CdSe QD emission becomes systematically larger in the QD-QW structures as the spacer thickness narrows to 20, and then to 12 nm. This clearly eliminates the possibility of direct spin injection from the DMS QW (which is predominantly populated by spin-down carriers) to CdSe QDs, such as that observed in other quantum structures [106]. This enhancement of P in the double-layer geometry can be ascribed to spin-dependent

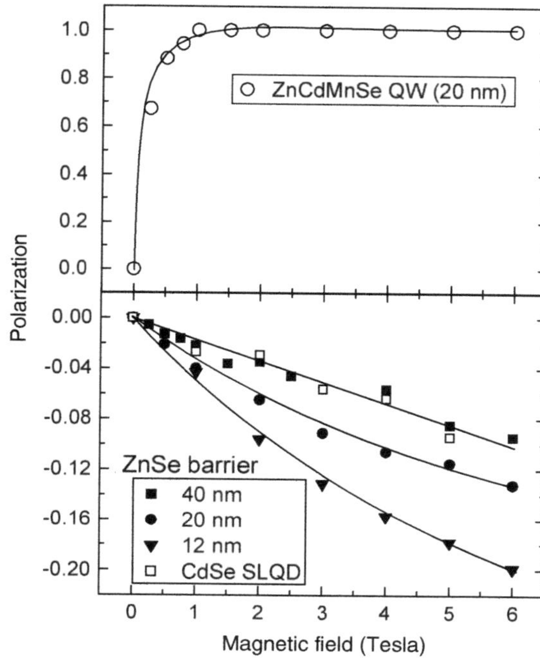

Fig. 4.27 Magnetic field dependence of PL polarization P for CdSe QDs in three DMS QD-QW structures. The *solid and open squares* represent, respectively, the values of P observed on CdSe QDs from the QD-QW system separated by a 40 nm ZnSe barrier, and on a single-layer of CdSe QD structure grown without the DMS QW, used as a reference sample. The *solid circles* and *triangles* are observed on the QD-QW structures with 20 and 12 nm barriers, respectively, showing that the magnitude of P for these specimens increases rapidly relative to that of the single-layer of CdSe QD structure as the barrier between the QDs and the DMS QW becomes thinner. *Open circles* are the values of P for the CdZnMnSe QW, showing 100% σ^+ polarization for fields above 1.0 T. The *lines* are guides for the eye. Reproduced with permission from [105], © (2005) Wiley – VCH Verlag GmbH & Co. KGaA

exchange interaction between carriers in the QDs and in the magnetic QW. Since the strength of the exchange interaction in the QD-QW system depends primarily on the overlap of wave functions from the QD and the QW subsystems, the spin-up polarization of carriers in non-DMS QDs is expected to increase with decreasing barrier thickness separating the DMS QW and the non-DMS QDs – in excellent agreement with the behavior of P seen in Fig. 4.27.

4.7.3 General Comments on Spin–Spin Interaction in Multiple Quantum Dot Systems

The possible mechanism for the spin-dependent correlation between carriers localized in coupled QDs is given below. Figure 4.28 shows schematically the possible

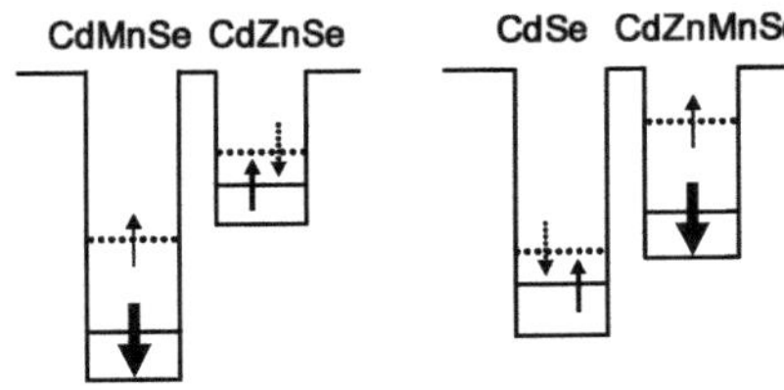

Fig. 4.28 Possible spin states shown schematically for the conduction bands of coupled non-DMS and DMS QD pairs. As a consequence of the large Zeeman splitting in the DMS QDs the carriers in those QDs are strongly polarized to spin-down states, as indicated by the *thick arrows*. The anti-parallel spin interaction between coupled QD pairs then increases the spin-up carrier population in the non-DMS QDs, resulting in an enhancement of the σ^- polarization, as seen in Figs. 4.25 and 4.26. Reproduced with permission from [100], © (2006) Wiley – VCH Verlag GmbH & Co. KGaA

spin states in coupled non-DMS and DMS QDs when a magnetic field is applied. (For simplicity, only the conduction band is shown in Fig. 4.28, but the picture can be readily generalized to both bands). The presence of coupling between dots indicates that there is carrier tunneling between states localized in the pairs of interacting dots. When a carrier from one QD virtually hops to another, the energy states and wave functions of the system can be calculated perturbatively. Detailed calculations in [94, 96] reveal that in a pair of QDs with inter-QD interactions the energy levels are lower than they would be in the same configuration if the inter-QD coupling were absent. Such virtual hopping process between coupled dots can occur only in the anti-parallel spin configurations due to the Pauli exclusion principle. This means that the spin states of carriers localized in one QD of a coupled pair prefer to align *anti-parallel* to the spins of the other QD.

Since the Zeeman splitting is much larger in DMS QDs than in non-DMS QDs, the spin polarization of carriers in DMS QDs is correspondingly much more pronounced [97]. Once the carriers in the DMS QD structures relax to their lowest-lying states (which in this case corresponds to the spin-down orientation, as indicated by thick arrows in Fig. 4.28), the preference for anti-parallel spin alignment of carriers in coupled systems *accelerates* the spin-flip process of the spin-down carriers in the non-DMS QDs into the lower-lying spin-up states. This process increases the number of spin-up carriers in the non-DMS QDs, resulting in an enhancement of the σ^- polarization, as is observed experimentally. This behavior will take place when (a) the quantum structures in adjacent layers are sufficiently strongly coupled to interact; and (b) the interaction is such as to align the carrier spins in adjacent quantum structures antiferromagnetically. This effect is clearly manifested in the coupled QD-QW system involving DMS QWs, in which antiferromagnetic interaction between the carriers in the QDs and the QW was unambiguously identified by the enhanced degree of polarization of the QD emission as the QD and the QW layers were brought closer together.

4.8 Spin Tracing

In our discussion of quantum structures in the preceding sections, we have focused on the effects of wave function leakage beyond the nominal size of a specific QW or QD structure. For example, the Zeeman splitting observed in non-DMS CdSe QDs embedded in a ZnMnSe matrix was ascribed primarily to the fact that, due to such wave function leakage, electrons and holes localized in the non-DMS CdSe QD 'feel' the DMS material that surrounds the confining structure, and their behavior would thus reflect to some degree the surrounding DMS properties. Similarly, in the case of non-magnetic QWs between DMS barriers (as, e.g., in the ZnCdSe/ZnMnSe combination) the large Zeeman splitting which characterizes the QW states is primarily determined by the degree to which wave functions of the QW states penetrate into the DMS barriers. This then affords the possibility of mapping how these wave functions are distributed throughout the system.

In our discussion, however, we have used idealized systems, i.e., we have assumed that in systems such as the ZnCdSe/ZnMnSe QW structure just mentioned, the boundaries between the layers are perfectly abrupt, with no Mn ions in the ZnCdSe QW layers, and no Cd ions in the ZnMnSe layers. In real systems, however, there is significant atomic intermixing that occurs near the interfaces. We can divide such atomic inter-penetration into two distinct processes. First, there is atomic segregation that occurs near the interface during the growth that is directly related to the *direction* of growth, and thus results in an *asymmetric composition profile* of the structure. And second, there can occur diffusion across the interfaces – a process that can be strongly enhanced by postgrowth annealing carried out at elevated temperatures. As will be seen, this process typically leads to symmetric atomic distributions near the interfaces, i.e., to symmetric 'smearing out' of the boundaries between adjacent materials.

To establish the distribution of 'alien' ions near interfaces is a difficult task. For example, in the well-known case of InAs QDs embedded in the GaAs matrix (or CdSe in ZnSe), it is generally assumed that the QDs are in reality InGaAs (or CdZnSe), in which the amount of Ga (or Zn) can be estimated from the photoluminescence (PL) energy emitted by the QD. This, however, is a very approximate process, since the PL energy also depends on the QD size and shape, and it is difficult to separate the effects caused by morphological parameters and by the composition.

In this context, the exchange interaction between Mn ions and band electrons provides an extremely powerful tool for mapping the distribution of 'unintentional' ions that make their way across the idealized interfaces. While the mapping process which we will describe applies specifically to Mn ions, it will also automatically provide insights into processes of across-interface migration of *all* atomic species that occur during the growth, or that are produced intentionally by post-growth annealing.

4.8.1 Spin Profiles Formed During Growth

We will first consider intermixing at interfaces that occurs during the growth, and is therefore unavoidable in any heterostructure. For simplicity, let us first consider the simplest case of a CdTe/MnTe superlattice. We recall that the MnTe layer is an antiferromagnet, with a zero net magnetic moment, and therefore with no exchange interaction of the type described in the introduction. Similarly, the 'idealized' CdTe layers can be assumed to contain no Mn ions. Thus, the exchange-induced spin splitting should be completely absent in this extreme case. The experiments, however, provide a very different picture. First, electron paramagnetic resonance (EPR) measurements on CdTe/MnTe (and on the closely related ZnTe/MnTe) superlattices clearly show EPR signals characteristic of dilute Mn distributions that could only come from Mn ions present in dilute amounts in the nominally pure CdTe (or ZnTe) layers [107, 108]. Furthermore, extensive experiments on spin entanglement carried out on similar (nominal) CdTe/MnTe superlattices also show clear Zeeman splitting of excitons that could not be produced in either CdTe or MnTe layers of these structures, and must therefore be ascribed to the presence of dilute amounts of Mn ions in the CdTe layers [109–111].

The process by which Mn ions make their way into the adjacent nonmagnetic layers has been studied extensively by Gaj et al. [112] and Grieshaber et al. [113] in a series of ingenious experiments on QW structures specifically designed to reveal the effect of the spin profile on the splitting of exciton states. This seminal research, along with subsequent extended investigations [114, 115], has not only provided an invaluable picture of the mechanisms which govern near-interface Mn segregation, but has also served to develop new techniques which can be used to map such distribution profiles.

Before looking at the details of these specific investigations, let us first return to the idealized example of the MnTe/CdTe superlattice to qualitatively describe the process of atomic segregation, which occurs at interfaces during the growth. To facilitate discussion, we will refer to the MnTe layers of the MnTe/CdTe superlattice as the barriers, and to the CdTe layers as the wells. Consider now what happens when the final MnTe monolayer is deposited upon completing the growth of the MnTe barrier, i.e., at the point when the Mn source is closed off and the Cd source is switched on. It is essential to appreciate that the final Mn layer is not truly 'fixed' – its bonds are only half-complete – and it is therefore partly volatile. To recognize that the physical properties of atoms in that last-deposited layer are different from those fully encapsulated in the bulk, we will refer to it as the 'interlayer'. As Cd atoms arrive at the MnTe interlayer, some of the Mn atoms from the interlayer can leave their expected positions of the idealized MnTe lattice and be replaced by Cd, while the Mn atoms will move to occupy the Group-II cation positions in subsequent layers.

While very accurate mathematical modeling has been established for this atomic redistribution [112, 114], it is useful to first discuss the process in terms of a simplified qualitative picture, which can serve as a useful guide to what physically occurs, as follows. As the growth transforms from MnTe to CdTe deposition, part of the

Mn atoms in the interlayer remain attached at the original interface layer, and the other part is exchanged for Cd and move to form the next interlayer. For simplicity of the following discussion, let us consider the special case in which half of the Mn atoms moves to the next layer and hence a complete exchange of Mn and Cd between two layers takes place. The same process repeats further on: half of the Mn atoms in the *new interlayer* remain in place, and half of that participate in forming the subsequent interlayer; and so on. In the case of the MnTe/CdTe interface chosen for this illustration, we then have 50% of Mn atoms at the original ('ideal') interface, 25% in the next monolayer, 12.5% in the next, and so on. Taking roughly 0.3 nm for the II–VI monolayer thickness, it is easy to see that after switching off the Mn source we will have just under 1.0% of Mn coverage by the time the 6th monolayer is formed, i.e., after ca. 2.0 nm of CdTe have been deposited on the nominal interface position.

The same reasoning can be applied to the Cd distribution after deposition of the CdTe well has been completed, and the MnTe deposition is resumed, i.e., when the final monolayer *of CdTe* acts as the interlayer. The composition profile of the MnTe/CdTe system is thus expected to be as shown in Fig. 4.29b. The essential feature of this picture is that the profile of the Mn distribution in the CdTe well (and, correspondingly, of the Cd distribution in the MnTe barrier) is *unidirectional*, i.e., there is no 'leaking' of Cd back into the MnTe barrier when the MnTe deposition is terminated, nor is there any leaking of Mn back into the CdTe well when the MnTe deposition is resumed. We have used the MnTe/CdTe superlattice for simplicity, but identical arguments will apply to ternary systems, and it is those systems that have been successfully used to establish the profile of Mn distribution by 'spin tracing' in [112,113], and [114]. Consider, e.g., a system with $Cd_{1-x}Mn_xTe$ barriers and nominal CdTe wells. Upon completion of the $Cd_{1-x}Mn_xTe$ barrier growth, the interlayer contains an atomic fraction x of Mn, of which approximately half will go to form the next interlayer, and so on, exponentially, as described above.

Measurements of the actual spin tracing have been based on the Zeeman splitting of excitons in $Cd_{1-x}Mn_xTe$/CdTe/$Cd_{1-x}Mn_xTe$ (symmetric) or $Cd_{1-x}Mn_xTe$/

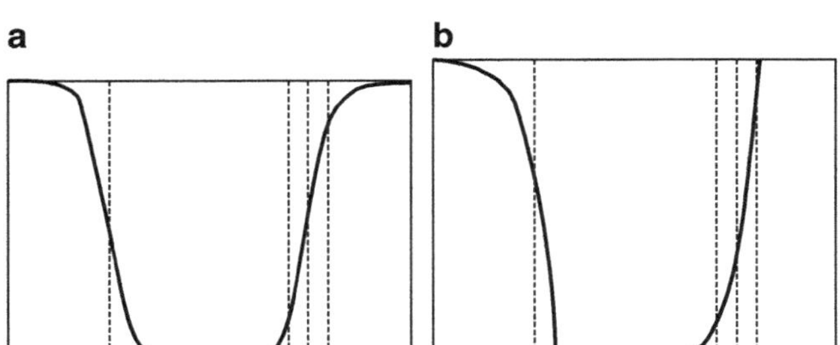

Fig. 4.29 Examples of quantum well potentials arising from interface profiles with a symmetric (**a**) and asymmetric (**b**) atomic intermixing. The asymmetric case (**b**) is a good representation of the potential expected for a $Cd_{1-x}Mn_xTe$/CdTe/$Cd_{1-y}Mg_yTe$ QW with the DMS barrier grown first, and with the growth proceeding from right to left in the figure. Reprinted with permission from [112], © (1994) by the American Physical Society

$CdTe/Cd_{1-y}Zn_yTe$ and $Cd_{1-y}Zn_yTe/CdTe/Cd_{1-x}Mn_xTe$ (asymmetric) quantum well structures, usually detected in reflectivity experiments [112–114]. To facilitate discussion of interfaces occurring in these structures, the authors use the following terminology: 'normal interface' is used to describe $Cd_{1-x}Mn_xTe$ grown on CdTe; and 'inverted interface' applies to CdTe grown on $Cd_{1-x}Mn_xTe$. In analyzing the Zeeman shift of exciton states in QWs with an asymmetric distribution of Mn ions such as that illustrated in Fig. 4.29b, two additional features must be considered. The first is that the spin splitting of an exciton in a particular region of a $II_{1-x}Mn_xVI$ alloy is proportional to the magnetization of the alloy in that region [5]. As is well known, the magnetization of $II_{1-x}Mn_xVI$ alloys does not scale with the Mn concentration x, but shows a maximum in the range near $x \simeq 0.10 - 0.15$ (depending on temperature), as illustrated by the solid line and full circles in Fig. 4.30 [19, 116], since at large values of x the magnetization of the Mn subsystem is reduced by antiferromagnetic interactions of the Mn ions with their surrounding neighbors, as discussed in [114]). The second feature to be considered is that the splitting of a particular state within the quantum well will also depend strongly on the profile of the wave function of that state, i.e., on the degree to which the exciton wave function overlaps with the (possibly graded) magnetization of the $II_{1-x}Mn_xVI$ layer comprising the well. These effects have been carefully modeled in a series of systematic spin tracing experiments [112, 114], and have allowed these authors to obtain a comprehensive quantitative picture of the Mn concentration profile of quantum wells and

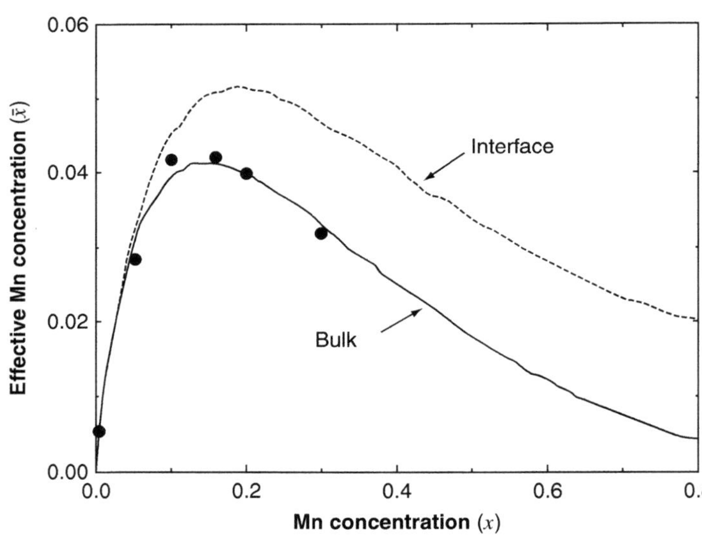

Fig. 4.30 Calculated effective magnetic concentration (which determines magnetization) as a function of real Mn concentration x for a bulk DMS (*solid curve*) and for an idealized DMS interface (*dotted curve*). Points are experimental for $Zn_{1-x}Mn_xSe$. Calculation is from [116]; experimental points are from [19]. Reprinted with permission from [116], © (1994) by the American Physical Society

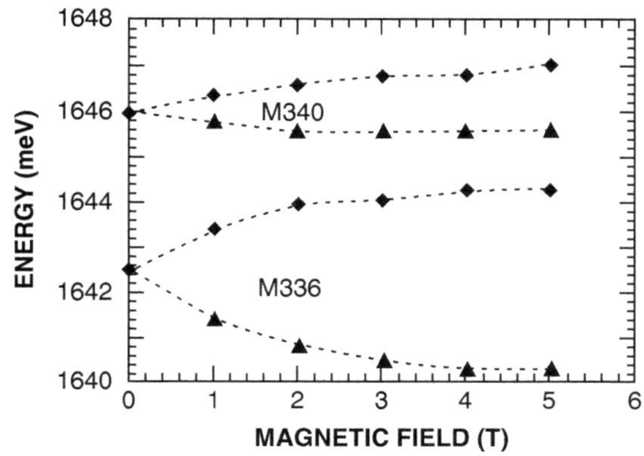

Fig. 4.31 Energies of excitonic transitions observed in magnetoreflectivity observed in two circular polarizations on asymmetric $Cd_{1-x}Mn_xTe/CdTe/Cd_{1-y}Zn_yTe$ QW systems. Data marked M336 involves a system with an inverted $Cd_{1-x}Mn_xTe/CdTe$ interface. The data marked M340 is for a sample where the $Cd_{1-x}Mn_xTe/CdTe$ interface is normal. Reprinted with permission from [112], © (1994) by the American Physical Society

superlattices. We will illustrate this by Fig. 4.31, which shows the exciton splitting in an asymmetric structure grown in the sequence $Cd_{1-x}Mn_xTe/CdTe/Cd_{1-y}Zn_yTe$, where Mn has segregated into the QW interior (marked M336 in the figure); and by the structure grown in the sequence $Cd_{1-y}Zn_yTe/CdTe/Cd_{1-x}Mn_xTe$ (marked M340), where the amount of Mn in the well is much lower, the Zeeman splitting being determined largely by wave function leakage into the DMS barrier. Focusing specifically on the DMS/non-DMS interfaces, we can refer to these two structures as 'inverted' and 'normal', respectively. For detailed mathematical modeling of exciton splitting in the presence of Mn segregation at normal and inverted interfaces, we refer the reader to [112] and [114].

The spin tracing technique just described provides valuable physical insights into the composition profile of structures grown by molecular beam epitaxy (and possibly by other epitaxial processes as well). In addition to the general understanding of spin segregation, it also offers some very practical uses. For example, this was exploited to achieve quantum dots in which only one Mn atom is contained – a feature that may become important in designing q-bits for quantum computing – as was described in Sect. 4.8.1 above. Using [117] and [118] as examples, this was achieved by growing a $Zn_{0.94}Mn_{0.06}Te$ barrier followed by a ZnTe layer of a selected thickness (in this case 10 ML). A layer of CdTe was deposited at this point to form CdTe QDs by the Stranski-Krastanow method, and the QDs were finally capped by ZnTe. Adjusting the thickness of ZnTe deposited after $Zn_{0.94}Mn_{0.06}Te$ has allowed the authors to choose the Mn concentration very precisely at the time when the QDs were formed [114]. In this way, the process of Mn segregation was used to achieve the situation where there is a high probability of forming dots that contain a single Mn atom, as indeed has been verified by the spectacular success of the experiments described in [117].

4.8.2 Inter-diffusion at Interfaces Mapped by the Spin Tracing Approach

From the discussion in Sect. 4.8.1, one sees that the compositions of quantum wells are of their very nature asymmetric, the near-exponential atomic segregation being imposed by the directional growth dynamics. This contrasts sharply with the process of *inter-diffusion* arising primarily from forces which tend to equilibrate the concentrations of different atomic species across an interface. This has been studied extensively by several groups in $Cd_{1-x}Mn_xTe/CdTe/Cd_{1-x}Mn_xTe$ quantum wells and in the more complicated $Cd_{1-x}Mn_xTe/CdTe/Cd_{1-y}Mg_yTe$ asymmetric materials combinations [119–121]. During the growth, such diffusion has been shown to be negligible compared to the segregation process described in the preceding section. However, when a given system is annealed at a temperature considerably higher than the growth temperature, the process of inter-diffusion has been shown to become increasingly important, resulting in interface broadening with symmetric compositional profiles, such as that seen in Fig. 4.29a [119, 120]. Again, spin tracing applied to the Mn ions has been invaluable in mapping the atomic distribution in the interface region of heterosotructures subjected to such high temperature annealing. We mention parenthetically that in the studies just referred to it was also shown that capping of heterostructures can be used to modify the atomic redistribution induced by annealing. Thus, owing to the development of the spin tracing technique we now have a rather clear picture of how interfaces look in heterostructures – information that is essential for the understanding of many of the electronic and opto-electronic properties of semiconductors multilayer systems, such as quantum confinement of excitons, the profile of their wave function, and their linewidth characterisitics.

4.8.3 General Remarks on Spin Tracing

In closing the discussion of spin tracing, it is essential to emphasize that the success of this approach arises from the unique properties by which the location of Mn atoms can be ‘flagged’ due to the Zeeman splitting which they produce; but the conclusions obtained by this ingenious approach apply equally well to other atoms, magnetic or nonmagnetic. For completeness, one should also remark here on two other features that characterize interfaces, not discussed in Sects. 4.8.1 and 4.8.2. First, we had argued that at concentrations x above about 0.15, the magnetization of $II_{1-x}Mn_xVI$ alloys (and thus the Zeeman splitting of excitons, of interest in spin tracing) decreases with increasing x due to antiferromagnetic interactions of the Mn ion with other Mn ions that surround it. This automatically implies that at interfaces between, e.g., CdTe and $Cd_{1-x}Mn_xTe$ with a large value of x, the contribution to the Zeeman splitting of the *intrinsic interface* would be larger than what it is in the $Cd_{1-x}Mn_xTe$ interior, because Mn ions in the interface layer have fewer Mn neighbors on one side. This is illustrated by the dotted curve in Fig. 4.30, as calculated by Fatah et al. [116] and should be considered. These intrinsic interface effects have been investigated as part of the spin tracing studies carried out in [112] and [114], and were found to be small compared to the effects discussed in Sects. 4.8.1 and 4.8.2.

Similarly, the unavoidable effects of interface roughness [122], while they do affect the exciton linewidth, were found in these same studies to make a negligible contribution to the Zeeman splitting of excitons in DMS/non-DMS quantum well structures compared to the interface mixing (atomic segregation) and inter-diffusion effects. We refer the interested reader to [112] and [114] for detailed modeling of these two issues.

4.9 Spin-polarized Devices Based on Band-offset Tuning

Recently, the field of spin-based electronics (or 'spintronics') has been growing in practical importance. Fundamental studies in this area focus on controlling the spin of the electron, with eye on making useful devices based on manipulating its population, orientation, or phase in solid state systems [123]. In this context, the investigations of spin transport and spin relaxation in electronic materials have recently acquired intense interest of the research community. Typical questions arising in the pursuit of these goals involve the identification of effective ways to polarize the spin system. Here, the generation of a spin-polarized condition usually means creating a nonequilibrium spin population (e.g., by spin injection). Although in general this can be achieved in several ways, for device applications the most desirable approach is to obtain spin injection by electrical means. Recently, electrical spin injection devices have been realized using II-Mn-VI DMS materials as spin aligners, with injection efficiencies of spin-polarized carriers into nonmagnetic semiconductor devices exceeding 50% [124, 125].

4.9.1 Spin-Polarized Light-Emitting Diodes

As noted above, experiments using II-Mn-VI DMSs as spin aligners, including CdMnTe [126], BeMnZnSe [124], and ZnMnSe [125], have demonstrated efficient electrical injection of spin-polarized carriers into semiconductor heterostructures. For example, carriers with impressive net spin polarizations have been electrically injected from a DMS contact into semiconductor light-emitting diode (LED) structures, where radiative recombination of spin-polarized carriers results in the emission of σ^+ or σ^- polarized light normal to the LED surface, forming in effect a 'spin-LED'. The observation of circularly polarized electroluminescence from such structures thus demonstrates successful electrical spin injection.

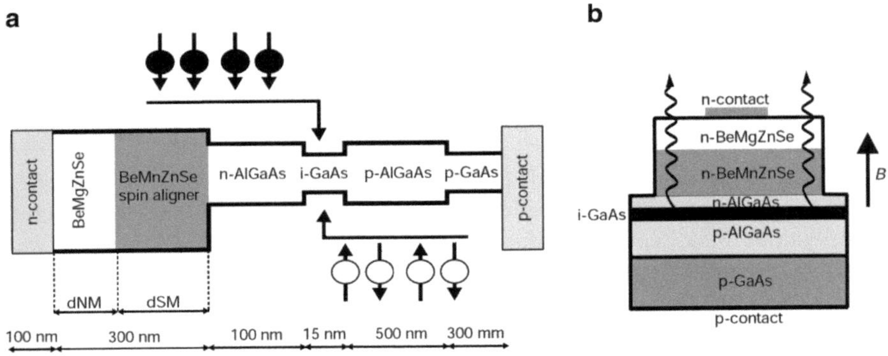

Fig. 4.32 (**a**) Schematic band structure of the spin-aligner and the adjacent GaAs-based light-emitting diode ('spin-LED'). Spin-polarized electrons are injected from the left into the active GaAs layer, unpolarized holes from the right. (**b**) Side view of the spin-LED device, showing directions of the magnetic field and of emitted light (after Fiederling et al., [124]). Reprinted by permission from Macmillan Publishers Ltd: Nature, [124], © (1999)

Figure 4.32 represents a spin-LED structure for demonstrating electrical spin-injection into GaAs reported by Fiederling et al. in [124]. The structure consists of a *p-i-n* diode with a 15-nm-wide GaAs layer embedded in $Al_{0.03}Ga_{0.97}As$ barriers. The top *n*-contacts consist of the DMS $Be_{0.07}Mn_{0.03}Zn_{0.90}Se$ with thickness d_{SM} and a nonmagnetic BeMgZnSe layer of thickness d_{NM}. The total thickness of the II–VI top layer ($d_{SM} + d_{NM} = 300$ nm) is kept constant to maintain the same overall injection conditions. In the top *n*-type DMS layer (n-$Be_xMn_yZn_{1-x-y}Se$) at low Mn concentration and at low temperatures, the *sp–d* exchange interaction results in a Zeeman splitting of the band edges when a small external magnetic field is applied. This splitting aligns the spins of all injected carriers to the energetically favored lower Zeeman level. As a result, electrons entering from the n contact are almost completely polarized in the spin-down state as they leave the spin aligner and are injected across the II-Mn-VI/AlGaAs interface. The degree of spin polarization of the injected electrons naturally depends on the thickness of the magnetic layer d_{SM}. The electrons then travel (by drift and diffusion) to the intrinsic GaAs quantum well, where they recombine with the unpolarized holes injected from the bottom *p*-GaAs substrate across the p-$Al_{0.03}Ga_{0.97}As$ barrier shown on the right in Fig. 4.32. To minimize the dephasing and scattering of the injected carrier spin, the distance between the injecting $Be_xMn_yZn_{1-x-y}Se$ layer and the GaAs layer is kept small (100 nm). One should note that the left-hand-side $Al_{0.03}Ga_{0.97}As$ spacer is only lightly doped (10^{16} cm^{-3}) to maximize the preservation of electron spin over as long a distance as possible [127].

In Fig. 4.33, Fiederling et al. [124] show the degree of circular polarization P_{opt} of electroluminescence under forward bias as a function of magnetic field. A strongly polarized electroluminescence is detected, with a maximum P_{opt} of 43% for the structure with thick magnetic injection layer (squares in Fig. 4.33). P_{opt} increases with magnetic field until it saturates for fields above 3.0 T. This

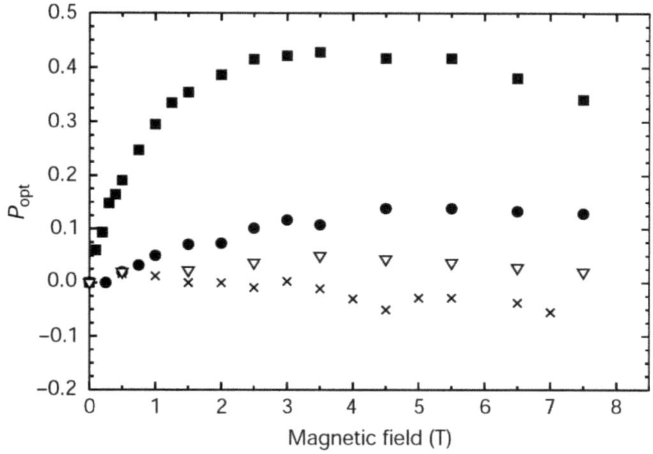

Fig. 4.33 Degree of circular polarization of the electroluminescence emitted by a series of LEDs. The results are shown as a function of the magnetic field for electron-injection across n-$Be_xMn_yZn_{1-x-y}Se$ spin-aligners with $d_{SM} = 300$ nm (*squares*), $d_{SM} = 3$ nm (*circles*); and for injection across a nonpolarizing n-$Be_xMg_yZn_{1-x-y}Se$ electrode ($d_{NM} = 300$ nm, $d_{SM} = 0$ nm, *triangles*). *Crosses* represent experimental data for the intrinsic degree of polarization of the GaAs layer (after Fiederling et al., [124]). Reprinted by permission from Macmillan Publishers Ltd: Nature, [124], © (1999)

behavior reflects the degree of spin-polarization of the injected carriers achieved in the $Be_xMn_yZn_{1-x-y}Se$ spin-aligner, where the Zeeman splitting follows a typical Brillouin function, and saturates when all carrier spins are aligned along the direction of the external magnetic field. In addition, Fiederling et al. also fabricated structures with magnetic injectors of reduced thickness $d_{SM} = 3$ nm as well as structures with a fully nonmagnetic injector ($d_{NM} = 300$ nm). The structures with $d_{SM} = 3$ nm show a reduced degree of optical polarization of 11% at high magnetic fields (circles in Fig. 4.33), indicating that the layer is too thin to achieve complete relaxation of all spins into the lower Zeeman level. The structure with a nonmagnetic injector ($Be_xMg_yZn_{1-x-y}Se$) shows almost no polarization of the electroluminescence (triangles in Fig. 4.33). Fiederling et al. have also measured the degree of optical polarization of the emission from the GaAs layer (crosses in Fig. 4.33) obtained by photoexcitation (using unpolarized light) in the AlGaAs barrier layers, below the energy gap of the spin-aligner DMS layers. The results of this experiment show a degree of photoluminescence polarization that is very small *and opposite in sign* to that obtained for the experiment with spin-polarized electron injection, thus ruling out any contribution of field-induced optical dichroism in the electroluminescence of the diode. This behavior can be fully attributed to the intrinsic g-factors of the GaAs layer. In the electroluminescence of the diodes, the intrinsic polarization of the GaAs layer becomes important only when the degree of polarization of the spin-injected electrons saturates. This is in fact seen in the form of decreasing P_{opt} observed for the electroluminescence at high magnetic fields (squares in Fig. 4.33).

The polarized electroluminescence emitted by the spin-LEDs observed by Fiederling et al. provides clear evidence of injection of spin-polarized carriers into the GaAs-based quantum well structures. Similar results using spin-LEDs based on ZnMnSe/GaAs have also been demonstrated by other groups, in which electron spin polarizations of 50%–85% were directly measured [125, 128, 129]. Regardless of any differences between the various structures used in the different laboratories these results have invariably demonstrated robust low-temperature spin injection from a DMS spin aligner into a non-magnetic region of the device.

4.9.2 DMS-based Resonant Tunneling Diodes

As demonstrated in the preceding section, spin injection can be realized by transferring the majority spin from a magnetic material (a DMS spin aligner [124, 130] or a metallic ferromagnetic metal contact [131]) into a nonmagnetic layer structure. In these structures, the spin injection directly depends on an externally applied magnetic field used for switching the magnetization of the magnetic material. For efficient electronic device, however, it is desirable to use systems where the polarization of the injected spin can be controlled by an applied voltage. In this regard, magnetic resonant tunneling diodes (RTDs) based on DMS materials have been proposed and realized for voltage-controlled spin polarized injection and detection [132, 133].

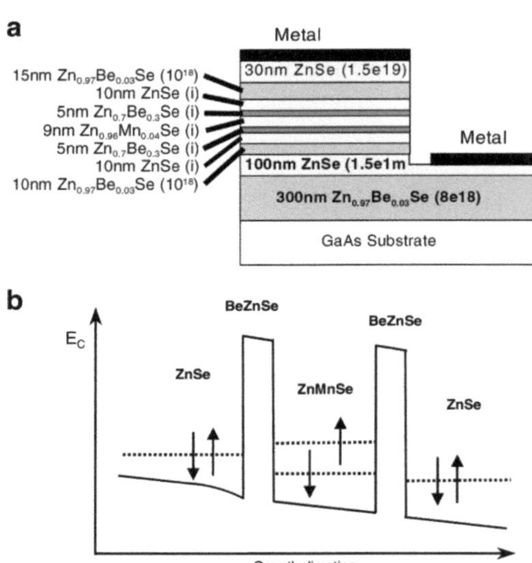

Fig. 4.34 (**a**) Structure of the DMS-based resonant tunnel diode (RTD) device; and (**b**) schematic view of the RTD band structure under finite bias (after Slobodskyy et al. [133]). Reprinted with permission from [133], © (2003) by the American Physical Society

Slobodskyy et al. investigated an all-II–VI semiconductor RTD based on ZnBeSe and a ZnMnSe DMS quantum well [133]. The complete structure of this type is shown in Fig. 4.34. The active region of the device consists of a 9 nm thick undoped $Zn_{0.96}Mn_{0.04}Se$ quantum well, sandwiched between two 5 nm thick undoped $Zn_{0.97}Be_{0.03}Se$ barriers. A schematic of the potential energy profile of the double-barrier structure under bias is also shown in Fig. 4.34. Slobodskyy et al. have observed typical RTD-like current-voltage ($I - V$) characteristics in such heterostructures with peak-to-valley ratios of over 2.5–1. When a magnetic field is applied to DMS-based RTDs of this type, these authors observed that the transmission resonance of the $I - V$ curve splits into two peaks, with a splitting corresponding to the separation of the energy levels in the ZnMnSe well. The $I - V$ characteristics of such DMS-based RTDs were measured by Slobodskyy et al. in magnetic fields ranging from 0 to 6.0 T, applied either perpendicular to or in the plane of the quantum well. The results observed in the perpendicular geometry are shown in Fig. 4.35 (lines). For clarity, curves corresponding to consecutive fields are offset by 10 μA. It is clear from the figure that the resonance peak seen a $B = 0$ is split into two peaks, and that the splitting grows as a function of the field. At 6.0 T, the separation between the split maxima is 36.5 and 42 mV for DMS layers with 4% and 8% Mn, respectively.

To explain the magnetic-field-induced behavior of the resonance, Slobodskyy et al. developed a model based on the giant Zeeman splitting of the spin-split levels in the DMS quantum well, as follows. First, they obtained the series contact resistance of the RTD from the measured $I - V$ curve at zero-magnetic field. They then

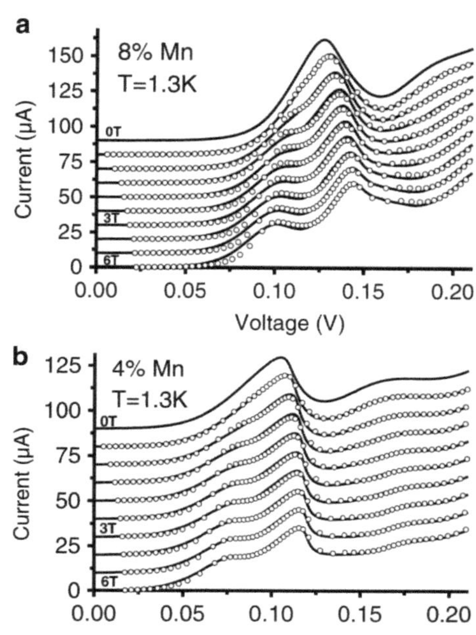

Fig. 4.35 Experimental (*solid curves*) and modeled (*circles*) $I - V$ characteristics for a resonant tunnel diode with (**a**) $Zn_{0.92}Mn_{0.08}Se$ and (**b**) $Zn_{0.96}Mn_{0.04}Se$ in the DMS quantum well. The successive curves are taken in 0.5 T intervals from 0 to 3.0 T and in 1.0 T intervals between 3.0 and 6.0 T, and are vertically displaced for clarity (after Slobodskyy et al. [133]). Reprinted with permission from [133], © (2003) by the American Physical Society

assumed that each of the two spin-split levels has the same conductance, and that each therefore transmits half of the total current across the device. In the presence of a magnetic field, the conductivity associated with each spin-polarized level as a function of applied voltage can then be given by a simple translation of the zero-field curve by the voltage corresponding to the energy shift of the spin-up and spin-down states in the well, respectively. After this translation, the authors added the conductivity contributions of the spin-up and spin-down curves, obtaining a modeled $I - V$ curve for each magnetic field. Figure 4.35 shows the modeled $I - V$ curves as the circles, which compare very well with the experimental data shown as the solid curves in the figure.

In addition, the data described above also allowed Slobodskyy et al. to extract the voltage splitting of the levels ΔV as a function of magnetic field. The authors found remarkable agreement between the magnetic field dependence of the experimental values ΔV and the Brillouin function of the spin level splitting δE characteristic of DMS materials, suggesting that their spin splitting model captures the essential properties of the device.

Acknowledgements

We thank our numerous co-workers and co-authors (see References) for their stimulating and fruitful discussions, and their very significant contributions to many aspects of DMSs reviewed in this article.

The contributors are grateful to the Authors and Publishers for permission to reproduce figures that appear in this chapter.

References

1. R.R. Galazka, J. Kossut, *Lecture Notes in Physics*, vol 132 (Springer, Berlin, 1980), p. 245
2. J. Mycielski, in *Recent Developments in Condensed Matter Physics*, ed. by J.T. Devreese (Plenum, New York, 1981), p. 257
3. N.B. Brandt, V.V. Moshchalkov, Adv. Phys. **33**, 193 (1984)
4. O. Goede, W. Heimbrodt, Phys. Status Solidi B **146**, 11 (1988)
5. J.K. Furdyna, J. Appl. Phys. **64**, R29, (1988)
6. J.A. Gaj, in *Diluted Magnetic Semiconductors*, ed. by J.K. Furdyna, J. Kossut. Semiconductors and Semimetals, vol 25, chap. 7 (Academic, Boston, 1988)
7. T. Dietl, *Diluted Magnetic Semiconductors*, in *Handbook of Semiconductors*, ed. by S. Mahajan, vol. 3B, (North Holland, Amsterdam, 1994)
8. J.K. Furdyna, J. Vac. Sci. Technol. A **4**, 2002 (1986)
9. J.A. Gaj, J. Ginter, R.R. Galazka, Phys. Status Solidi **89**, 655 (1978)
10. N.T. Khoi, J. Ginter, A. Twardowski, Phys. Status Solidi B **117**, 67 (1983)
11. A. Twardowski, M. Nawrocki, J. Ginter, Phys. Status Solidi B **96**, 497 (1979)
12. G. Rebman, C. Rigaux, G. Bastard, M. Menant, R. Triboulet, W. Giriat, Physica **117B & 118B**, 452 (1983)
13. A. Twardowski, P. Swiderski, M. von Ortenberg, R. Pauthenet, Solid State Commun. **50**, 509 (1984)

14. A. Twardowski, T. Dietl, M. Demianiuk, Solid State Commun. **48**, 845 (1983)
15. A.V. Komarov, S.M. Ryabchenko, O.V. Terletskii, Phys. Status Solidi B **102**, 603 (1980)
16. A.V. Komarov, S.M. Ryabchenko, Y.G. Semenov, B.D. Shanina, N.I. Vitrikhovskii, Zh. Eksp. Teor Fiz. **79**, 1554 (1980) [Sov. Phys. JETP **52**, 783 (1980)]
17. R.L. Aggarwal, S.N. Jasperson, J. Stankiewicz, Y. Shapira, S. Foner, B. Khazai, A. Wold, Phys. Rev. B **28**, 6907 (1983)
18. R.L. Aggarwal, S.N. Jasperson, P. Becla, J.K. Furdyna, Phys. Rev. B **34**, 5894 (1986)
19. S. Lee, M. Dobrowolska, J.K. Furdyna, H. Luo, L.R. Ram-Mohan, Phys. Rev. B **54,** 16939 (1996)
20. W.Y. Yu, A. Twardowski, L.P. Fu, A. Petrou, B.T. Jonker, Phys. Rev. B **51**, 9722 (1995)
21. E.R. Glaser, R. Magno, B.V. Shanabrook, J.G. Tischler, Phys. Rev. B **74**, 235306 (2006)
22. J.-J. Chen, B.P. Gila, M. Hlad, A. Gerger, F. Ren, C.R. Abernathy, S.J. Pearton, Appl. Phys. Lett. **88**, 042113 (2006)
23. M. Galluppi, L. Geelhaar, H. Riechert, M. Hetterich, A. Grau, S. Birner, W. Stolz, Phys. Rev. B **72**, 155324 (2005)
24. M. Muñoz, H. Lu, X. Zhou, M.C. Tamargo, F.H. Pollak, Appl. Phys. Lett. **83**, 1995 (2003)
25. Q.S. Zhu, S.M. Mou, X.C. Zhou, Z.T. Zhong, Appl. Phys. Lett. **62**, 2813 (1993)
26. J. Siviniant, F.V. Kyrychenko, Y.G. Semenov, D. Coquillat, D. Scalbert, J.P. Lascaray, Phys. Rev. B **59**, 10276 (1999)
27. P. Chen, J.E. Nicholls, J.H.C. Hogg, T. Stirner, W.E. Hagston, B. Lunn, D.E. Ashenford, Phys. Rev. B **52**, 4732 (1995)
28. P. Harrison, T. Stirner, S.J. Weston, S.R. Bardorf, S. Jackson, W.E. Hagston, J.H.C. Hogg, J.E. Nicholls, M. O'Neill, Phys. Rev. B **51**, 5477 (1995)
29. B. Kuhn-Heinrich, W. Ossau, T. Litz, A. Waag, G. Landwehr, J. Appl. Phys. **75**, 8046 (1994)
30. S.-K. Chang, A.V. Nurmikko, J.-W. Wu, L.A. Kolodziejski, R.L. Gunshor, Phys. Rev. B **37**, 1191 (1988)
31. P.J. Klar, D. Wolverson, J.J. Davies, W. Heimbrodt, M. Happ, Phys. Rev. B **57**, 7103 (1998)
32. M. Kim, C.S. Kim, S. Lee, J.K. Furdyna, M. Dobrowolska, J. Cryst. Growth **214/215**, 325 (2000)
33. N. Dai, L.R. Ram-Mohan, H. Luo, G.L. Yang, F.C. Zhang, M. Dobrowolska, J.K. Furdyna, Phys. Rev. B **50**, 18153 (1994)
34. E.O. Kane, J. Phys. Chem. Solids **1**, 294 (1973)
35. L.R. Ram-Mohan, K.H. Yoo, R.L. Aggarwal, Phys. Rev. B **38**, 6151 (1988)
36. A. Wasiela, P. Peyla, Y. Merle d'Aubigne, J.E. Nicholls, D.E. Ashenford, B. Lunn, Semicon. Sci. Technol. **7**, 571 (1992)
37. A.C. Gossard, R.C. Miller, W. Wiegmann, Surf. Sci. **174**, 131 (1986)
38. T. Wojtowicz, M. Kutrowski, M. Surma, P. Kossacki, Nguyen The Khoi, K. Kopalko, G. Karczewski, J. Kossut, M. Godlewski, Appl. Phys. Lett. **68**, 3326 (1996)
39. T. Wojtowicz, G. Karczewski, J. Kossut, Thin Solid Films **306**, 271 (1997)
40. T. Wojtowicz, M. Kutrowski, G. Cywiński, E. Dynowska, G. Karczewski, J. Kossut, R. Fiederling, G. Mackh, U. Zehnder, W. Ossau, Acta Phys. Polon. A **90**, 977 (1996)
41. T. Wojtowicz, M. Kutrowski, G. Karczewski, J. Kossut, Acta Phys. Polon. A **94**, 199 (1998)
42. M. Kutrowski, T. Wojtowicz, G. Cywiński, G. Karczewski, E. Janik, E. Dynowska, J. Kossut, P. Kossacki, R. Fiederling, A. Pfeuffer-Jeschke, W. Ossau, Acta Phys. Polon. A **92**, 887 (1997)
43. T. Wojtowicz, M. Kutrowski, G. Cywiński, G. Karczewski, E. Janik, E. Dynowska, J. Kossut, R. Fiederling, A. Pfeuffer-Jeschke, W. Ossau, J. Cryst. Growth **184/185**, 936 (1998)
44. W. Pötz, D.K. Ferry, Phys. Rev. B **32**, 3863 (1985)
45. W. Ossau, R. Fiederling, B. König, T. Wojtowicz, M. Kutrowski, G. Karczewski, J. Kossut, Physica E **2**, 209 (1998)
46. F. Kyrychenko, J. Kossut, Acta Phys. Polon. A **94**, 406 (1998)
47. F. Kyrychenko, J. Kossut, Sem. Sci. Technol. **13**, 1076 (1998)
48. H. Mathieu, P. Lefebvre, P. Christol, Phys. Rev. B **46**, 4092 (1992)
49. T.M. Rusin, J. Phys: Condens. Matter **12**, 576 (2000)
50. I.A. Merkulov, D.R. Yakovlev, A. Keller, W. Ossau, J. Geurts, A. Waag, G. Landwehr, G. Karczewski, T. Wojtowicz, J. Kossut, Phys. Rev. Lett. **83**, 1431 (1999)

51. B. König, I.A. Merkulov, D.R. Yakovlev, W. Ossau, S.M. Ryabchenko, M. Kutrowski, T. Wojtowicz, G. Karczewski, J. Kossut, Phys. Rev. B **61**, 16870 (2000)
52. W.P. Shen, M.L. Rustugi, J. Appl. Phys. **74**, 4006 (1993)
53. R.C. Myers, K.C. Ku, X. Li, N. Samarth, D.D. Awschalom, Phys. Rev. B **72**, 041302 (2005)
54. T. Kamizato, M. Matsuura, Phys. Rev. B **40**, 8378 (1989)
55. R. Dingle, A.C. Gossard, W. Wiegmann, Phys. Rev. Lett. **34,** 1327 (1975)
56. J.F. Smyth, D.D. Awschalom, N. Samarth, H. Luo, J.K. Furdyna, Phys. Rev. B **46,** 4340 (1992)
57. H. Kawai, J. Kaneko, N. Watanabe, J. Appl. Phys. **58**,1263 (1985)
58. C.C. Phillips, R. Eccleston, S.R. Andrews, Phys. Rev. B **40**, 9760 (1989)
59. A. Alexandrou, J.A. Kash, E.E. Mendez, M. Zachau, J.M. Hong, T. Fukuzawa, Y. Hase, Phys. Rev. B **42**, 9225 (1990)
60. L.V. Butov, A. Zrenner, G. Abstreiter, A.V. Petinova, K. Eberl, Phys. Rev. B **52**, 12153 (1995)
61. Y.J. Chen, E.S. Koteles, B.S. Elman, C.A. Armiento, Phys. Rev. B **36**, 4562 (1987)
62. G.W. Bryant, Phys. Rev. B **47**, 1683 (1993)
63. M.M. Dignam, J.E. Sipe, Phys. Rev. B **43**, 4084 (1991)
64. M. Bayer, V.B. Timofeev, F. Faller, T. Gutbrod, A. Forchel, Phys. Rev. B **54**, 8799 (1996)
65. S. Lee, G. Yang, M. Dobrowolska, J.K. Furdyna, J. Korean Phys. Soc. **39**, 447 (2001)
66. A.B. Dzyubenko, A.L. Yablonskii, Phys. Rev. B **53**, 16355 (1996)
67. S. Gasiorowicz, *Quantum Physics* (Wiley, New York, 1984), p. 85
68. S. Lee, M. Dobrowolska, J.K. Furdyna, L.R. Ram-Mohan, Phys. Rev. B **59**, 10302 (1999)
69. S. Lee, M. Dobrowolska, J.K. Furdyna, L.R. Ram-Mohan, Phys. Rev. B **61**, 2120 (2000)
70. L.R. Ram-Mohan, S. Saigal, D. Dossa, J. Shertzer, Comput. Phys. **4**, 50 (1990)
71. F.C. Zhang, N. Dai, H. Luo, N. Samarth, M. Dobrowolska, J.K. Furdyna, L.R. Ram-Mohan, Phys. Rev. Lett. **68**, 3220 (1992)
72. H. Luo, N. Dai, F.C. Zhang, N. Samarth, M. Dobrowolska, J.K. Furdyna, C. Parks, A.K. Ramdas, Phys. Rev. Lett. **70**, 1307 (1993)
73. S. Lee, M. Dobrowolska, J.K. Furdyna, L.R. Ram-Mohan, Opt. Mater. **23**, 79 (2003)
74. G. Prechtl, W. Heiss, A. Bonnani, W. Jantsch, S. Maćkowski, E. Janik, G. Karczewski, Phys. Rev. B **61**, 15617 (2000)
75. G. Prechtl, W. Heiss, A. Bonnani, W. Jantsch, S. Maćkowski, E. Janik, Phys. Rev. B **68**, 165313 (2003)
76. Ł. Kłopotowski, A. Gruszczyńska, E. Janik, M. Wiater, P. Kossacki, G. Karczewski, T. Wojtowicz, Phys. Rev. B **77**, 235312 (2008)
77. B.T. Jonker, X. Liu, W.C. Chou, A. Petrou, J. Warnock, J.J. Krebs, G.A. Prinz, J. Appl. Phys. **69**, 6097 (1991)
78. N. Dai, H. Luo, F.C. Zhang, N. Samarth, M. Dobrowolska, J.K. Furdyna, Phys. Rev. Lett. **67**, 3824 (1991)
79. W.C. Chou, A. Petrou, J. Warnock, B.T. Jonker, Phys. Rev. Lett. **67**, 3820 (1991)
80. R.B. Bylsma, W.M. Becker, J. Kossut, U. Debska, D. Yoder-Short, Phys. Rev. B **33**, 8207 (1986)
81. M. Dobrowolska, J.K. Furdyna, H. Luo in *The Encyclopedia of Advanced Materials*, eds. by D. Bloor, M.C. Flemings, R.J. Brook, S. Mahajan (Pergamon, Oxford, 1994), p. 2790
82. R.L. Gunshor, L.A. Kolodziejski, A.V. Nurmikko, N. Otsuka, Ann. Rev. Mater. Sci. **18**, 325 (1988)
83. X. Liu, A. Petrou, J. Warnock, B.T. Jonker, G.A. Prinz, J.J. Krebs, Phys. Rev. Lett. **63**, 2280 (1989)
84. E. Deleporte, J.M. Berroir, G. Bastard, C. Delalande, J.M. Hong, L.L. Chang, Phys. Rev. B **42**, 5891 (1990)
85. K. Leo, J. Shah, O. Göbel, T.C. Damen, S. Schmitt-Rink, W. Schäfer, Phys. Rev. Lett. **66**, 201 (1991)
86. N. Debbar, S. Hong, J. Singh, P. Bhattacharya, R. Sahai, J. Appl. Phys. **65**, 383 (1988)
87. R.G. Alonso, Eunsoon Oh, A.K. Ramdas, H. Luo, N. Samarth, J.K. Furdyna, L.R. Ram-Mohan, Phys. Rev. B **44**, 8009 (1991)
88. J.E. Zucker, A. Pinchuk, D.S. Chemla, A. Gossard, W. Wiegmann, Phys. Rev. B **29**, 7065 (1984)

89. J.J. Song, Y.S. Yoon, A. Fedotowsky, Y.B. Kim, J.N. Schulman, C.W. Tu, D. Huang, H. Morkoc, Phys. Rev. B **34**, 8958 (1986)
90. J.K. Furdyna, Solid State Electron. **37**, 1065 (1994)
91. F.C. Zhang, H. Luo, N. Dai, N. Samarth, M. Dobrowolska, J.K. Furdyna, Phys. Rev. B **47**, 3806 (1993)
92. H. Luo, W.C. Chou, N. Samarth, A. Petrou, J.K. Furdyna, Solid State Commun. **85**, 691 (1993)
93. D. Loss, D.P. DiVincenzo, Phys. Rev. A **57**, 120 (1998)
94. S. Itoh, S. Muto, Y. Ebiko, H. Sasakura, K. Shiramine, Jpn. J. Appl. Phys. **38**, L917 (1999)
95. A. Takeuchi, T. Kuroda, Y. Nakata, M. Murayama, T. Kitamura, N. Yokoyama, Jpn. J. Appl. Phys. **4**(2), 4278 (2003)
96. S. Lee, S.-R. Eric Yang, M. Dobrowolska, J.K. Furdyna, Semicond. Sci. Technol. **19**, 1125 (2004)
97. E. Oh, K.J. Yee, S.M. Soh, J.U. Lee, J.C. Woo, H.S. Jeon, D.S. Kim, S. Lee, J.K. Furdyna, H.C. Ri, H.S. Chany, S.H. Park, Appl. Phys. Lett. **83**, 4604 (2003)
98. W. Heller, U. Bockelmann, Phys. Rev. B **55**, R4871 (1997)
99. K. Chang, J.B. Xia, F.M. Peeters, Appl. Phys. Lett. **82**, 2661 (2004)
100. S. Lee, M. Dobrowolska, J.K. Furdyna, Phys. Stat. Sol. (B) **243**, 799 (2006)
101. S. Lee, M. Dobrowolska, J.K. Furdyna, Appl. Phys. Lett. **86**, 33114 (2005)
102. S. Lee, M. Dobrowolska, J.K. Furdyna, Phys. Rev. B **72**, 075320 (2005)
103. S. Lee, M. Dobrowolska, J.K. Furdyna, J. Appl. Phys. **99**, 08F702 (2006)
104. S. Lee, K. Park, M. Dobrowolska, J.K. Furdyna, J. Appl. Phys. **95**, 7184 (2004)
105. S. Lee, J.K. Furdyna, M. Dobrowolska, Phys. Stat. Sol. (C) **2**, 1085 (2005)
106. W.M. Chen, I.A. Buyanova, G.Yu. Rudko, A.G. Mal' Shukov, K.A. Chao, Phys. Rev. B **67**, 125313 (2003)
107. M. Qazzaz, G. Yang, S.H. Xin, L. Montes, H. Luo, J.K. Furdyna, Solid State Commun. **96**, 405 (1995)
108. J.K. Furdyna, M. Qazzaz, G. Yang, L. Montes, S.H. Xin, H. Luo, Acta Physica Polonica A **88**, 607 (1995)
109. J. Bao, A.V. Bragas, J.K. Furdyna, R. Merlin, Nat. Mater. **2**, 175 (2003)
110. J.M. Bao, A.V. Bragas, J.K. Furdyna, R. Merlin, Solid State Commun. **127**, 771 (2003)
111. J.M. Bao, A.V. Bragas, J.K. Furdyna, R. Merlin, Phys. Rev. B **71**, 045314 (2005)
112. J.A. Gaj, W. Grieshaber, C. Bodin-Deshayes, J. Cibert, G. Feuillet, Y. Merle d'Aubigné, A. Wasiela, Phys. Rev. B **50**, 5512 (1994)
113. W. Grieshaber, J. Cibert, J.A. Gaj, Y. Merle d'Aubigné, A. Wasiela, Phys. Rev. B **50**, 2011 (1994)
114. W. Grieshaber, A. Haury, J. Cibert, Y. Merle d'Aubigné, A. Wasiela, J.A. Gaj, Phys. Rev. B **53**, 4891 (1996)
115. P. Kossacki, N.T. Khoi, J.A. Gaj, G. Karczewski, T. Wojtowicz, J. Kossut, K.V. Rao, Solid State Commun. **94**, 439 (1995)
116. J.M. Fatah, T. Piorek, P. Harrison, T. Stirner, W.E. Hagston, Phys. Rev. B **49**, 10341 (1994)
117. H. Mariette, L. Besombes, C. Bougerol, D. Ferrand, Y. Leger, L. Maingault, J. Cibert, Phys. Stat. Sol. **243**, 3709 (2006)
118. L. Maingault, L. Besombes, Y. Léger, C. Bougerol, H. Mariette, Appl. Phys. Lett. **89**, 193109 (2006)
119. S. Mackowski, N.T. Khoi, P. Kossacki, P.A. Golnik, J.A. Gaj, A. Lemaitre, C. Testelin, C. Rigaux, G. Karczewski, T. Wojtowicz, J. Kossut, J. Cryst. Growth **184/185**, 966 (1998)
120. S. Mackowski, J.A. Gaj, Defect and Diffusion Forum **174**, 47 (1999)
121. D. Tönnies, G. Bacher, A. Forchel, A. Waag, G. Landwehr, Appl. Phys. Lett. **64**, 766 (1994)
122. W. Grieshaber, C. Bodin, J. Cibert, J. Gaj, Y. Merle d'Aaubigné, A. Wasiela, G. Feuillet, Appl. Phys. Lett. **65**, 1287 (1994)
123. I. Zutic, J. Fabian, S. Das Sarma, Rev. Mod. Phys. **76**, 323 (2004)
124. R. Fiederling, M. Keim, G. Reuscher, W. Ossau, G. Schmidt, A. Waag, L.W. Molenkamp, Nature (Lond.) **402**, 787 (1999)

125. B.T. Jonker, Y.D. Park, B.R. Bennett, H.D. Cheong, G. Kioseoglou, A. Petrou, Phys. Rev. B **62**, 8180 (2000)
126. M. Oestreich, J. Hübner, D. Hägele, P.J. Klar, W. Heimbrodt, W.W. Rühle, D.E. Ashenford, B. Lunn, Appl. Phys. Lett. **74**, 1251 (1999)
127. D.D. Awschalom, J.M. Kikkawa, Phys. Today **6**, 33 (1999)
128. Y.D. Park, B.T. Jonker, B.R. Bennett, G. Itskos, M. Furis, G. Kioseoglou, A. Petrou, Appl. Phys. Lett. **77**, 3989 (2000)
129. R.M. Stroud, A.T. Hanbicki, Y.D. Park, A.G. Petukhov, B.T. Jonker, Phys. Rev. Lett. **89**, 166602 (2002)
130. Y. Ohno, D.K. Young, B. Beschoten, F. Matsukura, H. Ohno, D.D. Awschalom, Nature (Lond.) **402**, 790 (1999)
131. P.R. Hammar, M. Johnson, Phys. Rev. Lett. **88**, 066806 (2002)
132. Th. Gruber, M. Keim, R. Fiederling, G. Reuscher, W. Ossau, G. Schmidt, L.W. Molenkamp, A. Waag, Appl. Phys. Lett. **78**, 1101 (2001)
133. A. Slobodskyy, C. Gould, T. Slobodskyy, C.R. Becker, G. Schmidt, L.W. Molenkamp, Phys. Rev. Lett. **90**, 246601 (2003)

Chapter 5
Diluted Magnetic Quantum Dots

F. Henneberger and J. Puls

Abstract Epitaxial quantum dots either made of DMS or placed in an environment made of DMS are subjects of this chapter. Emphasis is put on (Cd,Mn)Se- and (Cd,Mn)Te-based heterostructures. The growth by self-assembly is discussed. Dynamic aspects of the interaction between carriers and magnetic ions are reviewed and a close analogy between DMS quantum dots and bound magnetic polarons is pointed out. Single dot spectroscopy and, in particular, spectroscopy of a single dot containing a single magnetic ion is described and discussed.

5.1 Introduction

This chapter summarizes the electronic, optical, and magnetic properties of zero-dimensional diluted magnetic semiconductor structures. Key aspects are the self-assembled growth of such structures, the modification of the confined exciton states by the presence of magnetic ions, the specific dynamical scenarios of the carrier–ion interaction, including magneto-polaron formation as well as spin-lattice relaxation, up to the limit where a single magnetic ion is coupled to the discrete few-particle state of a single quantum dot.

Quantum dots (QDs) are semiconductor structures where the carriers are confined in all three space dimensions. Their discrete energy spectrum, tunable in a wide range by structure design, in combination with the ability to embed QDs in a practical device environment, predestines these nanostructures for various applications in opto-electronics, spintronics, and quantum information processing. In this context, DMS QD structures represent an interesting research field. Strong localization of the carriers means increased overlap of their wavefunction with the moments of the magnetic ions and hence enhanced coupling. Localization plays also a role in donor or acceptor states of bulk DMS or in DMS quantum wells. However, the localization

F. Henneberger (✉) and J. Puls
Institute of Physics, Humboldt-University, 12489 Berlin, Germany
e-mail: fh@physik.hu-berlin.de, puls@physik.hu-berlin.de

J. Kossut and J.A. Gaj (eds.), *Introduction to the Physics of Diluted Magnetic Semiconductors*, Springer Series in Materials Science 144,
DOI 10.1007/978-3-642-15856-8_5, © Springer-Verlag Berlin Heidelberg 2010

is here weak and not temperature robust. Moreover, QDs are morphologically defined and can be selectively addressed optically or electrically.

This chapter concentrates on self-assembled II–VI QD structures grown by epitaxial techniques. There have been also attempts to dope nanocrystals, mostly chemically synthesized, with magnetic ions. However, these efforts do not seem to have reached a systematic level so far, in particular regarding the magnetic properties. The same holds true for work on III–V materials, where only recently progress has been made. Compared to the long history of bulk DMS, research on their QD structures is a very young field that started only in the beginning of this decade. It is thus only natural that not all findings and conclusion have yet come to a mature state. We have tried to include all work we are aware of to give the reader a comprehensive overview on the current status of this quickly developing branch of DMS research.

5.2 Epitaxial Growth of II–VI Quantum Dot Structures

The self-assemblage of QD structures by heteroepitaxy comprises a variety of morphologies. Localization in quantum well structures on interface roughness or compositional variations are referred to as "natural" QDs. Monolayer (ML) or submonolayer growth provides flat, disc-like islands of mostly kinetically determined shape, size, and density. Modes that work at least close to equilibrium are Volmer-Weber and Stranski-Krastanov growth [1]. While electronic localization is relatively easily attainable, the challenge in heteroepitaxy lies in the growth of well-defined or even tailored structures with high reproducibility.

Unlike their III–V counterparts, the epitaxial growth of II–VI QDs has been quite a puzzle for a while. Though the strain situation is very similar to the standard InAs/GaAs heterosystem, a Stranski-Krastanov transition can not be observed before pseudomorphic growth is lost at a critical thickness of typically 3-4 MLs only. Therefore, other procedures had to be developed. In what follows, we briefly outline these efforts for the molecular-beam epitaxy of (Cd,Mn)Se/ZnSe.

The optimum temperature for layer-by-layer growth of ZnSe is about $T_G = 310°C$. Deposition of CdSe on ZnSe at this temperature results merely in uncontrollable surface inhomogeneities. However, as revealed by in-situ reflection high-energy electron diffraction (RHEED), coherent Frank-van-der-Merwe growth of CdSe is accomplished when the temperature is sufficiently lowered. In a relatively narrow temperature window ($T_G = 210$–$230°C$), prominent intensity oscillations (Fig. 5.1a) appear and the RHEED pattern exhibits sharp diffraction rods from the reconstructed surface. Beyond 3 MLs, the onset of misfit-related plastic deformation is witnessed by a decrease of the distance between the streaks in the RHEED pattern and the disappearance of X-ray interferences. Raising quickly the temperature while capping the film with ZnSe, a quantum well geometry is formed. On the other hand, the as-grown pseudomorphic two-dimensional film undergoes a striking transformation into a QD morphology under postgrowth thermal activation [2]

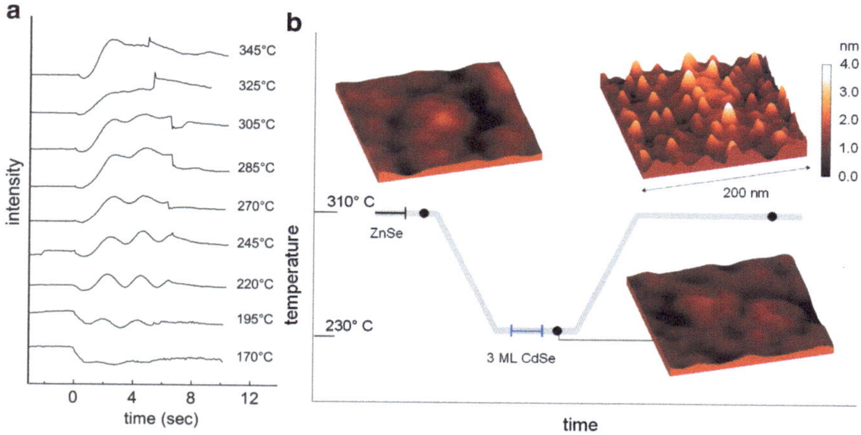

Fig. 5.1 (**a**) Specular beam RHEED intensity oscillations for growth of CdSe on a 1 μm thick strain relaxed ZnSe epilayer at different T_G. (**b**) AFM images revealing the reorganization of a two-dimensional CdSe film into an array of QDs. *Top left*: ZnSe grown at $T_G = 310°C$, *bottom right*: CdSe film grown on *top* at $T_G = 230°C$, *top right*: after thermal activation at growth temperature $T_G = 310°C$. GaAs(001) is used as substrate

as shown by the atomic-force microscopy (AFM) images summarized in Fig. 5.1b. To avoid desorption during activation, the surface is kept Se-rich. The reorganization of the film is also tracked in real-time by RHEED, where the otherwise streaky pattern becomes increasingly superimposed with distinct diffraction spots. Evaluation of the AFM data shows that only a fraction of an ML is reorganized into islands. Transmission electron microscopy (TEM) investigations demonstrate that the Stranski-Krastanov morphology is maintained after ZnSe overgrowth [3]. The islands are situated on top of a continuous wetting layer and exhibit an extended core of pure CdSe (Fig. 5.2a). AFM and TEM yield almost identical island heights (average 1.6 nm), in-plane extensions (below 10 nm), and QD density ($(1.1 \pm 0.3) \cdot 10^{11}\,\mathrm{cm}^{-2}$). Therefore, interdiffusion is not an essential point, neither during activation of the QD morphology nor during overgrowth.

CdSe films of 1 and 2 ML thickness do not undergo the reorganization confirming that elastic energy relaxation is involved in the QD formation. On the other hand, the thermally activated Stranski-Krastanov transition takes place just at critical thickness. Contrary to III–V materials, the interplay between strain relaxation and the formation of misfit dislocations is thus a critical issue for II–VI QDs. This point has been more explicitly elaborated by changing the surface energy through coverage of the two-dimensional film with an amorphous anion layer prior to reorganization. Reducing so the tendency of dislocation formation [5–7], the growth of CdTe/ZnTe Stranski-Krastanov QDs also succeeded by the thermal annealing procedure [5].

AFM is a standard tool to expose the formation of QDs in semiconductor heteroepitaxy. Regarding II–VI materials, caution is required when performing the measurements in ambient air. It has turned out that artifacts appear at the surface,

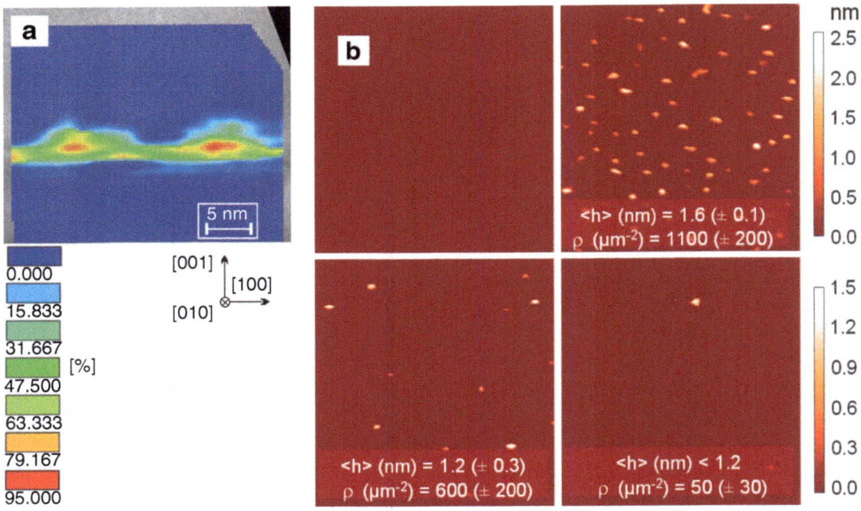

Fig. 5.2 (**a**) Chemical composition of capped ZnSe/CdSe/ZnSe QD structures evaluated by lattice fringe analysis (CELFA). The TEM signal is a 10–20 nm depth average, which reduces only seemingly the Cd concentration (given by the color scale) in the island center slightly below 100%. Reprinted with permission from [3]. Copyright (2002) American Institute of Physics. (**b**) AFM images of 3 ML (Cd,Mn)Se on ZnSe [4]. *Upper left*: two-dimensional surface without annealing ($x_{Mn} = 0.05$), other panels: after reorganization for $x = 0.0$ (*upper right*), $x_{Mn} = 0.05$ (*lower left*), $x_{Mn} = 0.1$ (*lower right*). The images are background corrected by subtracting the long-range roughness obtained within a local-mean-value algorithm. ϱ and $\langle h \rangle$ denote area density and average height, respectively, of the ensemble

which are not related to the intrinsic semiconductor growth. Relatively large islands of typically a few 10 nm extension undergoing rapid ripening are probably caused by oxidation processes. Such islands are absent if the AFM measurements are made under ultra-high-vacuum conditions [8].

The growth procedures described above for CdSe/ZnSe QD structures are maintained also under incorporation of Mn up to concentrations of about 0.10 [4]. The in-plane lattice constant of MnSe is closer to ZnSe so that the mismatch is reduced. In the standard strain-driven scenario, a tendency of forming larger QDs with smaller area density is expected. At variance, the QD size is only very little affected, whereas a dramatic decrease of the dot density occurs (Fig. 5.2b). That is, incorporation of Mn reduces further the small material amount aggregated in the islands. Therefore, strain is not the only factor that drives the QD formation. Surface processes controlled by the diffusibility of the atomic species play an important role. This conclusion is supported by the fact that a 3 ML thick CdSe film loses the capability of the QD transformation when kept a sufficiently long time at growth temperature [2, 3]. A recent study has addressed the kinetics between two-dimensional precursors for nonmagnetic CdSe/ZnSe structures and the mobile surface adatoms in more detail and showed that the QD density can be controlled by delaying the activation step [9].

Further work on the epitaxy of (Cd,Mn)Se- and (Cd,Mn)Te-based QD structures includes growth on Mn-passivated surfaces [10–12], atomic-layer epitaxy [13, 14], and submonolayer deposition [15]. Very diluted (Cd,Mn)Te QDs were fabricated by growing CdTe on ZnTe/(Zn,Mn)Te/ZnTe [16]. Segregation prior to thermal activation of the QD morphology takes Mn^{2+} ions to CdTe. The Mn concentration can be adjusted by the thickness of the ZnTe spacer down to the ultimate limit of a single Mn^{2+} ion per QD. Recently, also the direct incorporation of only a few Mn ions in the CdTe QD layer has succeeded [17].

(Cd,Mn)Se/ZnSe and (Cd,Mn)Te/ZnTe constitute structures where the magnetic ions are placed within the QDs. Another possibility is to embed nonmagnetic QDs in magnetic barriers. Penetration of the carrier wavefunctions in the outskirts and/or interdiffusion of Mn ions provides also an exchange coupling with the magnetic ions. DMS structures of that type have been grown by embedding CdSe QDs in (Zn,Mn)Se [18]. An alternative not relying on self-assembly has been demonstrated for CdTe/(Cd,Mn)Te [19, 20]. Selective ion implantation followed by rapid thermal annealing produces a lateral band-gap modulation in a CdTe quantum well through defect-related Cd-Mn intermixing, large enough to confine the carriers. The strong drop of the optical emission typically observed on QD structures fabricated by lithography does not arise here. The diameter of the dot-like regions is, however, about 250 nm.

5.3 Exchange Interactions Under Three-Dimensional Carrier Confinement

The strength of the $sp{-}d$ spin interaction is governed by the exchange integrals of the localized magnetic-ion states and the extended band-states of the semiconductor. These integrals are in general decreasing functions of the carrier wave vector. In a bulk DMS, only states in close vicinity of the band-edges are involved so that this fact is usually ignored. However, carriers strongly localized in space posses a large effective wave vector and the exchange parameters can be thus modified in QDs [21] (see also Chap. 3). In addition, heavy-light hole mixing related to the specific symmetry properties of the QD structure can reduce the exchange coupling [21, 22]. A further issue to be addressed is the location of the magnetic ions within the QD structure and how they overlap with carrier wave functions.

The $sp{-}d$ coupling is directly monitored by the Zeeman splitting of the exciton in an external magnetic field. Figure 5.3 summarizes photoluminescence (PL) data of (Cd,Mn)Se/ZnSe QD structures in a concentration range up to $x = 0.1$. The inhomogeneously broadened ensemble PL has a spectral half-width of about 100 meV. In accord with the Mn-induced band-gap increase, the PL maximum at zero field undergoes a marked hight-energy shift relative to pure CdSe/ZnSe QDs of more than 120 meV. Unlike spherical nanocrystals, epitaxial QD heterostructures exhibit a quantization axis for the hole spin defined through the growth direction, in most

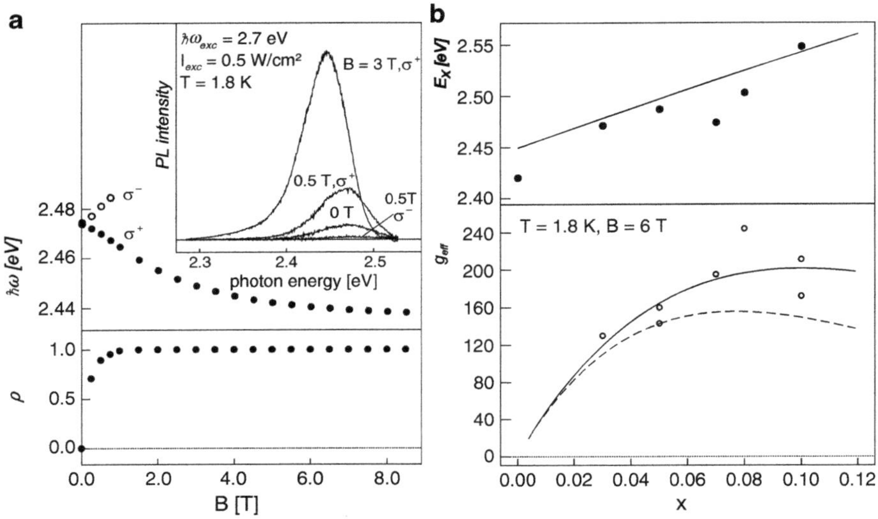

Fig. 5.3 Exciton PL of (Cd,Mn)Se/ZnSe QD structures [4]. (**a**) Magnetic-field dependence ($x = 0.07$). *Lower panel*: Circular polarization degree $\rho = (I^+ - I^-)/(I^+ + I^-)$ of the Zeeman components. *Upper panel*: PL maxima $\hbar\omega = E_X^\pm$ for $\sigma^\pm$ polarization detection. Inset: Ensemble PL for $B = 0\,\text{T}, 0.5\,\text{T}$ (σ^+ and σ^-), and $6.0\,\text{T}$ (σ^+ only). (**b**) Dependence on the Mn concentration. *Upper panel*: Zero-field exciton energies E_X (*solid dots*: experimental PL maxima, *line*: calculation). *Lower panel*: Effective exciton g factor $g_{X,\text{eff}}$ taken at $B = 6\,\text{T}$. *Open dots*: experiment, *solid* and *dashed lines*: calculation for r_{QW} and r_{bulk}, respectively, $g = 2.0$, $S = 5/2$. For details, see text. The Mn concentration is controlled by the calibrated flux pressure ratios during the growth of the initial (Cd,Mn)Se quantum well

cases, the [001] crystal axis. A field oriented along this axis (Faraday geometry[1]) results in diagonal magnetic coupling of the excitonic spin states. The PL of the QD structures splits up in two components with opposite circular polarization. The high-energy σ^- component disappears at larger field strength indicating rapid relaxation of the exciton spin. Almost complete spin polarization is reached at about $B = 1.0\,\text{T}$. The shift of the low-energy σ^+ component begins to saturate beyond 5 T. The effective g factor of the exciton is thus a function of the magnetic field. A natural definition is $g_{X,\text{eff}} = [E_X^+(0) - E_X^+(B)]/\mu_B B$, where E_X^+ denotes the energy position of the PL maximum. The g factor initially increases with the Mn concentration, but starts to decline close to $x = 0.1$. To obtain reliable values, spin heating effects have to be minimized by using below barrier excitation and sufficiently weak intensities ($I_{\text{exc}} < 1\,\text{W}\,\text{cm}^{-2}$). Below 1 T, where both components are visible and where their splitting is still linear in the field, g factors as large as $g_{X,\text{eff}} = 350$ are found. These findings resemble qualitatively those on DMS bulk materials [23, 24] and manifest the occurrence of a giant Zeeman splitting also in Stranski-Krastanov

[1] The light propagation is generally along the growth direction so that the field geometry can be also defined with respect to this axis.

QDs. In what follows, a theoretical analysis accounting for the specific features of the Stranski-Krastanov morphology is presented.

For zinc-blende crystal structure, the electron is an isotropic spin 1/2 particle, while the p-like hole has the angular momentum 3/2 or 1/2. When only heavy holes contribute significantly, the exciton ground-state comprises four spin configurations with total projections $s_z + j_z = \pm 1, \pm 2$. The ± 1 states are optically active in $\sigma^{\pm}$ photon polarization, whereas the other two represent dark excitons. The relevant exciton Hamiltonian reads then as

$$H = H_{\mathrm{X}} + H_{sp-d} \tag{5.1}$$

with

$$H_{\mathrm{X}} = E_{\mathrm{g}}(x) - \frac{\hbar^2}{2m_{\mathrm{e}}}\nabla_{\mathrm{e}}^2 + V_{\mathrm{e}}(\boldsymbol{r}_{\mathrm{e}}) - \frac{\hbar^2}{2m_{\mathrm{h}}}\nabla_{\mathrm{h}}^2 + V_{\mathrm{h}}(\boldsymbol{r}_{\mathrm{h}}) + V_{\mathrm{C}}(\boldsymbol{r}_{\mathrm{e}} - \boldsymbol{r}_{\mathrm{h}}) \tag{5.2}$$

and H_{sp-d} denoting the interaction with the magnetic ions. In the concentration range under investigation, the direct interaction of the carriers with the external magnetic field is much weaker and can be hence neglected. The same holds for the electron-hole exchange interaction, which produces energy corrections on the order of 1 meV and less [25] (editors note: see, however, Chap. 2 for cases when such corrections are important). In typical DMS QDs, these effects are outranged by the $sp-d$ interaction with energies in the range of some 10 meV. The major contribution to the exciton energy of some 100 meV originates from the single particle confinement potentials V_{e} and V_{h} as well as the Coulomb interaction V_{C}. It is thus justified to treat this part separately and to calculate the energy corrections due to the $sp-d$ interaction by perturbation theory.

For the ground-state with no orbital momentum, it is reasonable to start from single particle wave functions for electron and hole $\psi_j(\rho_j, z_j)$ $(j = \mathrm{e}, \mathrm{h})$ expressed in cylindrical coordinates related to the quantization axis z. These wave functions obey the Schrödinger equations

$$\left[-\frac{\hbar^2}{2m_j^{\star}}\left[\nabla_{\rho,j}^2 + \nabla_{z,j}^2 + \frac{1}{4\rho_j^2}\right] + V_j + V_{\mathrm{C},j} - E_j\right][\sqrt{\rho_j}\psi_j] = 0 \tag{5.3}$$

coupled by the Coulomb interaction. The latter is iteratively treated in Hartree approximation with $V_{\mathrm{C},j}$ being the effective potential of the electron (hole) in the averaged field of the hole (electron). The coupled eigenvalue problem can be solved numerically by standard finite-element methods.

A general difficulty in the calculation of carrier energies in self-assembled QDs is that the structural parameters are not precisely known and that these parameters even fluctuate across the ensemble. We present calculations for a prototype structure consisting of an island of spherical-lens shape situated on top of a wetting layer, both embedded in a material with higher band-gap (see Fig. 5.4). In the case of (Cd,Mn)Se/ZnSe, TEM and AFM data suggest an average island height of 1.2 nm,

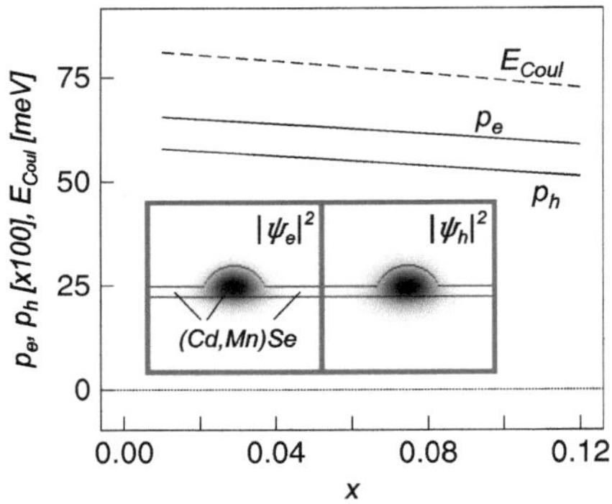

Fig. 5.4 Carrier confinement in (Cd,Mn)Se/ZnSe QDs. For details of the calculations, see text. Probabilities p_{e} and p_{h} (*solid lines*) to find electron and hole, respectively, in QD and wetting layer as well as electron-hole Coulomb energy E_{Coul} (*dashed*) vs. Mn content x. Inset: Density plots of squared electron and hole wave functions for $x = 0.05$ together with a cross-section of the QD morphology. Parameters: $m_{\mathrm{e}}^{\star} = 0.13\,m_0$ and $m_{\mathrm{h}}^{\star} = m_0/\gamma_1 = 0.22\,m_0$ [26] (m_0 – free electron mass), static dielectric constant $\varepsilon_0 = 9.6$ [27], an ML thickness increase of 10% due to the 7% compressive in-plane strain is accounted for in accordance with the elastic stiffness of CdSe. Care has been taken for a proper handling of the $1/\rho^2$ singularity. Using a grid with constant mesh distance h, the following replacement is used: $1/4\rho^2 \Rightarrow (2-\sqrt{1+h/\rho}-\sqrt{1-h/\rho})/h^2$, which reproduces the analytical solution of a cylindrical slab with infinite barriers

an aspect ratio of 1:3, and a two ML thick wetting film. The (Cd,Mn)Se band-gap follows a linear interpolation $E_{\mathrm{g}}(x) = 1.765\,\mathrm{eV}\,(1-x) + 3.3\,\mathrm{eV}\,x$ between zinc blende CdSe and MnSe [24]. A relative valence-band offset of 0.3 as known for (Zn,Cd)Se quantum wells [28] is used. Based on the weak interdiffusion found for CdSe/ZnSe QDs, abrupt band discontinuities are assumed at the interfaces. Strain-induced shifts of the band-edges can be neglected, since the contributions from hydrostatic and shear deformation largely compensate each other. The exciton energies computed in this way agree indeed fairly well with the experimental position of the PL bands in the whole composition range covered by Fig. 5.3b. An important point in the context of ion-carrier coupling is that the 120 meV high-energy shift of the exciton is only a 60% portion of the band-gap widening, indicating a significant penetration of the wave function into the ZnSe barriers. Integrating ψ_{e}^2 and ψ_{h}^2 over the space regions of QD and wetting layer provides the total probabilities p_{e} and p_{h} of finding the electron and the hole within regions containing the magnetic ions. The numerical results (Fig. 5.4) document a situation specific of Stranski-Krastanov systems: robust confinement yielding Coulomb energies about five times larger than the exciton binding energy of bulk CdSe [27], despite a substantial wave function leakage of 30–40%, quite independent on x.

The Hamiltonian of the $sp{-}d$-interaction can be written as

$$H_{sp-d} = -\sum_i \left[\mathcal{J}_{\mathrm{e}}(\boldsymbol{R}_i)\boldsymbol{s} + \frac{1}{3}\mathcal{J}_{\mathrm{h}}(\boldsymbol{R}_i)\boldsymbol{j}\right]\boldsymbol{S}_i. \tag{5.4}$$

$\boldsymbol{S}_i$ is the spin of the magnetic ion at site $\boldsymbol{R}_i$ and $\boldsymbol{s}$ and $\boldsymbol{j}$ are spin and angular momentum operator of electron and hole, respectively. The coupling parameters are given in effective-mass approximation by

$$\mathcal{J}_j(\boldsymbol{R}_i) = \int \mathrm{d}^3\boldsymbol{r} \int \mathrm{d}^3\boldsymbol{r}' \, \psi_j^*(\boldsymbol{r})\mathcal{J}_j(\boldsymbol{r},\boldsymbol{r}',\boldsymbol{R}_i)\psi_j(\boldsymbol{r}') \, ,$$
$$\mathcal{J}_j(\boldsymbol{r},\boldsymbol{r}',\boldsymbol{R}_i) = \sum_{\boldsymbol{k}}\sum_{\boldsymbol{k}'} \tilde{\mathcal{J}}_j(\boldsymbol{k},\boldsymbol{k}')\mathrm{e}^{ik(r-R_i)-ik'(r'-R_i)}, \tag{5.5}$$

where $\tilde{\mathcal{J}}_j(\boldsymbol{k},\boldsymbol{k}')$ are the scattering matrix elements of the extended band-states in the bulk material [21]. The s–d contribution of the electron is isotropic, while for a pure heavy hole only a component along the z-axis is present ($\boldsymbol{j}\,\boldsymbol{S} = j_z S_z$). A noticeable transverse component of the p–d coupling can be created by heavy-light hole coupling. If the spin of the magnetic ions is determined solely by the external magnetic field, the sp–d interaction can be treated in mean-field approximation. Here, all magnetic ions carry the same average spin. The correction to the exciton energy in a field oriented along the z-axis is then

$$\Delta E_{\mathrm{X}} = (\mathcal{J}_{\mathrm{e}} s_z + \frac{1}{3}\mathcal{J}_{\mathrm{h}} j_z)\langle S_z \rangle. \tag{5.6}$$

In thermal equilibrium at spin temperature T_{S}, the average spin projection of the magnetic ions can be expressed through the Brillouin function $B_{5/2}$

$$\langle S_z \rangle = -S\mathrm{B}_{5/2}[\frac{Sg\mu_{\mathrm{B}}B}{k_{\mathrm{B}}(T_{\mathrm{S}} + T_0)}] \tag{5.7}$$

(S and g: magnetic-ion electron spin and g factor, T_0: ion–ion superexchange correction [23]). For a QD containing a sufficiently large number of magnetic ions, it is easy to show that the coupling constants in (5.6) are given by

$$\mathcal{J}_j = \int \mathrm{d}\boldsymbol{r} \int \mathrm{d}\boldsymbol{r}' \psi_j^*(\boldsymbol{r})\mathcal{J}_j(\boldsymbol{r} - \boldsymbol{r}')\varrho(\boldsymbol{r}')\psi_j(\boldsymbol{r}'). \tag{5.8}$$

Here, ϱ is the density of the magnetic ions and $\mathcal{J}_j(\boldsymbol{r})$ comprises only diagonal scattering $\tilde{\mathcal{J}}_j(\boldsymbol{k},\boldsymbol{k})$. If the range of $\mathcal{J}_j(\boldsymbol{r})$ is much shorter than the scale on which the ψ_j change, a suitable approximation is

$$\mathcal{J}_j = \tilde{\mathcal{J}}_j(0,0) \int \mathrm{d}\boldsymbol{r}\,\varrho(\boldsymbol{r})|\psi_j(\boldsymbol{r})|^2, \tag{5.9}$$

where, in the usual notation, $\tilde{\mathcal{J}}_{\mathrm{e}}(0,0) = \alpha$ and $\tilde{\mathcal{J}}_{\mathrm{h}}(0,0) = \beta$. This is equivalent with using a δ-function $J_i(\boldsymbol{R}_i) \sim \delta(\boldsymbol{R}_i - \boldsymbol{r}_j)$ in (5.4) (see Chaps. 3 and 5). Such approach is indeed suggested by the relatively smooth wave functions for Stranski–Krastanov QDs (Fig. 5.4). Penetration of the wave functions into regions with no magnetic ions is here more important than the $\boldsymbol{k}$-dependence of the exchange matrix elements. Making finally use of the virtual-crystal approximation $\varrho = xN_0$ where N_0 is the density of cation sites, one obtains

$$\Delta E_X = -xr(x)[p_e \alpha N_0 s_z + \frac{1}{3} p_h \beta N_0 j_z] \langle S_z \rangle . \tag{5.10}$$

Besides the wave function leakage, a further quantity specifically modified in a QD geometry is the factor $r(x)$, accounting for antiferromagnetic pairing of magnetic ions on neighboring sites. Such pairing reduces the number of effective ion spins. Higher-order clusters can be neglected when x is smaller than 0.1. In a three-dimensional coordination, it holds $r_{\mathrm{bulk}}(x) = (1-x)^{12}$, where $(1-x)$ is the probability there is no ion on nearest-neighbor cation site and 12 is the number of those sites in the f.c.c. lattice. For a quantum well comprising N MLs, this changes to $r_{\mathrm{QW}} = (N-2)(1-x)^{12}/N + 2(1-x)^8/N$, as the number of nearest-neighbor cation sites at each interface is only 8. The true $r(x)$ of a Stranski–Krastanov morphology is intermediate between these cases.

The bulk exchange constants of (Cd,Mn)Se are $N_0\alpha = 0.23\,\mathrm{eV}$ and $N_0\beta = -1.26\,\mathrm{eV}$ [24]. The exciton state of lowest energy in a (Cd,Mn)Se/ZnSe QD is thus constituted by an electron with $s_z = -1/2$ and a hole with $j_z = 3/2$ giving rise to σ^+-polarized PL as observed experimentally. The coupling of the exciton and the magnetic ions is hence antiferromagnetic. A calculation of the effective g factors based on the magnetic-field shift given by formula (5.11) reproduces quite well the experimental values (Fig. 5.3b). It also shows that the saturation at about $x = 0.1$ is caused by the increasing number of antiferromagnetic pairs. Closer agreement is achieved when using the pairing probability r_{QW} of a quantum well. Obviously, this matches better to the flat wetting-layer-island morphology of these heterostructures. The saturation of the g factor represents also a tool that allows one to validate the magnetic-ion concentration in the QDs. In the present case, it can be concluded that diffusion or segregation of Mn play a minor role.

In nonmagnetic CdSe QDs embedded in (Zn,Mn)Se, a low-energy shift of the zero-field PL is observed [18]. A likely reason is a size increase induced by the presence of Mn in the substrate film [29, 30]. The Zeeman energies are about five times smaller than for (Cd,Mn)Se/ZnSe structures, but markedly larger than expected from the penetration of the wave function into the magnetic barriers [31]. Probably, interdiffusion creates also Mn^{2+} ions within the CdSe islands. Exciting above the (Zn,Mn)Se barrier, prominent PL from the barriers appears as well, likely a result of enhanced localization in the ternary material. At low magnetic fields $B < 1\,\mathrm{T}$, the circular polarization degree of the QD band is markedly smaller than for the barrier PL, indicating that injection of spin-polarized excitons from the barriers in the QDs is not a dominant process. Similarly to (Cd,Mn)Se/ZnSe, both barrier and QD degree reach a level of about 100% at $B \approx 1\,\mathrm{T}$.

So far, coupling between the hole states described by the 6×6 Kohn-Luttinger Hamiltoninan has been ignored. Typically, the 1/2 holes are sufficiently split-off by the spin–orbit interaction. The degeneracy of the 3/2 heavy- and light-hole states is lifted by confinement and strain in self-assembled QDs. In CdSe/ZnSe QDs, shape and strain anisotropies can produce a light-hole contribution in the heavy-hole ground-state up to 10% [32]. Such mixing only marginally influences the $sp{-}d$ interaction in the concentration and field range considered above. However, as pointed out in [22], the structure of the hole ground-state is quite sensitive to the aspect ratio. For a QD elongated along growth direction, i.e., a shape just inverse to the flat Stranski-Krastanov morphology, even a light hole-type ground-state is predicted. In this case, the $sp{-}d$ interaction is substantially reduced. In the limit of a pure light-hole state ($j_z = \pm 1/2$), the hole exchange interaction becomes three times smaller and, as a consequence of the signs of α and β, electron and hole contribution largely compensate.

Accounting also for the direct interaction with the external field, the total effective g factor of the exciton in a DMS QD is

$$g_{\mathrm{X,eff}} = g_{\mathrm{X}}^{0} + g_{\mathrm{X}}^{sp-d}. \tag{5.11}$$

Here, the sign of the g factor becomes important. The intrinsic contribution g_{X}^{0} is defined to be positive if the σ^- exciton is the one of lower energies. This is true for (Cd,Mn)Se and (Cd,Mn)Te while the negative hole exchange constant provides $g_{\mathrm{X}}^{sp-d} < 0$. The intrinsic value is of the order of one (CdSe/ZnSe: $g_{\mathrm{X}}^{0} = 1.6$ [25]). Therefore, a crossover from positive to negative g factor occurs at a very low concentration of magnetic ions. On the other hand, as g_{X}^{sp-d} is a function of the external field, the effective $g_{\mathrm{X,eff}}$ factor can be tuned to zero at critical field for proper structure design [33]. A sign change of g has been inferred from a sign reversal of the circular polarization degree on (Cd,Mn)Se/ZnSe QDs occurring at a critical concentration of about 1% Mn [34]. This is roughly one order of magnitude larger than the concentration where the crossover is expected. In natural (Cd,Mn,Mg)Te QDs, it has been found that the σ^+ PL occurs indeed high-energy shifted relative to the zero-field position already at $x = 0.005$ [35].

The magnetic interactions in (Cd,Mn)Se/(Zn,Mn)Se disc-like structures have been probed by Raman scattering [15]. Similarly to nonmagnetic CdSe/ZnSe QDs [25], Stokes and anti-Stokes features related to an electron spin-flip are observed. The Raman shift representing the Zeeman splitting of the electron directly displays the s–d interaction in the conduction band. Effective g factors of the electron up to $g_{\mathrm{e}} \sim 10$ are found. In addition, similarly to the findings in quantum wells, Raman transitions between ground- and first excited states of antiferromagnetically ordered nearest neighbor Mn^{2+} ions [36] and collective multiple spin-flip Raman lines [37] within the Zeeman-split Mn ground-state are detected.

The discussion so far has ignored the formation of magneto-polarons where a spin alignment of the magnetic ions is created without the need of an external magnetic field (see Sect. 5.4.2). The derivation of an effective g factor in the way described above is then misleading. For semiconductor structures with (yet partial) translational symmetry, there is a trade-off between the gain in energy due to the exchange interaction and the loss caused by the increase of the kinetic energy when the carriers localize [38]. The polaron is therefore only stable under certain parameter constellations. The situation becomes qualitatively different in QDs, where the polaron is always the state of lower energy. However, how close the carrier-ion system can actually evolve toward the energy minimum is controlled by the relation between the carrier lifetime and the time needed to align the magnetic moments. In the following section, we review the major dynamical scenarios.

5.4 Dynamic Processes

5.4.1 Interplay Between QD and Internal Mn States

A prominent finding is that incorporation of Mn in CdSe/ZnSe QD structures strongly quenches the PL yield [4, 18, 39, 40]. As an example, this is shown for (Cd,Mn)Se/ZnSe in Fig. 5.5. In a longitudinal external magnetic field, i.e., along the quantization axis, the yield recovers partially reaching a 0.1 fraction of the nonmagnetic structures at $B = 8\,\text{T}$. The mean energy of the exciton ground-state in CdSe/ZnSe QDs, as deduced from the maximum of the inhomogeneously broadened PL band, is about 2.4 eV and increases when Mn is incorporated (see Sect. 5.3). The energy required to promote a $3d^5$-shell electron of Mn^{2+} from the 5/2 ground-state ($^6A_1$) to the excited states $^4T_1$ and $^4T_2$ with 3/2 total spin amounts in (Cd,Mn)Se only to 2.1 eV. That is, the QD exciton can recombine nonradiatively by an Auger-processes (assisted by the emission of phonons) with simultaneous electron excitation in the magnetic ions [41, 42]. Participation of magnetic ions in the recombination dynamics is indeed demonstrated experimentally by a redistribution

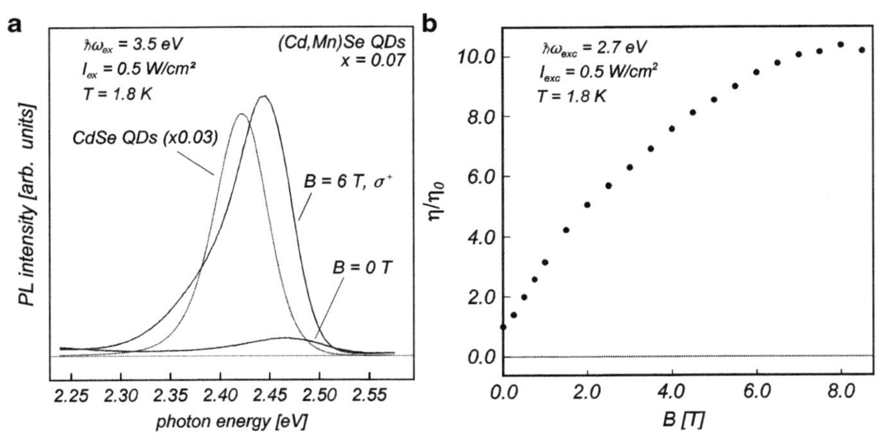

Fig. 5.5 PL yield recovery of (Cd,Mn)Se/ZnSe QD structures in an external magnetic field [4]: (**a**) PL spectra in comparison with nonmagnetic CdSe/ZnSe QDs (above barrier excitation). Note the scaling factor of 0.03. (**b**) Field dependence of the total yield $\eta = I^+ + I^-$ (below barrier excitation)

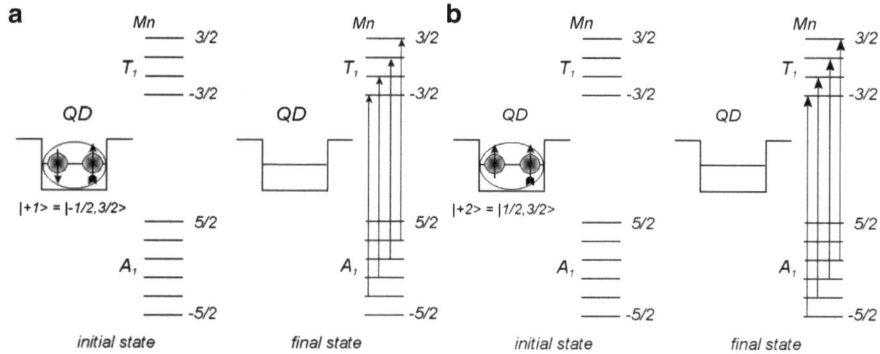

Fig. 5.6 Schematics of the Auger-processes in a magnetic field for bright (**a**) and dark (**b**) excitons. For explanations, see text

of the yield from the ${}^4T_1$–${}^6A_1$ emission at 2.1 eV toward the benefiting QD PL on the high-energy side with increasing magnetic field [18, 40, 43]. A magnetically induced shrinkage of the carrier wave functions suppressing nonradiative channels [44] is less likely, also in view of the fact that the yield recovery does not follow a quadratic field-dependence (Fig. 5.5b).

The Auger-process is subject to the condition that the total spin of the carrier-ion system is conserved providing the spin selection rule $s_z^e + s_z^h = \Delta S$, where s_z^e and s_z^h are the projections of electron and hole spin, respectively, and ΔS denotes the change of the spin projection for the magnetic ion. Note that the pure hole spin enters here, whereas the angular momentum j matters for the optical selection rules. That is, it holds $\Delta S = 0$ for the radiative excitons and $\Delta S = \pm 1$ for the dark excitons. The resultant Auger-processes are schematically drawn in Fig. 5.6. A rigorous treatment, properly accounting for the orbital momenta in the valence band and the $3d^5$-shell as well as spin–orbit interaction, confirms the above selection rules and shows that the direct Coulomb interaction is responsible for the processes involving radiative exciton states, while those of the dark excitons originate from the exchange part [39]. In particular, there is no need to invoke extra carriers (donor-bound excitons, trions) to guarantee spin conservation [45]. In a longitudinal external field, as stressed in the previous section, the σ^+ exciton state is predominantly occupied already at about 1 T. Further increasing the field strength, the $3d^5$-shell Mn electrons start to populate increasingly the -5/2 state from which Auger-processes are forbidden and, accordingly, the emission yield recovers (Fig. 5.5). Even at fields close to 10 T, the PL is still considerably weaker compared to nonmagnetic CdSe/ZnSe QD structures. A likely reason is the participation of Mn pairs in the Auger process, as larger fields are necessary to break up their antiferromagnetic ordering [23, 24] and to align these Mn spins, too.

The situation becomes entirely different in a transverse field, where the heavy hole is magnetically inactive [39]. Coupling of spin-up and spin-down states for the electron translates into a mixing of radiative and dark exciton states. The dark exciton component lifts the blocking of the Auger-transitions. Indeed, the recovery of the yield observed experimentally in this geometry is only a factor of 1.5 at $B = 11\,\mathrm{T}$ [39].

Time-resolved PL measurements on (Cd,Mn)Se/ZnSe QD structures have yielded an exciton lifetime of only a few 10 ps [43]. In contrast, the radiative lifetime in non-magnetic CdSe/ZnSe QDs is about 300 ps [46], which underlines the dominance of nonradiative recombination in the presence of Mn. The lifetime is too short for enabling the formation of a magneto-polaron state in this type of structures. On the other hand, the strong polarization degree of the PL signifies an exciton spin-flip time even shorter than the nonradiative lifetime. The same conclusion is drawn for CdSe/(Zn,Mn)Se QDs where, time-resolving both polarization components of the QD emission, no difference was observed within the resolution of the streak camera [31]. In contrast, the spin lifetime of the exciton in non-magnetic CdSe/ZnSe QDs observed under resonant excitation is much longer than its radiative lifetime [46]. It is tempting to assign the rapid spin relaxation in Mn-containing structures to flip-flop processes between carrier and magnetic ion spins. Such processes are mediated by the transverse part of the $sp-d$ interaction. However, the large detuning between the spin states by the giant Zeeman effect contradicts energy conservation. A possible explanation might be due to spin exchange processes with carriers in the continuum states of the structures generated by off-resonant excitation. Spin flip-flops become important for QDs containing only a single magnetic ion, where the energy scale of the spin splitting is strongly reduced (Sect. 5.5).

5.4.2 Magneto-Polaron Formation

To allow for a significant spin polarization of the exciton-magnetic-ion system without external magnetic field, the nonradiative Auger-processes described in the previous subsection have to be circumvented. Using specially tailored CdSe/(Zn,Mn)Se structures, the QD exciton energy could be moved below the ${}^4T_1$–${}^6A_1$ transition of the Mn^{2+} $3d^5$-shell [47]. The result is an increase of the exciton lifetime from 20 to 580 ps [48].

Formation of a magneto-polaron state is made visible by a transient low-energy shift of the ensemble PL of such a structure after short-pulse excitation in its energy continuum, as depicted in Fig. 5.7. In the inset, the energy position reached within the exciton lifetime is plotted as a function of temperature. As already mentioned, the robust carrier confinement in QDs makes the magneto-polaron formation different from that in higher-dimensional systems. Generally, the formation time τ_{MP} of the magneto-polaron includes both a transient change of the carrier wave function and the pure spin response τ_{S}. The first contribution becomes negligible in QDs so that the transients in Fig. 5.7 display the sole spin time ($\tau_{\mathrm{MP}} = \tau_{\mathrm{S}}$).

It is useful to describe the magneto-polaron state in terms of an exchange field $\boldsymbol{B}_{\mathrm{exc}} = \boldsymbol{B}^{\mathrm{e}}_{\mathrm{exc}} + \boldsymbol{B}^{\mathrm{h}}_{\mathrm{exc}}$ created by the carrier spins and the magnetization $\boldsymbol{M}$ of the ion system. For homogeneous coupling, i.e., using constant carrier wavefunctions

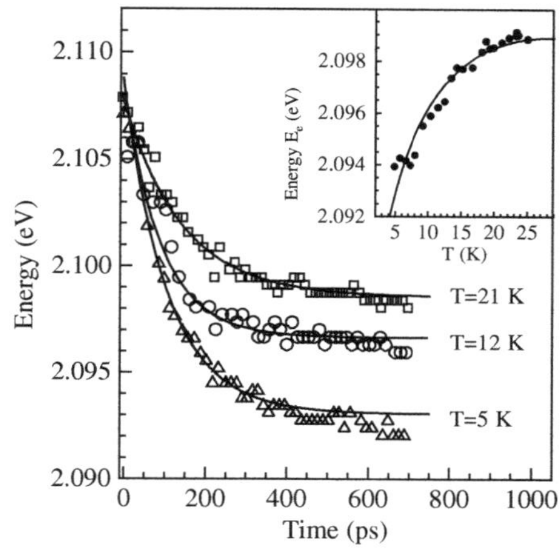

Fig. 5.7 Transient PL photon energy in CdSe/(Zn,Mn)Se QDs after short pulse excitation for various temperatures. Inset: Equilibrium energy position E_e vs. temperature. Reprinted with permission from [48]. Copyright (2002) American Physical Society

$\psi_j = \sqrt{p_j/V_{\text{mag}}}$ within the magnetic volume V_{mag}, the $sp{-}d$ Hamiltonian reads as

$$H_{sp-d} = -\boldsymbol{B}_{\text{exc}} V_{\text{mag}} \boldsymbol{M} \tag{5.12}$$

with

$$\boldsymbol{B}_{\text{exc}} = \frac{1}{V_{\text{mag}}} \frac{p_{\text{e}}\alpha \boldsymbol{s} + p_{\text{h}}\beta \boldsymbol{j}/3}{g\mu_{\text{B}}}, \quad V_{\text{mag}}\boldsymbol{M} = g\mu_{\text{B}} \sum_i \boldsymbol{S}_i. \tag{5.13}$$

It can be useful to replace V_{mag} by the number of cation sites $N_{\text{cat}} = V_{\text{mag}} \cdot N_0$ in the magnetic volume so that the exchange energies αN_0 and βN_0 of the bulk material appear in $\boldsymbol{B}_{\text{exc}}$. If populations of dark exciton states are completely negligible, the modulus of the exchange field is simply $B_{\text{exc}} = (p_{\text{e}}\alpha N_0 - p_{\text{h}}\beta N_0)/2g\mu_{\text{B}}N_{\text{cat}}$ for antiferromagnetic coupling ($\beta < 0$). Making use of the susceptibility $\chi = \text{d}M/\text{d}B$, the magneto-polaron energy can be represented by

$$E_{\text{MP}} = V_{\text{mag}}\chi B_{\text{exc}}^2. \tag{5.14}$$

There is a direct relation to the giant Zeeman shift in an external field[2] examined in Sect. 5.3, formula (5.10):

$$E_{\text{MP}} = \Delta E_{\text{X}}(B + B_{\text{exc}}) - \Delta E_{\text{X}}(B) = \frac{\text{d}}{\text{d}B}\Delta E_{\text{X}}(B)\, B_{\text{exc}} \tag{5.15}$$

or

$$V_{\text{mag}}\chi = xr(x)N_{\text{cat}}g\mu_{\text{B}}\frac{\text{d}}{\text{d}B}\langle S_z\rangle. \tag{5.16}$$

The derivative of $\langle S_z\rangle$ in (5.16) reflects the fact that no additional alignment of the ion spins can be induced when the saturation regime is already reached by the

[2] Here, the assumption is made that $\boldsymbol{B}_{\text{exc}}$ is not sufficient to drive $\boldsymbol{M}$ into saturation.

application of a sufficiently strong external field. At first glance, when inspecting the Hamiltonian in (5.12), the increase of the exchange field with shrinking QD volume seems to be counterbalanced by the decreasing number of ions contributing to the magnetization. However, at a given ion concentration, a larger B_{exc} generates a larger M so that the magneto-polaron energy becomes an inverse function of the magnetic volume.

Assuming that thermal equilibrium at the lattice temperature is reached, the emission energy of the exciton can be written as $\hbar\omega = E_{\text{X}}(T) - E_{\text{MP}}(B_{\text{exc}}, T)$, where $E_{\text{X}}(T)$ accounts for the temperature dependence of the band-gap. A fit to the data in Fig. 5.7 yields $B_{\text{MP}} = 2.7\,\text{T}$ at low temperatures. This relatively low value and the small magneto-polaron energies of only a few meV are related to the weak overlap of the carrier wave functions with the magnetic ions in the barriers. The built-up time depends much weaker on temperature or external magnetic field than for the (Cd,Mn)Se bulk material [48]. Steady-state single-dot measurements confirm these findings [47]. Though a magneto-polaron state is not formed in (Cd,Mn)Se/ZnSe QDs as a result of the short carrier lifetime, one can deduce its parameters in the case of direct wavefunction overlap from the external-field data providing $B_{\text{exc}} \sim 14\,\text{T}$ and $E_{\text{MP}} \sim 35\,\text{meV}$ close to saturation. That is, realizing structures with longer lifetimes would allow for the formation of more robust magneto-polaron states.

A system, where the condition for the absence of nonradiative Auger-processes is more easily fulfilled, is (Cd,Mn)Te, because of the lower band-gap. For QD structures, the high-energy shift caused by confinement may not become too large. In [17], (Zn,Cd)Te barriers are applied to ensure an exciton ground-state energy well below the $^6A_1 - {}^4T_{1,2}$ transitions. The magneto-polaron state has been studied quite early on natural (Cd,Mn)Te QDs [49, 50]. Evidence is again provided by a temperature-induced high-energy shift of the single-dot emission [49]. Magneto-polaron formation associated with a resident QD electron has been concluded from the presence of a donor-bound electron spin-flip Raman feature at zero-magnetic field in (Cd,Zn,Mn)Se/(Zn,Mn)Se disc structures [15]. A spin polarization memory assigned to magneto-polaron formation has been observed under quasi-resonant excitation, one LO-phonon energy above the exciton ground-state, of CdTe/(Cd,Mn)Te/ZnTe QD structures [51, 52]. Further aspects of the magneto-polaron formation are discussed in relation to single-dot measurements in Sect. 5.5.

5.4.3 Spin Temperature Dynamics

At the end, the spin temperature of the magnetic-ion system controls the degree of spin polarization that can be achieved. In this context, the spin-lattice relaxation (SLR) comes into play. It also defines the timescale on which the magnetic-ion population can be switched. A detailed overview on spin and energy transfer between magnetic ions, carriers, and the lattice in DMS structures is presented in Chap. 9 of this book. Here, the question for specific SLR mechanisms in DMS QDs is raised. Spin diffusion involving the wetting layers might be a peculiar process [53, 54].

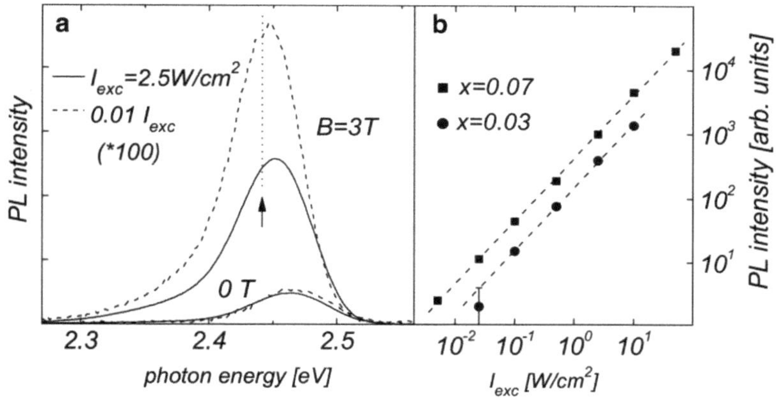

Fig. 5.8 Magnetic-ion spin heating under optical excitation monitored by PL (**a**) PL band of a (Cd,Mn)Se/ZnSe QD structure ($x = 0.07$) at low and high excitation. The *arrow* marks the spectral position for transient SLR studies (see Fig. 5.10a). (**b**) Spectrally integrated PL yield vs. excitation intensity I_{exc} for two Mn concentrations ($x = 0.03, 0.07$) at $B = 0$. The lower yield of $x = 0.03$-sample is not intrinsic to magnetic-ion system but due to stronger donor-acceptor pair recombination in the ZnSe buffer reducing the carriers reaching the QDs. Reprinted with permission from [55]. Copyright (2005) American Physical Society

Before the dynamical aspect of SLR will be treated, the stationary situation is examined.

Carriers excited optically heat directly or indirectly the magnetic-ion spin system. The resultant reduction of the spin polarization shows up in two ways. First, the magnetic-field-induced shift of the QD PL features shrinks and second, as the Auger rates grow up, the PL yield drops down. Figure 5.8 demonstrates both effects for (Cd,Mn)Se/ZnSe QD structures [55]. A mechanism that might also decrease the spin polarization is the creation of nonequilibrium populations in the excited Mn^{2+} states $^4T_1$ and $^4T_2$ steadily addressed by the Auger process. However, in a wide range, the PL yield is strictly a linear function of the excitation intensity. This excludes an essential contribution from Auger populations which would block the nonradiative rate. From the shift of the PL band as a function of the magnetic field, the spin temperature at a given excitation intensity can be derived using (5.7) and (5.10). Experimental care has to be taken to obtain reliable data. The superexchange parameter T_0 has to be deduced independently at excitation intensities as low as possible. In addition, local generation of photo-carriers as close as possible to the QD layer has to be ensured. In the measurements of Fig. 5.9a, the ZnSe barrier exciton is resonantly excited. The large absorption coefficient, in combination with a 1 μm ZnSe buffer, minimizes contributions from heating of the GaAs substrate.

The $T_S(I_{\text{exc}})$ relationship derived in this way is strongly nonlinear (Fig. 5.9b). It enables one to translate a certain excitation intensity into the spin temperature generated in the magnetic system.

For undoped DMS heterostructures, two main mechanisms are discussed so far for the magnetic-ion spin heating by photo-excited carriers: (1) the direct spin

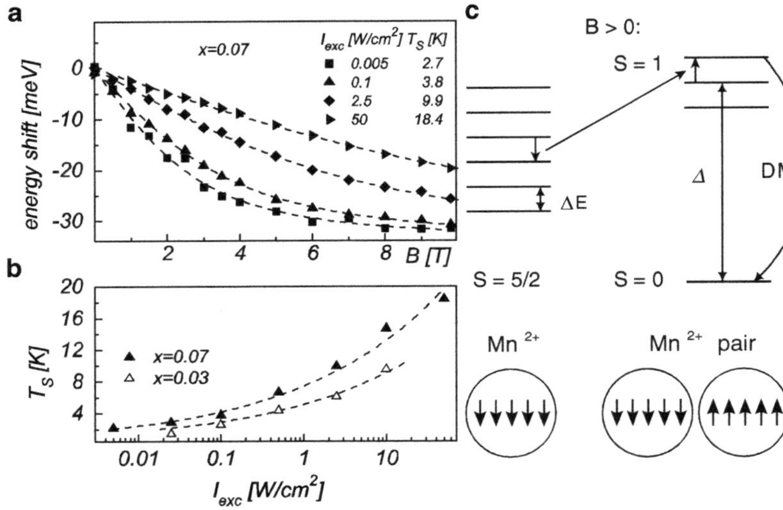

Fig. 5.9 Spin heating of (Cd,Mn)Se/ZnSe QD structures under photo excitation (**a**) Magnetic-field induced shift of the σ^+ PL at different excitation intensities ($x = 0.07$). Lines represent a fit of the experimental data by (5.7) and (5.10); (**b**) Derived spin temperatures for stationary heating ($x = 0.07, 0.03$). *Dashed lines* represent a fit of the data by $I_{\text{exc}} \sim T_S^4 - T_b^4$, $T_b = 1.6$ K. Note the logarithmic scale of I_{exc}. The excitation photon energy in excess is larger than the ZnSe band gap energy with an excess of 350 meV with respect to the PL maximum energy. $T_0 \approx 3.0$ K ($x = 0.07$) and 1.5 K (0.03); (**c**) Schematics of the spin cooling via the Dzyaloshinskii–Moriya coupled excited state of Mn pairs. The Zeeman splitting of all energy levels with $S > 0$ is $\Delta E = g\mu_B B$ and Δ is the antiferromagnetic exchange coupling. Figure 5.9a, b are reprinted with permission from [55]. Copyright (2005) American Physical Society

transfer via the $sp-d$ interaction [56, 57] and (2) indirect heating through phonons generated during relaxation of the carriers (see, e.g., [58]). Direct spin transfer, as evidenced by a linear relationship between spin temperature and excitation intensity, has been observed on quantum wells [56]. It can overcome the phonon heating, which increases exponentially with the concentration of the magnetic ions, only in very diluted structures [34, 57]. Since the inverse process, the spin cooling, is mediated solely by SLR, the spin system moves out of thermal equilibrium with the lattice, even under stationary excitation. In QDs, as already stressed in Sect. 4.1, energy conservation cannot be satisfied in the direct spin-exchange process as the energy states are discrete and largely Zeeman-split. Auger recombination increasing the amount of heat generated per photo-carrier favors phonon-mediated spin heating. In this case, a thermal equilibrium is established between the spin and the phonon system ($T_S = T_p$). The results found on (Cd,Mn)Se/ZnSe QD structures are fully consistent with this expectation. In the Debye limit and for one-dimensional heat conduction, it holds $I_{\text{exc}} \sim T_p^4 - T_b^4$ [58], where $T_b = 1.6$ K is the temperature of the He bath surrounding the sample. This dependence fits the experimental data indeed perfectly well (Fig. 5.9b).

Modulated PL excitation provides access to the time transients of both spin heating and cooling [55]. Here, the excitation intensity is periodically switched between two levels, I_{low} and I_{high}, respectively. When the switch is sufficiently fast, the spin temperature follows slowly and the adjustment of the new equilibrium is directly displayed by the transient PL change. Optical excitation is hence used to manipulate the spin temperature as well as to facilitate read-out of its dynamics. Detection on the low-energy side of the PL band has the advantage of better sensitivity, since both Zeeman shift and PL yield contribute in the same direction to the signal change. Data obtained on (Cd,Mn)Se/ZnSe QD structures are depicted in Fig. 5.10. The signal decrease during the high-excitation period reflects spin heating. Switch back to low excitation suddenly reduces the carrier density, which manifests in an instantaneous signal drop on the timescale of the figure. The subsequent PL increase displays spin cooling. In first order, when the changes are small, it holds $\Delta T_S \propto -\Delta I_{PL}(t)$ so that the transients directly reflect the time variation of the spin temperature. Both heating and cooling transients behave single-exponential with time constants (τ_{SLR}) in the 1 μs range. No indication of a long SLR component [54] is found on these DMS QD structures. Heating is always faster than cooling (a factor of 2 in the inset of Fig. 5.10a). This can be indicative of diverse mechanisms or, as the excitation level is different in the two periods, of a temperature-dependent SLR rate. Indeed, identical heating and cooling rates are found if one calibrates the transients to the same spin temperature by means of the $T_S(I_{exc})$ relationship derived from steady-state

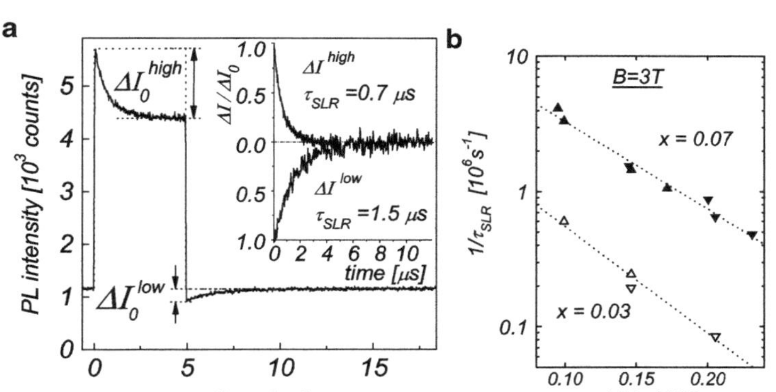

Fig. 5.10 Spin temperature dynamics revealed by PL excitation modulation. The sample is a (Cd,Mn)Se/ZnSe DMS QD structure ($x = 0.07$). (**a**) Typical PL transient (*solid line*) at $B = 3$ T and the underlying excitation modulation (*dashed line*). A high-speed acousto-optical modulator with a rise/fall time of 10 ns modulates the excitation laser light yielding a rectangular pulse profile superimposed on a lower background intensity. The PL is detected by a fast micro-channel photomultiplier and recorded by a multi-scaler photon counter. A spectral component on the low-energy side of the PL band is selected (see *arrow* in Fig. 5.8a). Pulse duration and observation window are long enough to ensure formation of an stationary state. Inset: Enlarged view of the normalized heating and cooling transients. Thick lines represent single-exponential fits. (**b**) SLR rate $1/\tau_{SLR}$ vs. inverse spin temperature $1/T_S$ for two different Mn concentrations. *Lines* are to guide the eye. Reprinted with permission from [55]. Copyright (2005) American Physical Society

measurements (Fig. 5.9b). When plotted versus the inverse temperature (Fig. 5.10b), an exponential behavior is revealed with an activation energy of 1.3 meV. While this energy is independent of the magnetic-ion concentration, the rates themselves slow down considerably for higher dilution. These experimental observations are clearly supportive of phonon-mediated spin heating.

Various experimental facts evidence a scenario, where the spin temperature T_S equilibrates with a local phonon temperature T_P established much quicker before. First, the single-exponential transients demonstrate that phonon dynamics and SLR proceed on well-separated timescales. Second, the distinct concentration dependence of the time constants ensures that indeed SRL and not phonon dynamics is displayed. The SLR times imply that the phonon system reaches a quasi-equilibrium state within a time shorter than 1 μs. This is faster than the phonon thermalization in bulk semiconductors [59]. Possibly, strain, alloy disorder, and the Stranski-Krastanov morphology itself lead to an increased anharmonic coupling of the phonon modes. This point deserves further investigations. To the best of our knowledge, no information on nonequilibrium phonons in self-assembled QDs is currently available.

The fast SRL rates require explanation. Relaxation within isolated Mn^{2+} ions is inefficient as there is no spin–orbit interaction due to the zero angular momentum ($L = 0$) of the 3d-electrons [60]. However, Mn^{2+} cluster, where the spin conservation is broken due to Dzyaloshinskii-Moriya coupling, can take over the excess spin of an isolated ion. The importance of Mn^{2+} pairs has been already stressed in relation to the saturation of the exciton g factor in Sect. 3. The observation of faster SLR at higher ion concentrations is then evident. However, more importantly, the activation energy governing the temperature dependence of the rates agrees very well with twice the antiferromagnetic exchange constant of (Cd,Mn)Se [61] ($2k_B J_1 = 1.4$ meV), which defines the energy separation Δ between ground- and first excited state of an ion pair. As sketched in Fig. 5.9c, the spin of an isolated ion can be transferred to the excited $S = 1$ state without violating energy conservation just by changing the spin projection from $M = 0$ to $M = 1$ (cooling) or $M = -1$ (heating). The spin is subsequently dissipated by relaxation in the Dzyaloshinskii-Moriya coupled $S = 0$ ground-state via phonon emission. The SLR rate of this process combines the Boltzmann occupation of the initial ($S = 1$, $M = 0$) state and the phonon emission factor, yielding $1/\tau_{SLR} \sim \exp(-\Delta/k_B T_P)[(\exp(\Delta/k_B T_P) - 1)^{-1} + 1] = 1/[\exp(\Delta/k_B T_P) - 1]$, similarly to the standard Orbach process [62, 63] and as observed experimentally.

In summary, no specific SLR scenario is found for (Cd,Mn)Se/ZnSe DMS QDs with clear Stranski-Krastanov morphology. As in DMS bulk materials or quantum wells, the relaxation takes place locally. The number of magnetic ions is obviously large enough to form a sufficient number of pairs. In the size and concentration range under investigation, we find 7–10 isolated ions and 1–4 Mn pairs. Mn^{2+} triades serving as the dominant channel in the bulk [60] play a minor role. Other QD morphologies can behave differently. Finally, it should be noted that SLR cannot be involved in the spin response τ_S during magneto-polaron formation which is orders of magnitude faster and which requires a transfer of the excess spin out of the magneto-polaron volume.

5.5 Single-Dot Spectroscopy: From Magnetic Fluctuations to Single Magnetic Moments

Micro-optical studies addressing single DMS QDs have uncovered a dramatically increased linewidth of the PL features compared to their nonmagnetic counterparts [12, 47, 64–66]. The cause are fluctuations of the magnetic-ion spins. The importance of such fluctuations has been outlined many years ago in studies on electrons bound to donors in bulk DMS [67,68]. When the carriers are excited optically, one has to distinguish whether magneto-polaron formation takes place within the carrier lifetime or not. We first consider the situation where this formation is suppressed by the fast Auger recombination.

Typical micro-optical PL spectra of (Cd,Mn)Se/ZnSe DMS QD structures are depicted in Fig. 5.11a [66]. The number of features is consistent with the areal dot density for the respective Mn concentration found by AFM (Sect. 5.2). The single-dot PL linewidth of nonmagnetic CdSe/ZnSe QDs is clearly below 100 μeV [46], i.e., more than two orders of magnitude smaller.

In the absence of an external field and magneto-polaron formation, the magnetic-ion spins fluctuate randomly. If the number of magnetic ions is large enough, the probability of a certain magnetization $\boldsymbol{M}$ obeys a Gaussian distribution

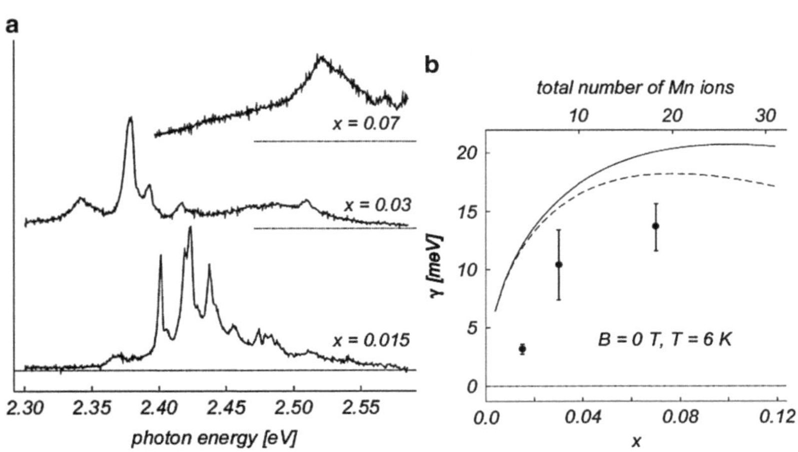

Fig. 5.11 Single-dot PL of (Cd,Mn)Se/ZnSe DMS QD structures. Mesas with lateral dimensions of 200 nm prepared by lithographic wet chemical etching are addressed by confocal microscopy (**a**) Spectra for three different Mn concentrations. (**b**) FWHM γ of the PL features vs. Mn concentration, *lines*: calculated (*full*: r_{QW}, *dashed*: r_{bulk}), *dots*: experimental, the *vertical bars* give the scattering of the width over the QDs examined. Reprinted with permission from [66]. Copyright (2004) American Physical Society

$$P(\boldsymbol{M}) = \frac{1}{(2\pi\langle M^2\rangle)^{3/2}} \exp\left[-\frac{M^2}{2\langle M^2\rangle}\right]. \tag{5.17}$$

The variance is the average of the squared projection along one space direction. For independent spins $\langle \boldsymbol{S}_i \boldsymbol{S}_j\rangle = \delta_{i,j} S(S+1)$, it follows (see (5.13))

$$\langle M^2\rangle = \frac{g^2\mu_B^2 N_{\text{mag}}}{V_{\text{mag}}^2} \frac{S(S+1)}{3}, \tag{5.18}$$

where N_{mag} is the number of magnetic ions in the magnetic volume. The average magnetization field scales thus like $\sqrt{\langle M^2\rangle} \sim \sqrt{N_{\text{mag}}}/V_{\text{mag}}$ reflecting the general fact that fluctuations become increasingly important when the system size shrinks. The photo-generated electron-hole pair samples a frozen spin configuration during its short lifetime which translates in certain energies of its spin components through the $sp-d$ interaction. The experimental data (both steady-state and time-resolved) are averages over a huge number of excitation events so that the distribution P shows up in the PL spectrum, here by a characteristic energy width of $\delta E_{\text{FM}} = B_{\text{exc}} V_{\text{mag}} \sqrt{\langle M^2\rangle}$. The consequences are two-fold. First, the lines become inhomogeneously broadened. Second, there is a zero-field splitting between the exciton spin states, though no macroscopic mean magnetization is present. For Mn^{2+} with $S = 5/2$, the energy width reads explicitly as

$$\delta E_{\text{MF}} = \frac{1}{N_{\text{cat}}} \sqrt{\frac{35}{12} N_{\text{Mn}}} \left[\frac{1}{2} p_{\text{e}} \alpha N_0 - \frac{1}{2} p_{\text{h}} \beta N_0\right]. \tag{5.19}$$

The above considerations are the high-temperature version of the dissipation-fluctuation theorem. At spin temperature T_S, the variance of thermodynamic fluctuations is related to the susceptibility by $\langle M^2\rangle = \chi k_B T_S / V_{\text{mag}}$. Using (5.14), one can express δE_{FM} through the magneto-polaron energy [67]

$$\delta E_{\text{MF}} = \sqrt{E_{\text{MP}} k_B T_S}. \tag{5.20}$$

This expression is also valid in presence of a longitudinal external magnetic field. The specific lineshape of the PL features depends on a number of factors. First, a pure heavy-hole state samples only fluctuations along the quantization axis, while the isotropic electron is sensitive on their modulus. The relation between the exchange parameters α and β is thus important. On the other hand, heavy–light hole mixing can produce a substantial in-plane component also of the $p-d$ coupling. Moreover, dark exciton states, gaining optical oscillator strength for transverse magnetization components, can contribute to the broadening. Finally, the degree of the exciton spin relaxation critically enters. In most experimental situations, spin coherences can be safely ignored. In other words, electron and hole spins align quickly to the direction of the relevant magnetization axis. However, how close the occupations can actually evolve toward equilibrium with the magnetic-ion system is only poorly investigated.

The experimental data of Fig. 5.11b follow reasonably well the square-root-like increase with the magnetic-ion concentration predicted by (5.19). Indeed, the size of these QD structures is quite independent on the Mn concentration (see Sect. 5.2). Using appropriate geometrical parameters, one finds $N_{\text{cat}} = 260$ and a number of Mn^{2+} ranging between 4 and 20 in the range covered. The resultant δE_{MF} overestimates the true broadening by about 40%. This is most likely the result of a partial exciton spin alignment as suggested by the strong polarization degree observed in external magnetic field.

An interesting question is to what degree the PL lines can be narrowed by an external magnetic field. The answer can be directly obtained from the field-dependence of the magnetic susceptibility. Using experimental data of magnetization measurements on bulk CdMnTe and CdMnSe, the analysis showed that the linewidth initially indeed declines down to the 1 meV range [69]. However, at fields above 10 T, the width starts to oscillate as antiferromagnetic Mn clusters break up. It will be thus highly unlikely that the much smaller line broadenings of nonmagnetic QDs in the 10 μeV range can be recovered.

In QD structures where the exciton resonances are energetically below the intra-Mn transitions, the change of the magnetization related to magneto-polaron formation has to be accounted for. The zero-field linewidth of CdSe/(Zn,Mn)Se QDs [65] is quite similar to what is shown in Fig. 5.11. Obviously, the polaron-induced magnetization is still markedly smaller than the fluctuating field. In a study on (Cd,Mn)Te/(Zn,Cd)Te structures, the lineshape has been examined in dependence on the QD size [17]. In consistent with volume dependence of the polaron energy, significant polaronic effects have been only observed for features situated on the center and high-energy side of the ensemble PL. Calculations based on the muffin-tin model for bound magneto-polarons allow for the extraction of the number of magnetic ions per QD from the experimental linewidths and energy positions as functions of a longitudinal external field. The effect of a transverse field on the line broadening has been studied in [71]. When heavy-light hole mixing is unimportant, the hole contribution $\boldsymbol{B}^{\text{h}}_{\text{exc}}$ to the exchange field is always directed along the z-axis. Then, the largest magnetization, i.e., the magneto-polaron state of lowest energy, is formed when the isotropic contribution $\boldsymbol{B}^{\text{e}}_{\text{exc}}$ of the electron aligns along $\boldsymbol{B} + \boldsymbol{B}^{\text{h}}_{\text{exc}}$. The magnetic-ion spins point then also in this direction. In such geometry, the exciton does not only sample the fluctuations of the amplitude of $\boldsymbol{M}$ but also those of its direction. The Gaussian variance of the latter scales like $\langle M_\perp^2 \rangle \propto M/B$. It decreases hence like the inverse field strength, that is, considerably smoother than $\mathrm{d}M/\mathrm{d}B$ governing the disappearance of the longitudinal fluctuations. Indeed, experimentally, an almost constant PL linewidth is found in Voigt geometry[3] up to 12 T [71].

While the role of magnetic fluctuations on the PL linewidth is well elaborated, to the best of our knowledge, no studies regarding a fluctuation-induced zero-field splitting of the exciton states of self-assembled DMS QDs is available. On donor-bound electrons in bulk CdMnSe, such splitting has been uncovered by Raman

[3] The field geometry is again defined with respect to the growth direction.

spin-flip scattering [68,72]. For localized excitons in DMS quantum well structures, a fluctuation-induced splitting of 2–3 meV has been estimated [73].

The above considerations are only valid for a sufficiently large number of magnetic ions in the QD. In the ultimate limit of a single magnetic ion, the exciton probes via the $sp-d$ interaction the discrete spin configurations of this ion. Such probing is only possible with a QD, since increasing dilution in the case of extended carrier states will merely cause the magnetic features to disappear. As already noted in Sect. 5.2, fabrication of such QD structures was succeeded for CdTe/(Zn,Mn)Te [16] allowing one to study various aspects of the $sp-d$ interaction in the single-ion/single-exciton limit [70, 74–77].

Figure 5.12 summarizes single-dot PL measurements on CdTe structures containing a single Mn^{2+} ion [70]. At zero field, the PL displays a fine structure consisting of six components, each twice degenerate. The relevant energies are found by diagonalization of the Hamiltonian (5.4), now with a single ion spin $\boldsymbol{S}$, in the common basis $|s_z\rangle|j_z\rangle|S_z\rangle$. The $2S+1$ multiplicity of the exciton PL is most easily understood when only the major Ising contribution of the heavy hole ($\propto j_zS_z$) is considered for simplicity. Because of the negative hole exchange parameter β, the states of lower energy are the three with antiferromagnetic coupling $|\mp 1/2\rangle|\pm 3/2\rangle|\mp|S_z|\rangle$,

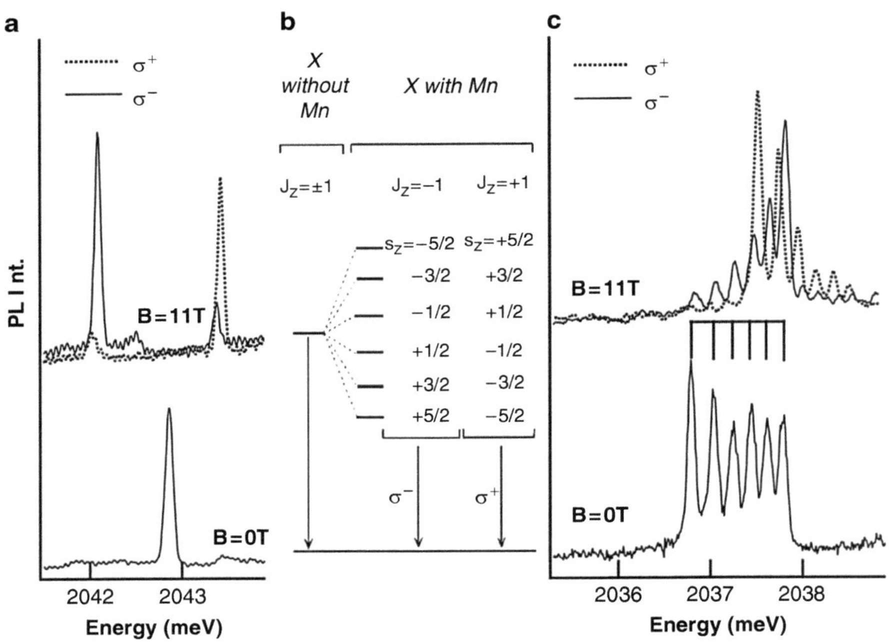

Fig. 5.12 PL spectra of an individual CdTe/ZnTe QD containing a single Mn ion. (**a**) Nonmagnetic reference QD. (**b**) Optical transitions of the Mn-exciton coupled system at zero-magnetic field. (**c**) QD with $N_{mag} = 1$. Reprinted with permission from [70]. Copyright (2004) American Physical Society

followed in opposite order by the ferromagnetic pairings $|\mp 1/2\rangle|\pm 3/2\rangle|\pm|S_z|\rangle$. In an external field, the degeneracy of each pair is removed providing 12 PL lines, six in each circular polarization orientation (Fig. 5.12c). The emission intensity is at zero field almost equally distributed over the lines. In external field, the two states $|-1/2\rangle|+3/2\rangle|-5/2\rangle$ and $|1/2\rangle|-3/2\rangle|-5/2\rangle$ dominate the emission comb of the respective polarization. These states correspond to the giant Zeeman splitting observed on QDs with many magnetic ions (Sect. 5.3). The energy separation between neighboring features in one of the polarizations is solely due to the exchange interaction at zero field. The intensity distribution at nonzero external field is indicative of an increasing population of the Mn^2+ spin ground-state when the Zeeman splitting widens. An effective spin temperature of $T_S = 15$ K is estimated from the field dependence. The absence of Mn pairs slows down the SLR and T_S becomes distinctly larger than the bath temperature (see Sect. 5.4.3).

The zero-field splitting caused by the $sp-d$ interaction is of the order of 1 meV. On this energy scale, the electron–hole exchange interaction cannot be longer ignored. It provides a characteristic fine structure of the exciton groundstate. The short-ranged and isotropic long-ranged part shifts the radiative states $J_X = \pm 1$ energetically upwards and separates them from the dark states $J_X = \pm 2$. In addition, when the in-plane symmetry of the QD is reduced by shape or strain anisotropies below D_2^d, the long-ranged exchange splits the radiative states in two linearly polarized components. In CdSe/ZnSe QDs, the energy separation between the dark and radiative doublet is about 2 meV, while the splitting between the radiative states scatters from a few tens to few hundreds of μeV [25, 46]. On the other hand, according to (5.9), the $sp-d$ coupling depends for a single magnetic ion on its position in the QD. Indeed, significant deviations from the bulk parameters α and β have been deduced from the line splitting in the single-ion case [70]. The interplay between the two exchange interactions – $sp-d$ and electron-hole – gives rise to geometrical effects that, depending on the in-plane anisotropy of the QD and the position of the magnetic ion, can lead to quite complex energy spectra and polarization properties [76]. In addition, the isotropic electron s–d interaction mixes the optically allowed and forbidden exciton states. In this way, the dark excitons gain optical oscillator strength and show up in the emission spectrum, even at zero external field.

At higher optical excitation density, a biexciton state consisting of two excitons is created in the QD [78, 79]. Its radiative decay leaves an exciton behind and is low-energy shifted relative to the single exciton PL by the two-exciton interaction energy. In the biexciton groundstate of zero total angular momentum, the spin of both the two electrons and holes is compensated. Hence, in first-order approximation, there is no $sp-d$ exchange interaction so that excitation of a second exciton in the QD removes the zero-field spin splitting of the magnetic-ion spin states. Experimental evidence of this fact is a mirror symmetry in the fine structure pattern of the biexciton and single-exciton PL [74]. Weak irregularities in the spacings are assigned to a contraction of the spatial carrier wavefunctions ψ_j, analog to the case of magneto-polaron formation.

Studies in a transverse external magnetic field have directly revealed the anisotropy of the $p-d$ exchange interaction of the hole in QDs [75]. The orientation of the magnetic-ion spin along the resultant direction of the external and the hole exchange field, already discussed above in the context of fluctuations, could be directly experimentally verified. In zero or longitudinal external field, the magnetic-ion spin projection is conserved during recombination of the exciton. In a transverse field, the quantization axis of the magnetic-ion spins is along the external field when no exciton is present. Creation of the exciton rotates this axis in the direction of $\boldsymbol{B} + \boldsymbol{B}^{\mathrm{h}}_{\mathrm{exc}}$. Magnetic-ion spin states of given S_z are no longer orthogonal in the two configurations through which the spin conservation in the recombination process is broken showing up experimentally by distinct satellite features. As known from investigations of nonmagnetic self-assembled structures [32], some QDs can exhibit a quite strong in-plane anisotropy providing a substantial transverse magnetic coupling of the hole ground state due to heavy–light hole mixing. In DMS QDs, the resultant transverse part of the p–d exchange provides spin flip-flops that have been also indicated experimentally [75].

The spin configuration of the magnetic ion is also very sensitive to the charge state of the QD. Using electrical injection or photo-depletion, the situation of neutral charge as well as of a single negative or positive charge could be realized for one and the same QD [77]. In the latter case, the PL originates from a trion state, either two electrons and one hole or two holes and one electron, respectively. For instance, in the case of negative charge, the PL emerges from a state where only the hole interacts with the magnetic ion, as the two electron spins are compensated. This provides for the Ising-type hole exchange the six, twofold generate initial carrier-ion states introduced in the context of Fig. 5.12. The final state is an electron coupled to the magnetic ion through an isotropic Heisenberg-type exchange ($\propto \boldsymbol{s}\boldsymbol{S}$) with ferromagnetic ordering. The 12 eigenstates of the electron-ion complex are differently organized: A groundstate septuplet with total spin $J = 3$ is followed by a fivefold degenerate manifold with $J = 2$. The optical transitions are imposed to the condition that the magnetic-ion spin is not subject to change as well as the polarization selection rules of the trion transitions. This gives rise to a rich emission pattern directly expressing the weights of the various spin configurations in the common carrier-ion states [77].

As long as the energy separation exceeds the homogeneous broadening, the fine structure of the optical transitions of a QD with a number of only very few magnetic ions reflects the ground- and excited state manifold formed by the collective spin configurations of both electron and hole as well as the ions. These manifolds increase like $6^{N_{\mathrm{mag}}}$ and $4 \times 6^{N_{\mathrm{mag}}}$, respectively, where the factor of 4 originates from the four possible electron–hole spin orientations. Exact numerical diagonalization of the standard DMS Hamiltonian was carried out up to $N_{\mathrm{mag}} = 4$ [80]. The spectrum is here already so complex that discrete features are hardly resolvable, on account of both the intrinsic homogenous width and the limited experimental resolution. The overall width of the envelope scales like $1/V_{\mathrm{mag}}$ consistent with the statistical approach outlined above. Recently, also the fine structure of excited carrier states in a QD with a single Mn ion has been examined [81]. In III–V DMS, Mn is not incorporated isoelectronically but forms an acceptor. The hole is thus strongly localized on the magnetic ion itself, while the confinement of the QD can be of minor importance. As the degeneracy of heavy and light hole is not lifted in this case, the ion-hole ground-state is given by a triplet corresponding to a total spin of $J = 1$ [82]. Very recently, respective spectra have been indeed obtained for a single Mn ion in an InAs QD applying an external magnetic field along growth direction. The data show a high-energy shift of the $J_z = 0$ state to due strain along growth direction as well as a fine structure of the $J_z = \pm 1$ states due to symmetry lowering in the QD plane [83].

A single magnetic ion in a QD can be also important in the context of quantum information [80, 84]. Conditional excitation and read-out by proper choice of the photon energy or polarization allows for preparation and manipulation of the magnetic-ion spin in a well-defined superpositions of quantum states.

5.6 Outlook: Novel Interactions and Configurations

Nanomagnetism is an emerging field with already many fascinating and unexpected findings over the last few years. DMS QDs exhibit a considerable potential in this regard. Though the paramagnetic response is most expressively seen through the giant Zeeman effect, already the formation of an excitonic magneto-polaron state is associated with ferromagnetic order in the magnetic-ion system. Still, the magnetization degree is weak and limited to low temperature. On the other hand, as has been extensively demonstrated so far on nonmagnetic structures, QDs can be easily charged with resident carriers by doping or even controllable by electrical injection via contacts. Because of the small volume size, huge doping levels are achieved, in particular for small-sized II–VI structures. Under such conditions, an RKKY-type interaction between the magnetic ions may overcome the weak antiferromagnetic superexchange [85–87]. The available magnetization is controlled by the structure of the energy shells and their filling with electrons. The theoretical treatments reveal a complex interplay between magnetism, quantum confinement, and Coulomb effects. For a single magnetic ion, its position is highly relevant [88]. These features make charged DMS QDs very versatile structures, the magnetic state of which can be tuned not only by the number of carriers but also by structural parameters like the confinement potential, gate voltage or Coulomb screening. Ferromagnetic behavior is predicted up to elevated temperature. An interesting point is whether holes with their markedly stronger s–p exchange may lead to an even more stable ferromagnetic state. On resonant tunneling diode structures containing self-assembled CdSe QDs embedded in an (Zn,Be,Mn)Se DMS host, a zero-field splitting of the electron ground-state has been observed in the current-voltage characteristics [89]. It has been tentatively ascribed to a local ferromagnetic interaction of the magnetic ions in the host mediated by the current-induced carriers of the QD. The theoretical analysis of spin-dependent transport through a single DMS QD predicts strongly polarized currents, even if the ferromagnetic source is not full polarized [90]. More details about the theoretical treatment of these phenomena can be found in Chap. 5.

QDs offer the possibility of spin storage and manipulation on the nanometer length scale. Spin injection from a DMS layer in a nonmagnetic QD has been demonstrated and used to evaluate the relevant spin relaxation processes [91, 92]. QD pairs are of particular interest, as they may serve as building blocks for quantum logical gates. Transport of carriers has been investigated on vertically stacked CdSe and (Cd,Mn,Zn)Se QD structures [93,94]. Strong tunneling coupling and the formation of an indirect exciton state have been observed on Cd(Mn,Mg)Te QD-molecules fabricated by interdiffusion [20].

Much work and progress is expected in the above directions in the next years. New QD structures, e.g., based on ZnO, might considerably enrich the research on DMS-related nanomagnetism.

Final Note

After completion of the manuscript, remarkable efforts have been undertaken toward fabrication of InAs QDs containing a single Mn atom and understanding their optical properties [83,95]. Since Mn acts as an acceptor in III–V materials, the resulting PL fine structure is expected to be analog to that of positively charged CdTe QDs studied previously [76, 96]. In contrast, a qualitatively different fine structure and a fan chart in an external magnetic field are found, which are not in accord with calculated spectra including all four contributing spins from the Mn ion, the two holes and the electron [97]. An explanation has been put forward within an effective two-spin model, where the acceptor hole and the Mn 3d electrons form an antiferromagnetic complex with a total angular momentum of $\boldsymbol{J} = 1$ interacting with the ± 1 momentum of the bright exciton states [83, 98].

The work on CdTe QDs with single Mn atom discussed in this chapter has demonstrated that polarization and photon energy of the PL carry the full information on the ion spin in the moment of exciton recombination. Very recently, the opposite mechanism, the orientation of the Mn spin via circularly polarized quasi-resonant photo-excitation has been demonstrated [99]. This opens a doorway toward all-optical manipulation of individual Mn spins. Indirect excitation of a single Mn spin via spin-polarized carriers created optically in a nonmagnetic QD and subsequent spin transfer to an adjacent DMS QD has been discussed in [100].

References

1. E. Bauer, Zeitschrift für Kristallographie **110**, 372 (1958)
2. M. Rabe, M. Lowisch, F. Henneberger, J. Cryst. Growth **184–185**, 248(1998)
3. D. Litvinov, A. Rosenauer, D. Gerthsen et al., Appl. Phys. Lett. **81**, 640 (2002)
4. P.R. Kratzert, J. Puls, M. Rabe et al., Appl. Phys. Lett. **79**, 2814 (2001)
5. F. Tinjod, B. Gilles, S. Moehl et al., Appl. Phys. Lett. **82**, 4340 (2003)
6. F. Tinjod, H. Mariette, Phys. Stat. Sol. (b) **241**, 550 (2004)

7. I.-V. Robin, R. André, C. Bougerol et al., Appl. Phys. Lett. **88**, 233103 (2006)
8. P.R. Kratzert, M. Rabe, F. Henneberger, Phys. Rev. Lett. **83**, 239(1999)
9. S. Mahapatra, T. Kiessling, E. Margopati et al., Appl. Phys. Lett. **89**, 043102 (2006)
10. L.V. Titova, J.K. Furdyna, M. Dobrowolska et al., Appl. Phys. Lett. **80**, 1237 (2002)
11. S. Maćkowski, S. Lee, J.K. Furdyna et al., Phys. Stat. Sol. (b) **229**, 469 (2002)
12. S. Maćkowski, H.E. Jackson, L.M. Smith et al., Appl. Phys. Lett. **83**, 3575 (2003)
13. X. Tang, B. Urbaszek, T.C.M. Graham et al., J. Cryst. Growth **251**, 586 (2003)
14. S. Kuroda, Y. Terai, K. Takita et al., J. Cryst. Growth **184–185**, 274 (1998)
15. I.I. Reshina, S.V. Ivanov, D.N. Mirlin et al., Phys. Rev. B **64**, 035303 (2001)
16. L. Maingault, L. Besombes, Y. Léger et al., Appl. Phys. Lett. **89**, 193109 (2006)
17. P. Wojnar, J. Suffczynski, K. Kowalik et al., Phys. Rev. B **75**, 155301 (2007)
18. C.S. Kim, M. Kim, S. Lee et al., J. Cryst. Growth **214–215**, 395 (2000)
19. T. Kümmell, G. Bacher, M.K. Welsch et al., J. Cryst. Growth **214–215**, 150 (2000)
20. S. V. Zsaitsev, M. K. Welsch, A. Forchel et al., Phys. Stat. Sol. (b) **24**, 3905 (2006)
21. A. K. Bhattacharjee, Phys. Rev. B **58**, 15660 (1998)
22. F.V. Kyrychenko, J. Kossut, Phys. Rev. B **70**, 205317 (2004)
23. J. K. Furdyna, J. Appl. Phys. **64**, R29 (1988)
24. O. Goede, W. Heimbrodt, Phys. Stat. Sol. (b) **146**, 11 (1988)
25. J. Puls, M. Rabe, H.J. Wünsche et al., Phys. Rev. B **60**, R16303 (1999)
26. H.J. Lozykowski, V.K. Shastri, Appl. Phys. Lett. **69**, 3235 (1991)
27. O. Madelung (ed.), *Numerical Data and Functional Relationships in Science and Technology – Landoldt-Börnstein, New Series, Group III*, vol. 17 (Springer, Berlin, 1982)
28. F. Liaci, P. Bigenwald, O. Briot et al., Phys. Rev. B **51**, 4699 (1995)
29. S. Lee, M. Dobrowolska, J.K. Furdyna, J. Cryst. Growth **292**, 311 (2006)
30. S. Lee, M. Dobrowolska, J.K. Furdyna, Solid State Commun. **141**, 311 (2007)
31. E. Oh, K.J. Yee, S.M. Soh et al., Appl. Phys. Lett. **83**, 4604 (2003)
32. A. Koudinov, I. Akimov, Y. Kusraev et al., Phys. Rev. B **70**, R241305 (2004)
33. K. Chang, J.B. Xia, F.M. Peeters, Appl. Phys. Lett. **82**, 2661 (2003)
34. T. Schmidt, M. Scheibner, L. Worschech et al., J. Appl. Phys. **100**, 123109 (2006)
35. G. Bacher, H. Schömig, J. Seufert et al., Phys. Stat. Sol. (b) **229**, 415 (2002)
36. J. Puls, F. Henneberger, J. Cryst. Growth **214/215**, 432 (2000)
37. J. Stühler, G. Schaack, M. Dahl et al., Phys. Rev. Lett. **74**, 2567 (1995)
38. A. K. Bhattacharjee, C. Benoit à la Guillaume, Phys. Rev. B **55**, 10613 (1997)
39. A.V. Chernenko, P.S. Dorozhkin, V.D. Kulakovskii et al., Phys. Rev. B **72**, 045302 (2005)
40. S. Lee, M. Dobrowolska, J.K. Furdyna, Phys. Rev. B **72**, 075320 (2005)
41. V.G. Abramishvili, A.V. Kamarov, S.M. Ryabchenko et al., Solid State Commun. **78**, 1069 (1991)
42. M. Nawrocki, Yu.G. Rubo, J.P. Lascaray et al., Phys. Rev. B **52**, R2241 (1995)
43. K. Shibata, E. Nakayama, I. Souma et al., Phys. Stat. Sol. (b) **229**, 473 (2002)
44. K. Shibata, T. Takabayashi, I. Souma et al., Physica E **10**, 358 (2001)
45. H. Falk, J. Hübner, P.J. Klar et al., Phys. Rev. B **68**, 165203 (2003)
46. T. Flisskowski, A. Hundt, M. Lowisch et al., Phys. Rev. Lett. **86**, 3172 (2001)
47. G. Bacher, H. Schömig, M.K. Welsch et al., Appl. Phys. Lett. **79**, 524 (2001)
48. J. Seufert, G. Bacher, M. Scheibner et al., Phys. Rev. Lett. **88**, 027402 (2002)
49. A.A. Maksimov, G. Bacher, A. McDonald et al., Phys. Rev. B **62**, R7767 (2000)
50. M. Godlewski, R. Narkowicz, T. Wojtowicz et al., J. Cryst. Growth **214–215**, 420 (2000)
51. S. Mackowski, T. Gurung, T.A. Nguyen et al., Appl. Phys. Lett. **84**, 3337 (2004)
52. S. Mackowski, T. Gurung, H.E. Jackson et al., Appl. Phys. Lett. **87**, 072502 (2005)
53. A.V. Scherbakov, D.R. Yakovlev, A.V. Akimov et al., Phys. Stat. Sol. (b) **229**, 723 (2002)
54. A.V. Scherbakov, A.V. Akimov, D.R. Yakovlev et al., Phys. Stat. Sol. (b) **241**, 361 (2003)
55. A. Hundt, J. Puls, A.V. Akimov et al., Phys. Rev. B **72**, 033304 (2005)
56. D. Keller, D.R. Yakovlev, B. König et al., Phys. Rev. B **65**, 035313 (2001)
57. D.R. Yakovlev, M. Kneip, A.A. Maksimov et al., Phys. Stat. Sol. (c) **1**, 989 (2004)
58. A.V. Koudinov, Yu.G. Kusrayev, I.G. Aksyanov, Phys. Rev. B **68**, 085315 (2003)
59. M.E. Msall, J.P. Wolfe, Phys. Rev. B **56**, 9557 (1997)

60. D. Scalbert, Phys. Stat. Sol. (b) **193**, 189 (1996)
61. D.U. Bartholomew, E.-K. Shuh, S. Rodriguez et al., Solid State Commun. **62**, 235 (1987)
62. R. Orbach, Proc. Roy. Soc. **A264**, 458 (1961)
63. A. Abragam, B. Bleaney, *Electron Paramagnetic Resonance of Transition Ions* (Clarendon Press, Oxford, 1970)
64. S. Kuroda, Y. Terai, K. Takita et al., J. Cryst. Growth **214–215**, 140 (2000)
65. G. Bacher, A.A. Maksimov, H. Schömig et al., Phys. Rev. Lett. **89**, 127201 (2002)
66. A. Hundt, J. Puls, F. Henneberger, Phys. Rev. B **69**, R121309 (2004)
67. T. Dietl, J. Spalek, Phys. Rev. Lett. **48**, 355 (1982), T. Dietl, J. Magn. Magn. Mater. **38**, 34 (1983)
68. M. Nawrocki, R. Planel, G. Fishman et al., Phys. Rev. Lett. **46**, 735 (1981)
69. R. Brazis, J. Kossut, Solid State Commun. **122**, 73 (2000)
70. L. Besombes, Y. Léger, L. Maingault et al., Phys. Rev. Lett. **93**, 207403 (2004)
71. P.S. Dorozhkin, A.V. Chernenko, V.D. Kulakovskii et al., Phys. Rev. B **68**, 195313 (2003)
72. D. Heiman, P.A. Wolff, J. Warnock, Phys. Rev. B **27**, 4848 (1983)
73. K.V. Kavokin, I.A. Merkulov, D.R. Yakovlev et al., Phys. Rev. B **60**, 16499 (1999)
74. L. Besombes, Y. Léger, L. Maingault et al., Phys. Rev. B **71**, 161307 (2005)
75. Y. Léger, L. Besombes, L. Maingault et al., Phys. Rev. B **72**, 241309 (2005)
76. Y. Léger, L. Besombes, L. Maingault et al., Phys. Rev. Lett. **95**, 047403 (2005)
77. Y. Léger, L. Besombes, J. Fernández-Rossier et al., Phys. Rev. Lett. **97**, 107401 (2006)
78. V.D. Kulakovskii, G. Bacher, R. Weigand et al., Phys. Rev. Lett. **82**, 1780 (1999)
79. I.A. Akimov, J.T. Andrews, F. Henneberger, Phys. Rev. Lett. **96**, 067401 (2006)
80. J. Fernández-Rossier, Phys. Rev. B **73**, 045301 (2006)
81. M.M. Glazov, E.L. Ivchenko, L. Besombes et al., Phys. Rev. B **75**, 205313 (2007)
82. A.O. Gorovov, Phys. Rev. B **70**, 035321 (2004)
83. A. Kudelski, A. Lemaitre, A. Miard et al., Phys. Rev. Lett. **99**, 247209 (2007)
84. A.O. Gorovov, A.V. Kalameitsev, Phys. Rev. B **71**, 035338 (2005)
85. F. Qu, P. Hawrylak, Phys. Lett. **95**, 217206 (2005); Phys. Lett. **96**, 157201 (2006)
86. J.I. Climente, M. Kursiński, P. Hawrylak et al., Phys. Rev. B **71**, 125321 (2005)
87. R.M. Abolfath, P. Hawrylak, I. Žutić, Phys. Lett. **98**, 207203 (2007)
88. Nga T.T. Nguyen, F.M. Peeters, Phys. Rev. B **76**, 045315 (2007)
89. C. Gould, A. Slobodskyy, D. Supp et al., Phys. Rev. Lett. **97**, 017202 (2007)
90. F. Qu, P. Vasilopoulos, Phys. Rev. B **74**, 245308 (2006)
91. J. Seufert, G. Bacher, H. Schömig et al., Phys. Rev. B **69**, 035311 (2004)
92. A. Muramaya, T. Furuta, K. Hyomi et al., Phys. Rev. B **75**, 195308 (2007)
93. S. Lee, H.S. Lee, J.Y. Lee et al., Appl. Phys. Lett. **86**, 033114 (2005)
94. S. Lee, D. G. Choi, T. K. Lee et al., Appl. Phys. Lett. **90**, 201916 (2006)
95. O. Krebs, E. Benjamin, A. Lemaître, Phys. Rev. B **80**, 165315 (2009)
96. Y. Léger, L. Besombes, L. Maingault et al., Phys. Rev. B **76**, 045331 (2007)
97. S. Chutia, A.K. Bhattacharjee, Phys. Rev. B **78**, 195311 (2008)
98. J. van Bree, P.M. Koenraad, J. Fernández-Rossier, Phys. Rev. B **78**, 165414 (2008)
99. C. Le Gall, L. Besombes, H. Boukari et al., Phys. Rev. Lett. **102**, 127402 (2009)
100. M. Goryca, T. Kazimierczuk, M. Nawrocki et al., Phys. Rev. Lett. **103**, 087401 (2009)

Chapter 6
Magnetic Ion–Carrier Interactions in Quantum Dots

Pawel Hawrylak

Abstract In this theoretical chapter, cylindrically shaped quantum dots are the center of attention. It is shown that $sp-d$ interaction does depend on exact position of the magnetic atoms in the dot and that the interaction strength changes when the dots are charged. Many body aspects of the carriers are, thus, of an importance for a discussion presented here. The dependence of $sp-d$ interaction on location of Mn and on charge state of the dot is also shown to influence the indirect coupling between magnetic ions mediated by charge carriers.

6.1 Introduction

Interaction of carriers with magnetic ions in semimagnetic semiconductors has been the subject of many investigations [1–3], well covered by other chapters in this book. This chapter reports mainly recent theoretical work on zero-dimensional semiconductor structures, quantum dots [4–6], carried out in the Quantum Theory group at the Institute for Microstructural Sciences, National Research Council of Canada. Quantum dots are nanoscale semiconductor structures, which allow for the engineering of carrier wavefunctions, their number, interaction among them, their total spin, and its coupling to magnetic ions. In the bulk diluted magnetic semiconductors, carrier localization by impurities enhances coupling of carrier spin to spins of magnetic ions. Carrier spin polarizes the spin of magnetic ions, leading to the formation of magneto-polarons [3, 7]. It was realized early on by one of us that such coupling could be enhanced if more carriers were localized in quantum dots [8]. Recent experiments on self-assembled quantum dots show that one can now control the number of carriers localized in quantum dots [6, 9], and separately, one can dope quantum dots with few magnetic ions, down to one [10–13]. Similar progress is reported for nanocrystals [14–17]. The control of carrier density in two-dimensional

P. Hawrylak
Quantum Theory Group, Institute for Microstructural Sciences, National Research Council of Canada, Ottawa, ON, Canada K1A 0R6
e-mail: pawel.hawrylak@nrc-cnrc.gc.ca

J. Kossut and J.A. Gaj (eds.), *Introduction to the Physics of Diluted Magnetic Semiconductors*, Springer Series in Materials Science 144,
DOI 10.1007/978-3-642-15856-8_6,

semimagnetic semiconductor systems has led to voltage control of their magnetic properties [18]. In parallel, the development of local probes bodes well for the future spatial control over the magnetic ions [19, 20]. It is the anticipation of the future control over both spins of carriers and spins of magnetic ions, and hence the development of spintronics at the nanoscale, which is the main motivation behind the work on magnetic quantum dots. To explore nano-spintronics, one needs to understand magnetic ion–carrier interactions in quantum dots. This topic has been a subject of significant experimental [10, 11, 13, 21–23] and theoretical [24–29] work, which we cannot hope to cover adequately. Instead, we focus on our own recent work with M. Korkusinski, F. Qu, R. Abolfath, and J.I. Climente [8, 13, 30–33].

6.2 One Electron States in a Quantum Dot

We describe here briefly the effective mass, $\boldsymbol{k} \cdot \boldsymbol{p}$ and atomistic tight-binding approaches to electronic states of a single carrier in self-assembled quantum dots [6].

6.2.1 Conduction Band Electron States

The simplest description of states of one conduction band electron in a quantum dot is that of a particle with effective mass m^* confined by an effective three-dimensional confining potential due to differences in energies of the conduction band minima in the dot and the barrier material. With strong confinement in the vertical direction, lens-shaped self-assembled quantum dots [6, 34, 35] can be approximated by a quasi-two-dimensional quantum dot with parabolic confinement, with a Hamiltonian given by:

$$\hat{H} = \frac{\boldsymbol{P}_x^2}{2m^*} + \frac{1}{2}m^*\omega_0^2 x^2 + \frac{\boldsymbol{P}_y^2}{2m^*} + \frac{1}{2}m^*\omega_0^2 y^2. \tag{6.1}$$

The Hamiltonian, the single-particle states and the energies of an electron in a parabolic QD correspond to two coupled harmonic oscillators with quantum numbers m and n, $\varphi_{mn}(x, y) = \varphi_m(x)\varphi_n(y)$ [34], where x and y are the electron coordinates in the plane of the quantum dot, with z being the growth direction. The lowest three one-dimensional harmonic oscillator states are $\varphi_0(\widehat{x}) = e^{-\widehat{x}^2/4}/(2\pi l_0^2)^{1/4}$, $\varphi_1(\widehat{x}) = \widehat{x}\, e^{-\widehat{x}^2/4}/(2\pi l_0^2)^{1/4}$, and $\varphi_2(\widehat{x}) = (\widehat{x}^2 - 1)e^{-\widehat{x}^2/4}/(8\pi l_0^2)^{1/4}$, with $\widehat{x} = x/\ell_0$ and $\ell_0 = 1/\sqrt{\omega_0}$. Here, ω_0 is the shell spacing, and length and energy are measured in effective Bohr radius $a_{\rm B} = \frac{\varepsilon}{m^* e^2}$ and effective Rydberg Ry $= \frac{1}{2m^* a_{\rm B}^2}$, respectively. Here, ε is the dielectric constant, e is electron charge, and $\hbar = 1$. The most important feature of the single-particle energy spectrum $E(nm) = (n + m + 1)\omega_0$ is the formation of degenerate shells, with degeneracy $g_s = s + 1$ whenever $s = n + m$, and shell spacing ω_0. This is

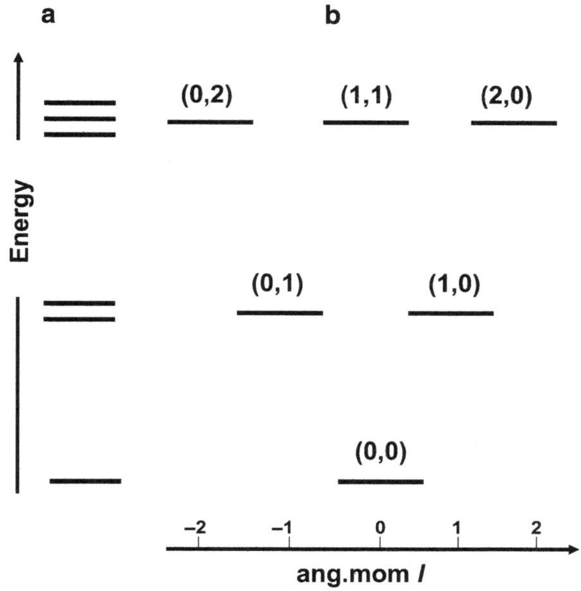

Fig. 6.1 (**a**) Single-electron energy spectrum, with degenerate energy levels separated in energy for visibility, and (**b**) single-electron energy spectrum as a function of the angular momentum l for a parabolic quantum dot

illustrated in Fig. 6.1a, which shows the electron energy levels as a function of energy. Fig. 6.1b shows the same spectrum but with energy levels classified by dimensionless angular momentum $l = m - n$ to visualize degeneracies. The s, p, and d shells are shown. Note the degeneracy of the third shell due to dynamical symmetry, i.e., the parabolic form of the confining potential.

6.2.2 Valence Band Hole States

The purpose here is to understand the confined levels of valence holes, and this can be accomplished by considering the simplest 4-band Luttinger-Kohn model [36].

The model considers a single valence hole on a quantum disk with radius R and height w (for a rectangular box, see [37]). The energy is measured from the top of the valence band, and increasing hole energy corresponds to states in the valence band with lower energy. To resolve degeneracies in the hole spectrum, we include external magnetic field B applied in parallel to the z-axis. The Luttinger-Kohn 4×4 Hamiltonian in cylindrical coordinates reads

$$H_L = \begin{pmatrix} P_+ & R & -S & 0 \\ R^* & P_- & 0 & S \\ -S^* & 0 & P_- & R \\ 0 & S^* & R^* & P_+ \end{pmatrix}, \tag{6.2}$$

where the basis of the above Hamiltonian consists of functions with total angular momentum $J = 3/2$ and $J_z = 3/2, -1/2, 1/2, -3/2$, respectively. The matrix elements of the Hamiltonian are

$$P_+ = \frac{\hbar^2}{2m_0}\left[(\gamma_1 - 2\gamma_2)k_z^2 + (\gamma_1 + \gamma_2)k_\rho^2\right], \tag{6.3}$$

$$P_- = \frac{\hbar^2}{2m_0}\left[(\gamma_1 + 2\gamma_2)k_z^2 + (\gamma_1 - \gamma_2)k_\rho^2\right], \tag{6.4}$$

$$R = \frac{\hbar^2}{2m_0}(-\sqrt{3})\gamma_{23}k_-^2, \tag{6.5}$$

$$S = \frac{\hbar^2}{2m_0}(2\sqrt{3})\gamma_3 k_- k_z, \tag{6.6}$$

where $\boldsymbol{k} = -i\nabla - (e/c\hbar)\boldsymbol{A}$ $(e > 0)$, $k_- = k_x - ik_y$, $k_\rho^2 = k_x^2 + k_y^2$ and γ_i are Luttinger parameters. We use the axial approximation: $\gamma_{23} = (\gamma_2 + \gamma_3)/2$. The vector magnetic potential is taken in the symmetric gauge $\boldsymbol{A} = B/2(-y, x, 0)$.

The disk confinement is introduced as a potential $V(\rho, z)$ so that the total Hamiltonian is

$$H = H_{\mathrm{L}} + V(\rho, z). \tag{6.7}$$

However, to investigate the importance of the heavy hole-light hole mixing, we assume a simple confinement, i.e., a disk with infinite walls ($V(\rho, z) = 0$ inside, and infinity outside the disk), yielding simple wave functions. A parabolic in-plane confinement can also be included.

In the absence of the mixing, the wave functions for holes can be written as

$$\langle \boldsymbol{r}|m, n, \nu, m_j\rangle = \frac{\sqrt{2}}{R|J_{m+1}(k_n^m R)|} J_m(k_n^m \rho)\frac{1}{\sqrt{2\pi}} \mathrm{e}^{im\phi}\xi^\nu(z)u_{m_j}(\boldsymbol{r}). \tag{6.8}$$

The above wave function consists of two parts: the periodic part of the Bloch function $u_{m_j}(\boldsymbol{r})$, for which $m_j = \pm 3/2$ for heavy holes and $m_j = \pm 1/2$ for light holes, and the envelope function. With rotational symmetry, the angular momentum m is a good quantum number, and the angular part of the envelope function is a simple exponential. The radial part of the envelope function is the Bessel function J_m for the given angular momentum channel m, with k_n^m being the hole wave vector defined in terms of the roots α_n^m of the Bessel function: $k_n^m = \alpha_n^m/R$, with n being the radial quantum number. Finally, the vertical part of the wave function $\xi^\nu(z)$ for a hard-wall disk can be expressed as $\xi^0(z) = \sqrt{2/w}\cos(\pi z/w)$ for the lowest subband, and $\xi^1(z) = \sqrt{2/w}\sin(2\pi z/w)$ for the second subband.

Upon the inclusion of mixing terms R and S in the Hamiltonian H_{L}, the angular momentum of the envelope function m and that of the Bloch function m_j are no longer good quantum numbers. However, the eigenstates of the total Hamiltonian can be classified by the z component of the total angular momentum: $L = m + m_j$. Moreover, the inversion symmetry of the system introduces yet another quantum number: the parity $\sigma = \downarrow$ or $\sigma = \uparrow$. Now the eigenstates of the Hamiltonian (6.2) take a form of spinors:

$$|L, N, \downarrow\rangle = \sqrt{\frac{2}{w}} \sum_n C_{n,m_j}^{L,N} \begin{pmatrix} f_n^{L-3/2}(\rho,\phi)\sin(2\pi z/w)|m_j = 3/2\rangle \\ f_n^{L+1/2}(\rho,\phi)\sin(2\pi z/w)|m_j = -1/2\rangle \\ f_n^{L-1/2}(\rho,\phi)\cos(\pi z/w)|m_j = 1/2\rangle \\ f_n^{L+3/2}(\rho,\phi)\cos(\pi z/w)|m_j = -3/2\rangle \end{pmatrix} \tag{6.9}$$

for parity $\sigma = \downarrow$, and

$$|L, N, \uparrow\rangle = \sqrt{\frac{2}{w}} \sum_n D_{n,m_j}^{L,N} \begin{pmatrix} f_n^{L-3/2}(\rho,\phi)\cos(\pi z/w)|m_j = 3/2\rangle \\ f_n^{L+1/2}(\rho,\phi)\cos(\pi z/w)|m_j = -1/2\rangle \\ f_n^{L-1/2}(\rho,\phi)\sin(2\pi z/w)|m_j = 1/2\rangle \\ f_n^{L+3/2}(\rho,\phi)\sin(2\pi z/w)|m_j = -3/2\rangle \end{pmatrix} \tag{6.10}$$

for parity $\sigma = \uparrow$. The quantum number N enumerates the states with given total angular momentum and parity, and the function f denotes the radial and angular part of the envelope function defined in (6.8).

An example of the eigenenergies of the 4×4 Luttinger-Kohn Hamiltonian is shown in Fig. 6.2 for model parameters corresponding to a SiGe quantum disk [6, 38].

The top of the valence band defines the zero energy, and the increasing energy of the hole implies hole states deeper in the valence band (lower energy). At zero-magnetic field, the states with the same L but opposite parity are degenerate (they form parity doublets). The energies of these states form a shell structure. The s shell is composed of the states with $L = 0$. The p shell consists of two states with parity $L = -1$ and $L = +1$; however, these doublets do not have the same energy, and the shell is not degenerate. Then there is a very broad d shell, consisting of the parity doublets with $L = -2$ and $L = 2$ similar in energy, and the parity doublet with $L = 0$, whose energy is larger.

The parity degeneracy is removed in the magnetic field. Moreover, the energy spectrum evolves with the magnetic field into Landau levels. The lowest Landau level comprises states with parity $\sigma = \uparrow$ and has a slope corresponding to holes of large mass. The second Landau level is composed of the states with parity $\sigma = \downarrow$ and has a more light-hole character.

The above considerations were carried out with an explicit assumption that the heavy- and light-hole band edges are degenerate. In self-assembled quantum dots such as InAs embedded in GaAs, strain removes this degeneracy, so that the heavy-hole band edge corresponds to a lower energy than the light-hole band edge. It means that as a result of the 4×4 Luttinger-Kohn approach the lowest hole states are expected to be predominantly of heavy-hole character. Moreover, in narrow-gap semiconductors such as InAs the conduction-valence band mixing is large and it is necessary to consider an 8×8 Luttinger-Kohn model [39]. However, the 4×4 Luttinger-Kohn model is sufficient to illuminate the differences between electrons and holes in quantum dots.

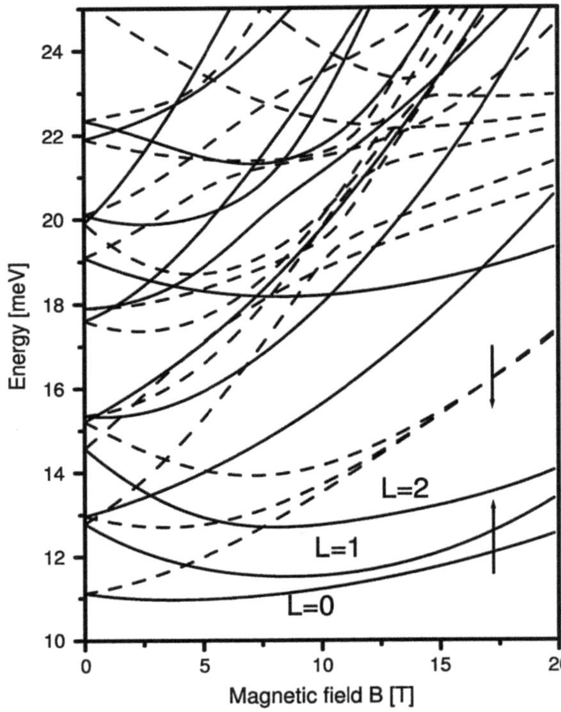

Fig. 6.2 Single-hole energy spectrum as a function of the magnetic field. *Solid lines* represent the parity $\sigma = \uparrow$ and *dashed lines* $\sigma = \downarrow$. From [6]

6.2.3 Tight-Binding Models

The one-band and multi-band $\boldsymbol{k} \cdot \boldsymbol{p}$ methods described in previous sections do not include the atomistic structure of the quantum dot, including the effect of interface between the dot and barrier materials. Yet, the results obtained from these effective mass approaches seem to work very well if one accepts the existence of very few effective parameters. This suggests that perhaps the details of atomic arrangement are washed out due to alloy fluctuations and only small number of effective parameters, such as effective mass and band-offset, suffice to describe the ground and excited states. Obviously, one would like to know how to extract these macroscopic parameters, including dielectric constant and confining potentials, from microscopic, i.e., atomistic structure of a quantum dot. A simulation of a structure containing an InAs self-assembled quantum dot embedded in GaAs material with a total size of 40 nm $\times$ 40 nm $\times$ 15 nm requires a simulation of $\approx$1 million atoms. With sp^3 orbitals per atom, i.e., eight electrons, this means dealing with $\approx$10 million electrons. Tight-binding methods, interpolating between first principle methods and $\boldsymbol{k} \cdot \boldsymbol{p}$ approaches, are capable of treating atomic systems of this size. In the tight-binding method, one selects most relevant orbitals $|a, i\rangle$ localized on the atom “a”. The wavefunction of a system is expanded on the basis of these localized orbitals. The expansion coefficients and the energy spectrum are obtained by diagonalization of the tight-binding Hamiltonian

$$H = \sum_{a,b,i,j} t_{i,j}^{a,b} c_{a,i}^{+} c_{b,j}, \qquad (6.11)$$

where $c_{a,i}^{+}$ creates an electron on orbital “i” of atom “a”. The energy spectrum is completely characterized by hopping matrix elements $t_{i,j}^{a,b}$. sp^3d tight-binding models have been implemented by Klimeck and co-workers [40] and Bryant and co-workers [41]. A simplified version of the tight binding model, the effective bond orbital model (EBOM), was developed by Y.C. Chang and co-workers [42]. This approach uses one s-like anti-bonding and three p-like bonding effective orbitals. The effective orbitals, or rather tight-binding hopping matrix elements, are chosen in such a way as to reproduce the $\boldsymbol{k} \cdot \boldsymbol{p}$ band structure of constituent materials near the center of the Brillouin zone. Thus, the EBOM model is a real space version of the $\boldsymbol{k} \cdot \boldsymbol{p}$ model not only with proper effective masses, but also with elements of atomistic description built in. The electronic structure obtained using EBOM reported in [42] appears, not surprisingly, to be in a good agreement with $\boldsymbol{k} \cdot \boldsymbol{p}$ calculations.

In Fig. 6.3, we show a comparison of the one electron energy spectrum of a lens-shaped InAs-GaAs self-assembled quantum dot obtained by the atomistic EBOM method, 8-band $\boldsymbol{k} \cdot \boldsymbol{p}$, and effective mass method by Sheng et al. [43]. The unknown parameters, e.g., effective masses, were fitted to the atomistic calculation. What the figure demonstrates is that the level ordering and the formation of electronic shells are present in all three approaches.

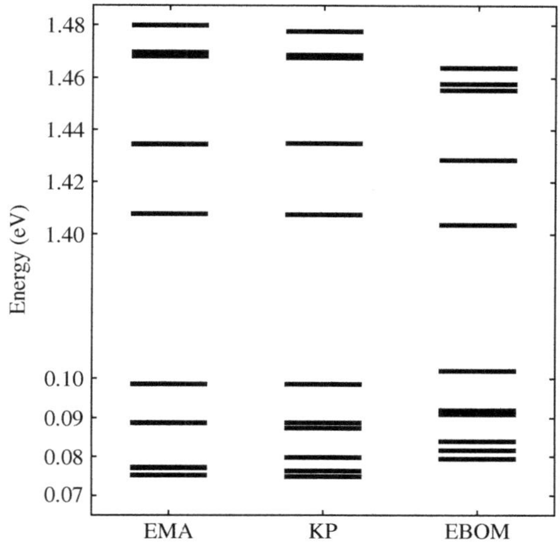

Fig. 6.3 Calculated energy levels of the $In_{0.5}Ga_{0.5}As/GaAs$ self-assembled quantum dot by the eight-band $k \cdot p$ (KP) method and EBOM. Also shown are the energy levels fitted for the effective-mass approximation (EMA). From [43]

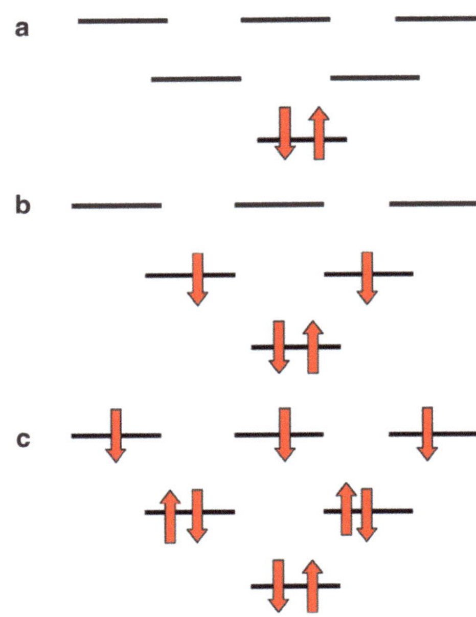

Fig. 6.4 Filling of single-electron energy levels with electrons: (**a**) s-shell with $S = 0$, (**b**) partially spin polarized p-shell with $S = 1$, and (**c**) partially spin polarized d-shell with $S = 3/2$

6.3 Total Spin of Many-Electron Quantum Dots

The shell structure, with its degeneracies, has important implications for the engineering of total spin S of electrons. The number of electrons can be varied using applied gates. As was shown early on by us using exact diagonalization techniques [44], and experimentally in vertical quantum dots [45], the total spin S of electrons can be engineered by changing the number of electrons. The spin is zero for closed electron shells with $N = 2, 6, 12, \ldots$ but is finite for partially filled shells. The filling of electronic shells is illustrated in Fig. 6.4. The $N = 2$ electrons fill the lowest s shell and with total spin $S = 0$. Additional two electrons have to be distributed over a shell of two degenerate levels. Exchange interaction leads to a spin-polarized state, a generalized Hunds rule for parabolic quantum dots. For d shell with threefold degeneracy, spin polarized state of three electrons is predicted.

Hence by changing the number of electrons with the gate voltage, the total electron spin S of the "artificial atom" can be engineered, as shown in Fig. 6.5. The stability of spin is governed by the exchange integral. For example, within the p shell this integral is $\approx 0.2\sqrt{\pi\omega_0}$. It is of the order of tens of meV, tunable by the dot size, level spacing, and external voltage.

6.4 Energy Spectrum of Magnetic Ions in Semiconductors

We have discussed the engineering of the total spin of electrons in a quantum dot, an artificial atom. This is possible because of degeneracies of the confining potential, and our ability to change the number of electrons in degenerate shells. The same

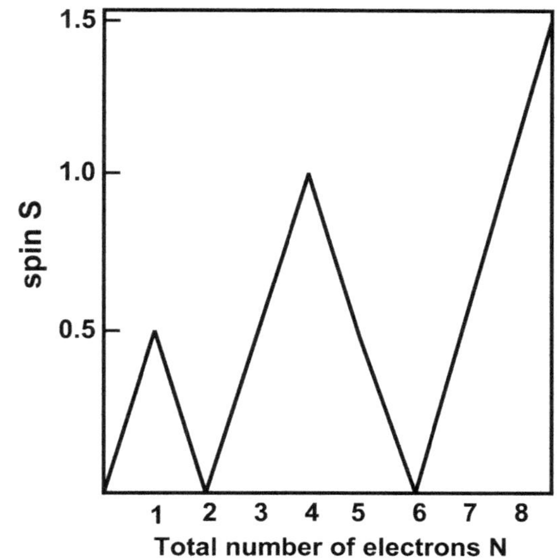

Fig. 6.5 Evolution of total spin S with increasing number of electrons N. Note $S = 0$ for filled shells and finite spin $S = 1/2, 1, 3/2$ for partially filled s, p, d shells

mechanism operates in natural atoms. In a Manganese Mn atom, $N = 5$ electrons fill the five-fold degenerate electronic d shell. The exchange energy is, however, of the order of eV, and the spatial extent is of the order of a fraction of nm. In II–VI semiconductors, Mn ion acts as isoelectronic impurity. Hence, the energy spectrum of electrons in a d shell of a single Mn ion is well approximated by a spectrum of a single angular momentum M with $M = 5/2$, and six states $|M_z\rangle$, with $M_z = -5/2, -3/2, -1/, +1/2, +3/2, +5/2$. However, in III–V semiconductors Mn ion acts as acceptor, adding both spin and a hole bound to a positively charged ion. A more microscopic approach can be found in, e.g., [46–48].

6.5 Mn–Mn Interaction

The interaction between two Mn ions is modeled by an effective spin exchange Hamiltonian $\hat{H}_{\text{Mn–Mn}} = J_{12}^{\text{AF}} \boldsymbol{M}_1 \cdot \boldsymbol{M}_2$. The antiferromagnetic exchange coupling parameter J_{12}^{AF} is a complex, strongly decaying function of the distance between the two ions [49]. In what follows [32] we have modeled the dependence of J_{12}^{AF} on $R_{12} = R_1 - R_2$ by $J_{12}^{\text{AF}} = J_{12}^{(0)} \exp[-\lambda(\frac{R_{12}}{a_0} - 1)]$. Here, a_0 is the lattice constant and $J_{12}^{(0)}$ is the nearest-neighbor interaction. The strength J_{12}^{AF} of exchange coupling between the Mn ion and its next nearest neighbor is reflected in the value of λ. We choose $\lambda = 5.1$ corresponding to J_{12} for the interaction between the next neighbor Mn ions being about 12% of $J_{12}^{(0)}$, which is consistent with experimental measurement [50].

6.6 Magnetic Ion–Electron Exchange Interaction

The electron–Mn $sp-d$ exchange interaction is modeled here by a contact ferromagnetic interaction [3, 51] $\hat{H}_{\text{e-Mn}} = -J_C^{2D} \boldsymbol{M} \cdot \boldsymbol{S} \delta(\boldsymbol{r} - \boldsymbol{R})$, where $\boldsymbol{S}$ ($\boldsymbol{M}$) is the electron (Mn) spin, $\boldsymbol{r}$ ($\boldsymbol{R}$) is the 2D electron (Mn) position in the plane of the quantum dot, and $J_C^{2D} = J_C 2/d$ is the interaction strength, with d being the thickness of the quantum dot. In what follows we adopt $J_C = 15\,\text{eV}$ [13], $d = 2\,\text{nm}$, $\varepsilon = 10.6$, $m^* = 0.106\,m_0$, Bohr radius $a_B = 52.9\,\text{Å}$, Ry $= 12.8\,\text{meV}$, with typical $\omega_0 = 4\,\text{Ry}$ and the effective width $l_0 = 26.45$, applicable to II–VI (Cd, Mn)Te semiconductor QDs. For additional information on electron–Manganese interaction, the reader is referred to the Chap. 3 by Merkulov and Rodina.

6.7 Hybrid System of Magnetic Ions and Carriers

We are now ready to consider a hybrid system composed of carriers and magnetic ions confined in a quasi-two-dimensional II–VI parabolic quantum dot. The Hamiltonian describing N interacting electrons and N_M magnetic Mn ions can be written as:

$$\hat{H} = \sum_{i=1}^{N} \left[\frac{\boldsymbol{P}_i^2}{2m^*} + \frac{1}{2} m^* \omega_0^2 \boldsymbol{r}_i^2 \right] + \sum_{i<j} V\left(\boldsymbol{r}_i - \boldsymbol{r}_j\right)$$
$$- J_C^{2D} \sum_{i=1}^{N} \sum_{j=1}^{N_M} \boldsymbol{M}_j \cdot \boldsymbol{S}_i \delta\left(\boldsymbol{r}_i - \boldsymbol{R}_j\right) + \frac{1}{2} \sum_{i=1}^{N_M} \sum_{j=1}^{N_M} J_{ij}^{AF} \boldsymbol{M}_j \cdot \boldsymbol{M}_i \qquad (6.12)$$

The first term in (6.12) corresponds to the sum of the one-electron Hamiltonians, the second term describes the electron–electron Coulomb interactions, the third term describes the $sp-d$ contact exchange interaction between jth Mn ion at position $\boldsymbol{R}_j$ and ith electron at position $\boldsymbol{r}_i = (x_i, y_i)$, and the fourth term describes short ranged antiferromagnetic mutual interaction of Mn ions.

The corresponding single-particle energies are those of two harmonic oscillators. The electron–electron interaction $V\left(\boldsymbol{r}_i - \boldsymbol{r}_j\right) = e^2/(\varepsilon|\boldsymbol{r}_i - \boldsymbol{r}_j|)$ is approximated by the 2D Coulomb interaction. The finite thickness of quantum dot is known to change the quantitative but not the qualitative effects of e–e interaction [6].

For computational purposes, it is convenient to transform the Hamiltonian into second quantization form. Let us denote single-particle harmonic oscillator states $|n, m\rangle = |i\rangle$, and transform the Hamiltonian into second quantization form by expanding the electron field operators $\hat{\psi} = \sum_{i,\sigma} \varphi_i(x, y) \chi_\sigma c_{i,\sigma}$ in orbital and spin eigenstates and annihilation (creation) $c_{i,\sigma}$ ($c_{i,\sigma}^+$) operators. We outline this procedure for the electron-magnetic ion scattering term, known from the Kondo and Anderson models:

$$H_{\text{e-Mn}} = -J_C^{2D} \sum_{i,j,\sigma\sigma'} \int \mathrm{d}\boldsymbol{r}\, \varphi_i^*(\boldsymbol{r}) \delta\left(\boldsymbol{r} - \boldsymbol{R}\right) \varphi_j(\boldsymbol{r}) \langle \chi_{\sigma'} \boldsymbol{S} \chi_\sigma \rangle \cdot \boldsymbol{M} c_{i,\sigma}^+ c_{j,\sigma'}. \qquad (6.13)$$

Integrating out position and spin degrees of freedom results in a Hamiltonian

$$H_{\text{e-Mn}} = -\sum_{i,j,\sigma\sigma'} J_{ij}(R) \boldsymbol{S}_{\sigma\sigma'} \cdot \boldsymbol{M} c^{+}_{i,\sigma} c_{j,\sigma'}, \tag{6.14}$$

which involves the exchange matrix elements $J_{ij}(R) = J_{\mathrm{C}}^{\mathrm{2D}} \varphi_i^*(R)\varphi_j(R)$, Pauli spin matrices $\boldsymbol{S}_{\sigma,\sigma'}$, and Mn spin operators. The exchange matrix elements $J_{ij}(R)$ are determined by the wave function of the two states (i, j) at the position $\boldsymbol{R}$ of the Mn ion. The electron–Mn exchange interaction combines two effects: flipping of electron spin simultaneous with flipping of Mn ion spin, and scattering of electrons between different orbitals, i.e., disorder. We can bring these effects out by rewriting the exchange Hamiltonian in terms of Mn spin rising and lowering operators M^{+}, M^{-}, with the final form of the Hamiltonian:

$$\begin{aligned}
\hat{H} &= \sum_{i\sigma} E_{i,\sigma} c^{+}_{i,\sigma} c_{i,\sigma} + \frac{1}{2} \sum_{ijkl\sigma\sigma'} \langle i,j|V_{ee}|k,l\rangle c^{+}_{i,\sigma} c^{+}_{j,\sigma'} c_{k,\sigma'} c_{l,\sigma} \\
&- \sum_{i,j}\sum_{k} \frac{J_{ij}(R_k)}{2} \left[(c^{+}_{i,\uparrow} c_{j,\uparrow} - c^{+}_{i,\downarrow} c_{j,\downarrow}) M_{k,z} + c^{+}_{i,\downarrow} c_{j,\uparrow} M^{+}_k + c^{+}_{i,\uparrow} c_{j,\downarrow} M^{-}_k \right] \\
&+ \frac{1}{2} \sum_{i=1}^{N_{\mathrm{M}}} \sum_{j=1}^{N_{\mathrm{M}}} J_{ij}^{\mathrm{AF}} \boldsymbol{M}_j \cdot \boldsymbol{M}_i .
\end{aligned} \tag{6.15}$$

Here, $E_{i,\sigma}$ is the energy of an electron on the single-particle orbital $|i\rangle$ with spin σ_i. The two-body Coulomb matrix elements $\langle i, j|V_{ee}|k, l\rangle$ [6] scatter electrons from states $|k\sigma', l\sigma\rangle$ to states $|i\sigma, j\sigma'\rangle$. The second line in (6.15) is the electron–Mn Hamiltonian, which consists of three terms. The first term measures the difference in spin up and down population and acts as Zeeman energy while the second and third term involve flipping of electron spin compensated by the flipping of Mn spin. The last line is the Mn–Mn interaction term. This Hamiltonian will now be analyzed in several limiting cases.

6.8 Magnetic Ion–Many Electron Interaction

The simplest problem is that of a single Mn spin interacting with an electron droplet.

Following [31], we now discuss a Hamiltonian of a single Mn impurity interacting with N mutually interacting electrons, from the point of view of the electronic system:

$$\begin{aligned}
\hat{H} &= \sum_{i\sigma} E_{i,\sigma} c^{+}_{i,\sigma} c_{i,\sigma} + \frac{1}{2} \sum_{ijkl}^{\sigma\sigma'} \langle i,j|V_{ee}|k,l\rangle c^{+}_{i,\sigma} c^{+}_{j,\sigma'} c_{k,\sigma'} c_{l,\sigma} \\
&- \sum_{i,j} \frac{J_{ij}(R)}{2} \left[(c^{+}_{i,\uparrow} c_{j,\uparrow} - c^{+}_{i,\downarrow} c_{j,\downarrow}) M_z + c^{+}_{i,\downarrow} c_{j,\uparrow} M^{+} + c^{+}_{i,\uparrow} c_{j,\downarrow} M^{-} \right].
\end{aligned} \tag{6.16}$$

The first line in (6.16) describes the electrons and the second line describes their scattering off magnetic impurity, with and without flipping the electron spin.

To calculate the electronic properties of the interacting electron–Mn system, the wave function is expanded on the following basis: $|i_1, i_2, \ldots, i_{N\uparrow}\rangle \times |j_1, j_2, \ldots, j_{N\downarrow}\rangle|M_z\rangle$, where $|i_1, i_2, \ldots, i_{N\uparrow}\rangle = c^+_{i_1,\uparrow} c^+_{i_2,\uparrow} \cdots c^+_{i_N,\uparrow}|0\rangle$, $|0\rangle$ is the vacuum, and $N\uparrow$ ($N\downarrow$) is the number of spin-up (down) electrons, $N\uparrow + N\downarrow = N$. The basis states are grouped into spin-up and spin-down electron states for each state of the Mn ion $|M_z\rangle$, with $M_z = \pm 5/2, \pm 3/2, \pm 1/2$. The number of possible configurations is determined by the number of single particle orbitals, the number of electrons, and the size of magnetic spin M. The Hamiltonian matrix is constructed on the basis of configurations, and upon diagonalization the eigenenergies and eigenstates of a single Mn and N interacting electrons were obtained. The measure of magnetic interaction was investigated in terms of energy shift Δ, defined by the difference between the ground state energy E_c of electrons coupled to a single Mn ion, and energy E_e of a QD without Mn ion. Because this energy splitting can be thought of as induced by a local magnetic field produced by the magnetic moment of a Mn ion, one can refer to it as the local Zeeman spin splitting.

Note that in the absence of electron–Mn exchange interaction the Coulomb interaction conserves the total angular momentum and z component of the total spin S of electrons. Hence, the diagonalization of the Hamiltonian $\hat{H}_{\text{elec}}$ can be performed separately for each L_e, S_z subspace. In a Mn ion-doped QD, however, electron–Mn exchange interaction induces coupling between electronic configurations with different total spin of electrons and/or different total angular momentum, breaking the symmetry. As a result, the diagonalization has to be performed in the entire Hilbert space, considerably increasing computational effort in comparison with that needed for an undoped QD.

6.8.1 *Engineering Mn–Carrier Exchange Interaction Matrix Elements*

The effect of the Mn ion is determined by the electron–Mn exchange interaction matrix elements $J_{ij}(R) = J^{2D}_C \varphi^*_i(R)\varphi_j(R)$. The scattering involves occupied and empty states, and hence predominantly states in the vicinity of the Fermi level are involved. This opens up a possibility of engineering matrix elements by moving the Fermi level with increasing the number of electrons, and by moving the Mn ion position with respect to the center of the quantum dot. Figure 6.6 shows the exchange parameter $\langle m', n'|J(R)|n, m\rangle = J^{2D}_C \varphi^*_{m',n'}(R)\varphi_{m,n}(R)$ as a function of Mn position R evaluated using harmonic oscillator states in Cartesian coordinates for s shell (a), p shell (b) and (c), and d shell (d–f).

In the s shell, the $J_{ss} = \langle 0, 0|J|0, 0\rangle$ matrix element decays as one moves Mn ion away from the center of the quantum dot. For the p shell, the two matrix elements $\langle 1, 0|J|0, 1\rangle$ and $\langle 0, 1|J|1, 0\rangle$ are zero at $R = 0$, have a maximum at finite R, and zeros along the X or Y axis. By choosing our coordinate system in such a way that

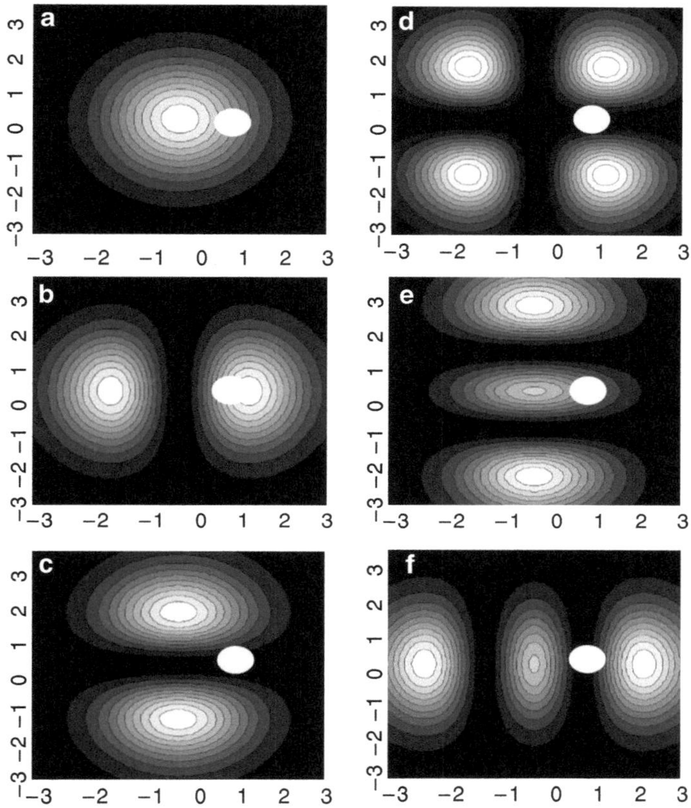

Fig. 6.6 Exchange matrix elements as a function of Mn position R for a QD with $\omega_0 = 4\,\text{Ry}$, $d = 20$ for s shell (**a**), p shell (**b**, **c**), and d shell (**d**–**f**). *Empty circle* indicates the position of Mn ion. From [31]

$\boldsymbol{R} = (X, 0)$, we see that Mn spin can be coupled to only one of the two p-orbitals. The situation is similar in the d shell, for which all three orbitals are shown. For a position of Mn $X = l_0$ we have coupling to only one of the d orbitals, as one of the orbitals has always a node in this direction, and a second one has a node at the same distance from the center of the dot. Therefore, one could engineer Mn position so it is coupled to only one of the three states of the d shell. Figure 6.6 clearly illustrates the potential for using Mn ion position as a tool in engineering magnetic interactions.

6.8.2 Mn–Carrier Spin Interaction

We now analyze the coupling of Mn spin to the spin of N electrons in a quantum dot. In the simplest approach, one would expect coupling of Mn spin $\boldsymbol{M}$ to the total electron spin $\boldsymbol{S}$, given by the effective Hamiltonian $H_{e-Mn} = -J_{eff}\boldsymbol{S} \cdot \boldsymbol{M}$, with effective exchange constant J_{eff} to be determined from microscopic calculation. This is, however, not what was found in [31]. We illustrate the nontrivial interaction of Mn ion with electrons by calculating the ground state energy of the full interacting system as a function of the number of electrons N for a fixed position of Mn, e.g., $\boldsymbol{R} = (l_0, 0)$. Figure 6.7 shows the energy shift Δ, $\Delta = |E_c(R) - E_e|$ calculated in [31].

The calculations were done in a limited Hilbert space of each partially filled shell. The solid line shows schematically the total spin of the ground state of the dot as a function of the number of electrons N, already discussed here. The total spin is zero for closed shells at $N = 2, 6, 12, \ldots$ and reaches maximum for a half-filled shell, i.e., $N = 1$ for the s shell, $N = 4$ for the p shell, and $N = 9$ for the d shell [6, 44, 45]. One might expect that Δ is proportional to the total spin

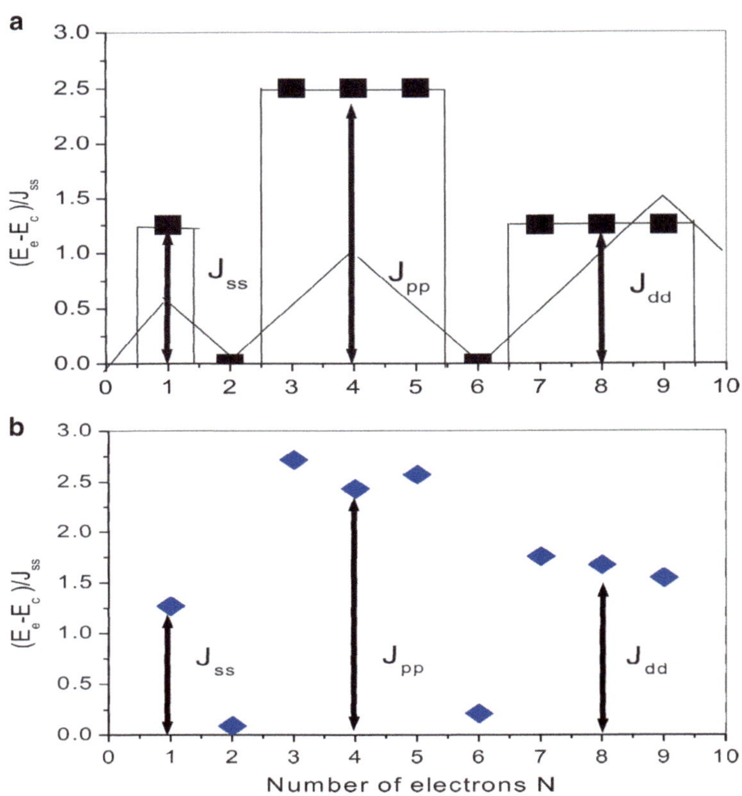

Fig. 6.7 Local Zeeman spin splitting $\Delta = |E_c - E_e|$ vs. number of electrons N for a QD with $\omega_0 = 4\,\text{Ry}, d = 2\,\text{nm}$ and $R = l_0$, calculated using a limited Hilbert space of corresponding partially filled shell. The corresponding energy scales, given by exchange parameters, are indicated by J_{ss}, J_{pp}, and J_{dd}, respectively. Inset shows results of numerical calculation including all configurations of s, p, and d shells. From [31]

S of electrons. However, Fig. 6.7a shows a different behavior of Δ with increasing number of electrons N. We find for closed shells $\Delta = 0$, which one expects for total electron spin $S = 0$ from the effective exchange interaction of the form $H_{\text{e–Mn}} = -J_{\text{eff}} \boldsymbol{S} \cdot \boldsymbol{M}$. However, for partially filled shells we find Δ to be independent of the filling of the shell despite the fact that the total calculated spin of the ground state, shown in Fig. 6.7, varies with shell filling. This was explained in [31] in two ways. One is to think of total spin S and effective exchange coupling J_{eff}. When total spin S increases with the filling up of a degenerate shell, effective J_{eff} decreases. An alternative point of view, supported by Fig. 6.6, is that the Mn spin couples effectively to only one electron of the electronic shell, irrespective of the shell filling, cancelling all many-particle contributions. This cancellation is no longer perfect when higher shells, and excitations from filled shells, are included. In Fig. 6.7b, we show the calculated shift Δ, including all configurations of the s, p, and d shells. It was found that Δ is no longer exactly zero for closed shells, and shows small variation across the states of the d shell. We would like to stress that the step-like behavior of Δ with increasing number of electrons N is independent of Mn position for up to $N = 6$ electrons. Of course, the amplitude of steps varies with Mn position. For higher electron numbers $N > 6$, the cancellation of many-particle contribution became more sensitive to Mn positions other than the one shown in Figs. 6.6 and 6.7, but the general trend persisted.

6.8.3 Addition Spectrum of N*-Electron Quantum Dot with a Mn Ion*

There are several methods to probe the coupling of electron and Mn spins, for example capacitance [9] and Coulomb [45] and spin blockade spectroscopies [52, 53]. In both cases, one measures electron addition energy equal to the chemical potential of the dot. The chemical potential is defined as the difference of total energies of a hybrid Mn–electron system $\mu(N) = E(N) - E(N-1)$. The effect of Mn ion on the chemical potential of N-electron QD can be extracted from the difference of chemical potentials of Mn ion doped- and undoped dots. Figure 6.8a shows the difference of the chemical potential $\Delta\mu = \mu_{\text{c}} - \mu_{\text{e}}$ of the QD with and without Mn ion as a function of N. The variation of the chemical potential with the number of electrons shows a behavior related to Fig. 6.7. One observes a negative shift for $N = 1$ and a positive and equal shift for $N = 2$. The $N = 1$ shift measures $-\Delta$.

For $N = 2$, the two-electron dot has a closed shell, and its energy is not modified by the presence of the Mn ion. Hence, the shift is now $+\Delta$. For $N = 3$, we measure $-\Delta$ of a $N = 3$ electron dot. For $N = 4$, i.e., in the half-filled shell the shift in the energy of the N electron and $N - 1$ electron dot is identical and cancels in the chemical potential. Hence, in the vicinity of half-filled shells we see no effect of Mn ion on the addition energy, the effects are only visible when electron number N crosses from one shell to the next one.

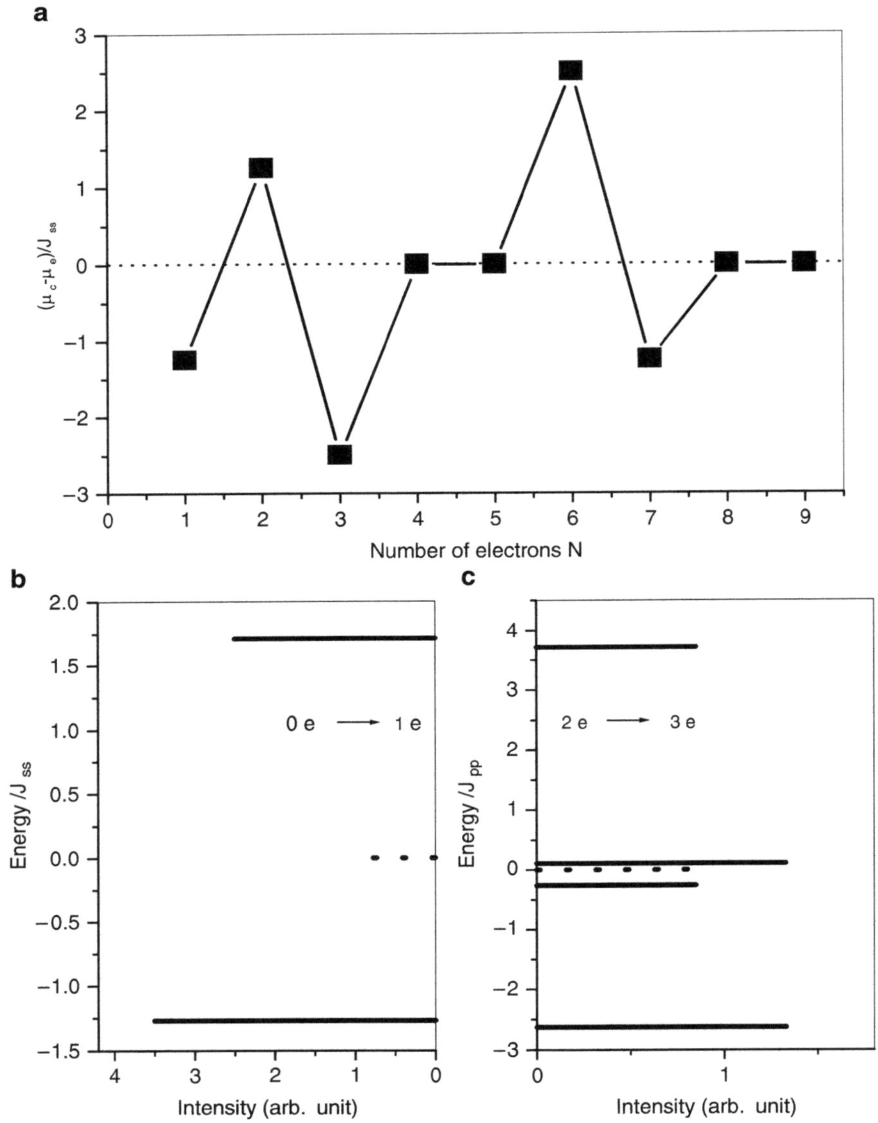

Fig. 6.8 (**a**) Difference of the chemical potential $\Delta\mu = \mu_c - \mu_e$ between a single Mn ion doped- and undoped quantum dot as a function of the number of electrons N. Spectral function for adding an electron with spin down to s-level of an empty QD (**b**), and to p-shell of N $= 2$ QD (**c**) with $\omega_0 = 4\,\mathrm{Ry}$ and $d = 2\,\mathrm{nm}$. For comparison, the spectral function of QD without Mn ion is shown as *dotted line*. From [31]

6.8.4 Electron Spectral Function of a N-Electron Quantum Dot with a Mn Ion

A more direct manifestation of the presence of Mn ion can be obtained from the high source and drain voltage bias spectroscopy [52, 53] of a quantum dot. With high bias applied to the drain an electron tunneling through a quantum dot probes QD excitation spectrum, proportional to the spectral function $A(m, \sigma, \omega) = \sum_i P_i \sum_f |\langle f, N+1|c^+_{m,n\sigma}|i, N\rangle|^2 \delta(E_f - E_i - \omega)$ of the quantum dot, where P_i is the probability that the dot is in initial state i. In the lower part of Fig. 6.8, we show examples of spectral functions. Figure 6.8b shows the spectral function corresponding to adding an electron with spin down to the s level of an empty quantum dot containing a single Mn ion. In the absence of the Mn ion, the spectral function, shown as dotted line, is a single peak $A(s, \downarrow, \omega) = \delta(\omega_0 - \omega)$ at ω_0. However, for a dot with Mn ion the spectral function (solid line) breaks into two pieces, $A(s, \downarrow, \omega) = A_- \delta(\omega_0 - J_{ss} - \omega) + A_+ \delta(\omega_0 + 2J_{ss} - \omega)$, separated by $3J_{ss}$. This can be understood by examining the effective exchange Hamiltonian $H_{\text{e–Mn}} = -J_{\text{eff}} \boldsymbol{S} \cdot \boldsymbol{M}$. For $S = 1/2$, there are two possible values of total spin $J_\pm = M \pm 1/2$. The two total spin values generate two degenerate groups, with degeneracy $2J_\pm + 1$. The two groups of degenerate states are probed by the added electron. When a magnetic field is applied, the degeneracies are removed and peaks split into groups of five and seven, a direct manifestation of the presence of the Mn ion.

A similar situation is encountered when adding an extra electron to the p shell of a dot with $N = 2$ electrons and Mn ion, as shown in Fig. 6.8c. The spectral function now splits into three pieces. The origin of the two extreme pieces, separated by $6J_{pp}$, is similar to the s shell, and the third, central piece represents the orbital not directly coupled to the Mn ion. The indirect coupling with closed s shell electrons does lead to a small splitting of degenerate Mn levels. The application of a weak magnetic field is expected to reveal proper degeneracies associated with the Mn ion spin.

We can now compare the effect of the Mn ion and quantum dot confinement on the electronic properties. Quantum confinement leads to quantization of energy levels with large spacing ω_0 but degenerate with respect to spin. The effect of Mn ion is to remove the spin degeneracy of energy levels of electron and of Mn ion. The scale of the effect presented here is measured in terms of exchange coupling J_{ss}. The scale of this coupling is proportional to level spacing, $J_{ss} \approx \omega_0$. The smaller the dot, the larger the spacing. Hence, the coupling of an individual electron to Mn ion in a quantum dot is enhanced due to electron confinement. For a model investigated in [31], J_{ss} was found to be 0.21 meV for $\omega_0 = 51.32$ meV.

6.8.5 Magnetic Ion in III–V Self-Assembled Quantum Dots

In III–V self-assembled quantum dots such as InAs-GaAs, Mn ions act as both spins and acceptors. Climente et al. [30] described a model device allowing voltage

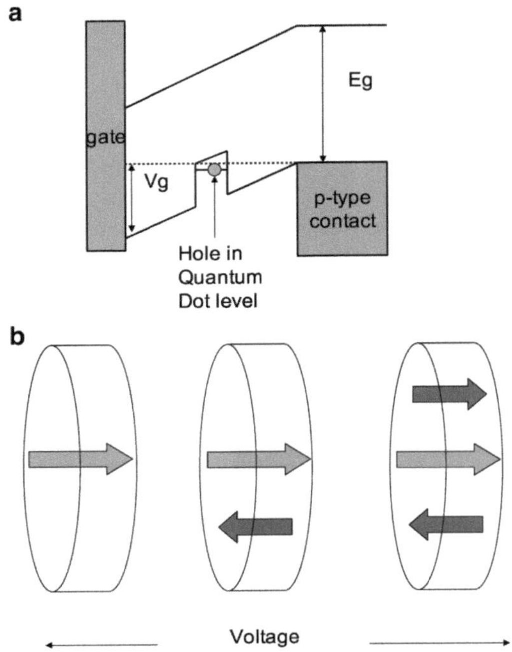

Fig. 6.9 (**a**) Schematic illustration of the device structure allowing voltage-control of the number of holes in a quantum dot. (**b**) Interacting spins in a III–V semiconductor quantum disk with one Mn ion under positive bias (*left*), zero bias (*center*), and negative bias (*right*). *Light-shaded arrow* corresponds to the Mn ion spin and *dark arrows* to valence holes spins. From [30]

control of the magnetic properties of magnetic ions in III–V self-assembled semiconductor quantum dots. The schematic operation of such a device is shown in Fig. 6.9.

The applied voltage, combined with the Coulomb blockade, allows the control of the number of holes in the quantum dot. It was shown that unlike for electrons in II–VI quantum dots, the spins of the holes are frozen in the growth direction, and interact with the spins of the magnetic ions via Ising-like $sp-d$ exchange interaction. The spectrum of a Mn ion in a p-type InAs quantum disk in a magnetic field was calculated as a function of the number of holes described by the Luttinger-Kohn Hamiltonian. For a neutral Mn acceptor, the spin of the hole leads to an effective magnetic field, which strongly modifies the magnetization of the ion. The magnetization can be modified further by charging the dot with an additional hole. The interacting holes form a singlet parity ground state, suppress the effective field, and modify the magnetic moment of the charged complex.

6.9 Mn–Mn Interactions Mediated by Interacting Electrons

In this section, we focus on the system of magnetic ions and how interaction with carriers induces interaction among magnetic ions.

Following [32], we focus on the understanding of two magnetic ions embedded in a quantum dot containing a controlled number of electrons N, with the Hamiltonian written as:

$$\hat{H} = g_{\mathrm{Mn}}\mu_{\mathrm{B}}BM_1^z + g_{\mathrm{Mn}}\mu_{\mathrm{B}}BM_2^z + J_{12}^{\mathrm{AF}}\boldsymbol{M}_1 \cdot \boldsymbol{M}_2$$
$$- \sum_{i,j,I} \frac{J_{ij}(R_I)}{2}\left[(c_{i,\uparrow}^{+}c_{j,\uparrow} - c_{i,\downarrow}^{+}c_{j,\downarrow})M_I^z + c_{i,\downarrow}^{+}c_{j,\uparrow}M_I^{+} + c_{i,\uparrow}^{+}c_{j,\downarrow}M_I^{-}\right]$$
$$+ \sum_{i\sigma} E_{i,\sigma}c_{i,\sigma}^{+}c_{i,\sigma} + \frac{1}{2}\sum_{ijkl}^{\sigma\sigma'} \langle i,j|V_{ee}|k,l\rangle c_{i,\sigma}^{+}c_{j,\sigma'}^{+}c_{k,\sigma'}c_{l,\sigma}. \quad (6.17)$$

The first line in (6.17) is the Hamiltonian of the Mn subsystem, including Mn–Mn interaction. The second line describes the interaction of Mn ions with electrons. The last line in (6.17) describes the electron Hamiltonian. Here, g_{Mn} (g_{e}) is the Mn (electron) g-factor, μ_{B} is the Bohr magneton, and B – the magnetic field along z-axis. In what follows we adopt [3] $g_{\mathrm{e}} = -1.67$, $g_{\mathrm{Mn}} = 2.02$, with all other parameters as in previous sections.

To calculate the electronic properties of the interacting Mn–electron system, the wave function was expanded on the following basis of configurations $|k\rangle$: $|k\rangle = |i_1, i_2, \ldots, i_{N\uparrow}\rangle|j_1, j_2, \ldots, j_{N\downarrow}\rangle|M_1^z M_2^z\rangle$, where $|i_1, i_2, \ldots, i_{N\uparrow}\rangle = c_{i_1,\uparrow}^{+}c_{i_2,\uparrow}^{+}\cdots c_{i_N,\uparrow}^{+}|0\rangle$, $|0\rangle$ is the vacuum, and $N\uparrow$ ($N\downarrow$) is the number of spin-up (-down) electrons, $N\uparrow + N\downarrow = N_{\mathrm{e}}$. The basis states were grouped into spin-up and spin-down electron states for each configuration of Mn ions $|M_1^z M_2^z\rangle$, with $M_I^z = \pm 5/2$, $\pm 3/2$, $\pm 1/2$. The number of possible electronic and magnetic configurations N_{C} is determined by the number of single particle orbitals N_S, the number of electrons, the size M of the magnetic ion spin, and their number N_{Mn}: $N_{\mathrm{C}} = (2M+1)^{N_{\mathrm{Mn}}} \sum_{N\downarrow=0}^{N} \binom{N_S}{N\downarrow}\binom{N_S}{N\uparrow}$. Using this basis states the Hamiltonian matrix was built, which upon diagonalization gave the eigenenergies and eigenstates of the interacting Mn and electron complex. When the electronic subsystem has partially occupied shells, presence of Mn ions removes angular momentum and spin conservation in the electronic system and significantly increases the computational effort. The purpose of this work was to determine the effective Mn–Mn interaction in the presence of electrons. The strength of carrier induced Mn–Mn interaction can be measured in terms of energy shift $\Delta = E_{\mathrm{c}} - E_{\mathrm{e}}$, defined by the difference between the ground state energy E_{c} of a quantum dot containing both N_{Mn} Mn ions and N electrons, and energy E_{e} of a quantum dot with only electrons.

6.9.1 RKKY Mn–Mn Interactions for Closed Shells

Let us first discuss Mn–Mn interaction for closed electronic shells, corresponding to electron numbers $N = 2, 6, 12, \ldots$, quantum dot analog of RKKY interaction in metals [54]. The simplest example is for two electrons in the s shell. If the shell spacing ω_0 is much larger than the corresponding Coulomb energy, the ground state of the electronic subsystem is well approximated by a single configuration with one spin-up and one spin-down electron occupying the s-shell. Due to singlet electronic ground state there is no direct coupling of total electron spin with Mn spins. The coupling of the two spin subsystems is through excited electronic and Mn states. To translate the coupling of the two spin subsystems into the effective Mn–Mn interaction, we evaluate the effect of electron–Mn interaction on the shift in ground state energy Δ in second-order perturbation theory as

$$\Delta = \sum_k \frac{\langle G|H_{\text{e-Mn}}|k\rangle\langle k|H_{\text{e-Mn}}|G\rangle}{E_G - E_k}, \tag{6.18}$$

where $|G\rangle$ is the noninteracting ground state and $|k\rangle$ is the noninteracting excited configuration of the two noninteracting subsystems. After some algebra, the effect of carriers can be expressed in terms of effective Mn–Mn interaction Hamiltonian $\Delta = J(\boldsymbol{R}_1, \boldsymbol{R}_2)\boldsymbol{M}_1 \cdot \boldsymbol{M}_2$, with effective RKKY-like interaction strength $J(\boldsymbol{R}_1, \boldsymbol{R}_2)$. Using analytical form of the single particle wave functions, an analytical form can be derived for the interaction strength as a function of Mn positions [32]:

$$J(\boldsymbol{R}_1, \boldsymbol{R}_2) = -\left(\frac{J_{\text{C}}^{\text{2D}}}{\pi l_0^2}\right)^2 \left(\frac{1}{4\omega_0}\right) e^{-\frac{R_1^2+R_2^2}{2}} \times \left[\alpha \boldsymbol{R}_1 \cdot \boldsymbol{R}_2 + \beta \frac{(\boldsymbol{R}_1 \cdot \boldsymbol{R}_2)^2 - (\boldsymbol{R}_1^2 + \boldsymbol{R}_2^2) + 2}{4}\right]. \tag{6.19}$$

Here, the Mn positions are measured in quantum dot length l_0 and α and β are parameters controlled by the number of shells and electron–electron interactions. For a quantum dot with only s and p shell $\beta = 0$ while for a quantum dot with s, p, and d shells, one finds $\alpha = 1.0$ and $\beta = 1.0$. Including e–e interactions renormalizes the many-particle energy spectrum and gives $\alpha = 3.59$ and $\beta = 2.57$. The characteristic energy scale is proportional to the square of the ratio of the 2D exchange constant (in units of energy × area) divided by the characteristic area of the quantum dot and divided by the shell spacing of the quantum dot, denoted as $J_0 = \left(\frac{J_{\text{C}}^{\text{2D}}}{\pi l_0^2}\right)^2 \left(\frac{1}{4\omega_0}\right)$. The smaller the quantum dot the stronger the RKKY interaction. The interaction decays exponentially with both Mn ions moving away from the center of the dot. The sign of interaction is either positive (antiferromagnetic) or negative(ferromagnetic) depending on Mn positions inside the quantum dot. If we place one Mn ion in the center of the dot, $R_1 = 0$, the RKKY interaction should vary as $J(0, \boldsymbol{R}_2) = -\left(\frac{J_{\text{C}}^{\text{2D}}}{\pi l_0^2}\right)^2 \left(\frac{1}{4\omega_0}\right) \left\{\frac{\beta}{4}\right\} \left\{2 - \boldsymbol{R}_2^2\right\} e^{-\frac{R_2^2}{2}}$ with position $\boldsymbol{R}_2$ of a second Mn ion. This isotropic interaction is ferromagnetic (negative) for $R_2 < \sqrt{2}$ and antiferromagnetic (positive) for $R_2 > \sqrt{2}$. The results of exact diagonalization of full

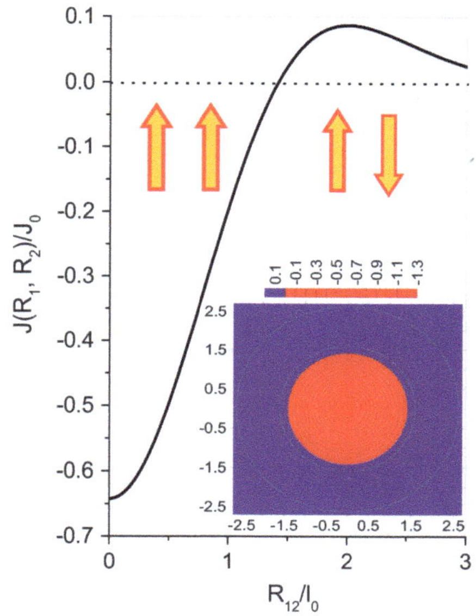

Fig. 6.10 RKKY interaction strength $J\ (R_1, R_2)$ as a function of Mn–Mn separation for a quantum dot containing two electrons and two Mn ions. One Mn ion is located at the center of the quantum dot R1 = 0, and the second one on the x-axis. The *dotted line* indicates the change of the sign of $J\ (R_1, R_2)$, and *arrows* schematically indicate preferred orientation of Mn spins. Inset shows the spatial map of RKKY interaction strength $J\ (0, R_2)$, with *red* indicating negative, i.e., ferromagnetic coupling. From [32]

Hamiltonian, (6.17), for $N = 2$ and two Mn ions, in the absence of the magnetic field, are shown in Fig. 6.10. This figure shows the effective RKKY-like interaction strength $J\ (\boldsymbol{R}_1, \boldsymbol{R}_2)$ as a function of the position of the second Mn ion with first ion in the center. The interaction strength $J\ (\boldsymbol{R}_1, \boldsymbol{R}_2)$ is very well described by the analytical expression discussed above, with ferromagnetic to antiferromagnetic transition at $R_2\ \sqrt{2}$. The inset shows a map of $J\ (\boldsymbol{R}_1, \boldsymbol{R}_2)$ as a function of R_2, showing isotropic dependence, and the regions with ferromagnetic and antiferromagnetic RKKY coupling. Similar complex maps can be obtained for different Mn ion positions. In Fig. 6.11a, the map of interaction strength $J\ (\boldsymbol{R}_1, \boldsymbol{R}_2)$ as a function of R_2 for $R_1 = (1, 0)$ is shown.

We see that the RKKY interaction is negative (ferromagnetic) when the two ions are close together but becomes positive (antiferromagnetic) when the second Mn ion is moved to the opposite side of the quantum dot, in agreement with analytical expression in (6.19).

To engineer Mn–Mn interactions, we can also change the number of carriers. Figure 6.11b shows the plot of the interaction strength $J\ (\boldsymbol{R}_1, \boldsymbol{R}_2)$ as a function of R_2 but for a quantum dot with two filled shells ($N = 6$). We see that in addition to the behavior shown in Fig. 6.11a there are new regions of ferromagnetic and antiferromagnetic coupling constant. This demonstrates the possibility of engineering carrier mediated RKKY interaction in quantum dots with Mn position, shell spacing, and number of filled shells.

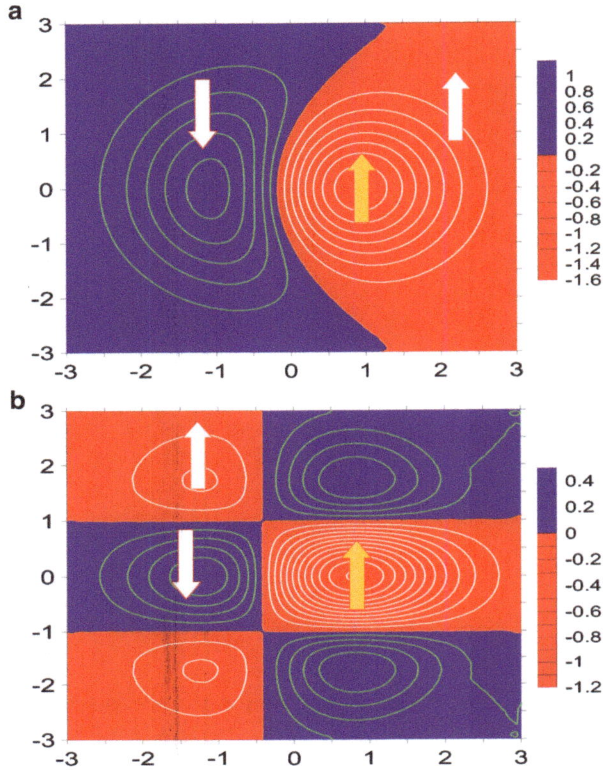

Fig. 6.11 Maps of RKKY interaction strength $J(R_1, R_2)$ as a function of position of the second magnetic impurity R_2 for $R_1 = 1$ in a quantum dot containing two (**a**) and six (**b**) interacting electrons. The *red color* indicates negative i.e., ferromagnetic coupling while *blue* indicates positive, i.e., antiferromagnetic coupling. The first Mn is indicated by *yellow arrow*. *White arrow* shows different orientations of the second Mn ion depending on its position. From [32]

6.9.2 Magneto-Polarons in Partially Filled Shells

Let us now turn to partially filled shells. For partially filled shells, electron spins couple directly to Mn spins, resulting in strongly coupled system, the magneto-polaron. The existence of magneto-polarons in quantum dots has been predicted already in [8].

The polaron effect can be illustrated by considering two Mn spins and a single electron in the s shell. Retaining only s-shell electronic state, assuming $R_1 = -R_2$, i.e., $J_{ss}(R_1) = J_{ss}(R_2)$ allows us to introduce total Mn spin $\boldsymbol{M} = \boldsymbol{M}_1 + \boldsymbol{M}_2$. Neglecting electron energy and setting magnetic field $B = 0$ we obtain a simplified one electron-two Mn ions Hamiltonian:

$$\hat{H} = \frac{J_{12}^{\mathrm{AF}}}{2}(\boldsymbol{M}^2 - \boldsymbol{M}_1^2 - \boldsymbol{M}_2^2) - \frac{J_{ss}}{2}\left[(c_{s,\uparrow}^+ c_{s,\uparrow} - c_{s,\downarrow}^+ c_{s,\downarrow})M^z + c_{s,\downarrow}^+ c_{s,\uparrow} M^+ + c_{s,\uparrow}^+ c_{s,\downarrow} M^-\right]. \tag{6.20}$$

The first term corresponds to the short-ranged Mn–Mn antiferromagnetic interaction classified by total Mn spin $0 \le M \le M_1 + M_2$. The second term describes ferromagnetic electron Mn interaction. Noting that total spin M is a good quantum number, expanding eigenstates in the basis $c_{s,\downarrow}^+|0\rangle|M, M^z\rangle$ and $c_{s,\uparrow}^+|0\rangle|M, M^z\rangle$, allows us to exactly diagonalize Hamiltonian, (6.20). One finds the degenerate ground state of the hybrid system with electron spin parallel to the total Mn spin, and energy

$$E(M,+) = -\frac{J_{ss}}{2}M + \frac{J_{12}^{\mathrm{AF}}}{2}\left[M(M+1) - \frac{35}{2}\right]. \tag{6.21}$$

While total Mn spin M is a discrete variable, we can gain the understanding by treating it as a continuous variable, and restoring its allowed discrete values at the end. In the absence of coupling to the electron spin, $J_{ss} = 0$, the ground state energy $E(M,+)$ is minimized for the total Mn spin $M = 0$, i.e., antiferromagnetic Mn arrangement. However, coupling to the electron spin gives minimum energy of the hybrid system for finite total Mn spin $M^* = (\frac{J_{ss}}{J_{12}^{\mathrm{AF}}} - 1)/2$. We can reach different spin alignments, i.e., ferromagnetic $M^* = 5$, canted $(0 < M^* < 5)$ and antiferromagnetic $(M^* = 0)$ by adjusting the ratio $\delta = J_{ss}/J_{12}$ with proper choice of the positions of a pair of Mn ions. For $M = M^* = (\delta - 1)/2$, the energy is found to be $E(M*,+) = -(\frac{J_{12}^{\mathrm{AF}}}{8})[(\delta-1)^2 + 70]$. It is negative, i.e., the interaction leads to lowering of energy and formation of a magneto-polaron, i.e., aligning of Mn spins. The binding energy of magneto-polaron is a nonmonotonic function of Mn–Mn separation due to two different length scales, the decay λ of short-range antiferromagnetic interaction and decay $\ell_0 = 1/\sqrt{\omega_0}$ of ferromagnetic carrier-induced interaction. This is illustrated in Fig. 6.12 where energy shift Δ of the ground state energy, the magnetopolaron energy, is shown for $N = 1$ electron from exact diagonalization of full Hamiltonian. The energy is negative, shows nonmonotonic dependence on the length scale where J_{12}^{AF} dominates, and a smooth decay on the length scale of a quantum dot where J_{ss} dominates. The formation of magneto-polarons in partially filled p and d shells is also shown in Fig. 6.12. The magnitude of polaron energy and its dependence on distance is found to depend strongly on the partially filled shell and a degree of filling.

To summarize this section, one can classify the effective interaction of magnetic ions mediated by electrons into two regimes: (a) closed shells with RKKY interactions, and (b) partially filled shells where nonperturbative magneto-polaron effects dominate. Explicit expressions for RKKY interaction has been derived in [32] and presented here, as well numerically obtained many-electron magneto-polaron energies for the first three quantum dot shells. This shows the way toward engineering magnetic properties on nanoscale using quantum dots containing magnetic ions and electrically tunable number of carriers.

Fig. 6.12 Energy shift $\Delta = E_c - E_e$, as a function of separation between two Mn ions, one located at the center of the quantum dot and the second one on the x-axis, for different electron numbers N. Results for partially filled s, p, and d electronic shells are shown. From [32]

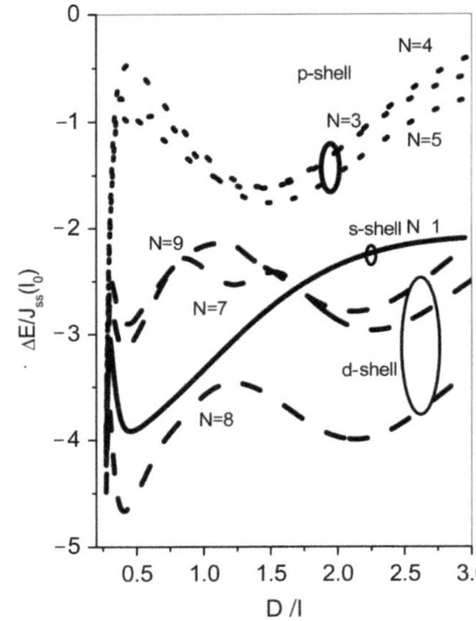

6.10 Control of Ferromagnetism in Quantum Dots

In previous sections, we discussed quantum dots with few – one and two – magnetic ions and a variable, controlled number of electrons. Here, we focus on a system of many magnetic ions, and a small number of electrons, as discussed by Abolfath et al. [33]. The question one would like to answer is whether one can enhance carrier-mediated ferromagnetism [18] by confining carriers into quantum dots. Fernandez-Rossier and Brey [29] investigated noninteracting electrons in quantum dots and suggested increased temperature of the ferromagnetic transition. The enhancement was predicted only for odd number of electrons N in a quantum dot, i.e., for nonzero total electron spin S. As we have discussed earlier, the total electron spin depends on degeneracies and e–e interactions. Since the transition temperature depends on the coupling of the electron spin to Mn ion spins, and this is also controlled by e–e interactions, the interactions may play a role in controlling the ferromagnetic transition temperature.

Abolfath et al. [33] focused on magnetic QD in zero applied magnetic field described by the Hamiltonian $H = H_e + H_m + H_{ex}$, with the electron contribution $H_e = \sum_{i=1}^{N}[-\frac{\hbar^2}{2m^*}\nabla_i^2 + U_{QD}(\mathbf{r}_i)] + \frac{1}{\epsilon}\sum_{i\neq j}\frac{e^2}{|\mathbf{r}_i - \mathbf{r}_j|}$, where $U_{QD}(\mathbf{r})$ was the confining potential of a three-dimensional QD. Further, the Mn Hamiltonian is given by $H_m = \sum_{I,I'} J^{AF}_{I,I'} \boldsymbol{M}_I \cdot \boldsymbol{M}_{I'}$. and the electron–Mn (e–Mn) exchange Hamiltonian is $H_{ex} = -J_{sd}\sum_{i,I} \boldsymbol{s}_i \cdot \boldsymbol{M}_I \delta(\mathbf{r}_i - \mathbf{R}_I)$, where J_{sd} is the exchange coupling between electron spin $\boldsymbol{s}_i$, at $\mathbf{r}_i \equiv (\boldsymbol{\rho}_i, z_i)$, and impurity spin $\boldsymbol{M}_I$, at $\mathbf{R}_I$.

Because of a large number of magnetic ions, exact diagonalization of Mn system is not possible. An effective mean field Hamiltonian describing electrons was obtained by replacing the Mn spins, which are randomly distributed, with a classical continuous field $H_e^{\rm eff} = H_e - \sum_i J_{sd} n_m \frac{\sigma_i}{2} \langle M_z(\mathbf{r}_i)\rangle$, where n_m is the density of Mn ions (number of ions per unit volume), and $\sigma = \pm 1$ for spin up ($\uparrow$), and down ($\downarrow$). The effective magnetic field seen by the electrons is the mean magnetic field produced by Mn ions. Assuming Mn impurities to be in equilibrium with thermal bath it follows that $\langle M_z(\mathbf{r}_i)\rangle = M B_{\rm M}(M b(\mathbf{r}_i)/k_{\rm B}T)$ where $B_{\rm M}(x)$ is the Brillouin function [3], and $k_{\rm B}$ is the Boltzmann constant. Here, $b(\mathbf{r}_i) = -Z_{\rm Mn} J^{\rm AF}\langle M_z(\mathbf{r}_i)\rangle + J_{sd}[n_\uparrow(\mathbf{r}_i) - n_\downarrow(\mathbf{r}_i)]/2$ is the effective field seen by the Mn ions [55]. The first term in $b(\mathbf{r}_i)$ describes the mean field of the direct Mn–Mn antiferromagnetic coupling [29]. $Z_{\rm Mn}$ is the averaged Mn coordination number, and $n_\sigma(\mathbf{r}_i)$ is spin-resolved electron density. The confining potential, $U_{\rm QD}$, is taken as a sum of a two-dimensional (2D) Gaussian $V_{\rm QD} = V_0 \exp(-\rho^2/\Delta^2)$ and one-dimensional parabolic potential $V_{\rm QD}^z = m^* \Omega^2 z^2/2$, where $\boldsymbol{\rho} \equiv (x, y)$. After expanding the QD wave functions in terms of its planar $\psi_{i\sigma}(\boldsymbol{\rho})$ and subband wave function $\xi(z)$, integrating out $\xi(z)$ gives the effective 2D Hamiltonian $H_e^{\rm eff}$. The two-body Coulomb interaction was written as a sum of Hartree potential $V_{\rm H}$ and spin-dependent exchange-correlation potential $V_{\rm XC}^\sigma$ [56], resulting in the Kohn-Sham Hamiltonian

$$H_{\rm KS} = \frac{-\hbar^2}{2m^*}\nabla_\rho^2 + V_{\rm QD} + V_{\rm H} + V_{\rm XC}^\sigma - \frac{\sigma}{2} h_{sd}(\boldsymbol{\rho}), \qquad (6.22)$$

with the effective magnetic field produced by magnetic ions

$$h_{sd}(\boldsymbol{\rho}) = J_{em} \int {\rm d}z |\xi(z)|^2 B_{\rm M}\left(\frac{M b(\boldsymbol{\rho}, z)}{k_{\rm B}T}\right). \qquad (6.23)$$

Here, $J_{em} = J_{sd} n_m M$ is the e–Mn exchange coupling. The Kohn–Sham eigenvectors and eigenvalues of (6.22), $\psi_{n\sigma}(\boldsymbol{\rho})$ and $\epsilon_{n\sigma}$, were calculated numerically for (Cd,Mn)Te QD with $J_{sd} = 0.015\,{\rm eV\ nm}^3$, $m^* = 0.106$, $\epsilon = 10.6$, and $Z_{\rm Mn} J^{\rm AF} = 0.02\,{\rm meV}$. The planar (x, y), and perpendicular (z) dimensions of the QD were taken as 42 and 1 nm with $n_m = 0, 0.025, 0.1\,{\rm nm}^{-3}$. In the central region of QD of area $4a_{\rm B}^{*2}$, where $a_{\rm B}^* = 5.29\,{\rm nm}$ is the effective Bohr radius in CdTe, $n_m = 0.1\,{\rm nm}^{-3}$ corresponds to ≈ 10 Mn atoms [3]. For a planar confinement, $V_{\rm QD}$, a Gaussian potential with $V_0 = -128\,{\rm meV}$, and $\Delta = 38.4\,{\rm meV}$, corresponding to $\omega_0 = 27\,{\rm meV}$, was used.

In Fig. 6.13, we show the magnetization $\langle M_z\rangle$ (a), electron polarization P (b), and the free energy difference ΔF between ferromagnetic and antiferromagnetic states (c) of $N = 4$ and $N = 8$ electrons as a function of temperature T. At low T, the spin triplet ground state of the $N = 4$ and $N = 8$ partially filled shells leads to finite electron polarization and finite Mn ion magnetization. With increasing temperature both $\langle M_z\rangle$ and P are decreasing, accompanied by a series of spin transitions in

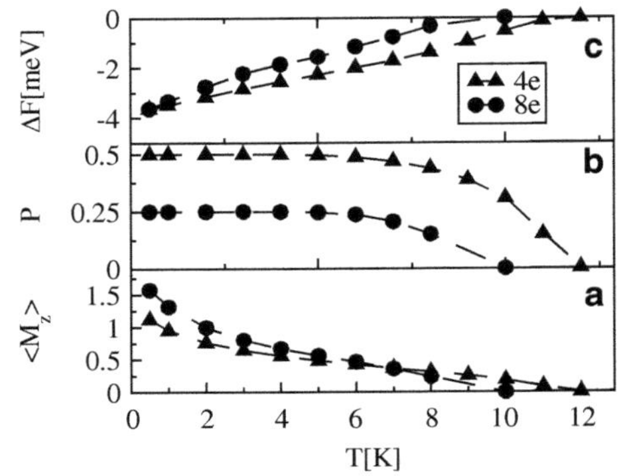

Fig. 6.13 Temperature evolution of Mn-magnetization per unit area $\langle M_z \rangle$ (**a**), the electron polarization P (**b**), and the free energy difference ΔF between ferromagnetic and antiferromagnetic QD (**c**) for $\gamma = 1$, and $n_m = 0.1\,\text{nm}^{-3}$. At low T, $N = 4, 8$ form half-filled shells with $P = 2/4, 2/8$. $T = T^*$, characterizes vanishing of $\langle M_z \rangle$, P, and ΔF. From [33]

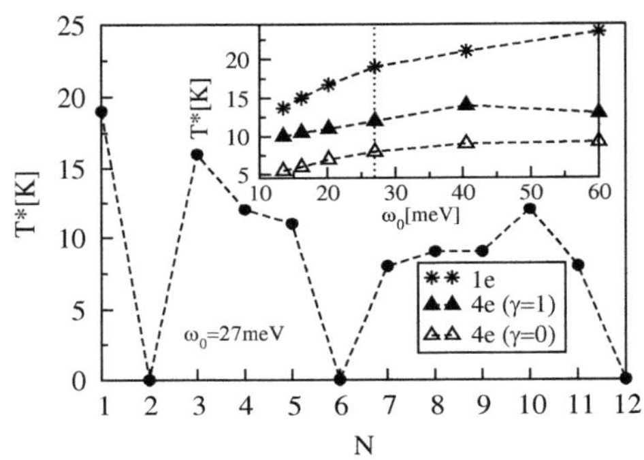

Fig. 6.14 T^* as a function of N for interacting electrons $\gamma = 1$, $n_m = 0.1\,\text{nm}^{-3}$, $\omega_0 = 27\,\text{meV}$, and $V_0 = -125\,\text{meV}$. Inset: The dependence of T^* on ω_0 for $N = 1$, and $N = 4$ with $\gamma = 0$ and $\gamma = 1$. $\omega_0 = 27\,\text{meV}$ is marked as a *dotted line*. For interacting system with $N = 4$ and $\gamma = 1$, there is an optimal confining potential which maximizes T^*. From [33]

electronic system. This allows us to define a characteristic temperature, T^*, at which $\langle M_z \rangle = P = \Delta F = 0$.

Figure 6.14 shows the critical temperature $T^*(N)$ as a function of the number of electrons N, for $\omega_0 = 27\,\text{meV}$ and $\gamma = 1$. We see that for a fixed ω_0, $T^*(N)$ decreases nonmonotonically with N. T^* is largest for $N = 1$, i.e., partially filled s-shell, there is no ferromagnetic transition for closed shells, i.e., $N = 2, 6, 12$, T^* decreases with increasing shell and electron number, and there is no significant variation across partially filled shells. The lack of enhancement of T^* for higher half-filled shells, corresponding to potentially higher total electronic spin, is somehow disappointing. However, results of [33] did indicate enhanced magnetization in the ferromagnetic phase corresponding to enhanced electron spin polarization.

The inset shows the dependence of T^* on shell spacing ω_0 for $N = 1$ and $N = 4$. The $N = 4$ case is shown for interacting electron system ($\gamma = 1$) and for noninteracting electron system ($\gamma = 0$). We see that at low ω_0 the e–e interaction strongly enhances T^*, while at large ω_0 the effect of confinement potential is dominant. Thus, we find $T^*(\gamma = 1) \rightarrow T^*(\gamma = 0)$ with increasing ω_0, which in turn gives rise to a small maximum in T^*.

Several trends in numerically calculated $T^*(N, \omega_0)$ can be obtained from a perturbative approach. Near $\langle M_z \rangle = P = 0$ for QD with one valence electron in s, p, or d shells, we find

$$T^* = J_{em} \sqrt{\frac{M+1}{3n_m M}} \left[\int \mathrm{d}\boldsymbol{r} |\psi_f(\boldsymbol{r})|^4 \right]^{1/2}, \tag{6.24}$$

where ψ_f is the wave function of the highest occupied orbital, and $J^{\mathrm{AF}} = 0$. For a given ω_0, T^* decreases with N, e.g., $T^*_{N=3} = 0.7T^*_{N=1}$ and $T^*_{N=7} = 0.6T^*_{N=1}$. One can also show that $T^* \propto \sqrt{\omega_0}$, consistent with bound magnetic polarons [57].

6.11 Summary and Outlook

We have described here our understanding of semiconductor quantum dots containing a controlled number of carriers and a controlled number of magnetic ions. Tuning the number of electrons leads to the variation of total spin of electrons and their exchange coupling with the Mn ions. The exchange coupling can be engineered by the choice of the electronic shell and Mn ion position. It can be switched off for closed electronic shells and maximized for partially filled shells. The presence of carriers mediates effective interaction among magnetic ions. For closed electronic shells, the interaction is RKKY-like while for partially filled shells it leads to the formation of many-electron magneto-polarons. The interaction leads to the stability of ferromagnetic state at increased temperatures.

Acknowledgements

This work has been carried out in collaboration with M.Korkusinski, F. Qu, R. Abolfath, and J.I. Climente. Partial support from Canadian Institute for Advanced Research, QuantumWorks, and NRC-CNRS CRP is acknowledged.

References

1. J.Gaj, R. Planel, G. Fishman, Solid State Comm. **29**, 435 (1979)
2. R.R. Galazka, J. Kossut, *Lecture Notes in Physics*, vol. 132 (Springer, Berlinm 1980), p. 245
3. J.K. Furdyna, J. Appl. Phys. **64**, R29 (1988)
4. L. Jacak, P. Hawrylak, A. Wojs, *Quantum Dots* (Springer, Berlin, 1998)

5. D. Bimberg, M. Grundmann, N.N. Ledentsov, *Quantum Dot Heterostructures* (Wiley, Chichester, 1999)
6. P. Hawrylak, M. Korkusinski, in *Topics in Applied Physics*, vol. 90, ed. by P. Michler (Springer, Berlin, 2003), pp. 25–92 and references therein
7. T. Dietl, J. Spalek, Phys. Rev. Lett. **48**, 355 (1982)
8. P. Hawrylak, M. Grabowski, J. J. Quinn, Phys. Rev. B **44**, 13082 (1991)
9. H. Drexler, D. Leonard, W. Hansen, J.P. Kotthaus, P.M. Petroff, Phys. Rev. Lett. **73**, 2252 (1994)
10. L. Besombes, Y. Léger, L. Maingault, D. Ferrand, H. Mariette, J. Cibert, Phys. Rev. Lett. **93**, 207403 (2004); Phys.Rev. B **71**, 161307 (2005)
11. Y. Léger, L. Besombes, J. Fernàndez-Rossier, L. Maingault, H. Mariette, Phys. Rev. Lett. **97**, 107401 (2006); Y. Léger, L. Besombes, L. Maingault, D. Ferrand, H. Mariette, Phys. Rev. Lett. **95**, 047403 (2005)
12. P. Wojnar, J. Suffczyński, K. Kowalik, A. Golnik, G. Karczewski, J. Kossut, Phys. Rev. B **75**, 155301 (2007)
13. C. Gould, A. Slobodskyy, D. Supp, T. Slobodskyy, P. Grabs, P. Hawrylak, F. Qu, G. Schmidt, L.W. Molenkamp, Phys. Rev. Lett. **97**, 017202 (2006)
14. W.K. Liu, K.M. Whitaker, K.R. Kittilstved, D.R. Gamelin, J. Am. Chem. Soc. **126**, 3910 (2006)
15. D.M. Hoffman, B.K. Meyer, A.I. Ekimov, I.A. Merkulov, A.L. Efros, M. Rosen, G. Couino, T. Gacoin, J.P. Boilot, Solid State Commun. **114**, 547 (2000)
16. S.C. Erwin, L. Zu, M.I. Haftel, A.L. Efros, T.A. Kennedy, D.J. Norris, Nature **436**, 91 (2005)
17. S.-J. Cheng, Phys. Rev. B **72**, 235332 (2005)
18. M. Ohno, D. Chiba, F. Matsukura, T. Omiya, E. Abe, T. Dietl, Y. Ohno, T. Ohtani, Nature **408**, 944 (2000)
19. A.J. Heinrich, J.A. Gupta, C.P. Lutz, D.M. Eigler, Science **306**, 466 (2004)
20. A.M. Yakunin, A.Yu. Silow, P.M. Koenraad, J.H. Wolter, W. Van Roy, J. De Boeck, J.-M. Tang, M.E. Flatte, Phys. Rev. Lett. **92**, 216806 (2004)
21. J. Seufert, G. Bacher, M. Scheibner, A. Forchel, S. Lee, M. Dobrowolska, J.K. Furdyna, Phys. Rev. Lett. **88**, 027402 (2002)
22. A. Hundt, J. Puls, F. Henneberger, Phys. Rev. B **69**, 121309 (2004)
23. S. Mackowski, T. Gurung, T.A. Nguyen, H.E. Jackson, L.M. Smith, G. Karczewski, J. Kossut, Appl. Phys. Lett. **84**, 3337 (2004)
24. A.K. Bhattacharjee, C. Benoit à la Guillaume, Phys. Rev. B **55**, 10613 (1997)
25. Al.L. Efros, M. Rosen, Ann. Rev. Mater. Sci. **30**, 475 (2000)
26. K. Chang, J.B. Xia, F.M. Peeters, Appl. Phys. Lett. **82**, 2661 (2003)
27. A.L. Efros, M. Rosen, E.I. Rashba, Phys. Rev. Lett. **87**, 206601 (2001)
28. A.O. Govorov, Phys. Rev. B **70**, 035321 (2004); Phys. Rev. B **72**, 075358 (2005); Phys. Rev. B **72**, 075359 (2005)
29. J. Fernández-Rossier, L. Brey, Phys. Rev. Lett. **93**, 117201 (2004); J. Fernández-Rossier, Phys. Rev. B **73**, 045301 (2006)
30. J.I. Climente, M. Korkusinski, P. Hawrylak, J. Planelles, Phys. Rev. B **71**, 125321 (2005)
31. F. Qu, P. Hawrylak, Phys. Rev. Lett. **95**, 217206 (2005)
32. F. Qu, P. Hawrylak, Phys. Rev. Lett. **96**, 157201 (2006)
33. R.M. Abolfath, P. Hawrylak, I. Zutic, Phys. Rev. Lett. **98**, 207203 (2007)
34. P. Hawrylak, Phys. Rev. Lett. **71**, 3347 (1993)
35. S. Raymond, S. Studenikin, A. Sachrajda, Z. Wasilewski, S.J. Cheng, W. Sheng, P. Hawrylak, A. Babinski, M. Potemski, G. Ortner, M. Bayer, Phys. Rev. Lett. **92**, 187402 (2004)
36. L.G.C. Rego, P. Hawrylak, J.A. Brum, A. Wojs, Phys. Rev. B **55**, 15694 (1997)
37. F.V. Kyrychenko, J. Kossut, Phys. Rev. **B 70**, 205317 (2004)
38. L.G.C. Rego, J.A. Brum, P. Hawrylak, Physica E **2**, 785 (1998)
39. C. Pryor, M.E. Pistol, L. Samuelson, Phys. Rev. B **56**, 10404 (1997); C. Pryor, Phys. Rev. B **57**, 7190 (1998)
40. G. Klimeck, F. Oyafuso, T.B. Boykin, R.C. Bowen, P. von Allmen, Comp. Model. Eng. Sci. **3**, 601 (2002)

41. G.W. Bryant, W. Jaskolski, Physica E **11**, 72 (2001)
42. S.J. Sun, Y.-C. Chang, Phys. Rev. B **62**, 13631 (2000)
43. W. Sheng, S.J. Cheng P. Hawrylak, Phys. Rev.B **65**, 235205 (2005)
44. A. Wojs, P. Hawrylak, Phys. Rev. **B53**, 10841 (1996)
45. S. Tarucha, D.G. Austing, T. Honda, R.J. van der Hage, L.P. Kouwenhoven, Phys. Rev. Lett. **77**, 3613 (1996)
46. P. Kacman, Semicond. Sci. Technol. **16**, R25 (2001) and references therein
47. J.-M. Tang, Michael E. Flatte, Phys. Rev. Lett. **92**, 047201 (2004)
48. P. Mahadevan, A. Zunger, D.D. Sarma, Phys. Rev. Lett. **93**, 177201 (2004)
49. B.E.Larson, K.C. Hass, H. Ehrenreich, Phys. Rev. **B 37**, 4137 (1988)
50. Y. Shapira Jr., N.F. Olivera, Phys. Rev. B **35**, 6888 (1987)
51. H. Kepa, V.K. Le, C.M. Brown, M. Sawicki, J.K. Furdyna, T.M. Giebultowicz, T. Dietl, Phys. Rev. Lett. **91**, 087205 (2003)
52. A. Sachrajda, P. Hawrylak, M.Ciorga, in *Transport in Quantum Dots*, chap. 3, ed. by J. Bird (Kluwer, Dordrecht, 2003)
53. M. Korkusinski, P. Hawrylak, M. Ciorga, M. Pioro-Ladrière, A.S. Sachrajda, Phys. Rev. Lett. **93**, 206806 (2004)
54. M.A. Ruderman, C. Kittel, Phys. Rev. **96**, 99 (1954)
55. S. Das Sarma, E.H. Hwang, A. Kaminski, Phys. Rev. B **67**, 155201 (2003)
56. C. Dharma-Wardana, F. Perrot, in *Density Functional Theory*, ed. by E.K.U. Gross, R.M. Dreizler (Plenum Press, New York, 1995)
57. D.R. Schmidt, A.G. Petukhov, M. Foygel, J.P. Ibbetson, S.J. Allen, Phys. Rev. Lett. **82**, 823 (1999)

Chapter 7
Magnetic Polarons

Dmitri R. Yakovlev and Wolfgang Ossau

Abstract Magnetic polaron, which is the subject of this chapter, is formed by exchange interaction of a carrier spin with localized spins of magnetic ions present in DMSs. It results in ferromagnetic alignment of the localized spins in vicinity of the carrier and can be treated as a magnetic molecule with large magnetic moment. There are stringent requirements if such a molecule is to be stable. The stability and dynamics of exciton magnetic polarons in systems of different dimensionality are considered in this chapter. We show that the stability of magnetic polarons in systems of lower dimensionalities is easier to achieve.

7.1 Introduction

The term "polaron" has been introduced first by Pekar for a correlated state of an electron and lattice vibrations in polar crystals [1]. Then it has been extended over all kinds of self-trapped carrier states. More generally, it is applied to the states of particles embedded in a relaxing medium, if a part of their binding energy is gained by polarization of the environment. In magnetic and diluted magnetic semiconductors (DMS), carriers are coupled with localized spins of magnetic ions by a strong exchange interaction [2]. This interaction causes the ferromagnetic alignment of the localized spins in the vicinity of the carrier (Fig. 7.1). The resulting cloud of polarized spins can be considered as a magnetic molecule with magnetic moment of hundreds Bohr magnetons [3]. Spin-organized systems of this kind are called magnetic polarons (MP), terms "ferron" [4] and "fluctuon" [5] were also suggested.

D.R. Yakovlev (✉)
Experimentelle Physik 2, Technische Universität Dortmund, 44221 Dortmund, Germany
and
Ioffe Physico-Technical Institute, Russian Academy of Sciences, 194021 St. Petersburg, Russia
e-mail: dmitri.yakovlev@tu-dortmund.de

W. Ossau
Experimentelle Physik 3, Universität Würzburg, 94074 Würzburg, Germany
e-mail: Wolfgang.Ossau@physik.uni-wuerzburg.de

J. Kossut and J.A. Gaj (eds.), *Introduction to the Physics of Diluted Magnetic Semiconductors*, Springer Series in Materials Science 144,
DOI 10.1007/978-3-642-15856-8_7,

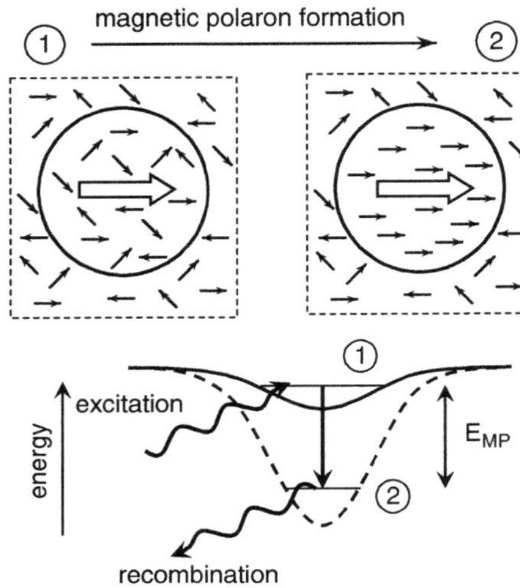

Fig. 7.1 Schematic presentation of the exciton magnetic polaron formation in diluted magnetic semiconductors. Spins of an exciton and Mn-ions are shown by *big* and *small arrows*, respectively. Being captured after photogeneration by a combined potential of alloy and magnetic fluctuations (state 1, fluctuation regime) the exciton aligns ferromagnetically the Mn spins inside its orbit and decreases its energy by forming a magnetic polaron in the collective regime (state 2). The latter can be accompanied by an autolocalization and respective reduction of the exciton size. E_{MP} is the energy of magnetic polaron. Exciton lifetime is limited by recombination

Theoretical studies of the magnetic polarons were pioneered by De Gennes in 1960 [6].

Existence of magnetic polarons in magnetic semiconductors such as EuO was established via transport and magnetization measurements [7]. But detailed information about the polaron energy spectrum and its spin structure was not available from these experiments. The situation changed after appearance of diluted magnetic semiconductors, where powerful optical methods can be applied to study magnetic polarons. II–VI DMS with magnetic Mn^{2+} ions, such as (Cd,Mn)Te, (Cd,Mn)Se and (Zn,Mn)Se, were widely used for these studies.

Most experimental and theoretical efforts before 1988 were concentrated on investigation of bound magnetic polarons (BMP). These polarons are formed by carriers localized on impurities. Therefore, the coordinate part of the carrier wave function in BMPs is controlled by the Coulomb potential of an impurity center, which simplifies considerably theoretical description. Also, the carrier lifetime on the center is usually much longer than the characteristic times of spin relaxation in the magnetic polaron, which allows polarons to reach their equilibrium states. Few types of bound magnetic polarons are reported: (1) Electron on a donor is mostly studied experimentally by Raman scattering [8–10]. It is suitable for detailed theoretical analysis due to simple spin structure [10–13]; (2) Hole on an acceptor was measured via donor–acceptor and electron–acceptor photoluminescence [14–16] and calculated with an account of complex structure of the valence band [17]; and (3) Exciton bound to either a neutral donor [18, 19] or a neutral acceptor [3, 20] was only studied experimentally. The main features of bound magnetic polarons, especially of the electron on a donor BMP, are understood fairly well (for review, see [11]).

Experimental findings on exciton magnetic polarons (EMPs) [20–24], whose formation is schematically displayed in Fig. 7.1, rose new questions related to localization conditions of excitons and dynamics of spin organization in the polaron formation process. These specific features distinguish the exciton magnetic polarons from the bound magnetic polarons:

1. Localization area and, respectively, the energy of EMP ground state are not determined by the Coulomb potential of the impurity center, but depend on the exciton localization or self-localization conditions. Moreover, the exciton autolocalization can play very important role, contributing to the polaron energy and affecting the polaron dynamics. In diluted magnetic semiconductors, exciton localization is determined by both magnetic and nonmagnetic fluctuating potentials. In turn the EMP formation, which is accompanied by a decrease of the exciton energy, can be considered as an additional localization process and, hence, can affect an exciton mobility in the crystal.
2. Exciton lifetime is limited by recombination processes (both radiative and nonradiative). In II–VI semiconductors, it is comparable with the EMP formation time. As a result, the process of the magnetic polaron formation can be interrupted by exciton recombination before the polaron reaches its equilibrium energy. The dynamics of magnetic polaron formation itself depends on exciton localization conditions, temperature, and external magnetic fields. It also depends strongly on concentration of magnetic Mn-ions. On the one hand, the exchange energy gained in the process of the polaron formation is also determined by the Mn concentration (this dependence is nonlinear and nonmonotonic because of antiferromagnetic coupling of nearest-neighbor Mn-spins [2]). On the other hand, a system of interacting Mn-spins plays an important role for energy and spin dissipation in the process of polaron formation (see [25] and Chap. 8). A modification of carrier wave functions in the process of polaron formation can affect the exciton radiative time [26].

Modern semiconductor technology, e.g., molecular-beam epitaxy, allows one to fabricate heterostructures with quantum wells (QWs) and superlattices (SLs) containing thin layers of diluted magnetic semiconductors and with quantum dots formed by a small amount of DMS material surrounded by nonmagnetic barriers. It extends the studies of the exciton magnetic polarons into the field of low-dimensional systems [27, 28] exploiting powerful methods of the band gap engineering (see Chap. 4).

One of the strong motivations to investigate magnetic polarons in low dimensional systems is the possibility to address the physics of the free magnetic polarons (FMPs), i.e., polaron states, which do not require initial localization for their formation. Theory predicts that the stability conditions for free magnetic polarons (i.e., self-localized magnetic polarons) are hardly realized in the bulk diluted magnetic semiconductors based on II–VI materials, but they are favored strongly by the reduction of dimensionality and the FMP is always stable in one-dimensional systems [26, 29, 30].

In this chapter, we give a survey of experimental and theoretical studies of exciton magnetic polarons in three-dimensional (3D) and two-dimensional (2D) systems realized in DMS epilayers, quantum wells, and superlattices. We briefly comment on magnetic polarons in zero-dimensional systems represented by DMS quantum dots, which are considered in Chap. 5. We focus our attention on the aspects where experiment and theory being complementary give rise to an unambiguous understanding of the underlying physics.

The chapter begins with the brief consideration of theoretical aspects of magnetic polaron stability and polaron dynamics (Sect. 7.2). In Sect. 7.3, the method of selective excitation used for the study of exciton magnetic polarons is described. Then, we give an overview of experimental results for exciton magnetic polarons in three-dimensional (Sect. 7.4) and two-dimensional (Sect. 7.5) systems. Finally, the dynamics of polaron formation is discussed and the hierarchy of relaxation processes controlling the polaron formation is given (Sect. 7.6).

7.2 Theoretical Aspects

It is not our intention to give here a comprehensive theoretical consideration of magnetic polarons, which can be found elsewhere [11, 31, 32], but rather comment on a few problems, which are important for understanding the experimental results. They are: (1) stability of free magnetic polarons and its dependence on the dimensionality, (2) dynamical aspects of exciton magnetic polarons, and (3) relation of the polaron parameters to the experimentally measured values.

There are two regimes for magnetic polarons: a fluctuation regime and a collective regime [11]. In the fluctuation regime, a carrier adjusts its spin along the magnetic moment of the fluctuation created by magnetic ions inside the carrier localization volume. This adjustment decreases the carrier energy. The value of the fluctuating magnetic moment is proportional to $\sqrt{N}$, where $N \gg 1$ is the number of localized magnetic moments interacting with the carrier. In DMS, carrier spin relaxation is usually much faster than the spin relaxation of the localized spins. In the fluctuation regime, the localized spins are not affected by the presence of the carrier. Namely, their orientation is not changed in the carrier exchange field. This is realized when the thermal energy $k_B T$ exceeds the Mn-spin energy in the carrier exchange field. The fluctuation regime is typical for the bound magnetic polarons formed by an electron on a donor [11], where the electron exchange field is relatively weak. For the exciton magnetic polarons, where the interaction with a hole represents the main part of the exchange energy, the fluctuation regime is realized at high temperatures. It is also actual for initial stages of the polaron formation for times shorter than the relaxation of Mn spins.

In the collective regime, the magnetic polaron energy increases due to Mn spin alignment forced by the carrier exchange field. The localized spins are driven to the energetically favorable state, which is however statistically unfavorable. This regime is realized at low temperatures. The polaron magnetic moment in the collective regime is proportional to N and for typical values of $N \approx 100$ exceeds by an order of magnitude the polaron magnetic moment in the fluctuation regime, as $\sqrt{N} \approx 10$. The collective regime is typical for the exciton magnetic polarons and for the case of BMP formed by a hole on an acceptor.

Evolution of the exciton magnetic polarons starts from the fluctuation regime. The collective regime is developed with time due to alignment of the localized spins. The fluctuation part of the polaron energy remains unchanged until the magnetization approaches the saturation. It is typically by an order of magnitude smaller than the collective part of the polaron energy.

In the most general case, the magnetic polaron Hamiltonian has the following form:

$$H_{\mathrm{MP}} = [K + U(\mathbf{r})] + [H_{\mathrm{dd}} + H_{\mathrm{mag}}] + H_{\mathrm{ex}}. \tag{7.1}$$

Here, $K + U(\mathbf{r})$ is the carrier Hamiltonian in the absence of exchange interaction with magnetic ions, K is the kinetic energy, and $U(\mathbf{r})$ is the potential which may include: (1) Coulomb potential of an impurity center for bound magnetic polarons, (2) localizing potential of alloy fluctuations or of the well width fluctuations in quantum wells, (3) carrier confinement potentials in heterostructures formed by band gap discontinuities. $H_{\mathrm{dd}} + H_{\mathrm{mag}}$ is the Hamiltonian of the magnetic ion spin system, where H_{dd} accounts for interactions between the localized spins including scalar interaction in a Heisenberg form, magneto-dipole interaction and Dzyaloshinskii–Moriya anisotropic interaction [32]. H_{mag} describes interaction of localized spins with an external magnetic field. H_{ex} is the Hamiltonian of exchange interaction of the carrier with the magnetic ions.

For a specific purpose of description of the autolocalized magnetic polaron, a nonlinear Hamiltonian is used

$$H_{\mathrm{MP}} = K(r_{\mathrm{loc}}, t) + U(\mathbf{r}) + V_{\mathrm{ex}}(\Psi^2(r_{\mathrm{loc}}, t), t). \tag{7.2}$$

Here, magnetic properties of the localized spins and their interaction with a carrier are included in the magnetic potential V_{ex}, which depends on the carrier wave function Ψ and changes in time during the magnetic polaron formation. The parameter r_{loc} is the carrier localization radius, which decreases due to autolocalization. Time dependence in (7.2) is important for the dynamics of the polaron formation and it can be omitted for analysis of the polaron stability and for calculation of polaron energy.

7.2.1 Stability of Magnetic Polarons in Systems of Different Dimensionality

The problem of the stability of free magnetic polarons was first analyzed by Kasuya et al. for 3D systems [33] and by Ryabchenko and Semenov for low-dimensional systems [34]. Later on after appearance of DMS quantum wells, an analysis of exciton free magnetic polarons in these structures was performed by A. Kavokin and K. Kavokin [26, 35] and by Benoit a la Guillaume et al. [29, 30].

Theoretical studies of the free magnetic polaron problem are generally based on the Schrödinger equation with an exchange potential that is proportional to a local magnetization. The magnetization is induced by the exchange interaction of magnetic ions with the carrier that forms the free magnetic polaron. The strength of the interaction for a magnetic ion is often expressed by an exchange field $B_{\mathrm{ex}}(\mathbf{r})$, which is proportional to the carrier density at the ion location. In case of holes, which play a dominant role in exciton magnetic polarons

$$B_{\mathrm{ex}}(\mathbf{r}) = \frac{1}{3\mu_{\mathrm{B}} g_{\mathrm{Mn}}} \beta J \, |\Psi(\mathbf{r})|^2. \tag{7.3}$$

Here, β is the hole exchange constant, $\Psi(\mathbf{r})$ is the spatial coordinate part of the hole wave function, J is the hole spin being equal to 3/2 in case of heavy holes. μ_{B} is the Bohr magneton and $g_{\mathrm{Mn}} \approx 2$ is the g-factor of the Mn^{2+} ion. Thus, the magnetization gained in this way depends on the carrier wave function, and the Schrödinger equation appears to be nonlinear.

This equation, even in the simplest case of the magnetization proportional to B_{ex}, has been solved exactly for one-dimensional (1D) systems only. The most reliable way, if one intends to fit experimental data, is to solve the equation numerically. A comprehensive method for the numerical procedure was suggested in [29]. To get less precise, but qualitatively clearer information, variational approaches were widely used [34, 35]. It should be noted here that, although the variational solution of a Schrödinger equation usually associates with minimization of energy, the functional that reaches its minimal value for the ground-state eigenfunction of a free magnetic polaron Hamiltonian, differs from the mean value of the Hamiltonian, i.e., from the carrier energy. The reason for this apparent controversy is the nonlinearity of the Schrödinger equation, or, in other words, dependence of the Hamiltonian on the wave function. It was shown that the eigenfunction provides the minimum to another functional, which has the sense of a work made to bring the system from the initial state, with a free carrier and zero magnetization, to the final state, where finite local magnetization exists and the carrier is autolocalized [30, 36, 37]. At a fixed temperature of the system, this functional coincides with the free energy, thus giving a simple relation to thermodynamics [37].

A criterion for the stable free magnetic polaron requires that the free energy of the magnetic ions in the carrier exchange field (F_{M}) should overcome the carrier kinetic energy gained via autolocalization (ΔK) [26, 34]. Therefore, it reads $\Delta K - F_{\mathrm{M}} < 0$. Equilibrium polaron state corresponds to the minimum of the

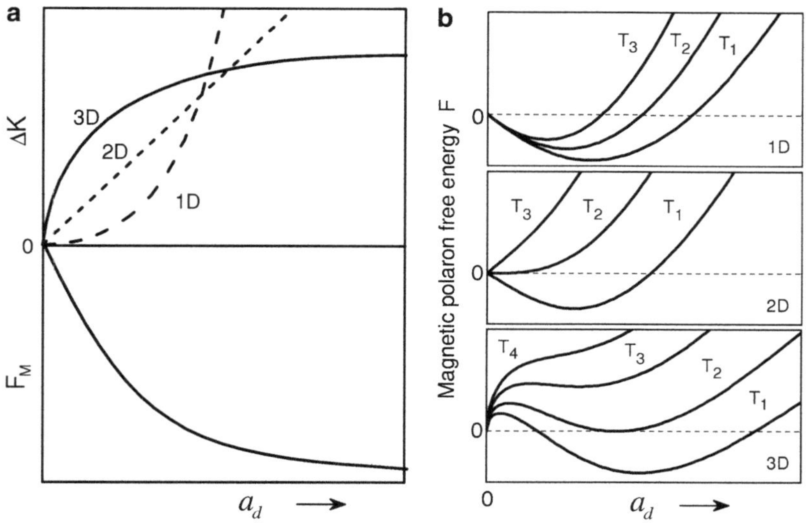

Fig. 7.2 Stability of magnetic polarons in systems of different dimensionality. (**a**) Dependencies of the kinetic energy ΔK and exchange part of the free energy F_M on the parameter $a_d \propto 1/(r_{loc}^d T)$. (**b**) Dependence of the polaron free energy on a_d. Represented temperatures have the following relation $T_1 < T_2 < T_3 < T_4$ [26,35]

free energy $F = \Delta K - F_M$. It is convenient to introduce a variational parameter $a_d \propto 1/(r_{loc}^d T)$, where d is the system dimensionality and T is the temperature [35]. Dependencies of ΔK and F_M on a_d for systems of different dimensionality are shown in Fig. 7.2a. Both ΔK and F_M increase in magnitude with a decrease of carrier localization volume $V \propto r_{loc}^d$. But their functional dependence on V depends on the dimensionality of the system. The dependence $F_M(V)$ is universal for any value of d an in weak exchange fields (i.e., for small a_d) $F_M \propto a_d \propto 1/V$. The kinetic energy is determined by the number of degrees of freedom accessible for the carrier motion $\Delta K \propto V^{-2/d} \propto r_{loc}^{-2}$ and therefore $\Delta K \propto (a_d T)^{2/d}$.

Typical dependencies of the magnetic polaron free energy on a_d for 1D, 2D, and 3D systems are given in Fig. 7.2b. The following conclusions on the formation process of the free magnetic polaron can be drawn [26, 34]:

1. It is always favorable (at any temperature) in 1D systems;
2. In 3D systems, it is always hindered by a potential barrier and an existence of a stable free magnetic polaron state behind this barrier depends on the behavior of $F_M(T)$ at stronger exchange fields B_{ex} corresponding to larger a_d;
3. There is no potential barrier in 2D systems, but existence of the free magnetic polaron depends on parameters of the system such as exchange constant and carrier effective mass and on the temperature.

These conclusions are confirmed by results of numerical calculations performed in [29] and shown in Fig. 7.3a. FMP energy is plotted as a function of the coupling

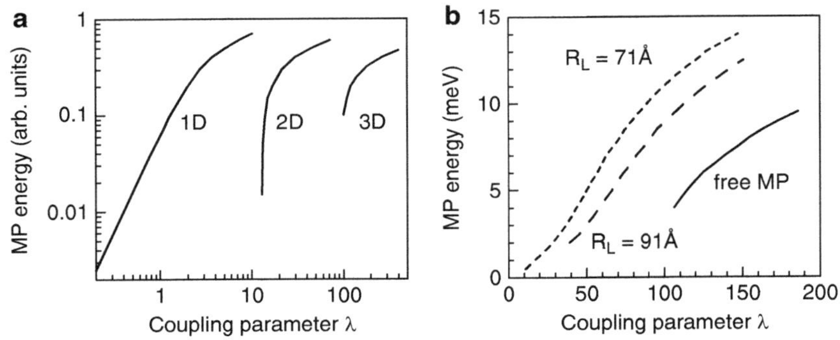

Fig. 7.3 (**a**) Energy of free magnetic polarons (normalized units) as a function of coupling parameter λ between a carrier and localized Mn spins for systems of different dimensionality. (**b**) Calculated energies of free (*solid line*) and localized (*dashed*) magnetic polarons in a 22-Å-thick $CdTe/Cd_{0.9}Mn_{0.1}Te$ quantum well. Radius of initial localization R_L is given in the panel [29]

parameter $\lambda \propto \beta^{5/2}(mxS_{\rm eff})^{3/2}/(T + T_0)$. Here, m is carrier effective mass, x is concentration of magnetic ions, $S_{\rm eff}$ and T_0 are parameters accounting for interactions between magnetic ions (see e.g., (8.15) in Chap. 8). One can see that the free magnetic polaron in one-dimensional system is indeed stable in the whole range of the coupling parameters. However, the polaron stability range is reduced with increasing dimensionality to $\lambda > 10$ for 2D case and $\lambda > 100$ for 3D systems.

To summarize, the free magnetic polaron formation is favored by the reduction of dimensionality and by the increase of exchange constant and carrier effective mass. It should be mentioned that the free magnetic polaron is a special case of the more general problem of a carrier self-trapping by any short-range interaction with the medium (for review, see [38]). A specifics of the magnetic polarons is related to the saturation of the polarization of localized spins in strong exchange field of the carrier. This saturation prevents collapse of the magnetic polarons. Estimations give values of the final radius of the free magnetic polaron, provided the latter exists, which exceed considerably the lattice constant.

It follows from the above considerations that in 2D and 3D systems the formation of magnetic polarons should be very sensitive to any random potential that can provide an additional localization of the carrier. From general consideration, one can suggest two possible sources of the localizing potential in 3D systems: variation of band edges due to alloy fluctuations, and spatially inhomogeneous magnetization. In 2D systems, such as quantum wells, monolayer variation of the quantum well width can play the main role in nonmagnetic localization [24, 26]. Contributions of all three mechanisms mentioned have been documented experimentally (see later), but no consistent statistical theory has been suggested so far.

Model calculations performed for a primary localization in rectangular potential wells have yielded a good agreement with experimental data on magnetic polaron energies at reasonable values of depth and radius of the wells [26, 29, 30]. One example for such calculations performed for $CdTe/Cd_{0.9}Mn_{0.1}Te$ quantum wells is given in Fig. 7.3b. Free magnetic polaron becomes unstable and is not formed for $\lambda < 100$, while an additional localization of exciton in-plane motion in the region with 71 Å radius shifts the stability range by an order of magnitude to $\lambda < 10$. Also for $\lambda > 100$, the localization leads to an increase of the polaron energy.

Let us summarize qualitative results of model calculations for magnetic polarons in diluted magnetic semiconductors [29, 30]:

1. Self-localization in 3D systems is very unfavorable and the initial localization of the carrier is of key importance for magnetic polaron formation. Such polarons formed from localized excitons we label as localized magnetic polarons (LMPs) [39]. The radius of the carrier localization is not expected to decrease significantly in the process of the LMP formation.
2. In experimentally available 2D systems, initial localization is still necessary to initiate the polaron formation. On the other hand, the localization radius r_{loc} can change during the polaron formation. Magnetic polarons of this kind can be considered as quasi-free polarons.
3. Since in 1D systems the free magnetic polarons are predicted to be stable, an additional localization is not so much important, although it can enhance the polaron effect to some extent.

The above conclusions on the free magnetic polaron properties done in terms of carriers are also valid for the exciton free magnetic polaron. A specifics of the exciton magnetic polarons in (Cd,Mn)Te is that the exchange constant and effective mass for holes are several times larger than for electrons. As a result, the hole contribution dominates the polaron formation. Simple considerations given below show that the exciton autolocalization can affect not only the polaron equilibrium energy, but also the dynamics of polaron formation. Henceforth, we will use the term autolocalization, as distinct from self-trapping, for the process of shrinkage of the exciton wave function, whether its initial localization is provided by nonmagnetic or magnetic potential.

7.2.2 Dynamics of Magnetic Polaron Formation

Evolution of exciton magnetic polarons has several stages controlled by exciton energy relaxation and recombination dynamics and by spin dynamics of the exciton and localized spins of magnetic ions:

1. Photogeneration of an exciton and its energy relaxation to the bottom of exciton band, which is assisted by emission of optical and acoustic phonons can take up to 100 ps [40]. The relaxation process is absent in case of selective excitation of localized excitons, the technique commonly used to study exciton magnetic polarons (see Sect. 7.3).
2. Establishing starting correlation between an exciton spin and localized spins inside the exciton. At this stage, the fluctuation regime of magnetic polarons is realized.

3. Ferromagnetic alignment of localized spins by an exciton exchange field (mainly hole exchange field), which is accompanied by an increase of polaron energy and polaron magnetic moment, i.e., by developing the collective regime. The magnetic polaron reaches its equilibrium state with an energy E_{MP}.
4. Polaron stays in its equilibrium state until exciton recombination. At this stage, the magnetic moment of the whole polaron can change its orientation, e.g., can be polarized by external magnetic fields.

An essential stage of the polaron formation, as it is shown in Fig. 7.1, is the ferromagnetic alignment of the localized spins along the exciton exchange field B_{ex}. However, just after the exciton photogeneration B_{ex} orientation should first be stabilized. We define this stage as a stage of establishing starting correlation between the exciton spin and the magnetic moment of localized spins inside the exciton volume, $\mathbf{M}_f$. $\mathbf{M}_f$ is controlled by fluctuations of magnetization [11, 41], and the starting correlation corresponds to the polaron fluctuation regime with parallel orientation of $\mathbf{M}_f$ and the exciton spin. Typical exchange energy in this case is about 1 meV, which exceeds the thermal energy $k_{\mathrm{B}}T \approx 0.1$ meV for $T = 1.6$ K.

Let us turn to the stage (3) of the evolution of a collective polaron regime. In wide band gap II–VI DMS free magnetic polarons are not stable and an initial localization is required to form exciton magnetic polarons. This is valid both for 3D system and quasi-2D systems, such as quantum wells. However, there is a considerable difference between these cases. In the 3D case, the magnetic polaron formation does not cause additional autolocalization and shrinkage of the exciton localization volume. In quantum wells, however, the latter can be reduced considerably, sometimes even a few times. Theoretical consideration of K. Kavokin [28, 36] shows that the polaron autolocalization in a 2D system not only increases the polaron energy, but also sufficiently modifies the dynamics of polaron formation.

In Kavokin's theory [28, 36], the evolution of the magnetization in the Mn-spin system $M(\mathbf{r},t)$ obeys a simple first-order differential equation with a unique relaxation time τ_s:

$$\frac{\mathrm{d}}{\mathrm{d}t} M(\mathbf{r},t) = -\frac{1}{\tau_s}[M(\mathbf{r},t) - M_{\mathrm{eq}}(B_{\mathrm{ex}}(\mathbf{r},t))], \tag{7.4}$$

where M_{eq} is the equilibrium magnetization that corresponds to the exchange field $B_{\mathrm{ex}}(\mathbf{r},t)$ defined by (7.3). For model considerations and for further comparison with experiment, it is convenient to introduce a magnetic polaron shift ΔE, which is the part of the polaron energy gained in the collective regime of the polaron formation. In the equilibrium, the polaron shift coincides with the polaron energy: $\Delta E(t \to \infty) = E_{\mathrm{MP}}$.

The dynamics of the polaron shift $\Delta E(t)$ is described by

$$\frac{\mathrm{d}}{\mathrm{d}t} \Delta E(t) = -\frac{1}{\tau_s}[\Delta E(t) - \Delta E_{\mathrm{eq}}(t)], \tag{7.5}$$

where $\Delta E_{\text{eq}}(t) \equiv \Delta E_{\text{eq}}(B_{\text{ex}}(t))$ is the equilibrium polaron energy in a given moment. It is the value to which the polaron energy formation tends at a given time moment. It is essential that for the autolocalization which increases B_{ex} the value of $\Delta E_{\text{eq}}(t)$ changes with time. For the equilibrium polaron state $\Delta E_{\text{eq}}(t \to \infty) = E_{\text{MP}}$. Suggesting a linear dependence of ΔE_{eq} on ΔE (see [28]) one can conclude from (7.5) that the time evolution of the polaron shift obeys the exponential law

$$\Delta E(t) = E_{\text{MP}}[1 - \exp(-t/\tau_f)], \tag{7.6}$$

with the polaron formation time

$$\tau_f = \tau_s \frac{E_{\text{MP}}}{\Delta E_{\text{eq}}(0)} = \tau_s(1 + \xi). \tag{7.7}$$

$\xi > 0$ is the part of the equilibrium polaron energy gained due to autolocalization. Note that for all studied structures the experimentally measured polaron dynamics closely follows an exponential law.

Analyzing (7.7) one can conclude that in 3D systems, where the autolocalization is negligible, $\xi = 0$, $\tau_f = \tau_s$. That means that the polaron formation is determined solely by spin dynamics of the Mn-spin system. In quantum wells, where the autolocalization contributes considerably to the polaron equilibrium energy, τ_f exceeds τ_s by a factor of $(1 + \xi)$.

Note that (7.7) should not be used for free magnetic polarons, where $\Delta E_{\text{eq}}(0) = 0$ would result in an infinite formation time. In this case, nonlinear character of the problem must be accounted at earlier stages of the model treatment. It was shown that the characteristic time of the free magnetic polaron formation also exceeds τ_s, but no unique exponential function can describe $\Delta E(t)$ [36].

We turn to mechanisms controlling the time τ_s. The polaron formation requires both the relaxation of energy by E_{MP} and the relaxation of spins of magnetic ions in the course of their ferromagnetic alignment. The slowest process among these two will control τ_s. However, the spin and energy transfers between carriers, magnetic ions, and lattice are often coupled with each other and have the very same mechanisms (see Chap. 8).

Experimental values for the polaron formation time τ_f fall in the range from 50 to 250 ps (Sect. 7.6.1), which is indicated by arrows in Fig. 7.4. Data points in this figure represent a collection of relaxation times for the Mn^{2+} spins in II–VI DMS [25]. The spin-lattice relaxation process, which assists energy transfer from the Mn spins to the phonon system, requires times in a range from 100 ns to 100 μs (see also Fig. 8.10 in Chap. 8). It is by several orders of magnitude longer than the spin–spin relaxation dynamics that covers a range from 1 ps to 1 ns. Both these relaxation processes are strongly accelerated with increasing Mn content in a sample due to an enhancement of the spin–spin interactions between magnetic ions. The polaron formation times correspond to the spin–spin relaxation dynamics. Spin-lattice relaxation is too slow to contribute to the fast polaron formation.

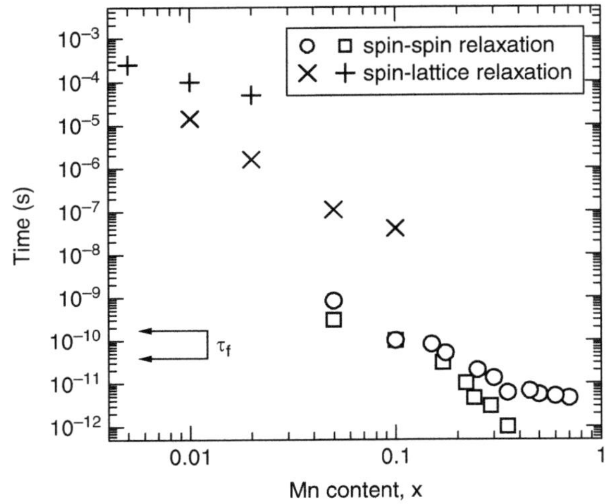

Fig. 7.4 A collection of spin-lattice and spin–spin relaxation times in $Cd_{1-x}Mn_xTe$ at 5 K [25]. *Arrows* show the range for typical magnetic polaron formation times

The polaron dynamics is governed by spin–spin interactions of magnetic ions [25]: (1) the system of interacting Mn spins has an energy reservoir of sufficient capacity, and energy transfer from the Mn spins to this reservoir happens with spin–spin relaxation rates; (2) the conservation of the spin momentum is broken by strong nonscalar spin–spin interactions, which allows generation of a nonzero magnetization inside the polaron volume.

7.2.3 Parameters of Exciton Magnetic Polaron

In this section, we discuss the relation of the exciton magnetic polaron parameters with the values that can be directly measured in magneto-optical experiments [42–46].

The polaron parameters of interest are: (1) polaron energy E_{MP}, (2) exchange field B_{ex}, (3) magnetic moment of the polaron in its equilibrium state M_{MP}, (4) polaron volume V_{MP}, and (5) fluctuation magnetic moment of localized spins in the polaron volume M_f, which contributes to establishing the starting correlation.

Among the experimentally measured values are:

1. Giant Zeeman splitting of heavy-hole exciton $\Delta E_z(B)$. According to (8.14) and (8.15), it reads

$$\Delta E_z(B) = (\alpha - \beta) N_0 x S_{\text{eff}} \mathrm{B}_{5/2}\left[\frac{5\mu_B g_{\text{Mn}} B}{2k_B(T + T_0)}\right]. \tag{7.8}$$

Here, $N_0\alpha$ and $N_0\beta$ are the exchange constants for the conduction and valence band. S_{eff} is the effective spin and T_0 is the effective temperature. These parameters permit a phenomenological description of the antiferromagnetic Mn–Mn exchange interaction. In the case of the exciton magnetic polaron, the heavy hole contributes dominantly to the polaron energy due to larger exchange constant and smaller localization volume compared with electron parameters. Therefore, we can limit our consideration to the holes. Their giant Zeeman splitting is directly related to that of the exciton

$$\Delta E_z^{\rm hh}(B) = \frac{|\beta|}{|\alpha| + |\beta|} \Delta E_z(B). \tag{7.9}$$

In the linear approximation of the modified Brillouin function $B_{5/2}$

$$\Delta E_z^{\rm hh}(B) = \left.\frac{{\rm d}\Delta E_z^{\rm hh}(B)}{{\rm d}B}\right|_{B=0} B = \gamma B. \tag{7.10}$$

The value of γ can be measured experimentally.

2. Energy of the exciton magnetic polarons $E_{\rm MP}$ can be measured by selective excitation of localized excitons as an energy shift of the polaron emission line from the laser energy. This is, however, not always possible. Interpretation problems appear, e.g., in situations when the polaron formation time exceeds that of exciton recombination, as is often the case of DMS quantum dots [47].
3. Circular polarization degree of the polaron luminescence $P_{\rm c}(B) = (I_+ - I_-)/(I_+ + I_-)$ induced by external magnetic field applied in the Faraday geometry. Here, I_+ and I_- are emission intensities detected in σ^+ and σ^- polarizations, respectively. Usually, the slope of $P_{\rm c}(B)$ dependence in low magnetic fields is of interest for us. Therefore, we introduce a parameter $\theta = {\rm d}P_{\rm c}(B)/{\rm d}B|_{B=0}$, which is related to the magnetic fluctuations [43]:

$$\theta = \frac{\sqrt{2\langle M_f^2\rangle}}{\sqrt{\pi} k_{\rm B} T}. \tag{7.11}$$

For evaluations, we use the "exchange box" approximation [12] which substitutes the inhomogeneous exchange field of a hole $B_{\rm ex}(\mathbf{r}) \propto |\Psi(\mathbf{r})|^2$ (see (7.3)) with a field $B_{\rm ex}$ homogeneous inside the hole localization volume $V_{\rm MP}$ and vanishing outside. We will treat $V_{\rm MP} = \frac{4}{3}\pi r_{\rm loc}^3$ as the polaron volume. Also, we consider heavy holes with $J = 3/2$. From (7.3), we get a relation between the exchange field and the polaron volume:

$$B_{\rm ex} = \frac{\beta}{2\mu_{\rm B} g_{\rm Mn}} \frac{1}{V_{\rm MP}}. \tag{7.12}$$

The experimental $E_{\rm MP}$ value can be linked with the hole giant Zeeman splitting

$$E_{\rm MP} = \frac{1}{2} \Delta E_z^{\rm hh}(B_{\rm ex}). \tag{7.13}$$

This gives a tool for experimental evaluation of $B_{\rm ex}$.

Finally, the magnetic moment of the polaron M_{MP} is by definition

$$M_{\mathrm{MP}} = \frac{E_{\mathrm{MP}}}{B_{\mathrm{ex}}} = \frac{\Delta E_z^{\mathrm{hh}}(B_{\mathrm{ex}})}{2B_{\mathrm{ex}}}. \tag{7.14}$$

Using the linear approximation (7.10) for the hole giant Zeeman splitting, one comes to the following relations for the polaron parameters:

$$E_{\mathrm{MP}} = \frac{1}{2}\gamma B_{\mathrm{ex}}, \tag{7.15}$$

$$B_{\mathrm{ex}} = \frac{2E_{\mathrm{MP}}}{\gamma}, \tag{7.16}$$

$$M_{\mathrm{MP}} = \frac{\gamma}{2}, \tag{7.17}$$

$$V_{\mathrm{MP}} = \frac{\beta}{4\mu_{\mathrm{B}} g_{\mathrm{Mn}}} \frac{\gamma}{E_{\mathrm{MP}}}, \tag{7.18}$$

$$\langle M_f^2 \rangle = \frac{k_{\mathrm{B}}T}{4} \frac{\gamma^2}{E_{\mathrm{MP}}}. \tag{7.19}$$

Concerning the polarization degree, the following set of relations can be written (assuming that autolocalization does not contribute significantly to the polaron energy):

$$E_{\mathrm{MP}} = \frac{1}{2\pi k_{\mathrm{B}}T} \frac{\gamma^2}{\theta^2}, \tag{7.20}$$

$$B_{\mathrm{ex}} = \frac{1}{\pi k_{\mathrm{B}}T} \frac{\gamma}{\theta^2}, \tag{7.21}$$

$$M_{\mathrm{MP}} = \sqrt{\frac{\pi k_{\mathrm{B}}T}{2}}\,\theta\sqrt{E_{\mathrm{MP}}}, \tag{7.22}$$

$$V_{\mathrm{MP}} = \frac{\pi k_{\mathrm{B}}T\beta}{2\mu_{\mathrm{B}} g_{\mathrm{Mn}}} \frac{\theta^2}{\gamma}, \tag{7.23}$$

$$\langle M_f^2 \rangle = \frac{\pi k_{\mathrm{B}}^2 T^2}{2}\theta^2. \tag{7.24}$$

It is worthwhile to note that all parameters in (7.20) can be measured experimentally. This provides a direct test of the model. Its validity was indeed confirmed in [42,45].

7.3 Optical Study of Exciton Magnetic Polarons by the Method of Selective Excitation

Several methods of optical spectroscopy, photoluminescence (PL), and polarized PL allow us to get detailed information about static and dynamic properties of magnetic polarons in diluted magnetic semiconductors. Polaron formation causes exciton localization in the exchange "potential" and the polaron energy can be evaluated from the low energy shift of the luminescence line respective to the free exciton energy. In DMS, the Stokes shift of luminescence line is due both to the polaron formation and to the exciton localization by a nonmagnetic random potential (e.g., alloy fluctuations), and it is necessary to distinguish between these two contributions. This can be done by the method of selective excitation, which was successfully used to study the exciton magnetic polarons in $Cd_{1-x}Mn_xTe$-based epilayers and heterostructures [21–24, 39, 48].

In this method, the excitons are excited selectively in the band of localized states at energies where the spectral diffusion due to phonon-assisted tunneling does not occur during the exciton lifetime. In this case, the Stokes shift between the luminescence line and the energy of selective excitation is determined by the magnetic polaron formation only, and can be associated with the polaron shift ΔE.

In Fig. 7.5, the method is presented for a 48-Å-period $Cd_{0.83}Mn_{0.17}Te$/ $Cd_{0.54}Mg_{0.46}Te$ superlattice. A tail of exciton localized states is depicted by PL excitation

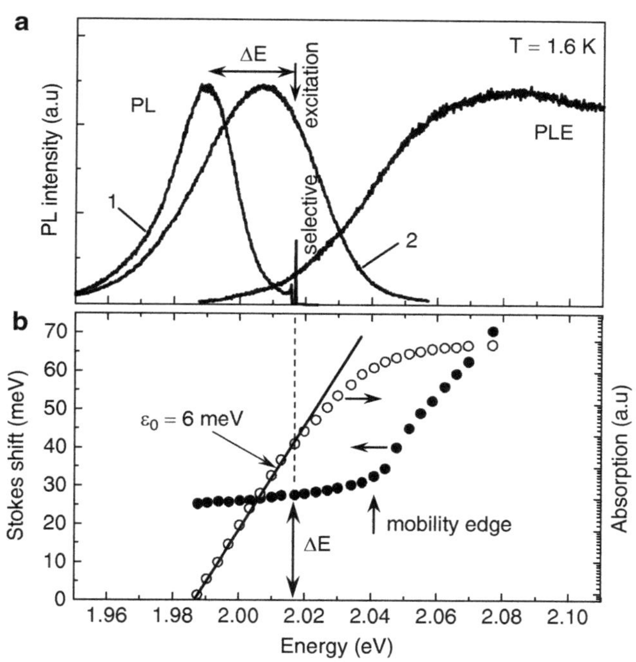

Fig. 7.5 Method of selective excitation presented for a 48-Å-period $Cd_{0.83}Mn_{0.17}Te/Cd_{0.54}Mg_{0.46}Te$ superlattice. (**a**) PLE spectrum and luminescence spectra measured under nonselective (curve 2) and selective (curve 1) excitations. (**b**) Absorption edge displays exponential tail of exciton localized states (*open circles*). Energy of 2.04 eV, where the dependence of the Stokes shift on the excitation energy changes its character, can be associated with the exciton mobility edge. Below this energy, spectral diffusion of excitons due to phonon-assisted tunneling does not occur during the exciton lifetime [49]

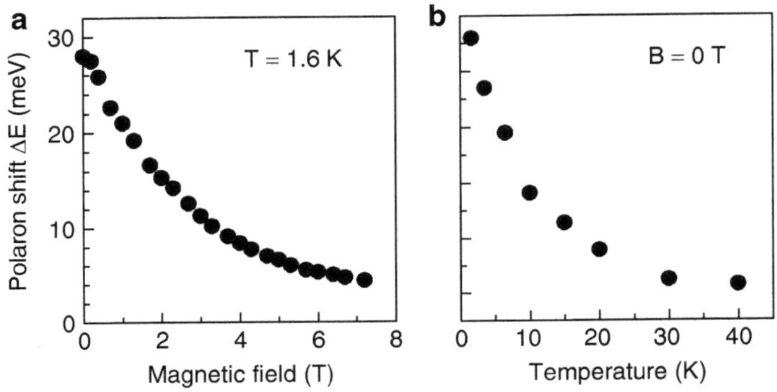

Fig. 7.6 Temperature and magnetic field (Faraday geometry) suppression of the magnetic polaron energy in a 48-Å-period $Cd_{0.83}Mn_{0.17}Te/Cd_{0.54}Mg_{0.46}Te$ superlattice

(PLE) spectrum in panel (a) and by the absorption edge shown in logarithmic scale in panel (b). The absorption edge was estimated by integrating luminescence intensity measured at different excitation energies in the band of localized states, as suggested in [50]. The edge is well described by an exponential law with a characteristic energy $\varepsilon_0 = 6$ meV [51]. The maximum of PLE spectrum at 2.08 eV corresponds to the free exciton energy. Luminescence line taken under nonselective excitation with energy exceeding 2.08 eV is shown by curve 2 in Fig. 7.5a. Nonmagnetic localization and polaron formation contribute to the Stokes shift ΔE. The Stokes shift shown by closed circles in panel (b) becomes nearly independent of the excitation energy below 2.04 eV, when only the excitons in the band of localized states are excited. Therefore, the observed Stokes shift value of 28 meV can be associated with the magnetic polaron energy.

The verification that the observed Stokes shift under selective excitation is of pure magnetic origin is provided by a characteristic suppression of ΔE either by application of an external magnetic field or by temperature increase shown in Fig. 7.6. Effective suppression of the polaron shift by temperature is due to the strong reduction of the magnetic susceptibility of the Mn-ion system and cannot be described by thermal activation as at $T = 20$ K thermal energy $k_B T \approx 2$ meV is much less than $\Delta E = 28$ meV.

We note here that the suggested method requires a certain initial localization of excitons, and strictly speaking, cannot be used for the study of the free magnetic polarons. But this limitation is not crucial for the structures reported in this chapter, as in most of them the free magnetic polarons are not stable.

Combination of the time-resolved spectroscopy with the selective excitation allows one to study dynamics of polaron formation with a picosecond resolution [21, 39, 52]. For experimental examples given in this chapter, the luminescence signal was recorded by a streak-camera with temporary resolution of 15 ps and selective photoexcitation was provided by 5 ps laser pulses.

Figure 7.7 shows the time evolution of magnetic polaron shift $\Delta E(t)$ in an 80-Å-thick $Cd_{0.89}Mn_{0.11}Te/Cd_{0.60}Mn_{0.11}Mg_{0.29}Te$ quantum well. The data are plotted in

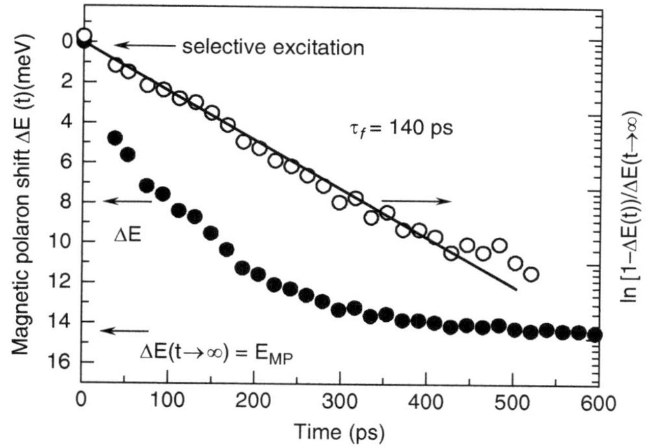

Fig. 7.7 Evolution of magnetic polaron shift $\Delta E(t)$ in an 80-Å-thick $Cd_{0.89}Mn_{0.11}Te/Cd_{0.60}Mn_{0.11}Mg_{0.29}Te$ quantum well plotted in linear and logarithmic scales. The strong dynamical effect on the magnetic polaron energy is clearly seen from a difference between ΔE measured under *cw* excitation and $\Delta E(t \to \infty)$

linear and logarithmic scales. The fast initial shift of the polaron line is followed by a saturation of $\Delta E(t)$ at the level of $\Delta E(t \to \infty) = 14$ meV, which corresponds to the equilibrium polaron energy E_{MP}. The formation process is well described by an exponential law $\Delta E(t) = \Delta E(t \to \infty)[1 - \exp(-t/\tau_f)]$, in line with model predictions. Linear fit to the experimental data yields a polaron formation time $\tau_f = 140$ ps. The polaron lifetime $\tau_0 = 110$ ps is determined from the decay of the spectrally integrated luminescence intensity.

In this structure, the polaron formation time exceeds the polaron lifetime. That means that the formation process is interrupted by recombination and the polaron shift $\Delta E = 8$ meV measured under continuous wave (*cw*) excitation is smaller than the polaron equilibrium energy $E_{MP} = \Delta E(t \to \infty) = 14$ meV received from the time-resolved experiments.

Here, we choose the structure with $\tau_0 < \tau_f$ to demonstrate that the magnetic polaron shift measured in steady-state experiments can differ strongly from the equilibrium polaron energy. Therefore, the time-resolved experiments are of importance not only for the study the polaron formation dynamics but also for the correct evaluation of the polaron energy.

Thereafter, we discuss the polaron formation in different structures, studied by the method of selective excitation. All structures reported in this chapter were grown by molecular-beam epitaxy on (100)-oriented substrates of CdTe or $Cd_{0.96}Zn_{0.04}Te$. Most of the experimental data are shown for a temperature of 1.6 K. External magnetic fields up to 8 T are generated by a split-coil superconducting magnet and are applied in the Faraday or Voigt geometries, i.e., with the field parallel or perpendicular to the structure growth axis (the *z*-axis), respectively.

7.4 Exciton Magnetic Polarons in 3D Systems

In this part, we discuss experimental results on exciton magnetic polarons in 3D systems represented by epilayers of $Cd_{1-x}Mn_xTe$ and $Cd_{1-x-y}Mn_xMg_yTe$. We are interested in the role of exciton localization and Mn concentration in the polaron stability and polaron binding energy. It will be shown that the polaron formation reduces exciton mobility and shifts exciton mobility edge.

As we discussed in the theory section, free magnetic polarons are not stable in 3D systems based on wide band gap II–VI DMS. Initial localization of an exciton becomes crucial for its further evolution to the magnetic polaron in the collective regime. In ternary and quaternary alloy semiconductors, the exciton localization is provided by potential fluctuations due to band gap variations [51,53]. Therefore, the exciton magnetic polarons in these alloys are often referred to as localized magnetic polarons (LMP). It should be noted that they are trapped by alloy fluctuations, which distinguishes them from the bound magnetic polarons (BMPs).

7.4.1 Magnetic Polarons in (Cd,Mn)Te

It is expected that the polaron energy in $Cd_{1-x}Mn_xTe$ should increase with growing Mn content. The starting point for this dependence is nonmagnetic CdTe, where magnetic polarons are absent by definition. The polaron energy increase is influenced by two factors: (1) by increasing number of localized spins in the exciton localization volume and (2) by enhancement of "nonmagnetic" exciton localization.

Exciton magnetic polarons were examined in a wide range of Mn contents $0 < x \leq 0.4$ by the selective excitation technique [39]. Results for polaron energies and polaron dynamics are summarized in Fig. 7.8. Open circles in panel (a) show the polaron shift measured under *cw* excitation. No shift is detected for $x < 0.10$. In the range $0.10 \leq x \leq 0.25$, the polaron energy increases with the Mn content and approaches a saturation value of about 28 meV for $x > 0.25$.

Polaron dynamics was examined to check to what extent the polaron shift under *cw* excitation corresponds to the equilibrium polaron energy E_{MP}. The polaron formation times and lifetimes are summarized in panel (b). The formation time decreases with increasing Mn content from 130 ps at $x = 0.13$ down to 55 ps at $x = 0.34$. Contrary to that, the lifetime increases in this range from 180 to 480 ps. For alloys with $x < 0.17$, the condition under which the exciton magnetic polaron can reach its equilibrium energy during the lifetime ($\tau_f \ll \tau_0$) is not met. For $x = 0.13$, the ratio between formation time and lifetime is $\tau_f/\tau_0 = 0.7$ and the saturation value $\Delta E(t \to \infty) = 12.8$ meV exceeds considerably the polaron shift $\Delta E_{cw} = 8.6$ meV. Equilibrium polaron energy evaluated from saturation of $\Delta E(t)$ in time-resolved experiment is shown by crosses in Fig. 7.8a. One can conclude that the dynamical effect on the polaron energy becomes important in $Cd_{1-x}Mn_xTe$ with $x < 0.17$. However, it does not change the general trend of the polaron energy

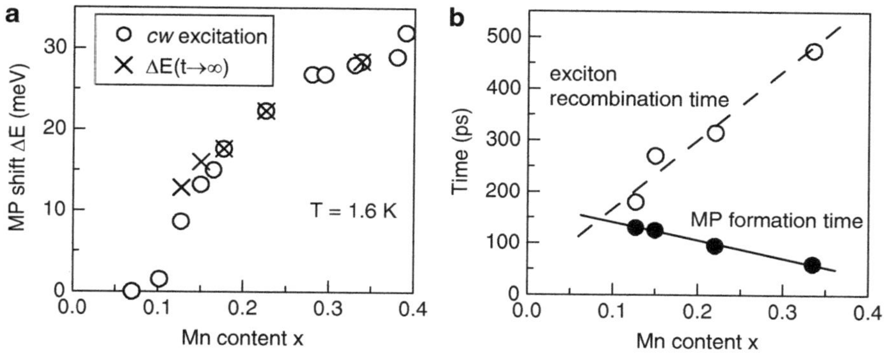

Fig. 7.8 Localized magnetic polarons in $Cd_{1-x}Mn_xTe$. (**a**) Polaron energies measured under continuous wave excitation (*circles*) and as the saturation value of the polaron shift under pulsed excitation (*crosses*). (**b**) Decay time of luminescence and polaron formation time. *Lines* are guides to the eye [39]

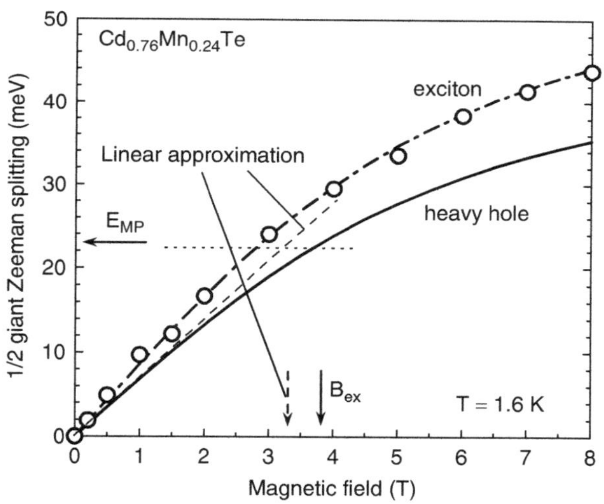

Fig. 7.9 Evaluation of the exchange field B_{ex} for localized magnetic polarons in $Cd_{0.76}Mn_{0.24}Te$. One half of the giant Zeeman splitting of the heavy-hole exciton (*circles* and *dashed-dotted line*) and of the heavy hole itself (*solid line*). For the polaron energy of 23 meV, the exchange field B_{ex} is 3.3 and 3.8 T for the linear approximation and real Zeeman splitting, respectively

concentration dependence, which is controlled by nonmagnetic localization on alloy fluctuations. Further arguments are given in Sect. 7.4.2.

Further polaron parameters can be evaluated following the routine suggested in Sect. 7.2.3. Figure 7.9 illustrates the evaluation procedure of the polaron exchange field B_{ex} on the base of the giant Zeeman splitting of the heavy-hole exciton that was measured experimentally. Data points show the energy shift of the lowest spin component, which corresponds to one half of the total giant Zeeman splitting for

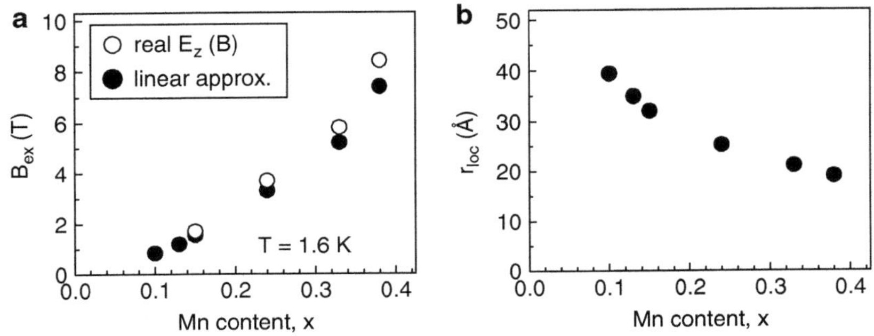

Fig. 7.10 Parameters of localized magnetic polarons in $Cd_{1-x}Mn_xTe$ as functions of Mn content. (**a**) Hole exchange field on the magnetic ions inside the polarons. (**b**) Hole localization radius [45]

the exciton in $Cd_{0.76}Mn_{0.24}Te$. Dashed-dotted line is a fit of experimental data with the modified Brillouin function (7.8) with parameters $T_0 = 8.4\,\text{K}$ and $S_{\text{eff}} = 0.42$. Contribution of the heavy-hole is taken as 80% of the exciton splitting on the base of the ratio of exchange constants for conduction ($N_0\alpha = 0.22\,\text{eV}$) and valence ($N_0\beta = -0.88\,\text{eV}$) bands: $|\beta|/(|\alpha| + |\beta|) = 0.8$. Polaron energy in this sample is $E_{\text{MP}} = 23\,\text{meV}$ and the polaron exchange field evaluated along (7.13) equals to 3.8 T. Use of the linear approximation along (7.16) gives a slightly lower value of 3.3 T.

The exchange field evaluated from the linear approximation and from the real shape of the giant Zeeman splitting of the heavy-hole is plotted in Fig. 7.10a as a function of the Mn content. Strong increase of the exchange field for large x is a result of enhanced exciton localization, as one can conclude from (7.12). The respective decrease of the hole localization radius from 40 down to 20 Å is illustrated in panel (b).

It is remarkable that the hole exchange field which addresses Mn spins inside the magnetic polarons can be as large as 8 T. Such a field can locally modify magnetic susceptibility of the Mn-spin system. Experimental evidence for that is discussed in Sect. 7.4.4.

7.4.2 Role of Nonmagnetic Localization

In ternary alloy $Cd_{1-x}Mn_xTe$, modification of the localized magnetic polarons with increasing Mn content is caused by changes of magnetic susceptibility and by an enhancement of exciton localization. For a deeper insight into polaron physics, separation of these contributions is required.

Let us study first the effect of magnetic susceptibility on the polaron binding energy. For that, we should fix polaron localization volume, which is the case for bound magnetic polarons. As the properties of the exciton magnetic polarons are

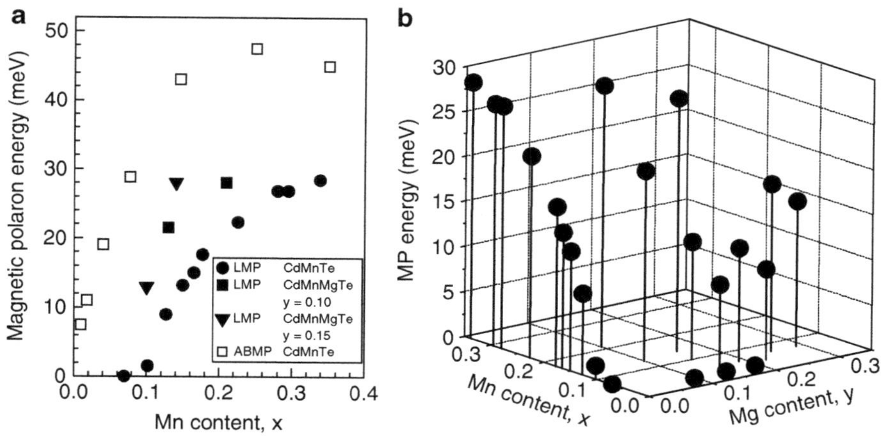

Fig. 7.11 Energies of magnetic polarons in 3D samples. (**a**) Localized magnetic polarons (*closed symbols*) [39, 54] and acceptor-bound magnetic polarons (*open symbols*) [15] vs. Mn content. (**b**) Localized magnetic polarons in $Cd_{1-x-y}Mn_xMg_yTe$ epilayers [55]. $T = 1.6\,K$

dominated by holes, acceptor-bound magnetic polarons (ABMPs) will be the closest analogy.

An initial hole localization in the acceptor-bound magnetic polarons is provided by the Coulomb binding of a hole to an acceptor [15]. The Coulomb potential controls the polaron volume, which is nearly independent of the x value. Therefore, the polaron energy dependence would reflect changes in the magnetic properties of the Mn-spin system with increasing Mn content.

Experimental data for acceptor-bound magnetic polarons in $Cd_{1-x}Mn_xTe$ are shown in Fig. 7.11a by open symbols. The polaron energy increases nearly linearly with an increasing Mn content up to $x \approx 0.15$ and, then, saturates. The initial increase coincides with the behavior of the magnetic susceptibility, which increases with the growing concentration of magnetic ions. However, the magnetic susceptibility reaches its maximum at about $x = 0.12$ and decreases for higher contents due to the antiferromagnetic coupling between neighboring Mn ions, which decreases the number of uncoupled spins. The saturation of the polaron energy at $x \geq 0.15$ is caused by two factors: (1) the antiferromagnetic coupling between neighboring Mn ions, which decreases magnetic susceptibility, and (2) the increasing input into the polaron energy from the clusters of antiferromagnetically coupled ions, which is provided by a strongly nonuniform exchange field of the localized hole [56].

Energies of localized magnetic polarons are given in panel (a) by closed circles. One can see that dependencies for acceptor-bound and localized magnetic polarons differ drastically in the whole range of Mn contents. In the case of localized magnetic polarons, the initial exciton localization is due to potential alloy fluctuations, which depend strongly on the Mn content. For $x < 0.10$, the exciton localization is not sufficient to favor the polaron formation. And for higher contents, the polaron energy of localized polarons is considerably smaller than the energy of the acceptor-bound magnetic polarons. Magnetic susceptibility dependence on the Mn content should give similar contribution to the energies of acceptor-bound and localized magnetic polarons. Therefore, we suggest that the increase of the localized magnetic polaron energy in $Cd_{1-x}Mn_xTe$ in the range of $0.15 \leq x \leq 0.33$ is due to the exciton localization enhancement.

Quaternary alloys $Cd_{1-x-y}Mn_xMg_yTe$ enhance the exciton localization on the potential fluctuations by increasing Mg content without changing the magnetic properties. It was shown from the analysis of the exciton Zeeman pattern that the partial substitution of Cd by Mg in $Cd_{1-x-y}Mn_xMg_yTe$ does not modify the exchange interaction of the band states with Mn ions or the exchange between neighboring Mn ions [54]. The data points depicted by closed squares and triangles in Fig. 7.11a clearly show the strong energy increase of the localized magnetic polarons caused by the enhancement of exciton localization in the quaternary alloys.

An extended set of experimental data for polaron energies in quaternary alloys is collected in Fig. 7.11b [55]. Strong effect of nonmagnetic Mg component on the polaron energy is seen for all studied samples. It is especially important for the low Mn contents of $0.10 < x$, where the localized magnetic polarons are otherwise unstable in ternary alloys.

Exciton effective mass is another factor that controls the nonmagnetic localization. The heavier the exciton, the more compact it is and the larger polaron energy is expected. The heavy-hole effective mass in $Cd_{1-x}Mn_xTe$ crystal is anisotropic, which may affect the polaron energy [57]. The effect was confirmed experimentally by a comparative study of (120) and (100)-oriented digital alloys $Cd_{1-x}Mn_xTe$ [58]. A systematic increase of the exciton magnetic polaron energy by about 20% in (120) samples was observed being in line with the heavier hole mass.

7.4.3 Magnetic Polaron Effect on Exciton Mobility

The polaron formation itself can increase the exciton localization and restrict the exciton mobility in the crystal. Exciton localization in random potential fluctuations was investigated in nonmagnetic semiconductors [51, 53]. It was shown that alloy fluctuations result in a tail of localized exciton states. Exciton spatial and spectral diffusion within the tail was traced down to be due to phonon-assisted tunneling between spatially separated localized states. The tunneling time is a strong function of localization energy, decreasing with the depth of the tail.

To describe the optical properties of the localized excitons, it is very instructive to introduce an effective mobility edge. It corresponds to a certain energy above which exciton motion is possible within the exciton lifetime and below which excitons are localized and recombine without spatial migration. The energy position of the mobility edge is determined by the tunneling time τ_t and by the exciton lifetime τ_0.

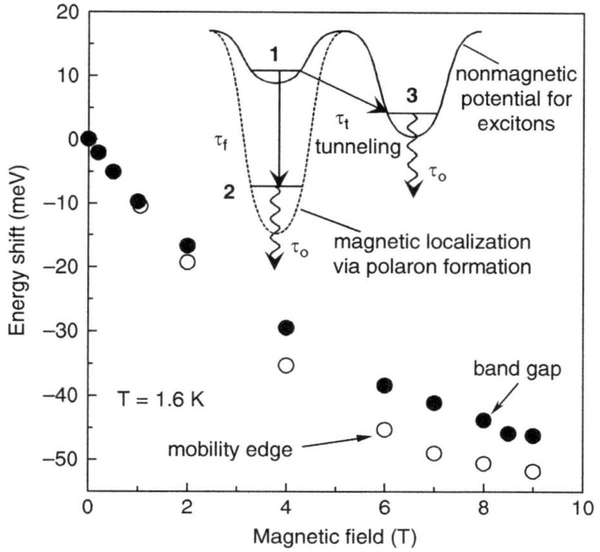

Fig. 7.12 Energy shifts of the band gap and the exciton mobility edge in a $Cd_{0.76}Mn_{0.24}Te$ epilayer in magnetic fields applied in the Faraday geometry. Schematical illustration of the competition between nonmagnetic spectral relaxation (1 → 3) and magnetic polaron formation (1 → 2) is displayed in the inset

In diluted magnetic semiconductors, the magnetic polaron formation enhances exciton localization. Therefore, the exciton mobility edge becomes sensitive to the magnetic localization [49,59]. Polaron formation time is usually shorter than that of exciton recombination, see, e.g., Fig. 7.8b. As a result, the fast magnetic localization interrupts spatial diffusion and nonmagnetic spectral relaxation to deeper localized states. In the inset of Fig. 7.12, the competition between polaron formation (1 → 2) and nonmagnetic spectral relaxation (1 → 3) is schematically illustrated. For the case of the nonmagnetic localization only, i.e., in the absence of magnetic polarons, and for $\tau_t < \tau_0$ exciton will tunnel from the state 1 to the state 3 and the energy of the state 3 will control the mobility edge. In the case of fast polaron formation, $\tau_f < \tau_t$ exciton will stay at the spatial position of the state 1 as after magnetic localization (1 → 2) its movement to the state 3 would be energetically unfavorable. In this case, the mobility edge will be controlled by the state 1. It is obvious that a suppression of the magnetic polaron, e.g., by an external magnetic field or by temperature increase, will shift the mobility edge.

The effect of magnetic polaron formation on the exciton mobility edge was demonstrated experimentally for a $Cd_{0.76}Mn_{0.24}Te$ epilayer [49, 59]. Here, the polaron formation time $\tau_f = 100$ ps is considerably shorter than the exciton lifetime $\tau_0 = 600$ ps [39]. External magnetic fields up to 9 T were used for the polaron suppression. A magnetic field of $B = 8$ T suppresses the polaron energy to less than 10% of its zero-field value. The mobility edge was determined as shown in Fig. 7.5

and measured as a function of magnetic fields. The giant Zeeman splitting of the band states prohibits consideration of the mobility edge in absolute energies. The mobility edge shift (open circles in Fig. 7.12) has to be analyzed with respect to the Zeeman shift of the free exciton energy (closed circles). In magnetic fields $B > 3$ T, the difference between the Zeeman shift and the mobility edge shift appears and it reaches 7 meV at 8 T. This shows that in magnetic fields, which suppress the polaron formation, the mobility edge moves deeper into the tail of localized states, where the density of states is smaller.

7.4.4 *Modification of Magnetic Susceptibility: Suppression of Spin Glass Phase*

It is interesting that the hole exchange field is sufficiently strong to modify the magnetic properties of the Mn-spin system. Magnetic ordering in the form of a spin glass formation in $Cd_{1-x}Mn_xTe$ with relatively high Mn concentrations $x > 0.1$ can be suppressed inside magnetic polarons. This is illustrated in Fig. 7.13 for $Cd_{0.67}Mn_{0.33}Te$. Here, the circular polarization degree of luminescence measured at low magnetic field of 0.4 T (open circles) is proportional to the magnetic susceptibility before the moment of exciton photogeneration, i.e., before the polaron formation [42]. A clear cusp is observed at the temperature of 8 K that is associated with the critical temperature of the spin glass formation [60]. This cusp is, however, absent for the temperature dependence of the polaron energy, which is also proportional to magnetization (closed circles) [61]. That means that the magnetization inside the magnetic polaron is modified by the exchange field of localized holes and, therefore, differs from the magnetization outside.

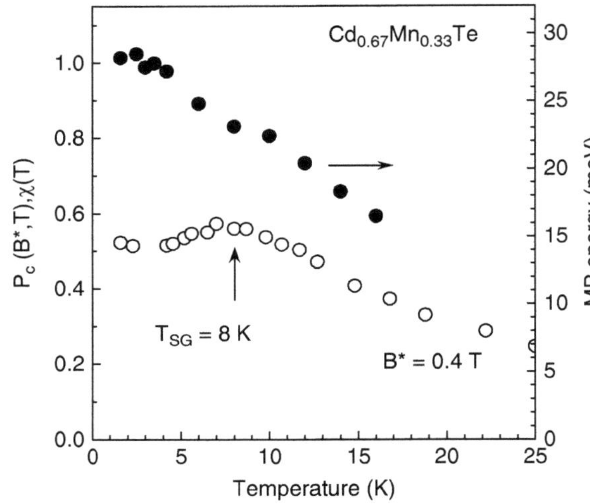

Fig. 7.13 Temperature dependencies of the localized magnetic polaron energy and magnetic-field-induced circular polarization of polaron emission. The polaron energy is measured at a zero-magnetic field [61]

7.5 Exciton Magnetic Polarons in Low-Dimensional Systems

Lowering dimensionality favors the magnetic polaron stability and increases the polaron binding energy. Therefore, the magnetic polaron effect plays an important role in quasi-two-dimensional quantum well heterostructures and in quasi-zero-dimensional quantum dots. Results for magnetic polarons in DMS quantum dots [47, 62–65] are discussed in Chap. 5, Sect. 5.4.2. Here, we concentrate on the polaron properties in quantum wells and superlattices.

7.5.1 Reduction of Dimensionality from 3D to 2D

The most direct way to study modification of the magnetic polaron properties with decreasing system dimensionality from 3D down to 2D is to decrease the quantum well width in DMS heterostructures. In such structures schematically shown in Fig. 7.14 excitons are confined in the layer with narrower band gap called quantum well (QW) with thickness L_z and their motion along the structure growth axis (z-axis) is blocked by the barriers having wider band gap. The exciton Bohr radius in CdTe is 70 Å. Therefore, the excitons in the wells with $L_z \geq 200$ Å can be treated as quasi-3D. The quasi-2D regime is realized for $L_z \leq 70$ Å.

One should be aware of several factors that might affect the polaron energy in quantum well structures. Apart from the dimensionality, it is the carrier wave function distribution between the well and the barrier layers, which changes with decreasing well width. It becomes so important that often either the wells or the barriers are made of nonmagnetic semiconductors (see structures labeled "type A"

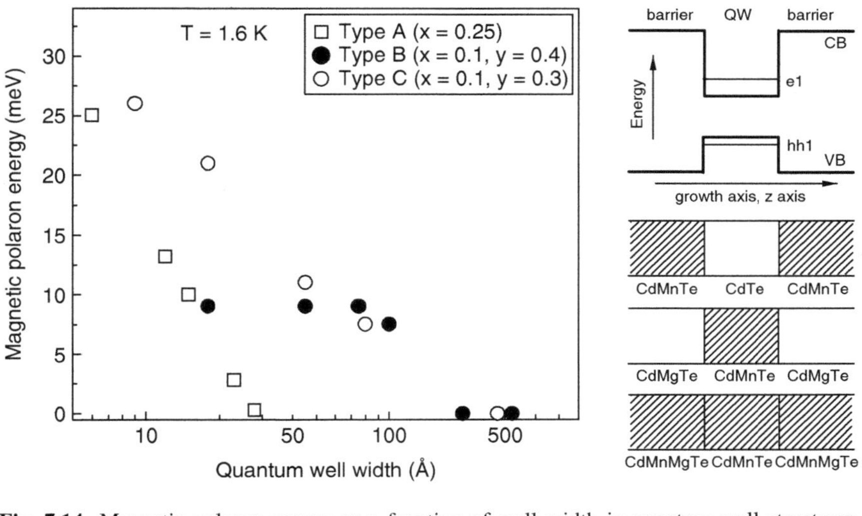

Fig. 7.14 Magnetic polaron energy as a function of well width in quantum well structures of different types [28]

and "type B" in Fig. 7.14). Also, the magnetic properties of interfaces between nonmagnetic and magnetic materials may differ from the bulk magnetic susceptibility [66, 67]. To account for and separate out these factors, different types of DMS quantum wells should be studied.

The right part of Fig. 7.14 displays three types of DMS quantum wells. Here, dashed patterns denote the presence of the Mn ions. CdTe/$Cd_{1-x}Mn_xTe$ structures (type A) have nonmagnetic wells embedded in magnetic barriers. Structures of type B and C both contain magnetic wells of $Cd_{0.9}Mn_{0.1}Te$, but their barrier materials are different. In the type C samples, Mn content is distributed homogeneously over the well and barriers, which makes them most suitable for the study of dimensionality effect on the polaron energy.

The polaron energies versus well width for all types of quantum wells are summarized in the left part of Fig. 7.14. Magnetic polaron formation is contributed by several factors. Their relative importance depends on the structure design and on the well width:

1. A considerable overlap of the exciton (hole) wave function with the Mn-containing parts of the structures is required for the polaron formation. This condition is always fulfilled in the structures of type B and C. In the type A structures, only the part of the wave function penetrating into the barriers contributes to the polaron formation. But this part is drastically reduced in thicker wells. As a consequence, the polaron energy decreases strongly with increasing well width and magnetic polarons are not formed in wells thicker than 30 Å (see open squares) [48].
2. Reduction of the dimensionality from 3D down to 2D strongly favors the polaron formation. The reason for that was discussed in the theoretical part. In the type B and C quantum wells with $L_z \geq 200$ Å, excitons can be considered as quasi-3D. These quantum wells exhibit the same polaron energy as found in bulk $Cd_{1-x}Mn_xTe$ [39]. The reduction of L_z from 200 Å down to 50 Å is accompanied by a strong increase of the polaron energy. For this range of well width, the role of the well width fluctuations in the exciton localization is relatively small and the exciton in-plane localization is dominated by alloy fluctuations [59, 68, 69]. Therefore, we conclude that the observed increase of the polaron energy with the reducing L_z manifests the effect of the reduction of the dimensionality on the stability of exciton magnetic polarons.
3. Enhancement of the in-plane exciton localization due to well width fluctuations becomes important for quantum wells thinner than 50 Å. We suggest that the increase of the polaron energy with decreasing $L_z < 50$ Å in the type C structures (open circles) is influenced by this factor. The constancy of the polaron energy in the type B structures (closed circles) is a result of two factors nearly compensating each other: enhancement of the localization, which increases the energy, and leakage of the hole wave function into the nonmagnetic barriers, which decreases it.
4. Modification of the magnetic properties of Mn-ion system at the interfaces between magnetic and nonmagnetic semiconductors can considerably increase the polaron energy in the type A and type B quantum wells. The type C structures are free from this effect. Mechanisms of such modification are discussed in detail in [27, 66, 67, 70, 71].

7.5.2 Magnetic Polaron in Spin Superlattice

Another spectacular way to realize smooth transition from 3D to 2D case by application of external magnetic field is offered by spin superlattices. The idea of such structures, first suggested by von Ortenberg [72], is illustrated in Fig. 7.15. The spin superlattice consists of alternating layers of diluted magnetic and nonmagnetic semiconductors with equal band gaps and zero band offsets. At zero-magnetic field, it is a flat-band structure and corresponds to 3D system. In external magnetic fields applied along the growth axis spatial separation of spin-down and spin-up carriers takes place and is accompanied by carrier confinement in magnetic and nonmagnetic layers, respectively.

Spin superlattice was realized in a 50 Å/150 Å $Cd_{0.9}Mn_{0.1}Te/Cd_{0.9}Mg_{0.1}Te$ heterostructure [73]. The asymmetric spin splitting of the exciton states shown in Fig. 7.15a documents the field-induced localization of the spin-down excitons, whose recombination contributes to the σ^+ polarized luminescence, in the 50-Å-thick layers of $Cd_{0.9}Mn_{0.1}Te$. The energy difference between σ^- and σ^+ components corresponds to the value of the magnetic-field-induced confinement potential. Magnetic polaron shift exhibits nonmonotonic behavior in magnetic fields

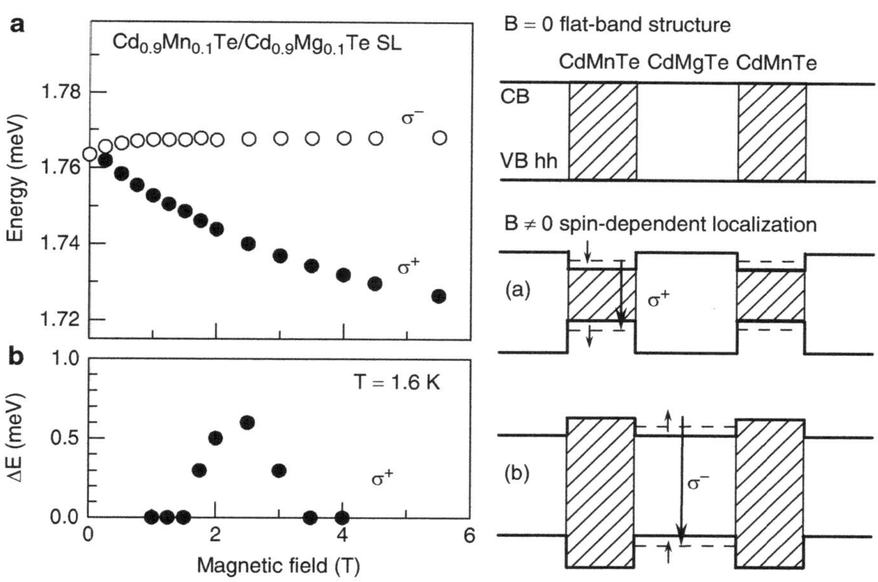

Fig. 7.15 (**a**) Magnetic-field-induced exciton spin splitting of a $Cd_{0.9}Mn_{0.1}Te/Cd_{0.9}Mg_{0.1}Te$ superlattice determined from the PL excitation spectra. *Closed* (*open*) *symbols* represent spin-down (spin-up) states. (**b**) Magnetic polaron energy measured for the spin-down exciton transition. The *right* part illustrates schematically a spin superlattice composed of alternating layers of diluted magnetic and nonmagnetic semiconductors with equal band gaps and with zero band offsets. The giant Zeeman splitting of band states in the magnetic material leads to a spatial separation of spin-up and spin-down states in external magnetic fields [28, 73]

(Fig. 7.15b). The exciton magnetic polaron is not stable at zero field, i.e., in 3D system. It appears at a field of 1.8 T for the σ^+ component, when excitons are confined sufficiently in the 50-Å-thick $Cd_{0.9}Mn_{0.1}Te$ layers and become quasi-2D. The polaron shift disappears with further field increase due to the polaron suppression (compare with Fig. 7.6a).

7.5.3 Anisotropic Spin Structure of 2D Magnetic Polaron

Anisotropy is an inherent feature of two-dimensional systems. In DMS quantum wells, an anisotropy of the exciton spin structure is revealed most brightly by the Zeeman patterns of free excitons [70], which exhibit a strong dependence on the orientation of external magnetic fields (Fig. 7.16b). This anisotropy results from strong spin–orbit interaction in the valence band. Size quantization lifts the degeneracy of Γ_8 valence band and splits it into two subbands with spin projections on the growth axis equal to $\pm 1/2$ (light holes) and $\pm 3/2$ (heavy holes). Both subbands demonstrate an anisotropic giant Zeeman splitting due to unequal spin components in the quantum well plane and normal to it [44, 74]. The spin of the heavy hole, which is the lowest in energy has zero in-plane spin component. This is confirmed by the absence of the exciton Zeeman splitting in Voigt geometry (open circles in panel (b)) in magnetic fields below 3 T.

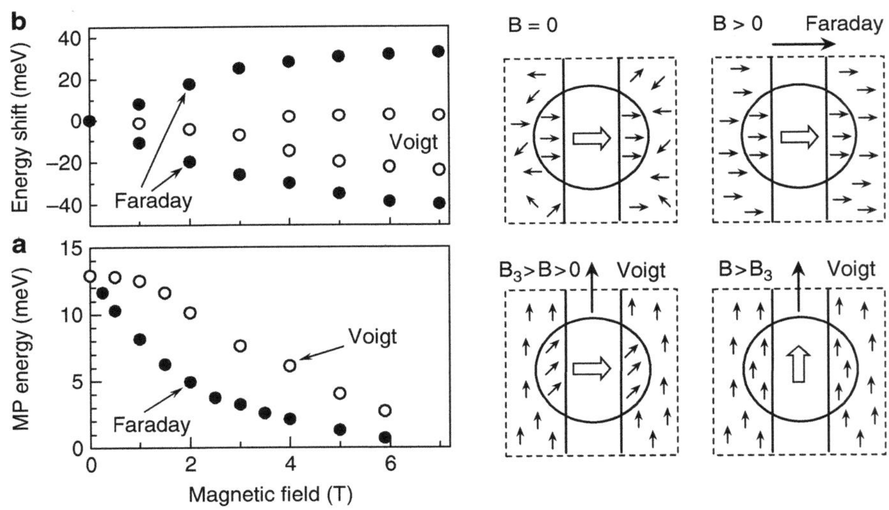

Fig. 7.16 (**a**) Suppression of magnetic polarons in external magnetic fields of different orientations in a 12-Å-thick $CdTe/Cd_{0.74}Mn_{0.26}Te$ quantum well [48]. (**b**) Anisotropic giant Zeeman splitting of the free exciton states in this structure. $T = 1.6$ K. Schematics of polaron suppression is given by the right part. The *open arrow* represents the hole spin in the exciton, small arrows represent Mn spins and the long *solid arrow* is the magnetic field vector

For exciton magnetic polarons, this anisotropy results first of all in a different suppression of the polaron energy in magnetic fields applied parallel (Faraday geometry) and perpendicular (Voigt geometry) to the z-axis. Such a behavior is very pronounced in a 12-Å-thick $CdTe/Cd_{0.74}Mn_{0.26}Te$ quantum well shown in Fig. 7.16a.

Schematic presentation of underlying mechanisms is displayed in the right part of the figure. When the external magnetic field is weak (for this structure – less than 2 T), the heavy-hole spin components in the quantum well plane are equal to zero, so that the hole exchange field is directed along the z-axis. The external field applied in the Faraday geometry does not change the orientation of the Mn spins inside the polaron, but aligns only those outside. This reduces the energy gain of the exciton in the polaron formation process. As a result, the polaron shift is suppressed, which reflects the saturation of Mn spin polarization described by (7.8):

$$E_{\mathrm{MP}}(B) = \frac{1}{2}\left[\Delta E_z^{\mathrm{hh}}(B + B_{\mathrm{ex}}) - \Delta E_z^{\mathrm{hh}}(B)\right]. \tag{7.25}$$

In the Voigt geometry, however, the polaron suppression is more complex because the polaron magnetic moment and the external field are not collinear. Detailed analysis of the hole spin structure in this geometry shows that there exists a critical value of a magnetic field B_3 at which the polaron magnetic moment flips into the direction of the external field [31, 44]. This process is controlled by the mixing of the heavy-hole and light-hole states. In magnetic fields exceeding B_3, the polaron suppression follows the scenario corresponding to the suppression in Faraday geometry. In the intermediate field range ($B_3 > B > 0$), the polaron suppression is due to the reorientation of Mn spins inside the polaron orbit, which is governed by the competition of the external and exchange fields.

The anisotropic spin structure of the two-dimensional magnetic polaron provides a new mechanism for the polaron formation in external magnetic fields applied in the Voigt geometry [75]. This new mechanism is governed by a transfer of the carrier-magnetic ions exchange energy to the energy reservoir determined by interaction of localized Mn-spins with an external magnetic field (i.e., to the Zeeman reservoir of magnetic spins). The energy transfer in this mechanism is not limited by the spin–spin relaxation rates and therefore can be much faster than the polaron formation at zero magnetic field. Therefore, this mechanism allows that the polaron comes closer to the equilibrium situation during the exciton lifetime. This can be observed experimentally in quantum wells with strong dynamical reduction of the polaron shift (see e.g., Fig. 7.17).

The scheme of the mechanism providing polaron formation via energy transfer to the Zeeman reservoir of localized spins is shown in the inset of Fig. 7.17b. We performed time-resolved experiments, where the localized excitons are photogenerated. The hole exchange field $\mathbf{B}_{\mathrm{ex}}$ is switched for Mn spins abruptly via 5 ps laser pulses. It is important to mention that the exchange field can reach values up to a few Tesla (Fig. 7.9a). Without this exchange field, the ensemble of localized spins is aligned by the external field $\mathbf{B}$ only, resulting in an average spin $\mathbf{I}(B, t \leq 0)$.

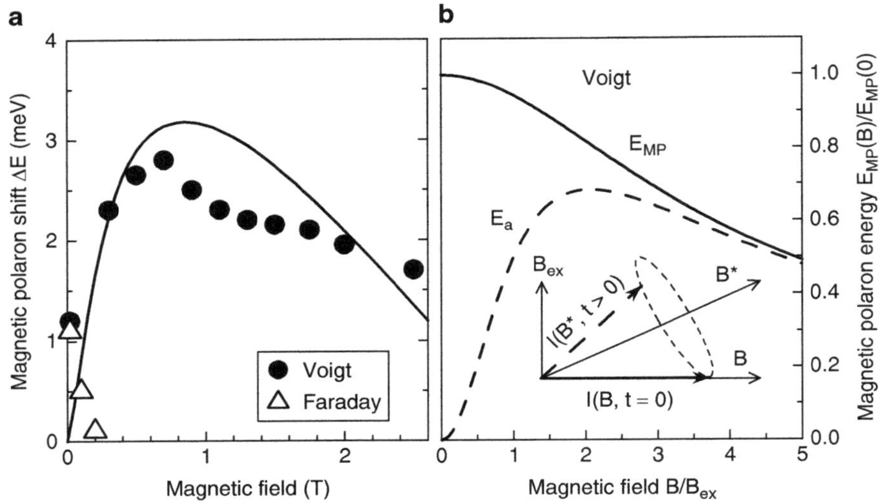

Fig. 7.17 Magnetic polaron formation in an 80-Å-thick $Cd_{0.93}Mn_{0.07}Te/Cd_{0.67}Mn_{0.07}Mg_{0.26}Te$ quantum well. (**a**) Magnetic polaron shift taken under *cw* excitation as a function of external magnetic fields applied in the Voigt and Faraday geometries. *Solid line* presents result of a model calculation. (**b**) Precession mechanism of the magnetic polaron formation in magnetic fields applied in the Voigt geometry. Calculated equilibrium polaron energy (E_{MP}) and the portion of the polaron energy gained by the precession (E_a) are shown by *solid* and *dashed lines*, respectively [75]

Within the orbit of the photocreated localized hole, this average spin is exposed to an effective field $\mathbf{B}^*$, which is the sum of the external and the exchange fields: $\mathbf{B}^* = \mathbf{B} + \mathbf{B}_{\mathrm{ex}}$.

If $\mathbf{B}_{\mathrm{ex}}$ and $\mathbf{B}$ are not parallel to each other, which is the case in the Voigt experimental geometry where $\mathbf{B}_{\mathrm{ex}} \perp \mathbf{B}$, the spin $\mathbf{I}$ starts to precess around $\mathbf{B}^*$ with the Larmor frequency $\omega_{\mathrm{L}} = \mu_{\mathrm{B}} g_{\mathrm{Mn}} B^*/\hbar$. This precessing spin $\mathbf{I}(B^*, t > 0)$ is represented by the dashed arrow. As a result of precession, the projection of $\mathbf{I}$ on $\mathbf{B}$ decreases and its projection on $\mathbf{B}_{\mathrm{ex}}$ increases. The magnetic polaron energy gained by this mechanism is equal to the projection of the average spin $\mathbf{I}$ on $\mathbf{B}_{\mathrm{ex}}$: $E_a = \mu_{\mathrm{B}} g_{\mathrm{Mn}} \mathbf{B}_{\mathrm{ex}} \mathbf{I}$ (dashed line in Fig. 7.17b).

The mechanism manifests itself in an increase of the magnetic polaron shift under applied magnetic fields. Figure 7.17a displays polaron energies in an 80-Å-thick $Cd_{0.93}Mn_{0.07}Te/Cd_{0.67}Mn_{0.07}Mg_{0.26}Te$ quantum well measured under continuous-wave (*cw*) excitation. For the Faraday geometry (triangles), the expected decrease of the polaron shift can be observed. Surprisingly, in the Voigt geometry the polaron shift (circles) increases first reaching its maximum at 0.7 T and, then, it decreases slowly. Such a behavior cannot be explained in any model that considers equilibrium magnetic polaron states only. It results from the strong dynamical effect on the polaron shift in the studied quantum well, where $\tau_f = 190$ ps and $\tau_0 = 110$ ps. Due to the fact that the polaron shift under *cw* excitation is 1.2 meV, which is considerably smaller than the polaron equilibrium energy $E_{MP}(t \to \infty, B = 0) = 8$ meV [75]. The new mechanism can gain part of the polaron energy during the precession time of about 10 ps, and therefore strongly shorten the polaron formation time. That moves the polaron shift closer to the equilibrium polaron energy. Modeling of the polaron shift that takes the energy transfer to the Zeeman reservoir into account, shown by the solid line in the panel (a), is in close agreement with the experiment.

Further effects caused by the anisotropic heavy-hole spin structure in quantum wells are the spontaneous symmetry lowering of the magnetic polarons [44] and the multiple Mn^{2+} spin-flip Raman scattering via the polaron states [76].

7.6 Spin Dynamics of Exciton Magnetic Polaron Formation

In this section, we discuss spin dynamics responsible for different stages of the exciton magnetic polaron formation. Evolution of collective polaron states in 3D and 2D systems is measured by means of time-resolved spectroscopy. Optical orientation allows to collect information about the very initial stages of the polaron formation, where the starting correlation between the exciton spin and the Mn spin fluctuation is established. Also, relaxation of the total polaron moment in external magnetic fields is analyzed. The comprehensive set of available data allows us to introduce a detailed hierarchy of relaxation processes controlling the magnetic polaron formation.

7.6.1 Magnetic Polaron Formation in 3D and 2D Systems

Let us discuss the formation dynamics of exciton magnetic polarons in 3D and 2D systems without going into details of microscopic mechanisms responsible for τ_s which are addressed in Sect. 7.2.2.

The polaron formation times measured in $Cd_{1-x}Mn_xTe$ epilayers (3D system) are shown in Fig. 7.18a. They shorten from 130 to 50 ps with x increase in the range 0.12–0.33. The increase of Mn content causes a change of magnetic properties and an increase of compositional disorder, which in turn leads to a stronger exciton localization. To distinguish between these two factors contributing the polaron formation time (see Sect. 7.4.2), the polaron formation was measured in $Cd_{1-x-y}Mn_xMg_yTe$ quaternary alloys with a nearly identical Mn content $x \approx 0.14$ (Fig. 7.18b). Despite the considerable increase of the equilibrium polaron energy from 12 meV in $Cd_{0.85}Mn_{0.15}Te$ to 28 meV in $Cd_{0.70}Mn_{0.14}Mg_{0.16}Te$, the formation times vary only slightly being in the range 90–110 ps. Therefore, the formation time does not scale with the polaron energy and with the degree of initial localization, but is controlled by the relaxation dynamics in the Mn-spin system.

Remember, that in 3D systems the polaron autolocalization is expected to be negligible and the experimentally measured polaron formation time τ_f can be directly

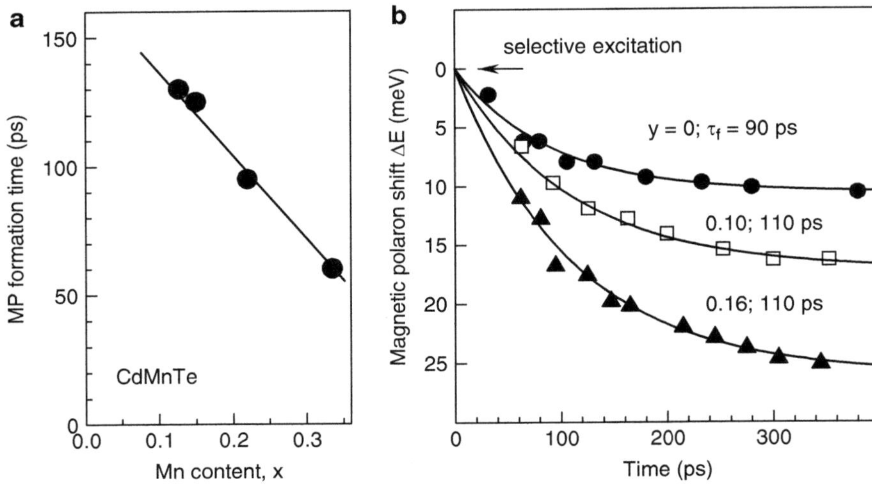

Fig. 7.18 Dynamics of localized magnetic polarons in DMS epilayers (3D case): (**a**) Polaron formation time vs. Mn content in $Cd_{1-x}Mn_xTe$ [39], (**b**) Time evolution of the polaron shift for $Cd_{1-x-y}Mn_xMg_yTe$ epilayers with $x \approx 0.14$ [54]. *Lines* are exponential fits along (7.6) with τ_f given in the figure

associated with the relaxation time τ_s, i.e., $\tau_f = \tau_s$ (Sect. 7.2.2). Therefore, the data in Fig. 7.18a can be treated as a functional dependence of τ_s on the Mn content.

A theoretical analysis of the magnetic polaron dynamics in 2D systems predicts a considerable contribution of the autolocalization, which is represented in (7.7) by the factor ξ resulting in $\tau_f = \tau_s(1 + \xi)$. Experimental data in Fig. 7.19a for $Cd_{0.93}Mn_{0.07}Te/Cd_{0.67}Mn_{0.07}Mg_{0.26}Te$ type C quantum wells confirm this prediction [77, 78]. Despite the uniform Mn content throughout the entire structure, the polaron formation time shortens from 185 to 135 ps with the well width decreasing from 80 to 20 Å. In narrow wells, an in-plane exciton localization by well width fluctuations is particularly strong. Which, in turn, suppresses the autolocalization and shifts τ_f closer to τ_s.

Polaron lifetime is displayed in Fig. 7.19a by open circles. Contrary to the τ_f behavior, τ_0 increases with narrowing the well width. If the well width is smaller than 30 Å, the lifetime τ_0 exceeds τ_f. With decreasing ratio τ_f/τ_0, the polaron shift increases and the difference between the polaron shift detected under *cw* conditions and the equilibrium polaron energy decreases. Both energies are plotted in panel (b). Indeed, not only the equilibrium energy increases for narrower wells, but also the polaron shift for *cw* excitation increases its relative contribution, which is in agreement with times from the panel (a).

Exciton lifetime can be increased by the polaron formation. The polaron-induced decrease of the exciton localization area decreases the coherence volume for the exciton radiative decay [79]. It may result in a sizable increase of the exciton radiative time. A model calculation of this effect is given in [80] and experimental evidences are published in [81].

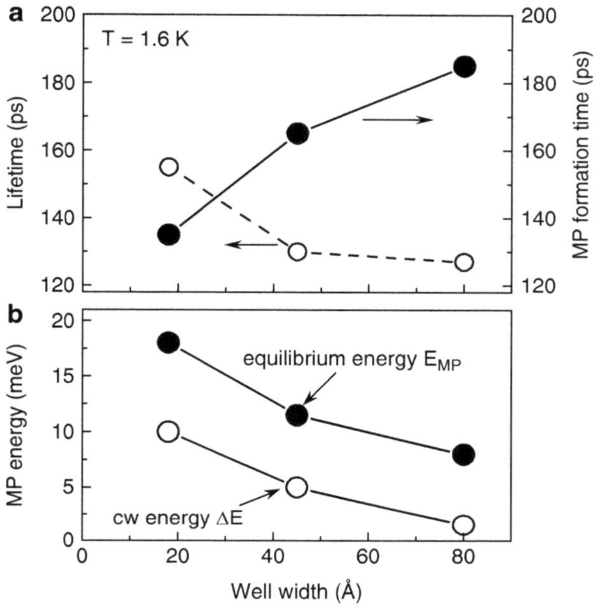

Fig. 7.19 Dynamics of magnetic polarons in $Cd_{0.93}Mn_{0.07}Te/Cd_{0.67}Mn_{0.07}Mg_{0.26}Te$ quantum wells: (**a**) Polaron formation time (*closed circles*) and lifetime (*open circles*) vs. well width. (**b**) Well width dependence of the polaron shifts measured under *cw* excitation (*open circles*) and the equilibrium polaron energies (E_{MP}) determined under pulsed excitation (*closed circles*) [77]

7.6.2 *Optical Orientation of Magnetic Polarons*

Circular polarization degree of polaron luminescence under *cw* excitation allows to address the spin dynamics at the very first stage of the polaron formation when the starting correlation is established. Two techniques are used for this study. In the *optical orientation technique*, spin polarized carriers are generated by a circular polarized light [82]. Optical orientation signal P_{oo} is controlled by the ratio of carrier lifetime and spin relaxation time. It approaches zero if spin relaxation takes place much faster than carrier recombination. Another technique is the *magnetic-field-induced polarization* P_c. Here, unpolarized light is used for excitation and thermal occupation of the carrier spin levels split by external magnetic field is analyzed. The maximal signal P_c corresponds to the equilibrium population and is achieved if the spin relaxation time is shorter than the carrier lifetime. Both techniques are widely used to study carrier and exciton spin dynamics in nonmagnetic semiconductors. In DMS, an interaction of carriers with magnetization fluctuations $\mathbf{M}_f$ and the magnetic polaron formation become dominating factors for the carrier spin dynamics [11, 41]. It will be demonstrated in this section.

We studied a 48-Å-period $Cd_{0.83}Mn_{0.17}Te/Cd_{0.54}Mg_{0.46}Te$ superlattice, whose optical properties and magnetic polarons were discussed in Sect. 7.3. Let us start with results on optical orientation, the signal of which is shown by circles in Fig. 7.20a. Under nonselective excitation, $P_{oo} \approx 0$, which means that the free excitons lose their initial spin orientation very efficiently due to the strong exchange interaction with the Mn spins [83]. However, a large optical orientation $P_{oo} = 0.27$ appears under selective excitation in the tail of localized exciton states. Panel (b) shows two polaron lines excited selectively with σ^+ polarized light (E^+) and detected in σ^+ (E^+A^+) and σ^- (E^+A^-) polarizations. A difference in the integral intensities of these lines gives

$$P_{oo} = \frac{I(E^+A^+) - I(E^+A^-)}{I(E^+A^+) + I(E^+A^-)}. \tag{7.26}$$

The energy shift of the E^+A^- line from the excitation energy (i.e., the polaron shift) is by $\Delta f = 3$ meV larger than the shift of the E^+A^+ line. It will follow from the model consideration that Δf is twice an energy of exciton exchange interaction with M_f. The exciton decreases its energy by this value when it makes spin flip from an unfavorable orientation to a favorable one.

The magnetic field dependence of the optical orientation is given in Fig. 7.21a. In nonmagnetic semiconductors, P_{oo} usually increases with magnetic fields applied in Faraday geometry due to the suppression of the carrier spin relaxation mechanisms [82, 84]. Contrary to that, in the DMS superlattice we observe a suppression of the optical orientation at fields as small as 0.5 T. Therefore, the nature of the optical orientation signal in DMS should differ from that in the nonmagnetic semiconductors.

A model of optical orientation of excitons in bulk DMS that accounts for Gaussian-distributed thermodynamical magnetic fluctuations inside the volume of exciton localization was suggested by Warnock et al. [41,85]. It was used to describe

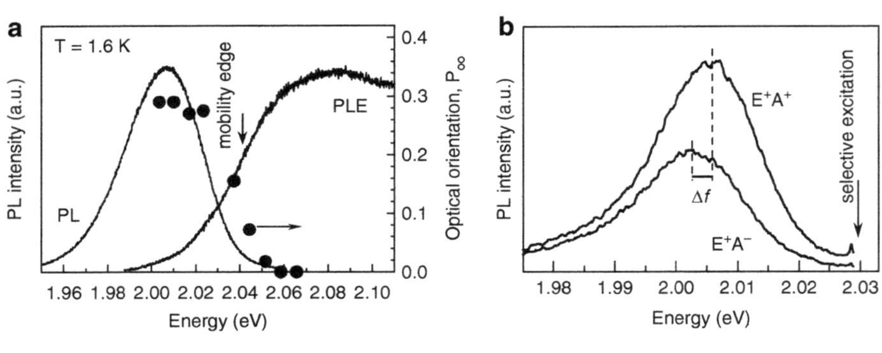

Fig. 7.20 Optical orientation of magnetic polarons in a 48-Å-period $Cd_{0.83}Mn_{0.17}Te$/ $Cd_{0.54}Mg_{0.46}Te$ superlattice. (**a**) Photoluminescence under nonselective excitation and PL excitation spectra. Optical orientation signal shown by *solid circles* appears only for selective excitation below the exciton mobility edge. (**b**) Photoluminescence spectra excited selectively with σ^+ polarized light (E^+) and detected in σ^+ (E^+A^+) and σ^- (E^+A^-) polarizations (see also Fig. 7.5) [50]

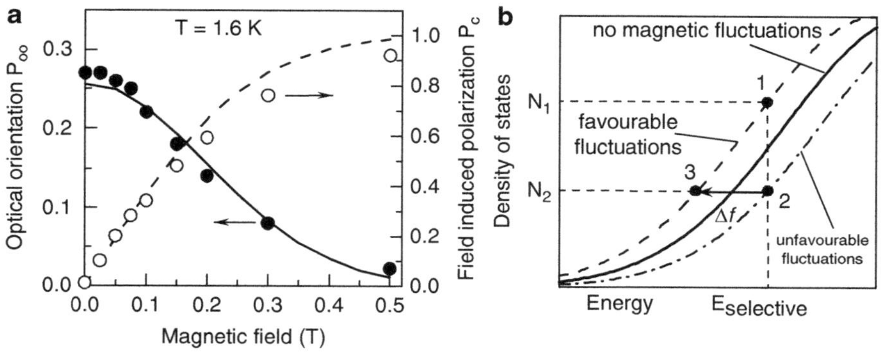

Fig. 7.21 (**a**) Degree of optical orientation (*closed circles*) and magnetic-field-induced polarization (*open circles*) as a function of magnetic field (Faraday geometry) in a 48-Å-period $Cd_{0.83}Mn_{0.17}Te/Cd_{0.54}Mg_{0.46}Te$ superlattice. *Lines* show the results of model calculations. (**b**) Model of optical orientation of excitons in DMS caused by magnetic fluctuations [50]

experimental data in $Cd_{1-x}Mn_xTe$ and $Cd_{1-x}Mn_xSe$. The model suggests that the optical orientation signal in DMS is not related to the exciton spin memory on the helicity of generating photons, but is caused by the specifics of the tail of localized states, which is contributed by nonmagnetic and magnetic fluctuations. Therefore, it is due for the excitons excited selectively in the tail of localized states at energies below the mobility edge.

One can easily modify this model for the 2D case suggesting that in superlattices the heavy-hole exciton spin has only two opposite orientations parallel to the z-axis. We assume that after photogeneration localized excitons first align their spins collinear with $\mathbf{M}_{f,z}$, which is a projection of $\mathbf{M}_f$ on the z-axis, and then form magnetic polarons keeping their spin orientation constant. In this case, P_{oo} is determined by the difference in the probabilities to excite excitons at the magnetic fluctuations with different orientations.

A scheme in Fig. 7.21b illustrated specifics of the optical orientation of localized excitons in DMS. Here, the tail of localized states formed by nonmagnetic fluctuations only is shown by a solid line. Two other lines show contribution of magnetic fluctuations. Here, favorable fluctuations (dashed line), where exciton spin is parallel to $\mathbf{M}_{f,z}$, shift the tail to lower energies. Respectively, unfavorable fluctuations (dashed-doted line), where exciton spin is antiparallel to $\mathbf{M}_{f,z}$, shift it to higher energies. As a result for the chosen energy of selective excitation (here $E_{\text{selective}}$), different numbers of excitons are excited in the favorable (N_1) and unfavorable (N_2) fluctuations.

In the favorable magnetic fluctuation (state 1), exciton has its minimal exchange energy. This exciton forms a magnetic polaron with spin orientation coinciding with the polarization of the exciting photon. For, e.g., σ^+ excitation, this polaron contributes to σ^+ polarized emission (E^+A^+ line). If the exciton is created in the unfavorable fluctuation (state 2), then it first makes a spin-flip (as shown by an arrow 2 → 3) and then forms a magnetic polaron. Such polaron has an orientation antiparallel to the initial exciton orientation and contributes to the σ^- polarized emission (E^+A^- line). The optical orientation is this case is controlled by a density of states in points 1 and 2

$$P_{oo} = \frac{N_1 - N_2}{N_1 + N_2}. \tag{7.27}$$

It increases for the stronger gradient in the tail of localized states controlled by nonmagnetic fluctuations and for larger Δf caused by magnetic fluctuations.

As we show in Fig. 7.5b, the tail of localized states formed by nonmagnetic fluctuations (they play dominating role here) can be well described by an exponential law

$$f_l(E) \propto \exp\left(\frac{E - E_0}{\varepsilon_0}\right). \tag{7.28}$$

Here, E_0 is a free exciton energy and ε_0 is a characteristic energy of the tail.

Following [11, 41], we suggest Gaussian distribution for the thermodynamical magnetic fluctuations [42]

$$f_m(M_f) = \frac{1}{\sqrt{2\pi\langle M_f^2\rangle}} \exp\left(-\frac{M_f^2}{2\langle M_f^2\rangle}\right). \tag{7.29}$$

Then the total density of states in the tail contributing by the nonmagnetic and magnetic fluctuations reads

$$F(E) = \int_{-\infty}^{+\infty} \mathrm{d}M_f\, f_l\left(E + \frac{M_f B_{\mathrm{ex}}}{2k_{\mathrm{B}}T}\right) f_m(M_f). \tag{7.30}$$

To obtain P_{oo} from (7.26) or (7.27), the integral intensities of σ^+ and σ^- polarized emission must be calculated by integrating the function $F(E)$ over the energy range of polaron emission.

The suggested approach can be used to calculate the magnetic field dependence of $P_{oo}(B)$. For that in (7.30) $f_m(M_f)$ should be substituted by its form accounting for the modification of the magnetic fluctuations in external magnetic field applied along the z-axis

$$f_m(M_f, B) = \frac{1}{\sqrt{2\pi\langle M_f^2\rangle}} \exp\left[-\left(\frac{M_f}{\sqrt{2\langle M_f^2\rangle}} - \frac{B\sqrt{2\langle M_f^2\rangle}}{2k_{\mathrm{B}}T}\right)^2\right]. \tag{7.31}$$

The solid line in Fig. 7.21a represents a result of the model calculation, which was obtained without fitting parameters. All required parameters were measured experimentally ($E_{\mathrm{MP}} = 28\,\mathrm{meV}$, $\gamma = 20\,\mathrm{meV\,T^{-1}}$, $\varepsilon_0 = 6\,\mathrm{meV}$) and evaluated using equations of Sect. 7.2.3 ($B_{ex} = 2.8$ T, $\langle M_f^2 \rangle = 0.5\,\text{meV}^2\,\text{T}^{-2}$). An excellent agreement between calculated and measured values of $P_{oo}(B)$ justifies the assumptions used in the model. The calculated value of $\Delta f(B = 0) = 2.8$ meV (see Figs. 7.20b and 7.21b) is also in a very good agreement with its experimental value of 3.0 meV, which independently confirms the model validity.

We remind that one of the model assumptions is the fast alignment of the exciton spin along the magnetic fluctuation. This should happen much faster than the polaron formation process, resulting in an energy gain comparable with Δf. The excellent model fit of the experimental data in Fig. 7.21a allows us to conclude that the directional stabilization of the exciton spin (and respectively, $\mathbf{B}_{ex}$), which occurs in the process of the starting correlation between exciton and the system of Mn spins, is controlled by the thermodynamical fluctuations of magnetization [43, 50]. The exciton spin must be aligned along the magnetic moment of the magnetic fluctuation within time, which is faster than picoseconds, and, then, does not change its orientation during the magnetic polaron formation until the exciton recombination.

Magnetic-field-induced polarization of polaron emission can be described in the model with the same assumptions as those used for the optical orientation. Open circles in Fig. 7.21a show experimental data for $P_c(B)$. Nonselective excitation was used in this experiment. The luminescence is unpolarized at zero field and the polarization degree gains its saturation level of 1.0 at a field of 0.5 T. The values of $P_c(B)$ are too small to be explained by the equilibrium thermal population of exciton spin sublevels subject of the giant Zeeman splitting. However, they can be well described by the model accounting for the thermodynamical magnetic fluctuations in the Mn-spin system [42, 83].

A result of the model calculation accounting for the equilibrium polarization of the fluctuation moments M_f is shown by a dashed line. It is in a very good agreement with experimental points. Note that the slope θ of $P_c(B)$ should be about ten times larger if the equilibrium polarization of the polaron magnetic moments M_{MP} is calculated [42]. Reason for that is that M_{MP} exceeds M_f about ten times. Experimental results are in a very good agreement with the equilibrium polarization of fluctuations M_f. That means that the polaron polarization just reflects the stage of the establishing starting correlation and not the equilibrium polarization of the polarons in their final state. It also means that the equilibrium polarization of M_{MP} is not established during the relatively short polaron lifetime. The later should be shorter than the directional relaxation of the polaron moment τ_{MP}. It was shown theoretically that $\tau_{MP} \sim \tau_f B_{ex}/B$ [42]. Therefore, for the superlattice in Fig. 7.20 where $B_{ex} = 2.8$ T, at $B = 0.2$ T the directional relaxation time is 14 times longer than the polaron formation time, and also much longer than the polaron lifetime. The measured times $\tau_f = 180$ ps and $\tau_0 = 330$ ps, and evaluated $\tau_{MP}(0.2\,\text{T}) \sim 2.5$ ns. Indeed $\tau_{MP} \gg \tau_0, \tau_f$.

The model with thermodynamical magnetic fluctuations offers a good description of experimental data for $Cd_{1-x}Mn_xTe$ with Mn contents in a range $x = 0.15$–0.30 [42, 45]. In samples with higher Mn contents ($x > 0.25$), "frozen" magnetic fields of Mn-spin clusters can modify the spectrum of magnetic fluctuations and contribute to the optical orientation signal [22, 46, 86].

7.6.3 Hierarchy of Spin Dynamics Contributing to Magnetic Polaron Formation

Summarizing the results on the dynamics of exciton magnetic polarons we suggest the following hierarchy of relaxation processes contributing to the polaron formation:

(a) Excitons can be generated either in localized states by means of selective excitation or being free with excess kinetic energy. In the later case, the exciton energy relaxation into localized states via emission of optical and acoustic phonons may require 30–100 ps.
(b) The starting correlation between the exciton spin and the Mn-spin system is established during a time in order of one picosecond ($\tau_{st} \leq 1$ ps). It is determined by the relaxation of the exciton spin on the direction of magnetic fluctuation $\mathbf{M}_f$, or its projection on the structure growth axis $\mathbf{M}_{f,z}$ in low-dimensional structures. The magnetic polaron in the fluctuation regime is formed.
(c) The evolution of the magnetic polaron energy covers the dynamical range $\tau_f = 50$–250 ps and is controlled by the spin–spin interactions in the Mn-spin system. In 3D systems, $\tau_f = \tau_s$. In low dimensional systems with significant autolocalization the polaron formation time exceeds spin-spin dynamics ($\tau_f \geq \tau_s$). The collective polaron regime is established.
(d) The directional relaxation time of the polaron magnetic moment exceeds the polaron formation time by at least an order of magnitude. It is also considerably longer than the polaron lifetime, which typically covers the range from 100 up to 600 ps ($\tau_{MP} \gg \tau_f, \tau_0$). Therefore, in external magnetic fields the equilibrium population of the polaron magnetic moments M_{MP} is not achieved.
(e) Exciton recombination in II–VI DMS typically covers the range $\tau_0 = 150$–600 ps. Usually $\tau_f < \tau_0$, but in some cases for $\tau_f \geq \tau_0$ the polaron formation is interrupted by recombination before the equilibrium polaron energy is achieved. The recombination terminates the existence of the exciton magnetic polaron as a quasi-particle. But it leaves behind a perturbation in the Mn-spin system in form of the cloud of ferromagnetically aligned Mn spins, whose magnetic moment equals to M_{MP}.
(f) Magnetic moment of the cloud relaxes with spin–spin relaxation time $\tau_s \approx \tau_f = 50$–250 ps keeping the excess energy in the Mn-spin system.
(g) The Mn-spin system reaches thermal equilibrium with the crystal lattice (i.e., with the phonon bath) during spin-lattice relaxation time $\tau_{SLR} = 10$–1000 ns.

7.7 Conclusions

Optical spectroscopy offers a comprehensive information on the properties of exciton magnetic polarons in II–VI DMS with Mn^{2+} magnetic ions. Initial localization of excitons plays a crucial role in the polaron stability both for three-dimensional epilayers and quasi-two-dimensional quantum wells. Reduction of the system dimensionality affects polaron stability, introduces anisotropy, and contributes to the polaron formation dynamics.

A variety of spin relaxation processes is responsible for the polaron formation dynamics. Their hierarchy is reconstructed in this chapter. It covers the temporal range from picoseconds for the establishing initial spin correlation with microseconds needed for spin-lattice relaxation of the Mn system after exciton recombination.

Acknowledgements

Results reviewed in this chapter originate from collaboration with I. A. Merkulov, K. V. Kavokin, A. V. Kavokin, G. Mackh, R. Fiederling, A. Waag, G. Landwehr, R. Hellmann, E. O. Göbel, T. Wojtowicz, G. Karczewski, and J. Kossut. Their contribution is greatly appreciated.

References

1. S.I. Pekar, Zh. Eksp. Teor. Fiz. **16**, 335 (1946)
2. J.K. Furdyna, J. Appl. Phys. **64**, R29 (1988)
3. A. Golnik, J.A. Gaj, M. Nawrocki, R. Planel, C. Benoit a la Guillaume, J. Phys. Soc. Jpn. Suppl. **A49**, 819 (1980)
4. E.L. Nagaev, J. Magn. Magn. Mat. **110**, 39 (1992)
5. M.A. Krivoglaz, A.A. Trushchenko, Sov. Phys. Solid State **11**, 2531 (1970) [Fiz. Tv. Tela **11**, 3119 (1969)]
6. P.G. De Gennes, Phys. Rev. **118**, 141 (1960)
7. T. Kasuya, A. Yanase, Rev. Mod. Phys. **40**, 684 (1968)
8. D.D. Alov, S.I. Gubarev, V.B. Timofeev, B.N. Shepel, Sov. Phys. JETP Lett. **34**, 71 (1981) [Pis'ma Zh. Eksp. Teor. Fiz. **34**, 76 (1981)]
9. M. Nawrocki, R. Planel, G. Fishman, R. Galazka, Phys. Rev. Lett. **46**, 735 (1981)
10. D. Heiman, P. A. Wolf, J. Warnock, Phys. Rev. B **27**, 4848 (1983)
11. P.A. Wolff, in *Semiconductors and Semimetals*, vol. 25, ed. by J.K. Furdyna, J. Kossut (Academic, London, 1988), p. 413
12. S.M. Ryabchenko, Yu.G. Semenov, Sov. Phys. JETP **57**, 825 (1983) [Zh. Eksp. Teor. Fiz. **84**, 1419 (1983)]
13. T. Dietl, J. Spalek, Phys. Rev. B **28**, 1548 (1983)
14. T.H. Nhung, R. Planel, C. Benoit a la Guillaume, A.K. Bhattacharjee, Phys. Rev. B **31**, 2388 (1985)
15. M. Bugajski, P. Becla, P.A. Wolff, D.Heiman, L.R. Ram-Mohan, Phys. Rev. B **38**, 10512 (1988)
16. Yu.G. Kusrayev, K.V. Kavokin, G.V. Astakhov, W. Ossau, L.W. Molenkamp, Phys. Rev. B **77**, 085205 (2008)
17. Yu.F. Berkovskaya, E.M. Vakhabova, B.L. Gelmont, I.A. Merkulov, Sov. Phys. JETP **67**, 750 (1988) [Zh. Eksp. Teor. Fiz. **94**, 183 (1988)]
18. C.A. Huber, A.V. Nurmikko, M. Gal, A. Wold, Solid State Commun. **46**, 41 (1983)
19. J.H. Harris, A.V. Nurmikko, Phys. Rev. Lett. **51**, 1472 (1983)
20. A. Golnik, J. Ginter, J.A. Gaj, J. Phys. C **16**, 6073 (1983)

21. T. Itoh, E. Komatsu, J. Luminescence **38**, 266 (1987)
22. B.P. Zakharchenya, Yu.G. Kusraev, Sov. Phys. JETP Lett. **50**, 225 (1989) [Pis'ma Zh. Eksp. Teor. Fiz. **50**, 199 (1989)]
23. V.F. Agekyan, A.Yu. Serov, Sov. Phys. Solid State **32**, 1951 (1990) [Fiz. Tv. Tela **32**, 3373 (1990)]
24. D.R. Yakovlev, W. Ossau, G. Landwehr, R.N. Bicknell-Tassius, A. Waag, I.N. Uraltsev, Solid State Commun. **76**, 325 (1990)
25. T. Dietl, P. Peyla, W. Grieshaber, Y. Merle d'Aubigne, Phys. Rev. Lett. **74**, 474 (1995)
26. A.V. Kavokin, K.V. Kavokin, Semicond. Sci. Technol. **8**, 191 (1993)
27. D.R. Yakovlev, in *Festkörperprobleme/Advances in Solid State Physics*, vol. 32, ed. by U. Rössler (Vieweg, Braunschweig, 1992), p. 251
28. D.R. Yakovlev, K.V. Kavokin, Comments Cond. Mat. Phys. **18**, 51 (1996)
29. C. Benoit a la Guillaume, Phys. Stat. Sol. (b) **175**, 369 (1993)
30. C. Benoit a la Guillaume, Yu.G. Semenov, M. Combescot, Phys. Rev. B **51**, 14124 (1995)
31. I.A. Merkulov, K.V. Kavokin, Phys. Rev. B **52**, 1751 (1995)
32. T. Dietl, in *Diluted Magnetic Semiconductors*, ed. by S. Mahajan. Handbook of Semiconductors, vol. 3b (North-Holland, Amsterdam, 1994), p. 1252
33. T. Kasuya, A. Yanase, T. Takeda, Solid State Commun. **8**, 1543 (1970)
34. S.M. Ryabchenko, Yu.G. Semenov, Sov. Phys. Solid State **26**, 2011 (1984) [Fiz. Tv. Tela **26**, 3347 (1984)]
35. A.V. Kavokin, K.V. Kavokin, Sov. Phys. Semicond. **25**, 1053 (1991)
36. K.V. Kavokin, Sov. Phys. Solid State **35**, 818 (1993)
37. K.V. Kavokin, A.V. Kavokin, Phys. Rev. B **53**, 2141 (1996)
38. E.I. Rashba, in *Excitons*, ed. by E.I. Rashba, M.D. Sturge (North-Holland, Amsterdam, 1982), p. 543
39. G. Mackh, W. Ossau, D.R. Yakovlev, A. Waag, G. Landwehr, R. Hellmann, E.O. Göbel, Phys. Rev. B **49**, 10248 (1994)
40. J. Shah, in *Hot Carriers in Semiconductor Nanostructures*, ed. by J. Shah (Academic Press, New York, 1992), ch. IV.1, p. 279
41. J. Warnock, R.N. Kershaw, D. Ridgely, K. Dwight, A. Wold, R.R. Galazka, J. Luminescence **34**, 25 (1985)
42. I.A. Merkulov, D.R. Yakovlev, K.V. Kavokin, G. Mackh, W. Ossau, A. Waag, G. Landwehr, Sov. Phys. JETP Lett. **62**, 335 (1995) [Pis'ma Zh. Eksp. Teor. Fiz. **62**, 313 (1995)]
43. I.A. Merkulov, G.R. Pozina, D. Coquillat, N. Paganotto, J. Siviniant, J.P. Lascaray, J. Cibert, Phys. Rev. B **54**, 5727 (1996)
44. I.A. Merkulov, K.V. Kavokin, G. Mackh, B. Kuhn-Heinrich, W. Ossau, A. Waag, G. Landwehr, D.R. Yakovlev, Phys. Solid State **39**, 1859 (1997) [Fiz. Tv. Tela **39**, 2079 (1997)]
45. K.V. Kavokin, I.A. Merkulov, D.R. Yakovlev, W. Ossau, G. Landwehr, Phys. Rev. B **60**, 16499 (1999)
46. A.V. Kudinov, Yu.G. Kusraev, V.N. Yakimovich, Phys. Solid State **37**, 359 (1995) [Fiz. Tv. Tela **37**, 660 (1995)]
47. A.A. Maksimov, G. Bacher, A. McDonald, V.D. Kulakovskii, A. Forchel, C.R. Becker, G. Landwehr, L. Molenkamp, Phys. Rev. B **62**, R7767 (2000)
48. D.R. Yakovlev, W. Ossau, G. Landwehr, R.N. Bicknell-Tassius, A. Waag, S. Schmeusser, I.N. Uraltsev, Solid State Commun. **82**, 29 (1992)
49. G. Mackh, D.R. Yakovlev, W. Ossau, A. Waag, G. Landwehr, Acta Phys. Pol. **87**, 265 (1995)
50. V.P. Kochereshko, I.A. Merkulov, K.V. Kavokin, D.R. Yakovlev, G. Mackh, W. Ossau, A. Waag, G. Landwehr, Proc. Int. Conf. on Semiconductor Heteroepitaxy, Montpellier, France 1995, ed. by B. Gil, R.L. Aulombard (World Scientific, Singapore, 1995), p. 218
51. S. Permogorov, A. Reznitsky, J. Luminescence **52**, 201 (1992)
52. D.R. Yakovlev, W. Ossau, G. Landwehr, R.N. Bicknell-Tassius, A. Waag, S. Schmeusser, I.N. Uraltsev, A. Pohlmann, E.O. Göbel, J. Cryst. Growth **117**, 854 (1992)
53. E. Cohen, M.D. Sturge, Phys. Rev. B **25**, 3828 (1982)

54. G. Mackh, M. Hilpert, D.R. Yakovlev, W. Ossau, H. Heinke, T. Litz, F. Fischer, A. Waag, G. Landwehr, R. Hellmann, E.O. Göbel, Phys. Rev. B **50**, 14069 (1994)
55. G. Mackh, W. Ossau, D.R. Yakovlev, M. Hilpert, A. Waag, G. Landwehr, J. Cryst. Res. Technol. **31**, 559 (1996)
56. P. Janiszewski, Proc. 20th Int. Conf. on the Physics of Semiconductors, Thessaloniki 1990, ed. by E. M. Anastassakis J.D. Joannopoulos (World Scientific, Singapore, 1990), p. 771
57. T.L. Linnik, Yu.G. Rubo, V.I. Sheka, Sov. Phys. JETP Lett. **63**, 222 (1996) [Pis'ma Zh. Eksp. Teor. Fiz. **63**, 209 (1996)]
58. R. Fiederling, D.R. Yakovlev, W. Ossau, G. Landwehr, I.A. Merkulov, K.V. Kavokin, T. Wojtowicz, M. Kutrowski, K. Grasza, G. Karczewski, J. Kossut, Phys. Rev. B **58**, R1766 (1998)
59. D.R. Yakovlev, G. Mackh, W. Ossau, A. Waag, G. Landwehr, in *Materials Science Forum*, vol. 182–184 (Trans Tech Publication, Switzerland, 1995), p. 537
60. D.R. Yakovlev, U. Zehnder, W. Ossau, A. Waag, G. Landwehr, T. Wojtowicz, G. Karczewski, J. Kossut, J. Magn. Magn. Mater. **191**, 25 (1999)
61. A.L. Chudnovskiy, R. Oppermann, B. Rosenov, D.R. Yakovlev, U. Zehnder, W. Ossau, Phys. Rev. B **55**, 10519 (1997)
62. G. Bacher, A.A. Maksimov, H. Schömig, V.D. Kulakovskii, M.K. Welsch, A. Forchel, P.S. Dorozhkin, A.V. Chernenko, S. Lee, M. Dobrowolska, J.K. Furdyna, Phys. Rev. Lett. **89**, 127201 (2002)
63. J. Seufert, G. Bacher, M. Scheibner, A. Forchel, S. Lee, M. Dobrowolska, J.K. Furdyna, Phys. Rev. Lett. **88**, 027402 (2002)
64. P.S. Dorozhkin, A.V. Chernenko, V.D. Kulakovskii, A.S. Brichkin, A.A. Maksimov, H. Schoemig, G. Bacher, A. Forchel, S. Lee, M. Dobrowolska, J.K. Furdyna, Phys. Rev. B **68**, 195313 (2003)
65. R. Beaulac, L. Schneider, P.I. Archer, G. Bacher, D.R. Gamelin, Science **325**, 973 (2009)
66. W. Ossau, B. Kuhn-Heinrich, Physica B **184**, 442 (1993)
67. J.A. Gaj, W. Grieshaber, C. Bodin-Deshayes, J. Cibert, G. Feuillet, Y. Merle d'Aubigne, A. Wasiela, Phys. Rev. B **50**, 5512 (1994)
68. G. Mackh, W. Ossau, D.R. Yakovlev, A. Waag, T. Litz, G. Landwehr, Solid State Commun. **88**, 221 (1993)
69. G. Mackh, W. Ossau, D.R. Yakovlev, R. Hellmann, E.O. Göbel, A. Waag, G. Landwehr, in *Materials Science Forum*, vol. 182–184 (Trans Tech Publication, Switzerland, 1995), p. 775
70. D.R. Yakovlev, G. Mackh, B. Kuhn-Heinrich, W. Ossau, A. Waag, G. Landwehr, R. Hellmann, E.O. Göbel, Phys. Rev. B **52**, 12033 (1995)
71. W. Ossau, B. Kuhn-Heinrich, A. Waag, G. Landwehr, J. Cryst. Growth **159**, 1210 (1996)
72. M. von Ortenberg, Phys. Rev. Lett. **49**, 1041 (1982)
73. W. Ossau, B. Kuhn-Heinrich, G. Mackh, A. Waag, G. Landwehr, J. Cryst. Growth **159**, 1046 (1996)
74. B. Kuhn-Heinrich, W. Ossau, E. Bangert, A. Waag, G. Landwehr, Solid State Commun. **91**, 413 (1994)
75. D.R. Yakovlev, K.V. Kavokin, I.A. Merkulov, G. Mackh, W. Ossau, R. Hellmann, E.O. Göbel, A. Waag, G. Landwehr, Phys. Rev. B **56**, 9782 (1997)
76. J. Stühler, G. Schaack, M. Dahl, A. Waag, G. Landwehr, K.V. Kavokin, I.A. Merkulov, Phys. Rev. Lett. **74**, 2567 (1995)
77. E.O. Göbel, R. Hellmann, G. Mackh, D.R. Yakovlev, W. Ossau, A. Waag, G. Landwehr, in *Materials Science Forum*, vol. 182–184 (Trans Tech Publication, Switzerland, 1995), p. 735
78. R. Hellmann, A. Euteneuer, E.O. Göbel, G. Mackh, D.R. Yakovlev, A. Waag, G. Landwehr, J. Cryst. Growth **159**, 976 (1996)
79. J. Feldmann, G. Peter, E.O. Göbel, P. Dawson, K. Moore, C. Foxon, R.J. Elliot, Phys. Rev. Lett. **59**, 2337 (1987)
80. A.V. Kavokin, J. de Physique IV, Colloque C5 **3**, 79 (1993)
81. M. Dahl, W. Gebauer, R. Prediger, A. Waag, Proc. Int. Conf. High Magnetic Fields in Semiconductor Physics, ed. by D. Heimann (World Scientific, Singapore, 1995), p. 670

82. F. Meier, B.P. Zakharchenya (eds.), *Optical Orientation* (North-Holland, Amsterdam, 1984)
83. V.P. Kochereshko, I.A. Merkulov, G.R. Pozina, I.N. Uraltsev, D.R. Yakovlev, W. Ossau, A. Waag, G. Landwehr, Solid State Electron. **37**, 1081 (1994)
84. E.L. Ivchenko, *Optical Spectroscopy of Semiconductor Nanostructures* (Alpha Science, Harrow, 2005)
85. J. Warnock, R.N. Kershaw, D. Ridgely, K. Dwight, A. Wold, R.R. Galazka, Solid State Commun. **54**, 215 (1985)
86. B.P. Zakharchenya, A.V. Kudinov, Yu.G. Kusraev, Sov. Phys. JETP **83**, 95 (1996) [Zh. Eksp. Teor. Fiz. **110**, 177 (1996)]

Chapter 8
Spin and Energy Transfer Between Carriers, Magnetic Ions, and Lattice

Dmitri R. Yakovlev and Igor A. Merkulov

Abstract Spin dynamics of free carriers and dynamics of the magnetization in diluted magnetic semiconductors are coupled to each other and mostly determined by the spin and energy transfer between the carriers, magnetic ions, and lattice. The efficiency of this transfer is examined here for low-dimensional heterostructures fabricated from II–VI semiconductor-based DMSs containing Mn^{2+}. We show here that the dynamics depends strongly on Mn content as well as on specific design of those structures. Therefore, it can be controlled by optical and electrical means.

8.1 Introduction

The magneto-optical, transport, and magnetic properties of diluted magnetic semiconductors (DMSs) and their heterostructures, which are treated nowadays as model materials for spintronics, are determined by three coupled subsystems: magnetic ions, lattice excitations (phonons), and free carriers.[1] Spin dynamics in DMS is

[1]Magnetic moments of d-electrons of Mn^{2+} ions are subject of hyperfine interaction with the Mn nuclei and nuclei of neighboring nonmagnetic ions. However, as the nuclear magneton is about three orders of magnitude smaller than the electron magneton, for the comparable concentrations of the electrons and the nuclei the hyperfine fields are screened by stronger random fields of neighboring magnetic ions. The effects of hyperfine interactions are observable at extremely low concentrations of magnetic ions [1]. They are not significant for the physics reported in this chapter and we do not discuss them here.

D.R. Yakovlev (✉)
Experimentelle Physik 2, Technische Universität Dortmund, 44221 Dortmund, Germany
and
Ioffe Physico-Technical Institute, Russian Academy of Sciences, 194021 St. Petersburg, Russia
e-mail: dmitri.yakovlev@tu-dortmund.de

I.A. Merkulov
Ioffe Physico-Technical Institute, Russian Academy of Sciences, 194021 St. Petersburg, Russia
and
Condensed Matter Sciences Division, Oak Ridge National Laboratory, Oak Ridge, TN 37831-6016, USA
e-mail: imp@mail.ornl.gov

J. Kossut and J.A. Gaj (eds.), *Introduction to the Physics of Diluted Magnetic Semiconductors*, Springer Series in Materials Science 144,
DOI 10.1007/978-3-642-15856-8_8,

controlled by the spin and energy transfer between these systems, where the free carriers (electrons and holes) and the magnetic ions have nonvanishing spins. Materials containing Mn^{2+} magnetic ions, such as (Cd,Mn)Te and (Zn,Mn)Se, have strong luminescence and sharp absorption edges formed by excitons, which make them very attractive for optical spectroscopy [2–4].

Giant magneto-optical effects are due to the strong exchange interaction of the free carrier spins with the localized magnetic moments of the Mn ions (Chaps. 1 and 3). As a result, the magnitudes of the spectroscopic responses are proportional to the magnetization of the Mn-spin system. Besides the strength of the external magnetic field, the magnetization is determined by the temperature of the Mn-spin system, which can differ from the bath temperature (i.e., the lattice temperature). Therefore, heating of the Mn-spin system can strongly influence magneto-optical properties, which offers an optical access to the magnetization dynamics. The Mn-spin temperature, T_{Mn}, can be evaluated from the giant Zeeman splitting of excitons, and Mn spin dynamics can be studied in a wide temporal range from picoseconds to seconds, and even hours. Various methods can be used to drive the Mn-spin system out of equilibrium with the lattice. Among those discussed in this chapter are injection of nonequilibrium phonons, heating of magnetic ions by microwaves in spin resonance conditions, heating by means of carriers, which in turn are either photogenerated with an excess kinetic energy or accelerated by electric fields.

Fast exchange scattering of the carriers on the magnetic ions (10^{-12}–10^{-11} s) provides efficient energy transfer into the Mn system [5–9], but cooling the Mn-spin system to the bath temperature is slowed down by the relatively long spin-lattice relaxation (SLR) time characteristic of low Mn concentrations of a few percent [10, 11]. The SLR time of the localized Mn spins is controlled by concentration dependent exchange interactions between the Mn ions and varies by several orders of magnitude from milliseconds down to nanoseconds with increasing Mn content from 0.4 to 11% [10, 11]. It has been shown that the presence of free carriers provided either by modulation doping or by applied gate voltages can accelerate SLR due to the additional channel for energy transfer from the Mn spins into the phonon system via free carriers [12, 13]. Also, the special design of heteromagnetic structures with modulated profile of Mn content has a strong effect on SLR dynamics [14, 15].

Most of the experimental data included in the chapter are for DMS quantum well heterostructures based on (Cd,Mn)Te and (Zn,Mn)Se fabricated by molecular beam epitaxy. The concentration of free carriers in these structures can be varied in a wide range by n-type or p-type modulation doping and by additional illumination. Static and dynamical properties can be controlled by the growth of heteromagnetic structures with a variable concentration of magnetic ions along the growth axis and by the digital alloy approach [14, 15]. Spin diffusion in the magnetic ion system becomes an important factor controlling spin-lattice relaxation rates in those heteromagnetic structures [16].

In this chapter, we overview the experimental data on spin and energy transfer, discuss the responsible mechanisms, and consider concepts allowing control over the transfer efficiency. The chapter is organized as follows: After an introduction to the systems responsible for spin dynamics in DMS (Sect. 8.2), we will consider the theoretical bases for spin and energy transfer between the systems, which often cannot be separated from each other (Sect. 8.3). Section 8.4 describes experimental approaches for the optical study of spin dynamics. Dynamics of the Mn-spin system is a focus of Sect. 8.5. Here, the SLR is considered together with the possibilities offered by heterostructures to tune the SLR dynamics by modulation doping and shaping the Mn profile. In Sect. 8.6, we turn to the energy and spin transfer between carriers and Mn ions. Transfer of spin coherence will not be considered in this chapter as Chap. 9 is devoted to that. For the information about carrier spin relaxation in DMS, we address readers to [8, 17] and references therein.

8.2 Systems Responsible for Spin Dynamics in DMS

Spin dynamics in DMS is controlled by the energy and spin transfer between the Mn-spin system, the free carriers, and the lattice (phonons). These interacting systems are shown schematically in Fig. 8.1, where the transfer channels are given by solid arrows.

The *Mn-spin system* plays a key role in spin dynamics. Its static and dynamic magnetization depends strongly on the Mn concentration due to the antiferromagnetic coupling between Mn ions located at the nearest and next-nearest sites of the crystal lattice (Chap. 1). It originates from the fact that the isolated Mn^{2+} ion has zero orbital moment and does not interact directly with the phonon system.[2] SLR

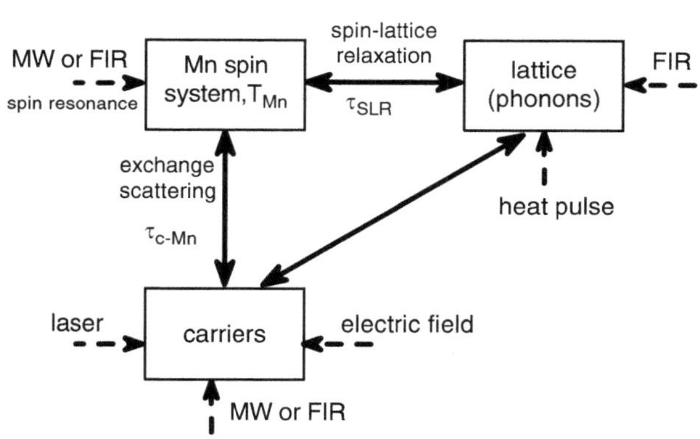

Fig. 8.1 Interacting systems of DMS. Channels for energy and spin transfer are shown by *solid arrows*. *Dashed arrows* correspond to different external influences causing heating of the systems. See Sect. 8.4 for details

[2] This statement is valid for the first-order effects on deformation of the Mn^{2+} local environment. In strained superlattices such as CdTe/MnTe and ZnTe/MnTe, a splitting of the d-shell multiplet has been observed [1]. It is caused by modification of the five electrons d-shell with S symmetry.

occurs with the help of Mn clusters and therefore the SLR time, τ_{SLR}, depends critically on the Mn content (Sect. 8.5). Engineering of the Mn profile in heterostructures offers an additional control over SLR dynamics by means of heteromagnetic structures and DMS digital alloys (see [14, 15] and Chap. 4). In external magnetic fields, the magnetization of the Mn system is characterized by the Mn-spin temperature, T_{Mn}, which can exceed considerably the bath temperature of the phonon system, T. Optical spectroscopy offers very sensitive methods for measuring T_{Mn}, enabled by a high sensitivity of the giant Zeeman splitting (GZS) of excitons (or individual carriers band states) to the polarization of Mn spins. We will exploit this feature widely in this chapter.

The *carrier system* includes the resident electrons or holes provided by n-type or p-type doping, respectively, and the photogenerated carriers. The latter are generated by light absorption and usually have an excess kinetic energy, which they share with the Mn system and the lattice. Photocarriers have a finite lifetime limited by various recombination processes, which for exciton recombination in wide-band-gap semiconductors falls in the subnanosecond range. Carriers are efficiently coupled with the lattice, e.g., the electron energy relaxation time by means of acoustic phonons is about 100 ps [18]. The carrier exchange scattering on the localized Mn spins is extremely efficient and happens with typical times, τ_{c-Mn}, faster than 10 ps [19]. This provides an efficient spin and energy exchange between the carriers and the Mn-spin system.

The *phonon system* due to the largest heat capacity serves as a bath. It is characterized by the lattice temperature, T, but can be also driven to a nonequilibrium situation by means of heat pulses (for review see [10]). Establishing equilibrium in the phonon system requires times in the range of microseconds.

The efficiency of different channels in the diagram of Fig. 8.1 can be studied by injecting energy and/or spin in different systems and measuring T_{Mn}. Various tools for doing so are shown by dashed arrows. We will consider them in detail in Sect. 8.4.

8.3 Theoretical View on Coupled Transfer of Spin and Energy

In this section, the theoretical basis for approaching the problem of spin and energy transfer between the coupled systems of DMS is formulated. In an equilibrium situation, the temperatures (in particular, spin temperatures) of the carriers, of the Mn spins, and of the lattice are equal. Therefore, spin and energy exchange between these three systems is absent. Here, we will consider electrons as typical representatives of the carriers. If the electrons or the Mn-spin system deviate from equilibrium, spin fluxes will be generated between these systems. Spin exchange is provided by

---

This modification appears as the first order of deformation in the perturbation theory. Mn^{2+} ion with the modified d-shell may interact with acoustic phonons but it is an effect of the second-order perturbation.

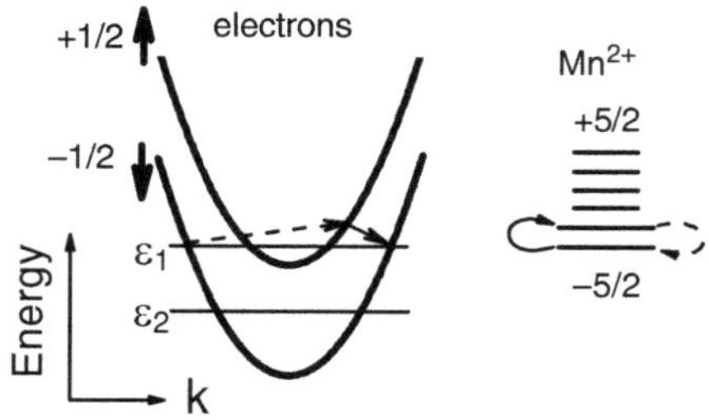

Fig. 8.2 Energy diagram of the exchange scattering of electron on Mn^{2+} ion in an external magnetic field. *Solid arrows* show the process of the electron scattering from spin up $(+1/2)$ to spin down $(-1/2)$ subband, which causes heating of the Mn spin. *Dashed arrows* show the reversed process. Because $\alpha_e > 0$, electron and Mn spins are parallel to each other in the lowest energy state. For degenerate 2DEGs, the number of active electrons contributing to the spin and energy transfer is independent of GZS if the Fermi level covers the upper spin subband ε_1. The transfer efficiency decreases exponentially when the Fermi level falls below the bottom of this subband ε_2

flip-flop transitions of the electron spin (**s**) and the Mn spin (**S**), which are described by $\mathbf{S}_+\mathbf{s}_- + \mathbf{s}_+\mathbf{S}_-$ terms in the Hamiltonian (12) introduced in Chap. 3. However, the energy of the Zeeman splitting for the electrons differs from that of the Mn spins considerably (see Fig. 8.2). For the electrons, it is due to the GZS of the conduction band in the exchange field of magnetic ions, and for the Mn spins it is just the splitting in an external magnetic field (neglecting the exchange field due to carriers). The energy mismatch for the flip-flop process is compensated by the electron kinetic energy. The scattering potential of a localized Mn spin for carriers has a δ-like shape, which permits with equal probability any momentum values of the difference between the initial and scattered electron state. Therefore, exchange scattering couples together spin systems of electrons and Mn ions and the reservoir of electron kinetic energy. Deviation from equilibrium causes energy and spin fluxes between these systems.

A quasi-equilibrium distribution of the electron system is established by means of electron–electron collisions. The timescale of electron–electron scattering is inversely proportional to the electron concentration (n_e) and to $E_e^{3/2}$, where E_e is the electron kinetic energy. For $E_e = 10\,\text{meV}$ and $n_e = 10^{15}\,\text{cm}^{-3}$, this time equals a few tens of picoseconds. The electron scattering time can be much faster than the time required to establish total equilibrium between all participating systems. The quasi-equilibrium electron distribution is characterized by an electron temperature, T_e, which may differ from the lattice temperature. T_e relaxes to the lattice temperature via the electron–phonon interaction, which in semiconductors typically requires times of 100 ps [20].

In the absence of magnetic ions, the electron spin relaxation time, τ_s, in semiconductors falls in the nanosecond time range. For example, for localized electrons in nonmagnetic CdTe/(Cd,Mg)Te quantum wells at a liquid helium temperature, it exceeds 30 ns [21]. As a result, at times longer than the electron–phonon interaction but shorter than τ_s the electron system is characterized by two parameters: the electron temperature and the nonequilibrium spin. The nonequilibrium spin is due to the difference of Fermi levels in spin subbands (see Chap. 2 in [22]).

$$\langle \mathbf{s}_{\mathrm{NE}} \rangle = \langle \mathbf{s} \rangle - \langle \mathbf{s}_{Te} \rangle . \quad (8.1)$$

Here, $\langle \mathbf{s}_{\mathrm{NE}} \rangle$ is the mean value of the nonequilibrium electron spin, $\langle \mathbf{s} \rangle$ is the mean value of the electron spin, and $\langle \mathbf{s}_{Te} \rangle$ is its equilibrium value at a temperature T_{e}.

The presence of the magnetic ions in DMS materials markedly reduces the spin relaxation time of electrons. However, spin exchange between the electrons and the magnetic ions in a magnetic field results in the transfer of both spin and energy. This process goes beyond simple spin relaxation and requires a more elaborate consideration, to be given in this chapter.

We turn now to the magnetic properties of the Mn-spin system. When the external conditions change (e.g., the magnetic field or the temperature) also the magnetization of the Mn-spin system relaxes to a new equilibrium. Characteristic relaxation times can cover a very wide range from picoseconds to minutes and even to hours. At vanishing external magnetic field and in the absence of spin ordering, components of the magnetization in all directions relax with the same rate. To be precise, this is valid for the crystals with cubic symmetry. Application of external magnetic field reduces the symmetry. Relaxation of longitudinal and transverse components of the magnetization is characterized by different times, T_1 and T_2, respectively. For this difference to be significant, the external magnetic field must exceed the local exchange field acting on a given Mn ion, produced by the neighboring Mn ions.

The energy relaxation of the Mn system is related to the relaxation of the longitudinal component of magnetization. It is determined by the energy exchange between the Mn-spin system and the lattice. Therefore, the longitudinal relaxation time T_1 is often called as spin-lattice relaxation time or energy relaxation time.

Transverse component of the magnetization is related to the coherent precession of many Mn spins around the magnetic field. Dephasing of this precession (see Chap. 9 for details) is caused by anisotropic spin–spin interactions and has a weak influence on the energy of spins. The transverse relaxation time T_2 is often called phase relaxation time or coherence time.

In weak magnetic fields, the energy of the spin system is mostly contributed by the energy of spin–spin interactions. At zero field limit, the equilibrium polarization of the Mn-spin system is equal to zero. In this case, all components of the nonequilibrium spin relax with the same time T_2. However, the energy (i.e., temperature) of the spin system relaxes to equilibrium with the time T_1, which is considerably longer than T_2 [23, 24].

In II–VI DMS, these times depend strongly on Mn content. The spin temperature, T_{Mn}, of the magnetic ion system is established on a timescale $T_2 = 10^{-12}$–10^{-9} s [19]. The assumption that this temperature is established will be valid throughout this chapter. Establishing energy equilibrium of the Mn system with the lattice requires times $T_1 = 10^{-8}$–10^{-4} s [19].

The main thermodynamical parameters of coupled systems in DMS are the electron temperature, the electron nonequilibrium spin, and the spin temperature of the magnetic ions. Let us analyze changes of these parameters under electron photogeneration.

For small deviations from equilibrium, the behavior of these parameters can be analyzed in the framework of linear approximation. It is convenient for a qualitative understanding of the general trends of the system behavior. However, for quantitative modeling of experimental results obtained at high excitation densities, this linear approximation is not sufficient [25].

The heat exchange between systems A and B with different temperatures T_A and T_B is commonly modeled by the Fourier law, which establishes a simple linear dependence between the heat flux, J_{AB}, and the temperature:

$$J_{AB} = q(T_A - T_B). \tag{8.2}$$

Here, $q > 0$ is thermal conductivity of the boundary separating the systems. In the approximation of small deviations of the temperatures from their equilibrium value, i.e., for $|T_A - T_B| \ll |T_{A,B}|$, the Fourier law due to its general character can be used to describe thermal fluxes between the magnetic ions and lattice. As a result, the rate of temperature change is described by a simple equation:

$$C_A \frac{\partial T_A}{\partial t} = -C_B \frac{\partial T_B}{\partial t} = -q(T_A - T_B), \tag{8.3}$$

where C_A, C_B are heat capacities of the systems.

At low temperatures, the small deviation approximation is valid only for very low levels of external influence. For example at the temperature of liquid helium (4.2 K), laser illumination with excitation density of about 1 W cm^{-2} may increase the carrier temperature a few times [20]. Accounting for the strong deviations requires the use of more elaborate equations, which often need numerical solution, but it does not bring qualitatively new effects.

The energy transfer between the electrons and the Mn-spin system cannot be considered solely in terms of equalization of temperatures as described in (8.3). The reason for this is that the energy transfer mediated by the exchange scattering of the electrons on the localized Mn spins is accompanied by the transfer of the nonequilibrium electron polarization to the Mn-spin system. For weak deviations from equilibrium, the spin flow into the Mn-spin system is proportional to the nonequilibrium electron polarization. Note the analogy to the Fourier law here

$$\left.\frac{\partial \mathbf{I}_{Mn}}{\partial t}\right|_{ex} \frac{N_{Mn}}{n_e} \approx \lambda_{JS} \langle \mathbf{s}_{NE} \rangle. \tag{8.4}$$

Here, $\mathbf{I}_{Mn}$ is the mean value of the magnetic ion spin, and N_{Mn} and n_e are the concentrations of the Mn ions and the electrons, respectively. λ_{JS} is the rate of electron spin relaxation on the magnetic ions. Index "ex" shows that $\mathbf{I}_{Mn}$ is changed by means of the electron exchange scattering.

In the presence of an external magnetic field **B**, the energy of the Mn-spin system, E_{Mn}, is changed by means of both spin and energy flow described by (8.4) and (8.2), respectively.

$$\left.\frac{\partial E_{\mathrm{Mn}}}{\partial t}\right|_{\mathrm{ex}} = C_{\mathrm{Mn}} \left.\frac{\partial T_{\mathrm{Mn}}}{\partial t}\right|_{\mathrm{ex}}$$
$$= q(T_{\mathrm{e}} - T_{\mathrm{Mn}}) + \mu_{\mathrm{B}} g_{\mathrm{Mn}} \lambda_{\mathrm{JS}} n_{\mathrm{e}} (\mathbf{B} \cdot \langle \mathbf{s}_{\mathrm{NE}} \rangle) \quad (8.5)$$

here μ_{B} is the Bohr magneton and $g_{\mathrm{Mn}} \approx 2$ is the g-factor of d-shell electrons of the Mn^{2+} ions. Therefore, the heating of the magnetic ions depends on the electron temperature and on the nonequilibrium electron polarization. In the absence of a nonequilibrium electron polarization, the heat flow is equal to zero for equal temperatures T_{e} and T_{Mn}. However, for a finite value of the nonequilibrium polarization, the energy flows are compensated under the condition $q(T_{\mathrm{e}} - T_{\mathrm{Mn}}) = -\mu_{\mathrm{B}} g_{\mathrm{Mn}} \lambda_{\mathrm{JS}} n_{\mathrm{e}} (\mathbf{B} \cdot \langle \mathbf{s}_{\mathrm{NE}} \rangle)$.

According to the Onsager relation [26], if the nonequilibrium electron polarization generates the energy flow into the Mn system, then in turn the energy flow, which is determined by the difference between T_{e} and T_{Mn}, should generate a nonequilibrium electron polarization

$$\left.\frac{\partial \langle \mathbf{s}_{\mathrm{NE}} \rangle}{\partial t}\right|_{\mathrm{ex}} = -\lambda_{\mathrm{JS}} \langle \mathbf{s} \rangle + \boldsymbol{\Lambda}_{\mathrm{ST}} (T_{\mathrm{e}} - T_{\mathrm{Mn}}). \quad (8.6)$$

The direction of the $\boldsymbol{\Lambda}_{\mathrm{ST}}$ vector is given by the magnetic field **B**. It is shown below in (8.12) that $\boldsymbol{\Lambda}_{\mathrm{ST}}$ is antiparallel to **B**. In the absence of the external field, the $\boldsymbol{\Lambda}_{\mathrm{ST}}$ value is equal to zero. According to the Le Chatelier principle [26], the flows generated by driving the systems out of equilibrium tend to recover the equilibrium state. Spin and energy transfer between the electrons and the Mn-spin system are described by (8.5) and (8.6). Phenomenological coefficients for these equations have been calculated in [25].

The electron system can be driven out of equilibrium either by the generation of nonequilibrium polarization, or by heating, which increases the kinetic energy of the electrons. The common way to generate nonequilibrium polarization of carriers is to use the optical orientation technique [22]. The sample is exposed to circularly polarized light, which generates spin-oriented carriers. Several ways to heat the electron system are shown in Fig. 8.1. All of them can be formally described by introducing a heat source Q. To solve the problem, all energy transfer channels between DMS systems shown by solid arrows in Fig. 8.1 should be taken into account. The task does not go far beyond the standard course of physics, but requires lengthy equations. For simplicity, we consider here only the effect of the nonequilibrium electron polarization on the energy transfer. A comprehensive theoretical description is given in [25].

The nonequilibrium electron polarization caused by the heat exchange between the electrons and the magnetic ions is described by the following balance equation

$$\frac{\mathrm{d}\langle \mathbf{s}_{\mathrm{NE}} \rangle}{\mathrm{d}t} = -\lambda_{\mathrm{JS}} \langle \mathbf{s}_{\mathrm{NE}} \rangle + \boldsymbol{\Lambda}_{\mathrm{ST}} (T_{\mathrm{e}} - T_{\mathrm{Mn}}) - \frac{\langle \mathbf{s}_{\mathrm{NE}} \rangle}{\tau_s} = 0, \quad (8.7)$$

then for the stationary condition one can find that

$$\langle \mathbf{s}_{\mathrm{NE}} \rangle = \frac{\mathbf{\Lambda}_{\mathrm{ST}} \tau_s (T_{\mathrm{e}} - T_{\mathrm{Mn}})}{\lambda_{\mathrm{JS}} \tau_s + 1}. \tag{8.8}$$

Substituting (8.8) in (8.5) we get the Fourier equation, but now with a renormalized coefficient of the boundary thermal conductivity

$$C_{\mathrm{Mn}} \frac{\partial T_{\mathrm{Mn}}}{\partial t}\bigg|_{\mathrm{ex}} = \left[q + \frac{\mu_{\mathrm{B}} g_{\mathrm{Mn}} \lambda_{\mathrm{JS}} \tau_s n_{\mathrm{e}}}{\lambda_{\mathrm{JS}} \tau_s + 1} (\mathbf{B} \cdot \mathbf{\Lambda}_{\mathrm{ST}}) \right] (T_{\mathrm{e}} T_{\mathrm{Mn}}) = \tilde{q}(T_{\mathrm{e}} - T_{\mathrm{Mn}}). \tag{8.9}$$

We are going to show that the nonequilibrium electron polarization decreases the effective thermal conductivity of the border between the electrons and magnetic ions. For that, the sign of $(\mathbf{B} \cdot \mathbf{\Lambda}_{\mathrm{ST}})$ should be found.

Spin and energy flows between the electrons and magnetic ions are coupled with each other in the presence of external magnetic fields. For simplicity, we consider the case of sufficiently strong magnetic fields that exceed the local fields of Mn–Mn interactions (for comprehensive consideration, see [25]). The transfer of the energy portion $\delta E = q(T_{\mathrm{e}} - T_{\mathrm{Mn}})\delta t$ from the electrons to the Mn-spin system is accompanied by the transfer of spin $\delta \mathbf{S}$

$$\delta E = \mu_{\mathrm{B}} g_{\mathrm{Mn}} (\mathbf{B} \cdot \delta \mathbf{S}) N_{\mathrm{Mn}}. \tag{8.10}$$

For the exchange scattering provided by the flip-flop transitions (terms $\mathbf{S}_+ \mathbf{s}_- + \mathbf{s}_+ \mathbf{S}_-$ of the Hamiltonian (11) in Chap. 3), the total spin is conserved. In other words, the variation of the Mn spin and the electron spin is equal in magnitude and has opposite signs, i.e., $\delta \langle \mathbf{s} \rangle n_{\mathrm{e}} = -\delta \mathbf{S} N_{\mathrm{Mn}}$. The energy transfer in (8.7) is described by the second term on the right-hand side. Therefore, one can write:

$$\begin{aligned} \delta \langle \mathbf{s} \rangle n_{\mathrm{e}} = \mathbf{\Lambda}_{\mathrm{ST}} (T_{\mathrm{e}} - T_{\mathrm{Mn}}) n_{\mathrm{e}} \delta t = -\delta \mathbf{S} N_{\mathrm{Mn}} &= -\frac{\mathbf{B}(\mathbf{B} \cdot \delta \mathbf{S}) N_{\mathrm{Mn}}}{B^2} \\ &= -\frac{\mathbf{B} \delta E}{\mu_{\mathrm{B}} g_{\mathrm{Mn}} B^2}. \end{aligned} \tag{8.11}$$

Then

$$\mathbf{\Lambda}_{\mathrm{ST}} = -\frac{\mathbf{B} q}{\mu_{\mathrm{B}} g_{\mathrm{Mn}} B^2 n_{\mathrm{e}}} \tag{8.12}$$

and $(\mathbf{B} \cdot \mathbf{\Lambda}_{\mathrm{ST}}) < 0$. Also, one obtains

$$\left[1 - \frac{\lambda_{\mathrm{JS}} \tau_s}{\lambda_{\mathrm{JS}} \tau_s + 1} \right] q = \tilde{q}. \tag{8.13}$$

In the case of fast spin relaxation of electrons, $\lambda_{\mathrm{JS}} \tau_s \ll 1$ and therefore the nonequilibrium electron polarization practically does not contribute to the energy transfer, i.e., $q \approx \tilde{q}$. If $\lambda_{\mathrm{JS}} \tau_s \gg 1$, then $\tilde{q} \approx q / \lambda_{\mathrm{JS}} \tau_s \ll q$. In this case, an induced nonequilibrium electron polarization suppresses the energy transfer from the electrons to the Mn-spin system.

The theoretical description for the case when the deviation from the equilibrium cannot be treated as weak is given in [25]. Numerical results of the spin and energy transfer between the electrons and the Mn ion system are shown in Fig. 8.14e for $Cd_{0.99}Mn_{0.01}Te/Cd_{0.76}Mg_{0.24}Te$ quantum wells (QW) with electron gases of different densities, n_e [27]. One can see that for small n_e the relaxation time increases exponentially already in with very weak magnetic fields, but with increasing n_e the growth of the relaxation time occurs at stronger fields. The reasons for that are shown schematically in Fig. 8.2. The GZS of electrons in the conduction band greatly exceeds the Zeeman splitting of the Mn ions. But for the coupled transfer of spin and energy, the energy conservation law should be satisfied. This condition restricts number of electrons participating in the transfer. For small n_e, when the 2DEG is nondegenerate, the number of active electrons is reduced with increasing GZS. For a degenerate 2DEG, the number of active electrons is independent of the GZS value if the Fermi level is in the upper spin subband. In this case, the strong magnetic field dependence of the transfer time appears when the Fermi level is leaving the upper spin subband.

Note that in the limiting case of vanishing magnetic field the energy and spin transfers between the carriers and Mn systems are decoupled. Since $(\mathbf{B} \cdot \delta\mathbf{S}) = 0$, the energy transfer does not induce spin polarization of the carriers or Mn ions. Equilibrium polarization of the Mn ions is equal to zero: $\langle \mathbf{I} \rangle (B = 0, T_{Mn}) \equiv 0$, and nonequilibrium polarization decays with the fast transverse relaxation time $T_2(B = 0)$.

In the case of hole exchange scattering on the Mn spins, the nonequilibrium hole polarization plays a minor role as it decays very fast due to the efficient spin relaxation of holes. It has been shown in (29, 30) of Chap. 3 that the anisotropic exchange interaction of two-dimensional holes opens a way for the direct energy transfer between the kinetic energy reservoir of the holes and the Mn-spin reservoir. The hole spin flip is not required for this process: the hole remains in the same spin subband while the Mn ion flips its spin. Experimentally, these processes have been found in [5] and are discussed in Sect. 8.6.2.

Note that in heterostructures where the carriers are localized in two or three dimensions and, therefore, interact locally with the Mn-spin system, the heating (cooling) of the Mn-spin system becomes spatially inhomogeneous. In this case, an evaluation of T_{Mn} should account for the spin diffusion within the Mn system between the regions interacting with carriers and the regions free of this interaction. An experimental example of such a situation is given in Sect. 8.5.2 for a QW structure with different Mn concentrations in the well and barrier layers.

8.4 Experimental Technique

In this section, we overview different experimental techniques based on optical detection of magnetization, and thus, of the Mn-spin temperature, and the application of these techniques to the investigation of the efficiency and the dynamics of spin and energy transfer between different systems of DMS, shown in Fig. 8.1.

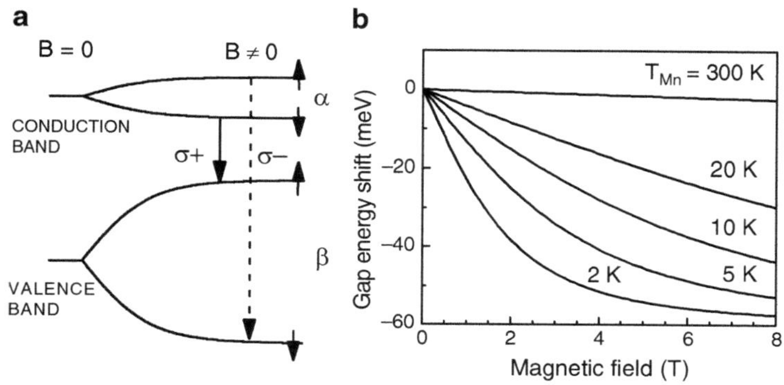

Fig. 8.3 (**a**) Optical transitions between the conduction and valence band states in DMS subject to the giant Zeeman splitting in an external magnetic field. Electron and hole spin states are indicated by *arrows*. (**b**) Band gap energy shift in $Zn_{0.05}Mn_{0.95}Se$ for different Mn-spin temperatures calculated with (8.15), using $S_{eff} = 1.5$, $T_0 = 1.8$ K

8.4.1 *Optical Detection of Mn-Spin Temperature*

The idea behind optical thermometry of the Mn-spin system is schematically illustrated by Fig. 8.3. Panel (a) shows optical transitions in vicinity of the band gap when the spin levels of electrons and holes are split by the GZS effect (Chap. 1). Two circularly polarized transitions corresponding to the heavy-hole exciton are given by arrows.

The giant Zeeman splitting of the heavy-hole excitons is proportional to the magnetization and thus to the average spin of the Mn ions $\langle S_z \rangle$:

$$\Delta E_z = (\delta_e \alpha - \delta_h \beta) N_0 x \langle S_z \rangle. \qquad (8.14)$$

Here, $N_0\alpha$ and $N_0\beta$ are the exchange constants for the conduction and valence band. In $Zn_{1-x}Mn_xSe$, they are equal to 0.26 and −1.31 eV, and in $Cd_{1-x}Mn_xTe$ they equal 0.22 and −0.88 eV, respectively [3]. x is the Mn mole fraction. Parameters $\delta_e, \delta_h \leq 1$ are introduced to account in heterostructures for the leakage of electron and hole wave functions into the nonmagnetic layers. A more precise description of the GZS effect in DMS heterostructures can be found in Chaps. 3 and 4. $\langle S_z \rangle$ is the mean thermal value of the Mn-spin component along the magnetic field $B = B_z$. It is expressed by the modified Brillouin function $B_{5/2}$:

$$\langle S_z \rangle = -S_{eff}(x) B_{5/2}\left[\frac{5\mu_B g_{Mn} B}{2k_B(T_{Mn} + T_0(x))}\right]. \qquad (8.15)$$

Here, $g_{Mn} = 2$ is the g-factor of the Mn^{2+} ions. S_{eff} is the effective spin and T_0 is the effective temperature. These parameters permit a phenomenological description of the antiferromagnetic Mn–Mn exchange interaction. For their values in, e.g.,

$Zn_{1-x}Mn_xSe$, we refer to Fig. 4 of [28]. One can see from (8.14), (8.15) that the GZS value is directly linked with T_{Mn}. Its decrease with the temperature growth is shown in Fig. 8.3b. The GZS value can be measured directly by different optical techniques: photoluminescence (PL), PL excitation (PLE) spectroscopy, reflection, and transmission. Its relative changes can be conveniently detected via the giant Faraday and Kerr rotation effects and via the circular polarization degree of PL.

8.4.1.1 Optical Spectra and Giant Zeeman Splitting of Excitons in (Zn,Mn)Se QWs

Giant magneto-optical effects in DMS take place at liquid helium temperatures, therefore, most of the experiments discussed in this chapter have been performed at $T = 1.6$ K. Figure 8.4a shows a survey of photoluminescence, PL excitation, and reflectivity spectra for a $Zn_{0.988}Mn_{0.012}Se/Zn_{0.94}Be_{0.06}Se$ QW. Two strong resonances corresponding to the exciton (X) ground states involving heavy holes (1s-hh) and light holes (1s-lh) are clearly seen in the reflectivity spectrum. They are split by 14 meV due to strain and quantum confinement effects. Both excitonic resonances are observable as strong lines in the PLE spectrum. The PL spectrum consists of two lines. The high energy line is due to recombination of the heavy-hole exciton; its small linewidth of 1.9 meV indicates a high structural quality of the sample. The low energy line corresponds to the negatively charged exciton (trion, T), which is a complex of two electrons bound to one hole [29]. The trion binding energy in this QW is 4.9 meV.

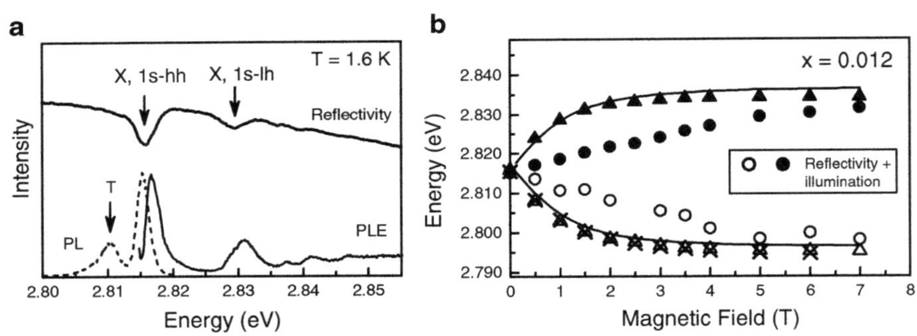

Fig. 8.4 Optical spectra of a 100-Å-thick $Zn_{0.988}Mn_{0.012}Se/Zn_{0.94}Be_{0.06}Se$ QW: (**a**) Photoluminescence, PL excitation, and reflectivity spectra in the absence of a magnetic field. (**b**) GZS of the heavy-hole excitons. Experimental data are shown by *symbols* (*open* for σ^+ and *closed* for σ^- polarization). *Triangles* are for reflectivity and *crosses* for PL data taken under a very low photoexcitation density of 0.016 W cm^{-2}. *Circles* represent exciton energies in reflectivity measurements with additional laser illumination ($\hbar\omega_L = 3.4$ eV, $P = 0.7$ W cm^{-2}). GZS reduction caused by the Mn system heating is clearly seen. *Solid lines* represent the best fit for the reflectivity data points measured without additional illumination (i.e., for the cold Mn system) using (8.14) with $S_{eff} = 2.21$ and $T_0 = 1.8$ K. Reprinted from [28]

When the magnetization does not vanish (e.g., in external magnetic field), the excitonic transitions experience a GZS. A set of the exciton GZS determined from reflectivity and PL spectra is presented in Fig. 8.4b. To avoid heating of the Mn system, PL spectra were measured under very low photoexcitation density $P = 0.016\,\mathrm{W\,cm^{-2}}$ (shown by crosses). These data points coincide very well with the reflectivity data detected without additional illumination (triangles). Both branches of the exciton GZS can be followed in the reflectivity spectrum, while only one branch is seen in PL due to exciton thermalization. Lines show a fit of the data to (8.14) from which the Mn content can be evaluated. Calculations show that in this 100-Å-thick QW $\delta_e = \delta_h = 0.96$, i.e., 96% of the carrier wave functions are localized in the QW.

8.4.1.2 Suppression of Giant Zeeman Splitting Under Continuous-Wave Excitation

The Mn-spin temperature can be increased even by moderate levels of continuous-wave (cw) photoexcitation. As shown by dots in Fig. 8.4b, an additional irradiation with laser power of $0.7\,\mathrm{W\,cm^{-2}}$ causes considerable GZS decrease evidencing the heating of the Mn-spin system. In more detail, this effect is seen in Fig. 8.5a, where the PL line position is shown for different excitation densities. Experimental data shown by symbols were fitted with (8.14) to evaluate T_{Mn}. It is seen that the Mn system can be heated to high temperatures of 23 K by means of relatively low excitation

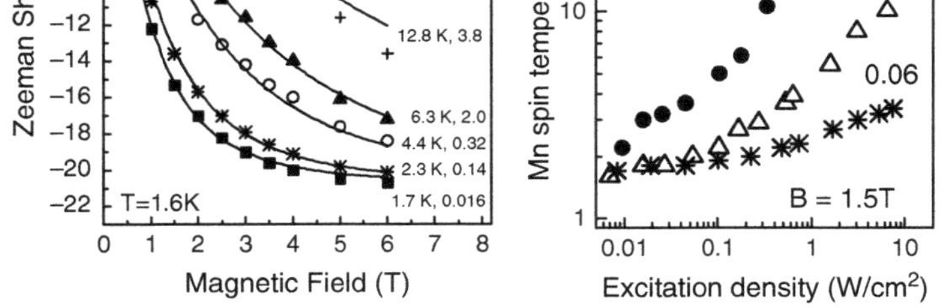

Fig. 8.5 Heating of the Mn-spin system in 100-Å-thick $Zn_{1-x}Mn_xSe/Zn_{1-y}Be_ySe$ QWs by cw laser excitation: (**a**) The giant Zeeman shift of excitons in a $Zn_{0.988}Mn_{0.012}Se/Zn_{0.94}Be_{0.06}Se$ QW evaluated from PL spectra measured under different excitation densities, given in $\mathrm{W\,cm^{-2}}$. *Lines* represent the fit which gives T_{Mn}. (**b**) Mn-spin temperature plotted against excitation density for structures with different Mn content x. Excitation energy $\hbar\omega_L = 3.4\,\mathrm{eV}$ exceeds the energy gap of barriers. $T = 1.6\,\mathrm{K}$. Reprinted from [28]

densities of $13\,\mathrm{W\,cm^{-2}}$. We note here that in this undoped QW in the whole range of magnetic fields up to 6 T very reliable fits can be achieved with a Mn temperature that is independent of the magnetic field. This differs from similar studies of QWs containing a two-dimensional electron gas, where the heating of the Mn system varies with the magnetic field (see Sect. 8.5.4).

The heating efficiency depends strongly on the Mn concentration. Figure 8.5b compares $T_{\mathrm{Mn}}(P)$ dependencies measured at $B = 1.5\,\mathrm{T}$ for the three samples with different Mn contents. For $x = 0.004$, T_{Mn} reaches 42 K at $P = 4.5\,\mathrm{W\,cm^{-2}}$, which corresponds to a GZS suppression down to 10% of its value at $T_{\mathrm{Mn}} = 1.6\,\mathrm{K}$. For higher Mn contents, the heating effect becomes less efficient. At $P \approx 4.5\,\mathrm{W\,cm^{-2}}$, T_{Mn} decreases from 42 to 3.2 K, when x increases from 0.004 to 0.06. We will show in Sect. 8.5 that this effect is due to the strong concentration dependence of the SLR time, which controls the cooling the Mn-spin system heated by carriers.

8.4.1.3 Time-Resolved Measurements of Spin Dynamics

Temporal characteristics of the spin dynamics can be measured by time-resolved optical techniques. For that, the external impacts which drive the Mn-spin system out of equilibrium with the bath should be pulsed (Fig. 8.1). Accordingly, changes of the optical response must be detected with time resolution. Here, an example of pulsed photoexcitation in DMS QWs is considered.

Photoluminescence spectra of the $Zn_{0.89}Mn_{0.11}Se/Zn_{0.89}Be_{0.11}Se$ DMS quantum well structure are given in Fig. 8.6a. The two bottom spectra were detected

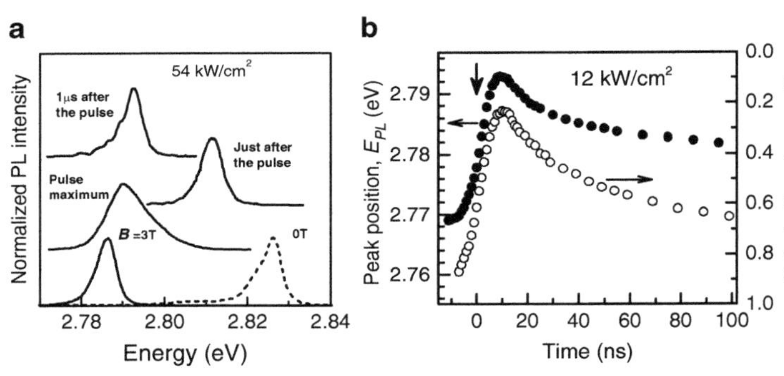

Fig. 8.6 Pulsed photoexcitation of a 100-Å-thick $Zn_{0.89}Mn_{0.11}Se/Zn_{0.89}Be_{0.11}Se$ QW. (**a**) The two lowest PL spectra are taken under very low *cw* laser excitation, the others are recorded at different time delays Δt with respect to the laser pulse maximum: $\Delta t \approx 0\,\mathrm{ns}$ (corresponds to the pulse maximum), 10 ns (just after the pulse), and 1 μs later. The excitation density $P = 54\,\mathrm{kW\,cm^{-2}}$, magnetic field $B = 3\,\mathrm{T}$ (*solid lines*) and $B = 0\,\mathrm{T}$ (*dashed line*), bath temperature $T = 1.6\,\mathrm{K}$. (**b**) Temporal variation of the PL spectral line position E_{PL} at a magnetic field $B = 3\,\mathrm{T}$ (*closed circles*) and of the circular polarization degree P_c at $B = 0.12\,\mathrm{T}$ (*open circles*). Excitation density $P = 12\,\mathrm{kW\,cm^{-2}}$; $T = 1.6\,\mathrm{K}$; the laser pulse maximum position is indicated by the *vertical arrow*. Reprinted from [11]

under *cw* laser illumination with very low excitation density to avoid heating of the Mn system above the bath temperature. The giant Zeeman shift of the emission line amounts to about 40 meV at $B = 3$ T. The three upper spectra show the emission line at different delays with respect to the impact laser pulse of 10 ns duration. PL spectra were detected with a gated charge-coupled-device (CCD) camera having a time resolution of 5 ns. To follow the magnetization evolution at time delays exceeding the exciton lifetime, an additional illumination with a weak *cw* laser has been provided [11]. Just after the pulse, the Zeeman shift is reduced to $\sim$14 meV, which corresponds to a heating of the Mn system up to $T_{\mathrm{Mn}} = 17$ K. After one microsecond, the line is shifted back to lower energies, reflecting the cooling of the Mn system.

There are two characteristics of the magneto-optical spectra that can be exploited for obtaining information about the temperature of the Mn-spin system. Both are related to the giant Zeeman splitting effect of the conduction and valence band states. The first characteristic is the energy shift of the emission line. It is convenient to use it in relatively strong magnetic fields (exceeding 0.5 T), for which the Zeeman shift can be clearly detected. For measurements in weak magnetic fields (below 0.5 T), the circular polarization degree of emission can be analyzed. It has been shown that both characteristics provide the same information about the Mn-spin temperature (see Fig. 10 in [28]).

Figure 8.6b shows the time evolution of both the Zeeman shift and the polarization degree induced by pulsed laser excitation. Fast heating of the Mn ions occurs during the laser pulse and is reflected by a high-energy shift of the PL maximum by 23 meV and a decrease of the polarization degree from 0.9 to 0.2. After the laser pulse, the Mn-spin temperature relaxes toward equilibrium with a relaxation time constant of 23 ns. However, it saturates at a level that exceeds the bath temperature. We will show below in Sect. 8.6.4 that this level is controlled by nonequilibrium phonons. To reach the equilibrium temperature of 1.6 K, a much longer time of a few μs is required. Remarkably, the Zeeman shift and the polarization degree show a very similar temporal behavior and, therefore, both of them are well suited for optical detection of the spin-lattice relaxation dynamics.

The magnetization dynamics in bulk DMS can be measured directly by a time-domain magnetic spectrometer based on a pickup electromagnetic coil [30, 31]. The coil being fixed around a DMS sample with the coil axis parallel to an external magnetic field detects the externally induced changes of the longitudinal magnetization. However, this technique is not sensitive enough for very thin magnetic layers and nanostructures because the net changes of magnetization in the whole sample are very small.

A very sensitive technique well suited for nanostructures is based on Faraday or Kerr rotation effects. Being combined with time-resolved pump-probe technique, it allows to study spin dynamics with a time resolution of 100 fs [7, 32], see also Chap. 9. Fast spin and energy transfer from photogenerated carriers to Mn system and long cooling time of the magnetic ions were reported [8, 33].

Recently, a technique based on short magnetic pulses created by a small coil was employed [34]. Sharp pulses with the rise and fall times of about 10 ns and the magnetic field amplitudes up to 40 mT were realized. Combined with optical detection this technique is capable to address magnetization dynamics in the absence of an external magnetic field by monitoring the optical response after the field pulse. This dynamics occurs with the fast transverse relaxation time T_2. For more details, the reader is referred to original publications [34–36].

8.4.2 Tools to Address the Mn-Spin System

To study the efficiency of the transfer channels between systems of DMS shown in Fig. 8.1, one should have a possibility to inject spin and energy in different systems to follow the changes of T_{Mn}. There is a rich set of tools that can be applied:

1. Microwave (MW) radiation. Its typical frequency range from 10 to 100 GHz corresponds to the photon energies from 0.04 to 0.4 meV, which allows exciting Mn^{2+} ions resonantly in conditions of electron spin resonance (ESR): $\hbar\omega_{\mathrm{MW}} = \mu_{\mathrm{B}} g_{\mathrm{Mn}} B$. Also, the ESR of free carriers and their nonresonant heating is possible [37–43].
2. Far-infrared (FIR) radiation covers the range of photon energy from 2 to 12 meV. Similar to MW radiation ESR for Mn^{2+} ions can be realized, but in very strong magnetic fields. Also, the nonresonant heating of carriers and their excitation in the cyclotron resonance condition is possible. Direct heating of the lattice under FIR radiation has been also reported [30, 31].
3. Heat pulse of nonequilibrium phonons injects energy directly into the phonon system. It can be induced by FIR or by means of a phonon generator, which is a thin metal film evaporated on the sample and excited either by electrical current or by laser pulses [10, 44].
4. Electrical current heating of free carriers in doped DMS [45, 46].
5. Laser light generates photocarriers with excess kinetic energy which then either interact directly with the Mn spins or generate nonequilibrium phonons [5, 6, 25, 47–54]. See also Sect. 8.6 and Chap. 5.

Let us consider a few examples from this list in a greater detail.

8.4.2.1 Microwave Heating

Optically detected magnetic resonance (ODMR) techniques with a 10 GHz MW generator allowed Ryabchenko and coworkers to discover in 1977 the strong exchange interaction of carrier spins with magnetic ions and announce a new class of DMS materials [38]. Faraday rotation in (Cd,Mn)Te was used for the optical detection of magnetization in this experiment. Later, ODMR was used for investigation of (Zn,Mn)Te [39] and (Cd,Mn)Te [40, 41, 43] bulk DMS and recently for (Cd,Mn)Te [42] and (Zn,Mn)Se QWs [37].

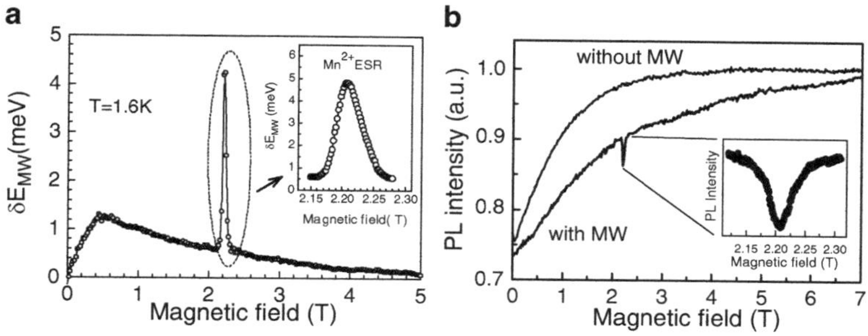

Fig. 8.7 Optically detected magnetic resonance of Mn^{2+} electron spin in a 100-Å-thick $Zn_{0.988}Mn_{0.012}Se/Zn_{0.94}Be_{0.06}Se$ MQW. (**a**) Microwave-induced changes of GZS. The largest changes are detected at the conditions of Mn^{2+} electron spin resonance (ESR). (**b**) Effect of MW on the peak PL intensity of excitons, which increases with the polarization of the Mn-spin system. Reprinted from [37]

An example in $Zn_{0.988}Mn_{0.012}Se/Zn_{0.94}Be_{0.06}Se$ QWs is given in Fig. 8.7. The GZS shifts the exciton PL line up to 20 meV. The line shift is reduced under MW radiation. The source operating at 60 GHz with power up to 300 mW was used and the sample was mounted in a cylindrical resonator [37]. The MW-induced shift δE_{MW} from panel (a) shows a strong ESR resonance of Mn^{2+} ion at 2.2 T. T_{Mn} at the resonance increases from 1.6 up to 5.2 K. Additionally, a nonresonant heating is observed for all magnetic fields. It is due to the MW interaction with free carriers, which then pass the energy to the Mn system.

In (Zn,Mn)Se-based QWs, a pronounced ODMR can be also detected through the changes of the PL peak intensity (panel (b)). This effect is related to the interplay between radiative and nonradiative recombination of excitons, which may depend on the Mn magnetization due to spin-dependent energy transfer from excitons into the Mn $3d^5$ shell [55, 56].

Pulsed excitation with MW allows one to measure SLR time, τ_{SLR}. In the case of long MW pulses exceeding characteristic times of magnetization relaxation, τ_{SLR} can be evaluated from the rise and decay dynamics of ODMR signals [37].

8.4.2.2 Heat Pulse of Nonequilibrium Phonons

Nonequilibrium phonons in [44, 57, 58] were generated by a heat pulse technique. A phonon generator (a 10-nm-thick Constantan film) with an area $0.5 \times 0.25\,\mathrm{mm}^2$ was evaporated on the narrow edge of the GaAs substrate and was heated by current pulses of duration ranging from 0.1 to 1 μs at a repetition rate of a few kHz. The pulse power density P_{h} was varied from 25 to 250 W mm^{-2}. The phonons from the generator propagate through the GaAs substrate, reaching a $Cd_{0.99}Mn_{0.01}Te$ QW and heating the Mn spins. The time-resolved kinetics in Fig. 8.8 was measured with a single-channel detector (photomultiplier). A spectrometer was set at a fixed energy, e.g., 1.634 eV (see inset), and the energy shift of the PL line was

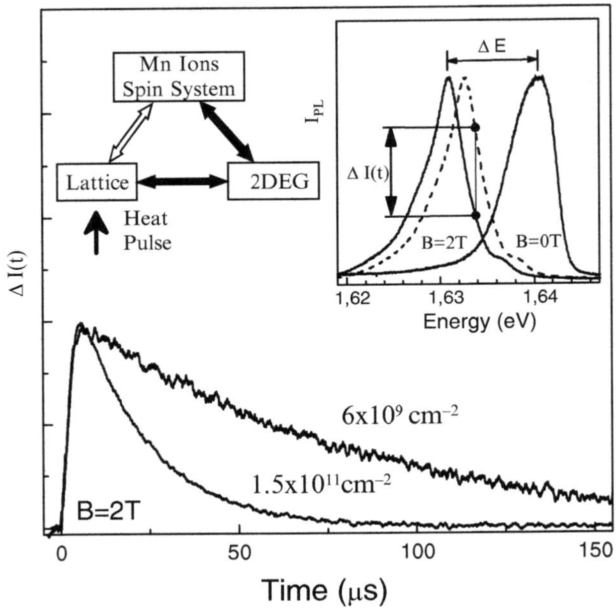

Fig. 8.8 Time evolution of the phonon-induced variation of the PL intensity (related to the variation of the Mn-spin temperature) in two 80-Å-thick $Cd_{0.99}Mn_{0.01}Te/Cd_{0.76}Mg_{0.24}Te$ QW samples with different n_e detected on the high energy side of PL line at 1.634 eV. Spectra are normalized to their peak intensity. The inset shows the stationary exciton PL spectra (*solid lines*) and the spectrum in the presence of nonequilibrium phonons (*dashed line*) for the sample with $n_e = 1.5 \times 10^{11}\,\text{cm}^{-2}$. The experiments were carried out at $T = 1.6$ K and photoexcitation density $P = 70\,\text{mW cm}^{-2}$. Reprinted from [27]

translated into the PL intensity variation. For small T_{Mn} changes, the measured phonon-induced signals $\Delta I(t) = I(t) - I_0$ are proportional to the variation of T_{Mn}. Here, I_0 and $I(t)$ are PL intensities before and after the heat pulse, respectively.

The leading edges of $\Delta I(t)$ signals have a width of about 2 μs, which is the duration of the phonon pulse reaching the QW. The decay represents the cooling of Mn system. In QWs, the phonon bottleneck is absent and, therefore, the decay corresponds to the SLR of the Mn ions. An acceleration of SLR in the presence of 2DEG is clearly seen in Fig. 8.8. The cooling of the Mn system is faster in the sample with higher electron density n_e. The spin relaxation time τ_{SLR} decreases from 83 μs in the nominally undoped sample with $6 \times 10^9\,\text{cm}^{-2}$ electrons to 20 μs in the sample with $n_e = 1.5 \times 10^{11}\,\text{cm}^{-2}$. The responsible mechanism will be discussed in Sect. 8.5.3.

8.4.2.3 Electric Field Heating via Free Carriers

In doped samples, free carriers that are accelerated by applied electric fields can pass part of their kinetic energy into the Mn-spin system. This heating mechanism

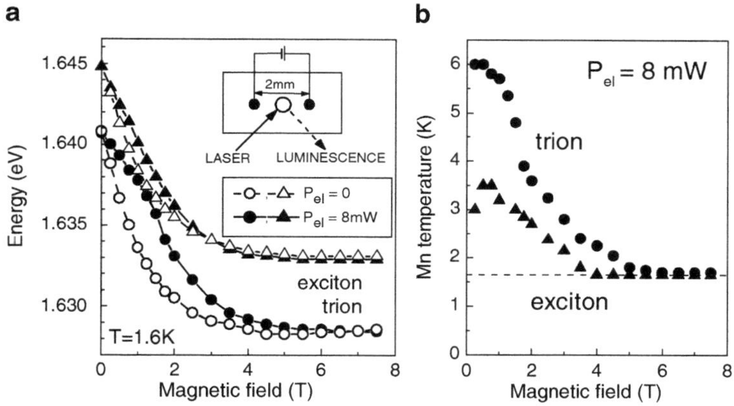

Fig. 8.9 Electric current heating of the Mn-spin system in an 80-Å-thick $Cd_{0.99}Mn_{0.01}Te/Cd_{0.76}Mg_{0.24}Te$ QW. (**a**) Giant Zeeman shift of exciton and negatively charged exciton (trion) photoluminescence lines with (*closed symbols*) and without (*open symbols*) electric current. (**b**) Mn-spin temperature dependence versus magnetic field evaluated for "current on" data from panel (**a**). Reprinted from [45]

of the Mn spins is very efficient due to the strong exchange interaction between the free carriers and localized Mn spins.

An example of the electric current heating is given in Fig. 8.9 for an 80-Å-thick $Cd_{0.99}Mn_{0.01}Te/Cd_{0.76}Mg_{0.24}Te$ QW containing a low density 2DEG with $n_e = 1.2 \times 10^{10}\,cm^{-2}$. At this electron density, both exciton and trion lines are observed in the PL spectrum. An electric field was applied in the plane of the QW through two contacts separated by 2 mm and PL was measured between the contacts, see inset in panel (a). The estimated electrical power was 8 mW. A pronounced shift of the PL lines to higher energy is seen when the magnetic and electric fields are applied, which confirms that the energy is transferred from the electrons to the Mn-spin system. The evaluated increase of T_{Mn} is shown in panel (b). The effect is stronger for trions, which can be explained by spatially inhomogeneous density of the diluted 2DEG with a Fermi energy comparable with localizing potential of the well width fluctuations. The trions are formed in sites containing the 2DEG, where the Mn heating should be stronger, and the excitons are photogenerated in sites free of background electrons, where a weaker heating effect is expected. In the latter case, the heating can be provided by energy diffusion inside the Mn-spin system from hot to cold regions [16].

8.5 Spin-Lattice Relaxation of Mn System

In this part, we discuss the magnetization dynamics of DMS related to the Mn-spin system. The spin-lattice relaxation of Mn ions, its concentration dependence, and possibility of its control in heterostructures will be considered. In the following, we discuss experimental data measured in relatively strong magnetic fields, i.e., under conditions when the magnetization dynamics is controlled by longitudinal relaxation time T_1.

8.5.1 Concentration Dependence of SLR Dynamics

SLR dynamics have been measured by time-resolved techniques described in Sect. 8.4.1.3. The interpretation of experimental results to separate magnetization dynamics from phonon dynamics is detailed below in Sect. 8.6.3. The results are summarized in Fig. 8.10, where the spin-lattice relaxation time is plotted as a function of Mn content. The full circles are data for (Zn,Mn)Se-based structures, while the open symbols show data for (Cd,Mn)Te. The SLR times of Mn ions in (Zn,Mn)Se cover five orders of magnitude from 10^{-3} down to 10^{-8} s as the Mn concentration varies from 0.004 up to 0.11. Such a strong dependence evidences that Mn–Mn interactions must play a key role in the SLR dynamics. Results for both materials closely follow each other allowing us to conclude that SLR of the Mn ions is rather insensitive to the ion host material.

We turn now to the physical mechanisms responsible for the strong dependence of SLR time on the Mn content. It is known that an isolated Mn^{2+} ion in a perfect II–VI semiconductor crystal has no coupling with the lattice [59] and, therefore, should have infinitely long SLR time. The reasons for this expectation are the following: The electric field of the phonons does not act on the magnetic moments, and also the magnetic field induced by temporal variation of the electric field is relativistically small. In principle, gradients of the electric field may interact with a quadrupolar

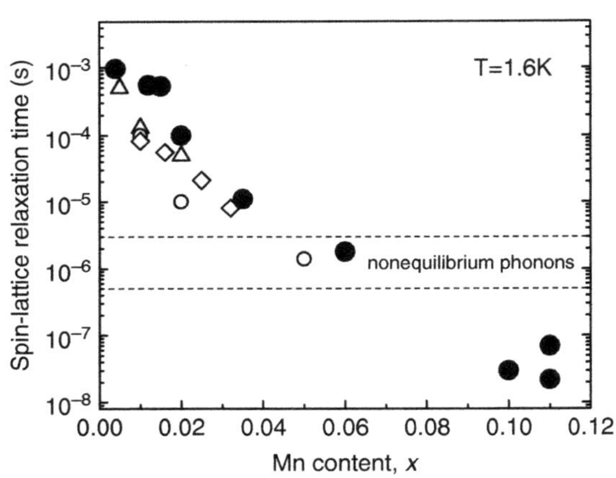

Fig. 8.10 Dependence of the spin-lattice relaxation time on Mn content for nominally undoped $Zn_{1-x}Mn_xSe/Zn_{0.94}Be_{0.06}Se$ structures (*closed circles*). Reprinted from [11]. *Open symbols* represent the data for $Cd_{1-x}Mn_xTe$ bulk samples (*triangles* and *circles*) [51,52] and $Cd_{1-x}Mn_xTe$-based heterostructures (*diamonds*) [58]. The *dashed lines* indicate typical lifetimes of nonequilibrium phonons

moment of the magnetic ion, which in turn interacts with the spin. But for the Mn^{2+} ions the magnetic d-shell is half-filled and its quadrupolar moment is equal to zero. In this case, the dominating mechanism of the spin–phonon interaction is due to phonon modulation of the spin–spin interaction between neighboring Mn ions. It is known as the Waller mechanism [59], which is obviously strongly dependent on the concentration of magnetic ions.

The spin–spin interactions for the Mn ions are provided by three mechanisms: the exchange interaction, the Dzyaloshinskii–Moriya interaction [60, 61], and the magneto-dipole interaction. All of them decrease with the increasing distance between the Mn ions, but the magneto-dipole interaction has the longest range. Therefore, despite being relatively weak, it plays a dominant role for SLR dynamics in the limit of very small Mn contents. The exchange and Dzyaloshinskii–Moriya interactions play a more dominant role at higher Mn contents, when Mn clusters are formed. The number of Mn spins coupled in clusters and the typical cluster size increase progressively with growing Mn concentration [62], and this is the reason for the strong dependence of the SLR time on the Mn concentration [51, 52, 63]. It can be shown in the framework of a simple model that accounts for spin diffusion from Mn ions to Mn clusters, where the spin has efficient relaxation, that the SLR rate (i.e., $1/\tau_{SLR}$) has a strong dependence on Mn content with a power law between x^3 and x^4 [64]. This is in a good qualitative agreement with experimental results of Fig. 8.10. SLR dynamics of the Mn ions is strongly accelerated by an increase of the lattice temperature [52,58] and also becomes faster in stronger external magnetic fields [31, 58].

It is worthwhile to note that in heterostructures with thin DMS layers of a few tens of nanometers the SLR dynamics of Mn ions is free from the phonon bottleneck effect, which might strongly modify (slow down) spin dynamics in bulk DMSs (see [58] and references therein).

One can conclude from Fig. 8.10 that for the fixed external conditions the Mn content x is the main factor determining the SLR dynamics. This establishes a relationship between the static and dynamic magnetization, e.g., between the GZS value and SLR time – both being dependent on the Mn content. This restricting relationship can be made less stringent (or even broken) by the growth of heterostructures with modulated Mn content and/or with modulation doping by donors or acceptors providing free carriers. Below, a few examples of such structures are considered, where the magnetization dynamics have been accelerated greatly without changing the static magnetization.

8.5.2 Mn Profile Engineering

8.5.2.1 Digital Alloys

The concept of "digital alloying" offered by molecular beam epitaxy is a very efficient tool for tailoring static and dynamic magnetic properties of DMSs [15].

Compared to common "disordered alloys" with the same Mn concentration, the spin-lattice relaxation dynamics of magnetic Mn ions has been accelerated by an order of magnitude in (Cd,Mn)Te digital alloys (DA), without any noticeable change in the giant Zeeman splitting of excitonic spin states, i.e., without any effect on the static magnetization. The basic idea is that the static and dynamical properties of DMS are governed by different mechanisms. Paramagnetic Mn spins provide the main contribution to the static magnetization, and their tendency to form antiferromagnetically coupled Mn–Mn clusters, when Mn content increases, is unfavorable in this respect. On the other hand, the SLR dynamics is controlled by anisotropic exchange interactions of Mn ions in such clusters. The digital alloying of DMS results in a considerably different degree of clustering which, in turn, modifies the magnetization dynamics, while keeping the number of paramagnetic spins (which controls the static magnetization) about constant.

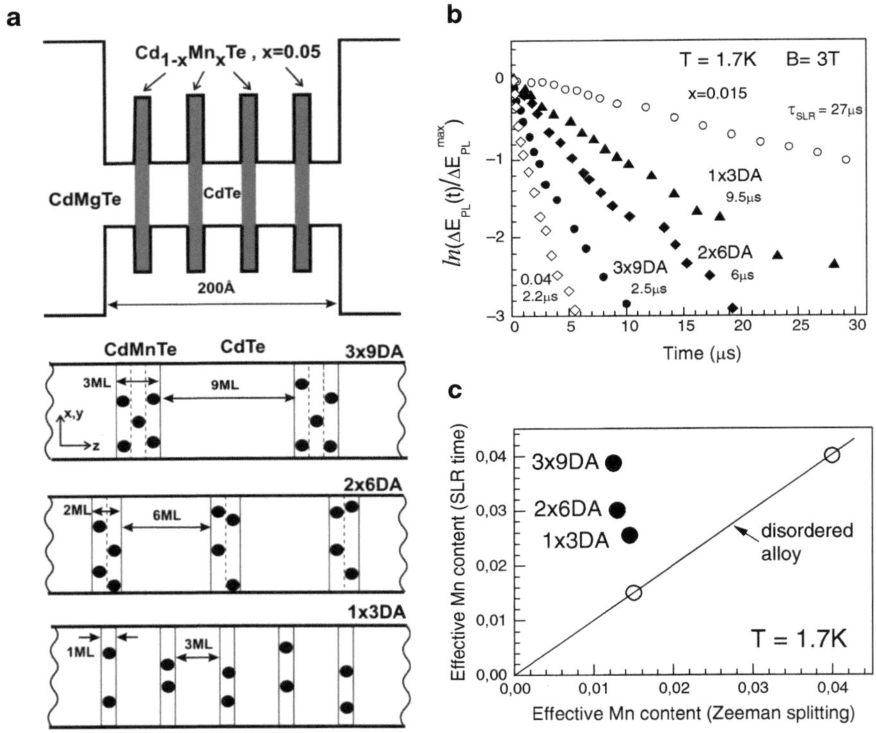

Fig. 8.11 (**a**) Schematic diagram of the conduction and valence band profiles and the Mn ion concentration profile in (Cd,Mn)Te digital alloy structures. All samples have the same averaged Mn content $x_{DA} \approx 0.013$. (**b**) Spin-lattice relaxation in three $Cd_{1-x}Mn_xTe$ digital alloy samples (*closed symbols*). *Open symbols* correspond to $Cd_{1-x}Mn_xTe$ disordered alloys. Sample parameters and characteristic times are given in the panel. (**c**) Diagram linking the static (giant Zeeman splitting) and dynamical (SLR time) magnetic characteristics in form of effective Mn contents for disordered (*open circles* and *solid line*) and digital (*closed circles*) $Cd_{1-x}Mn_xTe$ alloys. Reprinted from [15]

Instead of $Cd_{0.987}Mn_{0.013}Te$ ternary alloys, DMS digital alloys consisting of short-period superlattices $Cd_{0.95}Mn_{0.05}Te/CdTe$ with layers of one to ten monolayers have been grown (see schematics in Fig. 8.11a). Static magnetic properties measured by means of the exciton GZS were very similar for the disordered and for the digital alloys, but the SLR time in the digital alloys has been reduced significantly from 27 μs down to 2.5 μs. Corresponding SLR kinetics is shown in Fig. 8.11b by closed symbols. Open symbols display typical kinetics for the disordered alloys.

An instructive way to present the tunability of the DA parameters is given by Fig. 8.11c. Here, the correspondence diagram for the effective Mn concentrations determined by the static (i.e., the giant Zeeman splitting) and the dynamic (i.e., the SLR time) magnetic properties in DMS is shown. For the disordered alloys, the effective concentrations, extracted from both effects, are equal to the "real" Mn content. As a result, the disordered alloys are described in the diagram by the straight solid identity line. Open points here are experimental data for the disordered alloys. The DA data points do not fall onto this line, confirming that the static and dynamic parameters of magnetization in the DMS digital alloys can be tuned separately.

8.5.2.2 Heteromagnetic Structures

It is often unfavorable to increase the Mn content to values higher than $x \sim 0.01$ in the active layer of spintronic structures because alloy fluctuations and defects induce carrier scattering and open nonradiative recombination channels, deteriorating transport and optical properties. The dilemma is analogous to the optimization of transport properties in semiconductors, where the increase of carrier density obtained by intentional doping is opposed by higher scattering rates of the carriers on impurity centers. Modulation-doped heterostructures, where the layer of dopants is separated from the active layer with free carriers, turned out to be the solution for this problem. Recently, a magnetic counterpart of this concept has been suggested.

A concept of *heteromagnetic semiconductor structures* composed of neighboring layers with different Mn contents has been introduced and verified experimentally [14]. A strong decrease of the SLR time in the whole structure was observed, while keeping x in the active layer at a low level. The concept is based on the diffusion of Mn spins between magnetic layers with different dynamic characteristics. Such diffusion over distances of about 10 nm may occur much faster than the SLR process in layers with low x [16]. Hence in heteromagnetic nanostructures, the excited spin from the layer with low x diffuses quickly to the layer with high x value, where the SLR is fast. In other words, the presence of the layer with high x accelerates the overall relaxation of the Mn spin polarization in the structure. This idea is shown schematically in Fig. 8.12c. The "slow SLR" along the direct path from the material containing 1% Mn to the lattice is bypassed by the faster two-stage path "spin diffusion + fast SLR," which involves the material with 7% of Mn.

The concept has been tested by comparing SLR dynamics in two $Zn_{1-x}Mn_xSe/Be_{1-y}Mn_yTe$ multiple QW structures with a type-II band alignment [14]. In both

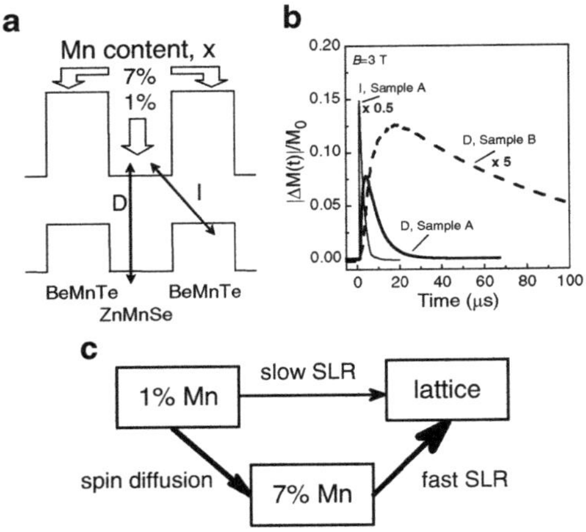

Fig. 8.12 (**a**) Illustration of spatially direct (D) and indirect (I) optical transitions in a $Zn_{0.99}Mn_{0.01}Se/Be_{0.93}Mn_{0.07}Te$ heteromagnetic structure with a type-II band alignment. (**b**) Dynamics of the relative changes in magnetization, $M(t)/M$ (M is equilibrium magnetization at $T = 1.6$ K, $B = 3$ T) measured in a 200 Å/100 Å $Zn_{0.99}Mn_{0.01}Se/Be_{0.93}Mn_{0.07}Te$ (sample A) and a 200 Å/100 Å $Zn_{0.99}Mn_{0.01}Se$/BeTe (sample B) multiple QW structures. In both samples, the electrons are confined in the $Zn_{0.99}Mn_{0.01}Se$ layer. *Thick lines* display the behavior of the direct optical transition corresponding to Mn spin dynamics in $Zn_{0.99}Mn_{0.01}Se$. The *thin line* traces the dynamics of the indirect optical transition, which is dominated by SLR dynamics in $Be_{0.93}Mn_{0.07}Te$. Reprinted from [14]. (**c**) Scheme explaining an acceleration of SLR

samples, electrons have been localized in $Zn_{0.99}Mn_{0.01}Se$ layers and holes have been spatially separated from the electrons in either DMS $Be_{0.93}Mn_{0.07}Te$ layers (sample A) or nonmagnetic BeTe layers (sample B). The type-II band alignment in these heterostructures gives rise to spatially direct (D) and indirect (I) optical transitions (Fig. 8.12a), which are spectrally separated [65]. Monitoring D and I photoluminescence lines resulting from these transitions allows one to measure the magnetization dynamics in each particular layer. The D luminescence probes the magnetization in the $Zn_{0.99}Mn_{0.01}Se$, while the I luminescence probes the $Be_{0.93}Mn_{0.07}Te$ layers in the sample A [65, 66].

The measured spin dynamics after a phonon-induced heating of the Mn-spin system are shown in Fig. 8.12b. The magnetization decay in sample B ($\tau_{SLR} = 90\,\mu s$) is close to the relaxation time in bulk $Zn_{0.99}Mn_{0.01}Se$. The presence of the magnetic $Be_{0.93}Mn_{0.07}Te$ layers in sample A leads to a drastic decrease of the SLR time by more than one order of magnitude down to $\tau_{SLR} = 6\,\mu s$. The thin solid line traces the spin dynamics in $Be_{0.93}Mn_{0.07}Te$ layers, which is limited by the dynamics of nonequilibrium phonons to about 1.5 μs. Experimental data clearly demonstrate that acceleration of the spin-lattice relaxation does indeed occur in heteromagnetic semiconductor structures.

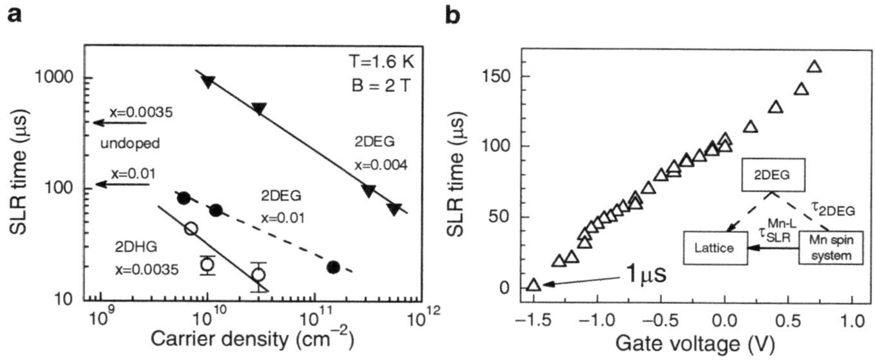

Fig. 8.13 (**a**) SLR time as a function of carrier density in $Cd_{0.9965}Mn_{0.0035}Te/Cd_{0.66}Mg_{0.27}Zn_{0.07}Te$ QWs with a 2DHG (*open symbols*). Reprinted from [12]. Data for $Cd_{0.99}Mn_{0.01}Te/Cd_{0.76}Mg_{0.24}Te$ QWs [27] and $Zn_{0.996}Mn_{0.004}Se/Zn_{0.94}Be_{0.06}Se$ QWs [10] with a 2DEG are shown by *closed symbols* for comparison. SLR times for undoped samples are shown by *arrows*. *Lines* are guides for the eye. (**b**) Gate voltage tuned n-type modulation-doped $Zn_{0.985}Mn_{0.015}Se/Zn_{0.94}Be_{0.06}Se$ QW: SLR time dependence on the gate voltage, with $B = 3\,T$ and $T = 1.7\,K$. Reprinted from [13]. Paths for spin-lattice relaxation are shown in the insert

8.5.3 Acceleration by Free Carriers

It is known for metals with magnetic impurities that the free carriers play an important role for SLR and for energy transfer away from magnetic ions [67]. The Korringa effect and the Knight shift are examples of the effects caused by such interaction. Similar effects have been reported for bulk narrow gap DMSs with a high concentration of free carriers [68].

Also in wide band gap DMSs, the SLR dynamics can be modified significantly by the presence of free carriers. The carriers are strongly coupled with both the magnetic ions and the phonons. Therefore, they may serve as a bypass channel for the slow direct spin-lattice relaxation (see scheme in Fig. 8.13b) with an efficiency controlled by the carrier concentration. Indeed, an acceleration of SLR has been found experimentally in (Cd,Mn)Te and (Zn,Mn)Se-based QWs modulation doped with either electrons or holes [10, 12, 27].

In QWs, the free carriers (electrons or holes) can be implemented by modulation doping of barrier layers with donors or acceptors. Their concentrations can be tuned additionally by above-barrier illumination and/or by the application of a gate voltage along the structure growth axis. Both methods have been used to study the effect of free carriers on SLR dynamics.

Results for the *n*-type and *p*-type doped samples are summarized in Fig. 8.13a. Here, the carrier density has been controlled by the doping level during the growth. In (Zn,Mn)Se-based QWs with $x = 0.004$ (triangles) the electron density varies from $10^{10}\,cm^{-2}$ up to $5.5 \times 10^{11}\,cm^{-2}$, which causes a shortening of the SLR times from 960 μs down to 70 μs. Therefore, in the highly doped sample the SLR dynamics is controlled by the free carriers. The same conclusion can be drawn for *n*-type as well as p-type doped (Cd,Mn)Te-based QWs. The effect is stronger in the QWs with free holes, which can be explained by the stronger $p-d$ interaction for holes with Mn ions.

Application of a gate voltage to the $Zn_{0.985}Mn_{0.015}Se/Zn_{0.94}Be_{0.06}Se$ modulation-doped QW allows one to tune the electron density in the range from $5 \times 10^{10}\,cm^{-2}$ to $3.1 \times 10^{11}\,cm^{-2}$ (gate voltages from 0.7 to -1.5 V, correspondingly) [13]. One can see in Fig. 8.13b that the tunability of the SLR time in this sample extends over two orders of magnitude from 1 μs up to 160 μs. Note that the static magnetization is not influenced by the electron density. This result shows that the magnetization dynamics in DMS can be controlled separately from the static magnetization by means of an electric field.

8.5.4 *Regime of Degenerate 2DEG*

At low temperatures and sufficiently large electron densities, when the thermal energy becomes smaller than the Fermi energy ($k_B T \ll \varepsilon_F$), a 2DEG becomes degenerate. The properties of such an electron system are described by Fermi statistics. Two circumstances should be taken into account when the interaction of a Mn-spin system with a degenerate 2DEG is analyzed. Both are related to the fact that in an external magnetic field, the flip-flop process of electron–Mn exchange should conserve not only spin but also the energy of the whole system. The Zeeman splitting between spin sublevels of Mn^{2+} ions is given by $\Delta E_{Mn}[meV] = \mu_B g_{Mn} B = 0.116 B[T]$, i.e., it is about 0.1 meV in a magnetic field of 1 T.

First, only a part of the electrons can interact with Mn. These are electrons in the vicinity of the Fermi level being separated from it by either $k_B T$ or ΔE_{Mn}. For other electrons, "deeper" in the 2DEG, the final states of the flip-flop process are already occupied. As a result, the SLR rate of the Mn system via free carriers should saturate with increasing carrier density for $\varepsilon_F > k_B T, \Delta E_{Mn}$. Also for the fixed carrier density the SLR rate should increase with the magnetic field. It is due to the increasing number of "active" electrons near the Fermi level while the Zeeman splitting of the Mn ions increases.

Second, the 2DEG becomes fully spin polarized when the GZS of the conduction band ($E_{Z,el}$) exceeds $2\varepsilon_F$. The spin-flip scattering for the electron is not possible inside the same spin subband and the subband for the opposite spin orientation is split too far away in energy (Fig. 8.14a). This should cause a strong reduction of the electron–Mn interaction efficiency with growing magnetic fields.

In QWs with a 2DEG, the SLR dynamics are governed by two channels of energy flow from the Mn-spin system, both of which depend on the magnetic field strength:

$$\frac{1}{\tau_{SLR}} = \frac{1}{\tau_{SLR}^0} + \frac{1}{\tilde{\tau}_{e-Mn}}. \tag{8.16}$$

Here, τ_{SLR}^0 is the SLR time caused by the direct coupling of the Mn ions with the lattice. Its magnetic field dependence measured for the nominally undoped sample

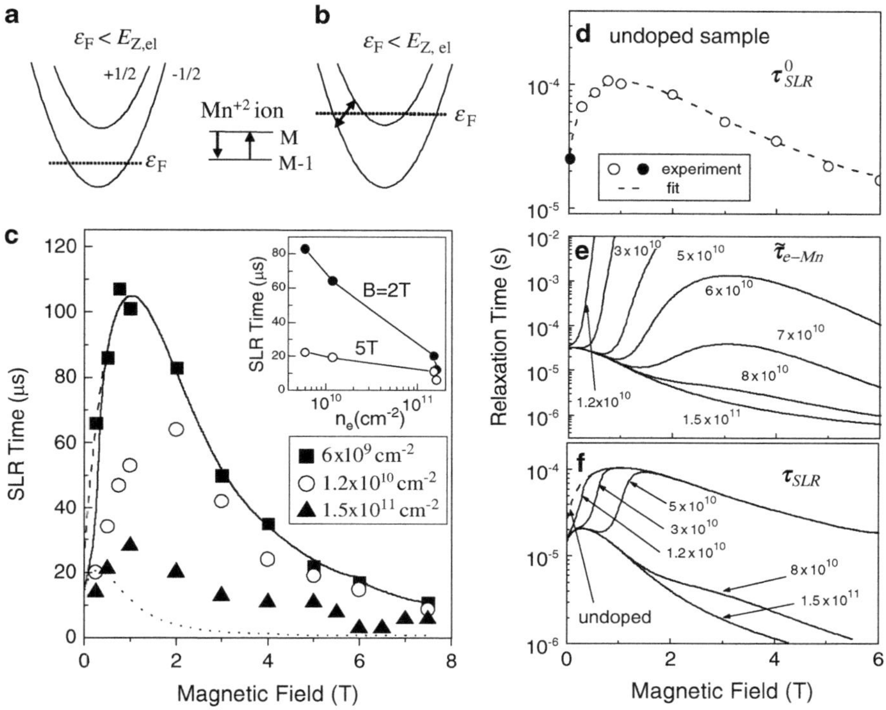

Fig. 8.14 Energy diagram illustrating spin-flip transitions in the presence of a magnetic field: (**a**) spin-flip transitions are not allowed at low n_e because of energy conservation; (**b**) the number of possible transitions increases with B when n_e is sufficiently high. (**c**) Dependence of SLR times on the magnetic field measured in an 80-Å-thick $Cd_{0.99}Mn_{0.01}Te/Cd_{0.76}Mg_{0.24}Te$ QWs with 2DEG of different densities (symbols). *Solid* and *dotted lines* are results of modeling for $n_e = 1.2 \times 10^{10}\,cm^{-2}$ and $1.5 \times 10^{11}\,cm^{-2}$, respectively. $T = 1.6\,K$. The $\tau^0_{SLR}(B)$ dependence is shown by a *dashed line*. Inset: SLR time as a function of electron density. Here, *lines* are guides for the eye. (**d**) Measured dependence in an undoped $Cd_{0.99}Mn_{0.01}Te$-based QW. (**e**) Modelling of $\tilde{\tau}_{e-Mn}(B)$ for different electron densities. (**f**) Calculation of $\tau_{SLR}(B)$. Reprinted from [27]

is given in Fig. 8.14d. It shows a nonmonotonic behavior with a maximum at about 1 T and it steadily decreases for higher magnetic fields [27, 31]. The characteristic time $\tilde{\tau}_{e-Mn}$ gives the efficiency of the Mn ion interaction with the 2DEG. It has been calculated for $Cd_{0.99}Mn_{0.01}Te$-based QWs by means of the formalism developed in [25], see panel (e) in Fig. 8.14. The resultant spin dynamics accounting for both channels, combines data from panels (e) and (d) via (8.16). It is plotted in panel (f). One can see that for $n_e < 7 \times 10^{10}\,cm^{-2}$, τ_{SLR} at low fields ($B < 1$–$2\,T$) is dominated by $\tilde{\tau}_{e-Mn}(B)$, but at high fields it is determined by the $\tau^0_{SLR}(B)$ dependence. The situation is different for $n_e \geq 7 \times 10^{10}\,cm^{-2}$, where $\tilde{\tau}_{e-Mn}$ contributes to τ_{SLR} most in the whole range of magnetic fields.

The modeling results are in fair agreement with experimental data on spin dynamics measured for $Cd_{0.99}Mn_{0.01}Te/Cd_{0.76}Mg_{0.24}Te$ QWs with different electron

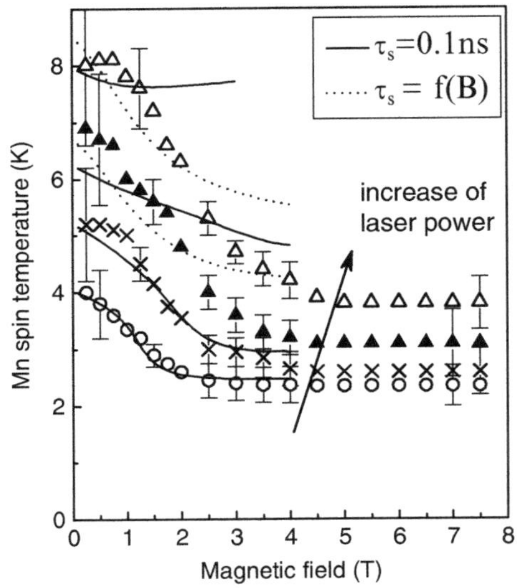

Fig. 8.15 Magnetic field dependence of T_{Mn} measured for different excitation densities of *cw* laser varied from 0.5 to 4 W cm^{-2} (*symbols*). Calculations for various electron spin relaxation times τ_s are shown by *lines*. 80-Å-thick $Cd_{0.99}Mn_{0.01}Te$/$Cd_{0.76}Mg_{0.24}Te$ QW with 2DEG densities varied from 1.5×10^{10} to 3.2×10^{10} cm^{-2} with increasing laser power. $T = 1.6$ K. Reprinted from [25]

densities (panel (c)). Here, the data for $n_e = 6 \times 10^9$ cm^{-2} closely follows the $\tau^0_{SLR}(B)$ dependence represented by the dashed line, which above 1 T coincides with the solid line. Although the general behavior for $n_e = 1.2 \times 10^{10}$ cm^{-2} is similar at low magnetic fields ($B < 1$ T), τ_{SLR} is much shorter than that for $n_e = 6 \times 10^9$ cm^{-2}. Then, τ_{SLR} continues to increase with B. For fields $B > 2$ T, the values of τ_{SLR} do not differ significantly from $\tau^0_{SLR}(B)$. A further increase of the electron density up to 1.5×10^{11} cm^{-2} leads to a decrease of τ_{SLR} in the whole range of magnetic fields. SLR time dependencies on the electron density are plotted in the inset. One can see that the variation of τ_{SLR} at $B = 2$ T is much stronger than that at the higher field of 5 T.

The presence of free carriers modifies not only the cooling dynamics of the Mn-spin system, but also its heating efficiency [25]. Results for the carrier heating of the Mn system under *cw* photoexcitation are depicted in Fig. 8.15. The Mn-spin temperature plotted for different excitation densities shows a pronounced decrease in magnetic fields exceeding 1 T. Two factors are responsible: (1) shortening of τ_{SLR}, which is controlled in this sample by $\tau^0_{SLR}(B)$ for $B > 2$ T, and (2) a decrease of the energy transfer efficiency from the 2DEG to the Mn ions when ε_F becomes smaller than the electron GZS.

To summarize the results of Sects. 8.5.3 and 8.5.4, the presence of free carriers significantly modifies the dynamic magnetic properties of DMS heterostructures. This offers a new method of controlling the spin dynamics, such as spin-lattice and spin-spin relaxation rates, by tuning the free carrier density, which is important for spintronic applications.

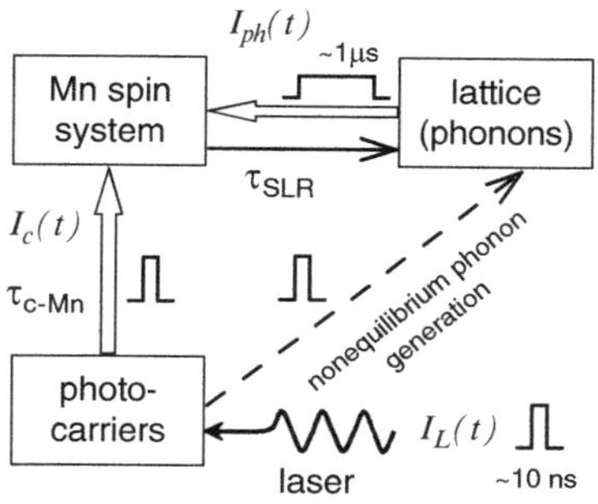

Fig. 8.16 Interacting subsystems in undoped DMS structure under pulsed laser excitation with duration of 10 ns. Channels for energy transfer in the process of heating (*open arrows*) and cooling (*solid arrows*) of the Mn-spin system are shown. The *dashed arrow* shows phonon generation due to energy relaxation of photocarriers. Laser light heats the Mn-spin system by carrier $I_c(t)$ and phonon $I_{ph}(t)$ impacts. The carrier impact is limited by the laser pulse duration, and the phonon impact is given by typical lifetimes of nonequilibrium phonons of about 1 μs. The Mn-spin system relaxes toward equilibrium (given by the lattice temperature) with the spin-lattice relaxation time

8.6 Spin and Energy Transfer from Carriers to Mn System

In this section, the spin and energy transfer from photogenerated carriers to the Mn-spin system will be discussed. For simplicity, model considerations and experimental illustrations are limited to undoped structures, which do not contain free carriers in the absence of laser illumination. Figure 8.16 is a scheme of Fig. 8.1 reduced to the specific case of pulsed laser excitation of undoped structures. Here, the carrier system consists of photocarriers only.

Photocarriers with excess kinetic energy are generated by laser light. The Mn-spin system receives energy from the photocarriers, i.e., is heated with respect to the bath lattice temperature, via two channels shown by open arrows. The direct path is provided by fast exchange scattering between the carriers and the Mn ions. The efficiency of this channel is characterized by a time τ_{c-Mn}. The indirect path involves at the first stage generation of nonequilibrium phonons due to carrier energy relaxation (dashed arrow). At the second stage, the phonons heat the Mn-spin system. This channel is characterized by the SLR time, τ_{SLR}. These two paths of energy transfer can be distinguished with high accuracy by means of a time-resolved spectroscopy with resolution considerably shorter than the phonon lifetimes (which are typically about 1 μs) [6, 11]. The direct heating of the Mn system by carriers takes place only during the carrier lifetime (or during the laser pulse if it is longer than the carrier lifetime), but the indirect phonon heating lasts considerably longer. Time-integrated (*cw*) spectroscopy may also offer the possibility to identify the dominant path, when, e.g., the heating efficiency is measured as function of Mn content (Fig. 8.5). A solid line arrow in the scheme corresponds to the cooling of the Mn-spin system to the bath temperature, which in this case is provided by direct SLR dynamics only.

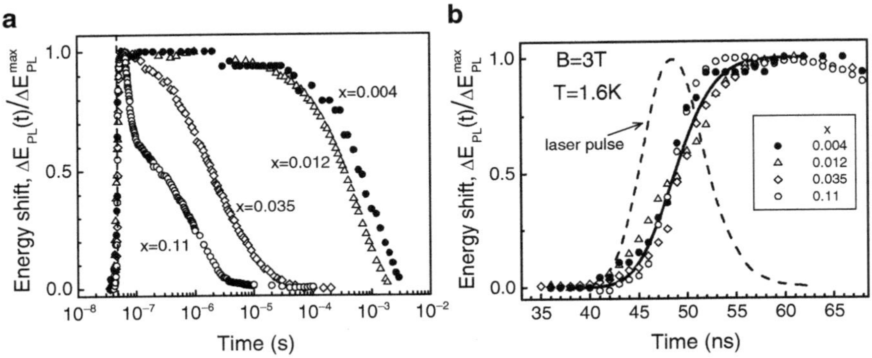

Fig. 8.17 (**a**) Normalized energy shifts of PL lines induced by 7 ns laser pulses in $Zn_{1-x}Mn_xSe$/$Zn_{1-y}Be_ySe$ QWs with different Mn content. The maximum shifts ΔE_{PL}^{max} are 5, 12, 23 and 26 meV for samples with $x = 0.004$, 0.012, 0.035 and 0.11, respectively. Time position of the laser pulse is shown by the *vertical dashed line*. For convenient comparison of the different samples, the data are plotted on a logarithmic time scale. $B = 3$ T. (**b**) Closeup of the data from panel (**a**) to highlight the initial energy shift (*symbols*) in comparison with the laser pulse integral (*solid line*). The *dashed line* shows the excitation laser pulse profile. Reprinted from [6]

8.6.1 Direct Spin and Energy Transfer

An efficient direct transfer from the carriers to the Mn system has been reported for (Zn,Mn)Se-based QWs [6], see also Chap. 9 and [8, 32]. Results of time-resolved experiments for $Zn_{1-x}Mn_xSe$/$Zn_{1-y}Be_ySe$ QWs with Mn concentrations varied from 0.004 up to 0.11 are collected in Fig. 8.17. In panel (a), the magnetization dynamics visualized via the energy shift of the emission line is shown over the full time range. To make the comparison of different samples convenient, the data are normalized to the maximum shift in each structure $\Delta E_{\rm PL}^{\rm max}$. A logarithmic scale for the time delay was chosen to highlight the huge dynamic range of SLR times from 20 ns up to 1 ms covered by the energy shift, which reflects the cooling of the Mn-spin system [11]. It is, however, remarkable that the heating of the Mn-spin system (the rise of the signal) is very fast and identical for all samples.

The rising parts are given in more detail in panel (b). A dashed line there traces the temporal profile of 7 ns laser pulse. Typical exciton lifetimes in these structures do not exceed 200 ps, therefore the carriers are present only during the laser action. The solid curve shows the integral of the laser pulse, which corresponds to the expected carrier impact to the direct Mn heating. Experimental data for all concentrations are grouped closely around this curve. This allows one to conclude that in (Zn,Mn)Se-based QWs the direct energy transfer dominates over the indirect one. This is in good agreement with the results obtained earlier for *cw* laser excitation [28]. Note that considerably longer heating times are expected for indirect heating involving nonequilibrium phonons.

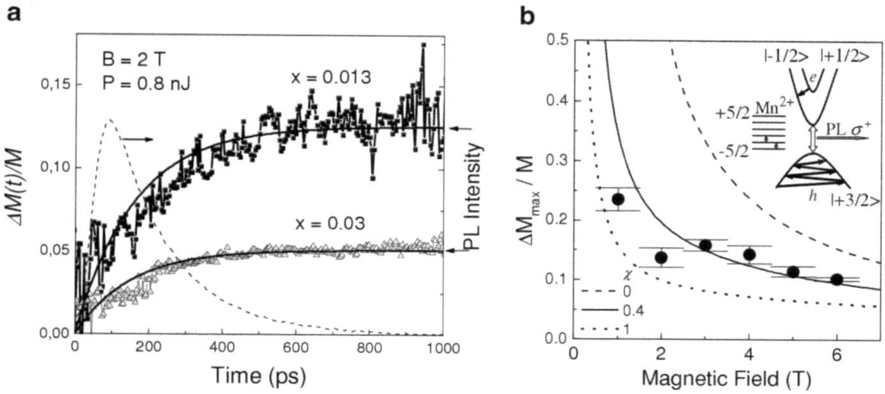

Fig. 8.18 (**a**) Time evolution of magnetization changes in two $Zn_{1-x}Mn_xSe/Zn_{0.94}Be_{0.06}Se$ QWs. *Solid lines* are single exponential fits to the experimental data with $\tau_M = \tau_{PL} = 160$ ps ($x = 0.013$) and 150 ps ($x = 0.03$) [5]. *Dashed line* shows photoluminescence kinetics for the $x = 0.013$ sample. $T = 1.6$ K. (**b**) Magnetic field dependence of $\Delta M_{\max}/M$, measured in the sample with $x = 0.013$ for $P = 0.7$ nJ. Dependencies of $\Delta M_{\max}/M$ on B for different fitting parameters χ are given by lines (see [5] for details). The inset is a scheme of transitions involved in multiple spin transfer from carriers to Mn^{2+} ions. There are only flip-flop transitions for electrons, while for holes the multiple transfer of spin may be possible during the relaxation inside the same Zeeman subband $|+3/2\rangle$. Reprinted from [5]

8.6.2 *Multiple Transfer of Angular Momentum Quanta from Holes*

The interaction of free carriers with magnetic ions in DMS nanostructures provides the basis for ultrafast manipulations of the magnetization on the timescale of picoseconds [7, 9]. We show in this part that the magnetization change can not only be very fast, but also very efficient due to the fact that one hole may flip about hundred Mn spins.

Picosecond kinetics of the magnetic-field-induced magnetization in $Zn_{1-x}Mn_xSe/Zn_{0.94}Be_{0.06}Se$ quantum wells (QWs) has been examined [5]. Samples were excited by 160 fs pulses with energy of 0.8 nJ at a repetition rate of 4 kHz and detected with a streak camera with a time resolution of 20 ps. The results were not sensitive to the polarization of the excitation light.

The transient curves for magnetization dynamics $\Delta M(t)/M$ for two samples with $x = 0.013$ and 0.03 are shown in Fig. 8.18a for $B = 2$ T. Here, M is the magnetization, which is proportional to the band gap GZS shift (see (8.14)):

$$\Delta E_Z = \frac{1}{2}\frac{\alpha - \beta}{\mu_B g_{Mn}} M(B). \tag{8.17}$$

The magnetization decreases under laser excitation and its photoinduced variation $\Delta M(t)$ is proportional to changes of the total GZS value. The value $\Delta M(t)/M$

increases and saturates in a time of several hundreds of picoseconds at a value $\Delta M_{\rm max}/M$, shown by horizontal arrows. The dashed curve shows the photoluminescence decay with a time $\tau_{\rm PL} = 160$ ps, which is identical to the time constant of the magnetization decrease $\tau_{\rm M} = 130$–180 ps. Therefore, the magnetization dynamics reflects the heating of the Mn-spin system when energy is transferred from the hot photogenerated carriers. The saturation value $\Delta M_{\rm max}/M$ depends linearly on the excitation power. The maximum obtained value for $\Delta M_{\rm max}/M$ was as high as 50% in a sample with $x = 0.005$ [5].

Let us discuss what transfer mechanism can provide the huge reduction of magnetization $\Delta M_{\rm max}$ within the short time $\tau_{\rm M}$. The analysis is based on the theoretical approach developed in [25], where the energy and spin transfer are due to spin flip-flop processes. A single flip-flop scattering event between a carrier and an ion is faster than a few picoseconds [8, 69], but the energy transferred to the Mn ion is rather small, e.g., $\mu_{\rm B} g_{\rm Mn} B = 0.23$ meV at $B = 2$ T. Transfer of a significantly larger amount of energy requires multiple scattering, which in principle is possible within the typical carrier lifetime of 100 ps. However, after the first flip-flop, the carrier has an unfavorable spin orientation and can no longer transfer energy to Mn ions until its spin relaxes back by any mechanism except exchange scattering with Mn. The latter would reverse the effect of the original flip-flop process and, therefore, cannot be considered as contribution to multiple spin transfer, as it would transfer spin and energy back from the Mn ion to the carrier.

The maximum relative change of the magnetization can be written as:

$$\frac{\Delta M_{\rm max}}{M} = \frac{\lambda_{\rm e-Mn} n_{\rm e} + \lambda_{\rm h-Mn} n_{\rm h}}{x d N_0 S_{\rm eff} {\rm B}_{5/2}[\frac{5\mu_{\rm B} g_{\rm Mn} B}{2k_{\rm B}(T+T_0)}]}, \tag{8.18}$$

where $n_{\rm e,h}$ are the electron (e) and hole (h) densities generated by a laser pulse in a QW of width $d = 100$ Å; $\lambda_{\rm e-Mn}(\lambda_{\rm h-Mn})$ is the number of irreversible exchange scattering events with Mn for a photoexcited electron (hole). Equation (8.18) allows one to estimate whether multiple spin transfer ($\lambda_{\rm e-Mn} + \lambda_{\rm h-Mn} \gg 1$) is required to explain the experimental results.

Let us first analyze the electron contribution. As we mentioned above, for multiple spin transfer the electron should relax its spin by any mechanism except exchange scattering with Mn. The electron spin relaxation time in nonmagnetic QWs varies between 100 ps and a few ns [70] and is about equal to or longer than the carrier lifetime $\tau_{\rm PL} \sim 150$ ps. Therefore $\lambda_{\rm e-Mn} \sim 1$, so that only a single transfer event on average can take place within $\tau_{\rm M} \sim \tau_{\rm PL}$. The estimate for the carrier density for $P = 0.8$ nJ excitation energy gives $n_{\rm e,h} = (3-7) \times 10^{11}$ cm^{-2}. This value is much less than the sheet density of Mn ions in a QW (1.3×10^{14} cm^{-2} for $x = 0.013$). From (8.18) with $x = 0.013$, $B = 2$ T, $\lambda_{\rm e-Mn} = 1$ and $n_{\rm e} = 5 \times 10^{11}$ cm^{-2} one can expect a magnetization change of $\Delta M_{\rm max}/M \sim 0.003$, which is almost two orders of magnitude smaller than the experimentally measured value of 0.12. Therefore, the electrons alone cannot explain the large change of the magnetization.

Contrary to electrons, the spin relaxation time of holes can be as fast as 0.1–1 ps even in nonmagnetic semiconductors. Due to the strong spin-orbit coupling, any momentum scattering of a hole likely also changes its spin state [22]. Therefore, holes may undergo multiple spin-flip transitions. But they need to be scattered to a subband with a different spin state. This becomes a problem in external magnetic fields as holes are subject to the GZS of the valence band. For example, in the sample with $x = 0.013$ at $B = 2\,\mathrm{T}$, the energy difference between the lowest spin state $|+3/2\rangle$ and next spin state $|+1/2\rangle$ is equal to 22 meV. Therefore, only very hot holes during the first 80 ps of energy relaxation in the QW may undergo spin-flip between different subbands. It is seen, however, from Fig. 8.18a that about 70% of the demagnetization occurs for $t \geq 80$ ps.

The mechanism suggested to explain these phenomena can provide spin and energy transfer from spin-polarized holes to Mn ions, while the initial and the final states of the holes belong to the same spin subband $|+3/2\rangle$ (see scheme in Fig. 8.18b). Thus, although at first sight very surprising, scattering of a hole at the contact potential of a Mn ion is possible due to the $(\mathbf{k}\hat{\mathbf{J}})^2$ term with the 3D hole spin operator $\hat{\mathbf{J}}(j = 3/2)$ in the Luttinger Hamiltonian [71]. Due to this coupling, at finite k the heavy-hole state $|+3/2\rangle$ mixes with the light-hole state $|+1/2\rangle$, resulting in nonzero matrix elements between even and odd subbands of hole quantization in quantum wells (see (29, 30) in Chap. 3). In lowest order perturbation theory, the scattering rate for a heavy hole $|+3/2\rangle$ at a Mn ion is proportional to the ratio of the hole kinetic energy to the splitting E_1 between the first heavy-hole and the second light-hole subbands. For a nondegenerate hole gas, this ratio is proportional to $k_\mathrm{B}T_\mathrm{h}/E_1$. Thus, multiple spin and energy transfer is not restricted to flip-flop transitions alone, but may continue during the entire relaxation cascade of holes in the same spin subband. The evaluations, which are detailed in [5] indicate that one hole can flip up to a hundred Mn spins.

The magnetic field dependence of $\Delta M_\mathrm{max}/M$ presented in Fig. 8.18b can be well described in the framework of the suggested model. The results of calculations shown by lines were achieved for different values of the parameter $\chi = \tau_\mathrm{h-ph}/\tau_\mathrm{h-Mn}$, which is the ratio of hole scattering times with acoustic phonons and Mn spins.

8.6.3 Double Impact of Laser Pulses for Mn Heating

Laser light that is absorbed in a DMS structure generates photocarriers (electrons and holes) with excess kinetic energy. This energy can be transferred to the Mn-spin system by two ways shown schematically in Fig. 8.16. The direct way is provided by exchange scattering of the free carriers on magnetic ions. It might also contain a contribution from excitation of the internal Mn^{2+} ion transition $^6A_1 \rightarrow {}^4T_1$ via energy transfer from excitonic states [56,72,73]. The *indirect* way involves nonequilibrium phonons generated by the free carriers during their energy relaxation. Therefore, the laser pulses have a *double impact* on the heating of the Mn-spin system. These two impacts may differ in duration, time profile and heating efficiency. The relative contributions of these impacts depend on the DMS material, the structure parameters and the excitation conditions. On the one hand it has been shown that in (Zn,Mn)Se QWs [6,28], in n-type doped (Cd,Mn)Te QWs [25] and in undoped (Cd,Mn)Te QWs under high excitation density [48, 49] the direct transfer is dominant. On the other hand, the indirect transfer has been suggested as the leading mechanism in bulk (Cd,Mn)Te [51], in (Cd,Mn)Te QWs [53] and in (Cd,Mn)Se quantum dots [54].

Time-resolved spectroscopy with time resolution considerably shorter than the phonon lifetimes allows one to distinguish the two ways of energy transfer in the time domain. The direct heating of the Mn system by carriers takes place only during the laser pulse action, because the laser pulse in the experiments discussed below is longer than the carrier lifetime. However, the indirect phonon heating lasts considerably longer. Two experimental examples given below in Sect. 8.6.4 for (Zn,Mn)Se and (Cd,Mn)Te QWs have been measured under 7 ns laser pulses, whose temporal profile $I_L(t)$ is shown by the dashed line in Fig. 8.17b. A gated charge-coupled-device detector with a time resolution of 2 ns has been used to trace optically the magnetization dynamics. The carrier impact $I_c(t)$ with a duration Δt_c almost coincides with $I_L(t)$, but the phonon impact $I_{ph}(t)$ with a duration Δt_{ph} differs from it. Nonequilibrium phonons are generated by the photocarriers and, therefore, the leading edge of $I_{ph}(t)$ does not exceed ~10 ns (i.e., integral of laser pulse), but the trailing edge is determined by the lifetime of acoustic phonons in crystals at low temperatures, which is on the order of 1 μs [10, 20]. As a result, the Mn system is exposed to a short carrier impact and a long phonon impact, as is shown schematically in Fig. 8.16. The relative efficiency of these impacts for Mn heating can be characterized by the maximum temperature of the Mn system that results from them: Θ_c for the carrier impact and Θ_{ph} for the phonon impact.

The dynamical response of the Mn temperature $T_{Mn}(t)$ to an impact will differ for $I_c(t)$ and $I_{ph}(t)$, as it is determined by the difference in characteristic times during which energy can be transferred from the carriers to the Mn ions (τ_{c-Mn}), from the phonons to the Mn ions (τ_{SLR}) and from the Mn ions back to the lattice (τ_{SLR}). Therefore, the response allows one to measure these times experimentally. However, the different impact contributions need to be extracted from the magnetization relaxation, which is not always trivial.

Several scenarios can be realized. When τ_{SLR} exceeds the duration of the impact pulses, Δt_c for the carrier impact and Δt_{ph} for the phonon impact, the SLR time can be measured from the decay of $T_{Mn}(t)$. This regime has been realized experimentally enabling the measurement of τ_{SLR} in (Cd,Mn)Te and (Zn,Mn)Se QWs with $x < 0.035$ (see Figs. 8.8 and 8.17a).

The situation becomes more complicated when the SLR dynamics is faster than the phonon impact. In Fig. 8.19, we analyze the case $\Delta t_c < \tau_{SLR} < \Delta t_{ph}$ for different relative contributions of carriers and phonons. The impact profiles are shown by solid lines and the expected $T_{Mn}(t)$ are given by dashed lines. Cases (a) and (b) are for single impact conditions, when one of the contributions strongly dominates the other. For carrier impact only (case (a)), $\Delta t_c < \tau_{SLR}$ and τ_{SLR} determines the decrease of $T_{Mn}(t)$ toward the lattice temperature. In case (b), the SLR time can be

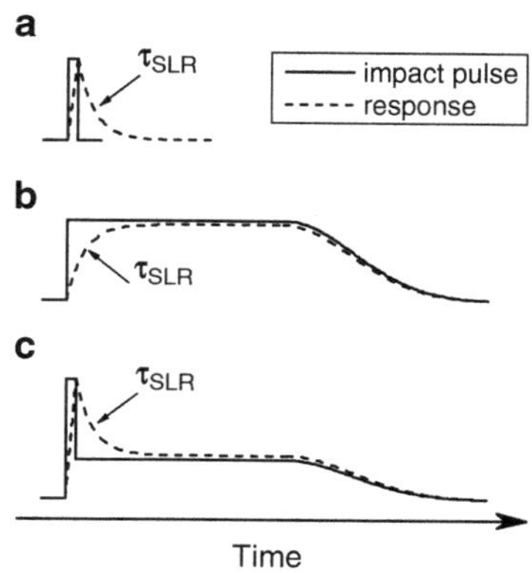

Fig. 8.19 Schematic presentation of the dynamical response of the Mn system to pulses under various experimental conditions: (**a**) carrier impact only, with a duration shorter than τ_{SLR}; (**b**) phonon impact only, with duration longer than τ_{SLR}; (**c**) double impact of carriers and phonons with $\Delta t_c < \tau_{SLR} < \Delta t_{ph}$ and $\Theta_c > \Theta_{ph}$. Reprinted from [11]

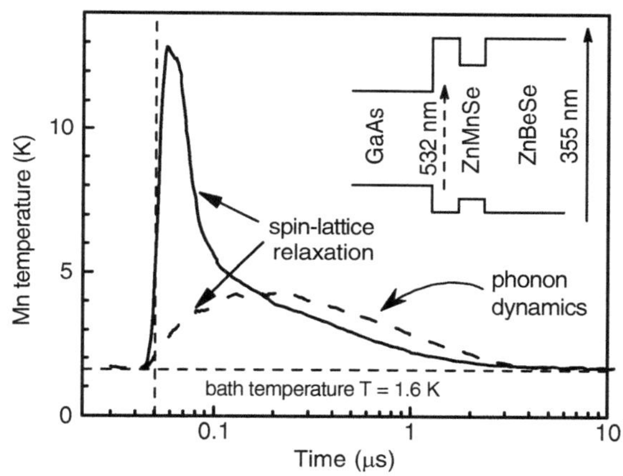

Fig. 8.20 Dynamics of the Mn temperature T_{Mn} for 355 nm ($P = 9\,\mathrm{kW\,cm^{-2}}$, double impact by carriers and phonons, *solid line*) and 532 nm ($P = 15\,\mathrm{kW\,cm^{-2}}$, single impact by phonons only, *dashed line*) laser excitations in 100-Å-thick $Zn_{0.89}Mn_{0.11}Se$/ $Zn_{0.89}Be_{0.11}Se$ QWs. The *vertical dashed line* indicates the maximum of the laser pulse at 50 ns. $B = 3$ T. Reprinted from [11]

measured from the rise of $T_{Mn}(t)$. This is possible due to the sharp rise of the phonon impact $I_{ph}(t)$, which is $\sim \Delta t_c$. The decrease of $T_{Mn}(t)$ follows the slow decay of the phonon impact. The double impact case (c) is realized for the condition $\Theta_c > \Theta_{ph}$. In this case, the decrease of $T_{Mn}(t)$ has fast and slow components, corresponding to the SLR and the phonon impact, respectively.

8.6.4 Competition of Direct and Indirect Transfer

The regime shown schematically in Fig. 8.19c is realized experimentally for $Zn_{0.89}Mn_{0.11}Se/Zn_{0.94}Be_{0.06}Se$ QWs grown on GaAs substrates. Relatively high Mn concentration is chosen to meet the condition $\Delta t_c < \tau_{SLR} < \Delta t_{ph}$. The Mn temperature obtained from the spectroscopic data at $B = 3$ T is shown in Fig. 8.20. Two different laser excitation wavelengths are used in these studies to distinguish carrier and phonon impacts, see scheme in Fig. 8.20.

Laser light of 532 nm wavelength (photon energy 2.33 eV) is not absorbed neither by the (Zn,Mn)Se QW nor by the (Zn,Be)Se barriers, and therefore does not generate carriers in the QWs. However, it is absorbed in the GaAs substrate and the Mn system in the (Zn,Mn)Se QW is heated by the phonon impact only. The phonons are generated by photocreated carriers in GaAs at distances of less than 1 μm from the (Zn,Mn)Se QW and thus the delay of phonon impact is less than ∼1 ns. The SLR time in this case controls the rise of the $T_{\mathrm{Mn}}(t)$ signal (case (b) in Fig. 8.19). The dynamics is given by a dashed line in Fig. 8.20. Here, $\tau_{\mathrm{SLR}} = 25$ ns has been extracted from the rise time. The decay of this curve with a time constant of about 0.6 μs is due to the dynamics of nonequilibrium phonons.

A 355 nm wavelength excitation (photon energy 3.49 eV) leads to absorption in the immediate region of the II–VI heterostructure and should cause a double impact. The dynamics of the Mn system shown by a solid line in Fig. 8.20 follows the scenario of case (c) from Fig. 8.19. The carrier impact drives up the Mn temperature to $\Theta_{\mathrm{c}} \approx 13$ K during the 10 ns of the laser pulse. Afterward, the Mn system relaxes on a time scale of $\tau_{\mathrm{SLR}} \approx 25$ ns to $\Theta_{\mathrm{ph}} \approx 4.2$ K, which is controlled by the phonon impact. At delay times longer than 100 ns $T_{\mathrm{Mn}}(t)$ follows the phonon impact $I_{\mathrm{ph}}(t)$. The data are plotted on a logarithmic scale to show a wide temporal range.

An independent confirmation of the assignment of the dynamical ranges has come from the laser power dependence of magnetization dynamics [6]. The SLR time decreases from 70 down to 20 ns with increasing power. This is in accord with the known trend of shorter SLR times at higher lattice temperatures [51, 58]. Simultaneously, the time characterizing the phonon dynamics increases from 350 to 1200 ns, which is due to a strong decrease of the mean free path of nonequilibrium

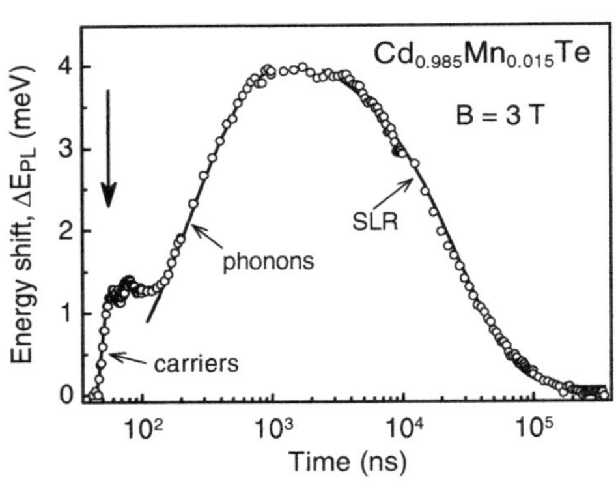

Fig. 8.21 Temporal behavior of the PL line energy shift ΔE_{PL} in 75 Å/75 Å $Cd_{0.985}Mn_{0.015}Te$/$Cd_{0.6}Mg_{0.4}Te$ multiple QW. The *indirect* (phonon) heating is stronger in this sample than the *direct* heating through carriers. The dynamics of Mn heating makes possible to distinguish contributions from carriers and nonequilibrium phonons. The solid lines represent at early times the integral of laser pulse (carriers), then the exponential growth with time constant of 0.3 μs due to phonons, and finally the monoexponential decay with an SLR time of 28 μs. The *vertical arrow* shows the laser pulse maximum position. $T = 1.6$ K. Reprinted from [6]

phonons as their average frequency increases [74]. Thus, under higher optical excitation, the propagation of phonons is hindered and becomes slower [74,75], resulting in longer lifetimes for nonequilibrium phonons inside the sample [20].

In (Cd,Mn)Te QWs, the situation can differ qualitatively from that in (Zn,Mn)Se QWs. As one can see in Fig. 8.21, in a $Cd_{0.985}Mn_{0.015}Te/Cd_{0.6}Mg_{0.4}Te$ QW sample the indirect heating is more efficient than the direct one: $\Theta_{ph} \approx 4.8$ K and $\Theta_c \approx 2.8$ K. Here, $\tau_{SLR} = 28\,\mu$s exceeds the phonon lifetimes and is therefore associated with the decay of the signal. The rise of the signal has two distinct parts. The fast but smaller in amplitude component is due to the direct energy transfer from photocarriers and the slow component is contributed by the indirect transfer involving nonequilibrium phonons. Figure 8.21 demonstrates clearly that the chosen experimental conditions, namely the short excitation pulses and the nanosecond time resolution, allow one to isolate the direct carrier contribution even for the cases when $\Theta_c < \Theta_{ph}$.

In most experimental situations, the distinction between direct and indirect mechanisms of Mn heating by hot photocarriers is not a trivial task. It requires extended sets of experimental data (i.e., for different Mn concentrations, excitation densities, temperatures, and magnetic fields) and careful interpretation. For example, the regime of long laser pulses exceeding τ_{SLR} offers, at first glance, a straightforward interpretation of the rise dynamics. In such experiments performed for bulk (Cd,Mn)Te [51,76] and (Cd,Mn)Se quantum dots [54], the heating times after switching on the laser pulse were shorter but of the order of cooling times of the Mn system after the end of the pulse. It was also shown that the heating time depends on the excitation density and shortens by an order of magnitude with its increase [76]. The heating times were associated with the SLR times and on this basis the conclusion of a dominating role of indirect energy transfer involving phonons was drawn. However, the analysis presented in [6] shows that a long heating time alone, even when it becomes comparable with SLR times, is not sufficient to conclude about the dominant role of the indirect energy transfer.

In experiments using high-density laser excitation performed for (Cd,Mn)Te/(Cd,Mg)Te QWs very fast heating during 0.5 ns and very high peak temperatures of the Mn-spin system (up to 300 K) have been found, which proves a dominating contribution from direct energy transfer [48–50]. Under these conditions, the coupled system of carriers and Mn ions becomes strongly inhomogeneous. It is separated into spatial domains having very different Mn temperatures. The positive feedback that drives this process is due to the strong dependence of the GZS of the band states on the Mn-spin temperature.

Note that time-resolved spectroscopy of magnetization in DMS can be also used to study the dynamics of nonequilibrium phonons. In the case when SLR dynamics is much faster than those of the phonons, the Mn-spin system has been used as a detector of subterahertz acoustic phonons [10].

8.7 Conclusions

Spin dynamics of DMS materials and their nanostructures are controlled by interacting systems of free carriers, magnetic ions, and lattice excitations (phonons). Spin and energy transfer between the free carriers and magnetic ions is provided by the very same process of flip-flop exchange scattering. A difference in the spin temperatures of the carriers and magnetic ions may cause a nonequilibrium polarization of the carriers, which in turn influences the energy transfer between these systems. This leads to a great variety of spin-related phenomena and their strong dependence on sample parameters (concentrations of free carriers and magnetic ions, design of DMS heterostructures) and external conditions (temperature, magnetic field strength, strength of external impact via laser, microwave radiation, electrical field and phonon injection).

Spin-lattice relaxation of the Mn^{2+} spin system in II–VI DMS is controlled by interactions between magnetic ions and, therefore, is a strong function of the Mn concentration. At liquid helium temperatures, the spin-lattice relaxation time covers a dynamic range of more than five orders of magnitude varying from 10^{-3} to 10^{-8} s for the Mn content increase from 0.4 to 11%. As a result, in bulk DMSs static and dynamical magnetizations become strongly related to each other, as both are controlled by the Mn content. DMS nanostructures offer several ways to overcome this natural limitation and to realize independent tunability of the static and dynamical magnetization. Digital growth of DMS, design of heteromagnetic structures with a complicated Mn profile, implementation of free carriers via modulation doping and control of the free carrier concentration by means of gate voltage and laser illumination are among the methods which have been successfully used to provide a control over the spin dynamics in DMS nanostructures.

Acknowledgements

Results reviewed in this chapter originate from collaboration with M. Kneip, A.A. Maksimov, I.I. Tartakovskii, A.V. Scherbakov, A.V. Akimov, R.A. Suris, V.Yu. Ivanov, S.M. Ryabchenko, M. Bayer, D. Keller, B. König, W. Ossau and G. Landwehr. Structures for these studies have been grown in the University of Würzburg by L. Hansen, A. Waag, T. Slobodskyy, G. Schmidt and L. W. Molenkamp, and in the Institute of Physics, Polish Academy of Sciences by G. Karczewski, T. Wojtowicz and J. Kossut. We appreciate greatly fruitful collaboration with these colleagues.

References

1. M. Qazzaz, G. Yang, S.H. Xin, L. Montes, H. Luo, J.K. Furdyna, Solid State Commun. **96**, 405 (1995)
2. *Diluted magnetic semiconductors*, ed. by J.K. Furdyna, J. Kossut. Semiconductors and Semimetals, vol. 25 (Academic, London, 1988)

3. J.K. Furdyna, J. Appl. Phys. **64**, R29 (1988)
4. T. Dietl, in *Diluted Magnetic Semiconductors*, ed. by S. Mahajan. Handbook of Semiconductors, vol. 3b (North-Holland, Amsterdam 1994), p. 1252
5. A.V. Akimov, A.V. Scherbakov, D.R. Yakovlev, I.A. Merkulov, M. Bayer, A. Waag, L.W. Molenkamp, Phys. Rev. B **73**, 165328 (2006)
6. M.K. Kneip, D.R. Yakovlev, M. Bayer, A.A. Maksimov, I.I. Tartakovskii, D. Keller, W. Ossau, L.W. Molenkamp, A. Waag, Phys. Rev. B **73**, 035306 (2006)
7. J.J. Baumberg, S.A. Crooker, D.D. Awschalom, N. Samarth, H. Luo, J.K. Furdyna, Phys. Rev. B **50**, 7689 (1994)
8. S.A. Crooker, D.D. Awschalom, J.J. Baumberg, F. Flack, N. Samarth, Phys. Rev. B **56**, 7574 (1997)
9. D.D. Awschalom, J. Warnock, IEEE J. Quant. Electron. **25**, 2570 (1989)
10. A.V. Akimov, A.V. Scherbakov, D.R. Yakovlev, in *Spin Lattice Relaxation in Magnetic Semiconductor Nanostructures*, ed. by A.A. Balandin, K.L. Wang. Handbook of Semiconductor Nanostructures and Nanodevices, chap. 2 (American Scientific Publishers, CA, 2006), pp. 45–93
11. M.K. Kneip, D.R. Yakovlev, M. Bayer, A.A. Maksimov, I.I. Tartakovskii, D. Keller, W. Ossau, L.W. Molenkamp, A. Waag, Phys. Rev. B **73**, 045305 (2006)
12. A.V. Scherbakov, A.V. Akimov, D.R. Yakovlev, W. Ossau, L.W. Molenkamp, S. Tatarenko, J. Cibert, Solid State Commun. **120**, 17 (2001)
13. M.K. Kneip, D.R. Yakovlev, M. Bayer, T. Slobodskyy, G. Schmidt, L.W. Molenkamp, Appl. Phys. Lett. **88**, 212105 (2006)
14. A.V. Scherbakov, A.V. Akimov, D.R. Yakovlev, W. Ossau, L. Hansen, A. Waag, L.W. Molenkamp, Appl. Phys. Lett. **86**, 162104 (2005)
15. M.K. Kneip, D.R. Yakovlev, M. Bayer, G. Karczewski, T. Wojtowicz, J. Kossut, Appl. Phys. Lett. **88**, 152105 (2006)
16. A.A. Maksimov, D.R. Yakovlev, J. Debus, I.I. Tartakovskii, A. Waag, G. Karczewski, T. Wojtowicz, J. Kossut, M. Bayer, Phys. Rev. B 82, 035211 (2010)
17. C. Camilleri, F. Teppe, D. Scalbert, Y.G. Semenov, M. Nawrocki, M. Dyakonov, J. Cibert, S. Tatarenko, T. Wojtowicz, Phys. Rev. B **64**, 08533 (2001)
18. J. Shah, *Ultrafast Spectroscopy of Semiconductors and Semiconductor Nanostructures* (Springer, Berlin, 1999)
19. T. Dietl, P. Peyla, W. Grieshaber, Y. Merle d'Aubigne, Phys. Rev. Lett. **74**, 474 (1995)
20. J. Shah, R.F. Leheny, W.F. Brinkman, Phys. Rev. B **10**, 659 (1974)
21. E.A. Zhukov, D.R. Yakovlev, M. Bayer, G. Karczewski, T. Wojtowicz, J. Kossut, Phys. Stat. Sol. (b) **243**, 878 (2006)
22. F. Meier, B.P. Zakharchenya (eds.), *Optical Orientation* (North-Holland, Amsterdam, 1984)
23. A. Abragam, *The Principles of Nuclear Magnetism* (Clarendon Press, Oxford, 1961)
24. M. Goldman, *Spin Temperature and Nuclear Magnetic Resonance in Solids* (Oxford University Press, London, 1970)
25. B. König, I.A. Merkulov, D.R. Yakovlev, W. Ossau, S.M. Ryabchenko, M. Kutrowski, T. Wojtowicz, G. Karczewski, J. Kossut, Phys. Rev. B **61**, 16870 (2000)
26. L.D. Landau, E.M. Lifschitz, *Statistical Physics*, vol. 5, 3rd edn. (Pergamon, NY, 1980)
27. A.V. Scherbakov, D.R. Yakovlev, A.V. Akimov, I.A. Merkulov, B. König, W. Ossau, L.W. Molenkamp, T. Wojtowicz, G. Karczewski, G. Cywinski, J. Kossut, Phys. Rev. B **64**, 155205 (2001)
28. D. Keller, D.R. Yakovlev, B. König, W. Ossau, Th. Gruber, A. Waag, L.W. Molenkamp, A.V. Scherbakov, Phys. Rev. B **65**, 035313 (2002)
29. G.V. Astakhov, D.R. Yakovlev, V.P. Kochereshko, W. Ossau, W. Faschinger, J. Puls, F. Henneberger, S.A. Crooker, Q. McCulloch, D. Wolverson, N.A. Gippius, A. Waag, Phys. Rev. B **65**, 165335 (2002)
30. T. Strutz, A.M. Witowski, R.E.M. de Bekker, P. Wyder, Appl. Phys. Lett. **57**, 831 (1990)
31. T. Strutz, A.M. Witowski, P. Wyder, Phys. Rev. Lett. **68**, 3912 (1992)
32. S.A. Crooker, D.D. Awschalom, N. Samarth, IEEE J. Sel. Top. Quant. Electron. **1**, 1082 (1995)

33. J.J. Baumberg, D.D. Awschalom, N. Samarth, J. Appl. Phys. **75**, 6199 (1994)
34. M. Goryca, D. Ferrand, P. Kossacki, M. Nawrocki, W. Pacuski, W. Maslana, S. Tatarenko, J. Cibert, Phys. Stat. Sol. (b) **243**, 882 (2006)
35. M. Goryca, D. Ferrand, P. Kossacki, M. Nawrocki, W. Pacuski, W. Maslana, S. Tatarenko, T. Wojtowicz, G. Karczewski, J. Cibert, Phys. Stat. Sol. (c) **4**, 307 (2007)
36. P. Kossacki, D. Ferrand, M. Goryca, M. Nawrocki, W. Pacuski, W. Maslana, S. Tatarenko, J. Cibert, Physica E **32**, 454 (2006)
37. V.Yu. Ivanov, M. Godlewski, D.R. Yakovlev, M.K. Kneip, M. Bayer, S. Ryabchenko, A. Waag, Phys. Rev. B **78**, 85322 (2008)
38. A.V. Komarov, S.M. Ryabchenko, O.V. Terletskii, I.I. Zheru, R.D. Ivanchuk, Sov. Phys. JETP **46**, 318 (1977); Zh. Eksp. Teor. Fiz. **73**, 608 (1977)
39. A.V. Malyavkin, Phys. Stat. Sol. (b) **115**, 353 (1983)
40. A.V. Malyavkin, A.A. Dremin, JETP Lett. **42**, 114 (1985); Pis'ma Zh. Eksp. Teor. Fiz. **42**, 95 (1985)
41. S.J.C.H.M. van Gisbergen, M. Godlewski, R.R. Galazka, T. Gregorkiewicz, C.A.J. Ammerlaan, N.T. Khoi, Phys. Rev. B **48**, 11767 (1993)
42. M.L. Sadowski, M. Byszewski, M. Potemski, A. Sachrajda, G. Karczewski, Appl. Phys. Lett. **82**, 3719 (2003)
43. S. Zeng, L.C. Smith, J.J. Davies, D. Wolverson, S.J. Bingham, G.N. Aliev, Phys. Stat. Sol. (b) **243**, 887 (2006)
44. A.V. Scherbakov, A.V. Akimov, D.R. Yakovlev, W. Ossau, G. Landwehr, T. Wojtowicz, G. Karczewski, J. Kossut, Phys. Rev. B **60**, 5609 (1999)
45. B. König, Ph. D. Thesis, Würzburg, Germany, (2000)
46. Y.S. Gui, C.R. Becker, J. Liu, M. König, V. Daumer, M.N. Kiselev, H. Buhmann, L.W. Molenkamp, Phys. Rev. B **70**, 195328 (2004)
47. S.M. Ryabchenko, Yu. G. Semenov, O.V. Terletskii, Sov. Phys. JETP **55**, 557 (1982); Zh. Eskp. Teor. Fiz. **82**, 951 (1982)
48. V.D. Kulakovskii, M.G. Tyazhlov, A.I. Filin, D.R. Yakovlev, A. Waag, G. Landwehr, Phys. Rev. B **54**, R8333 (1996)
49. M.G. Tyazhlov, A.I. Filin, A.V. Larionov, V.D. Kulakovskii, D.R. Yakovlev, A. Waag, G. Landwehr, Sov. Phys. JETP **85**, 784 (1997); Zh. Eksp. Teor. Fiz. **112**, 1440 (1997)
50. M.G. Tyazhlov, V.D. Kulakovskii, A.I. Filin, D.R. Yakovlev, A. Waag, G. Landwehr, Phys. Rev. B **59**, 2050 (1999)
51. W. Farah, D. Scalbert, M. Nawrocki, Phys. Rev. B **53**, R10461 (1996)
52. D. Scalbert, Phys. Stat. Sol. (b) **193**, 189 (1996)
53. A.V. Koudinov, Yu.G. Kusraev, I.G. Aksyanov, Phys. Rev. B **68**, 085315 (2003)
54. A. Hundt, J. Puls, A.V. Akimov, Y.H. Fan, F. Henneberger, Phys. Rev. B **72**, 033304 (2005)
55. M. Nawrocki, Yu.G. Rubo, J.P. Lascaray, D. Coquillat, Phys. Rev. B **52**, R2241 (1995)
56. H. Falk, J. Hubner, P.J. Klar, W. Heimbrodt, Phys. Rev. B **68**, 165203 (2003)
57. A.V. Akimov, A.V. Scherbakov, A.L. Zhmodikov, V.P. Kochereshko, D.R. Yakovlev, W. Ossau, G. Landwehr, T. Wojtowicz, J. Kossut, Phys. Rev. B **56**, 12100 (1997)
58. A.V. Scherbakov, A.V. Akimov, D.R. Yakovlev, W. Ossau, G. Landwehr, T. Wojtowicz, G. Karczewski, J. Kossut, Phys. Rev. B **62**, R10641 (2000)
59. A. Abragam, B. Bleaney, *Electron Paramagnetic Resonance of Transition Ions* (Clarendon Press, Oxford, 1970)
60. I. Dzyaloshinskii, J. Phys. Chem. Solids **4**, 241 (1958)
61. T. Moriya, Phys. Rev. **120**, 91 (1960)
62. B.E. Larson, K.C. Haas, R.L. Aggarwal, Phys. Rev. B **33**, 1789 (1986)
63. X. Wang, M. Dahl, D. Heiman, P.A. Wolf, P. Becla, Phys. Rev. B **46**, 11216 (1992)
64. R.A. Suris, private communication
65. D.R. Yakovlev, C. Sas, B. König, L. Hansen, W. Ossau, G. Landwehr, L.W. Molenkamp, Appl. Phys. Lett. **78**, 1870 (2001)
66. D.R. Yakovlev, A.V. Platonov, E.L. Ivchenko, V.P. Kochereshko, C. Sas, W. Ossau, L. Hansen, A. Waag, G. Landwehr, L.W. Molenkamp, Phys. Rev. Lett. **88**, 257401 (2002)

67. S.E. Barnes, Adv. Phys. **30**, 801 (1981)
68. T. Story, C.H.W. Swuste, P.J.T. Eggenkamp, H.J.M. Swagten, W.J.M. De Jonge, Phys. Rev. Lett. **77**, 2802 (1996)
69. R. Akimoto, K. Ando, F. Sasaki, S. Kobayashi, T. Tani, J. Appl. Phys. **84**, 6318 (1998)
70. J.M. Kikkawa, I.P. Smorchkova, N. Samarth, D.D. Awschalom, Science **257**, 1284 (1997)
71. R. Ferreira, G. Bastard, Phys. Rev. B **43**, 9687 (1991)
72. D. Some, A.V. Nurmikko, Phys. Rev. B **48**, 4418 (1993)
73. D. Leinen, Phys. Rev. B **55**, 6975 (1997)
74. A.A. Maksimov, I.I. Tartakovskii, JETP Lett. **42**, 568 (1985)
75. D.V. Kazakovtsev, A.A. Maksimov, D.A. Pronin, I.I. Tartakovski, JETP Lett. **49**, 61 (1989)
76. D. Scalbert, W. Farah, M. Nawrocki, Proc. 23 Int. Conf. Physics of Semiconductors, ed. by M. Scheffler, R. Zimmermann, Berlin, Germany, 1996 (World Scientific, Singapore 1996), p. 433

Chapter 9
Coherent Spin Dynamics of Carriers and Magnetic Ions in Diluted Magnetic Semiconductors

Scott A. Crooker

Abstract The strong $sp-d$ exchange interaction in DMS materials gives rise to a rich array of electron, hole, and Mn-ion spin dynamics that is broadly tunable with applied magnetic fields. These are discussed in-depth in this chapter. Terahertz spin precession, as also demonstrated in this chapter, of coherent electrons, ultrafast relaxation of hole spins, and the long-lived free-induction decays of the embedded Mn spin population can be directly initiated and probed in the time domain from femtosecond to microsecond timescales using ultrafast optical techniques based on Faraday rotation.

9.1 Introduction

In diluted magnetic semiconductors (DMSs), the strong J_{sp-d} exchange interaction between the electronic carriers (electrons and holes) and the embedded magnetic ions generates not only the well-known enhancements of static magneto-optical properties such as Zeeman splitting, but also gives rise to a rich array of *dynamic* and *spin-coherent* phenomena. Using ultrafast optical experiments based primarily on time-resolved Faraday rotation, this chapter describes how ensembles of spin-polarized carriers and embedded magnetic ions can be created in coherent superpositions of the Zeeman-split spin eigenstates that exist in applied transverse magnetic fields B. The time evolution of these superpositions can be monitored directly in the time domain. Semi-classically, this corresponds to time-domain measurements of a spin ensemble precessing at its Larmor frequency, $\omega_L = g\mu_B B/\hbar$, where g is the relevant g-factor and μ_B is the Bohr magneton. The decay of the oscillatory precession signals provides a direct measure of T_2^*, the *ensemble's* transverse spin relaxation time. T_2^* is effectively a measure of how long the net ensemble

S.A. Crooker
National High Magnetic Field Laboratory, Los Alamos National Laboratory, Los Alamos, NM 87545, USA
e-mail: crooker@lanl.gov

J. Kossut and J.A. Gaj (eds.), *Introduction to the Physics of Diluted Magnetic Semiconductors*, Springer Series in Materials Science 144,
DOI 10.1007/978-3-642-15856-8_9, © Springer-Verlag Berlin Heidelberg 2010

spin remains in a well-defined coherent superposition of spin eigenstates. Note that T_2^* of the measured ensemble may be considerably shorter than the true spin decoherence time T_2 of any *one* spin within the ensemble, particularly if spins within the ensemble precess with different frequencies due to an inhomogeneous magnetic environment – a process known as "ensemble dephasing." As will be discussed in Sect. 9.6 of this chapter, recent studies suggest that this latter situation is especially relevant to carrier spins in DMS materials.

In addition to the timescales T_2^* and T_2 that characterize the transverse spin relaxation of spin ensembles and individual spins, respectively, the longitudinal spin relaxation time T_1 describes the timescale for spins in an upper spin state to relax to a lower spin state. This process necessarily involves energy dissipation, and a comprehensive discussion of spin and energy transfer between carrier spins, magnetic ions, and the lattice in DMS materials can be found in Chap. 8 by Yakovlev and Merkulov.

In this chapter, transverse magnetic fields and time-resolved Faraday rotation studies directly reveal the influence of the magnetic ions on carrier spin coherence and also, in the converse process, reveal the influence of the carrier spins on the coherence properties of the magnetic ions. On the one hand, the presence of the magnetic ions leads to very fast (terahertz) precession of electron spins, due to the $s{-}d$ exchange-enhanced Zeeman splitting of the conduction band. However, the very same magnetic ions are also responsible for the rapid loss of this electron spin coherence in comparison with nonmagnetic semiconductors. Conversely, the strong $p{-}d$ exchange also drives a separate and (usually) long-lived spin coherence within the sublattice of embedded Mn moments. This coherence is established via the transient exchange field from optically injected holes, and can be exploited for time-domain paramagnetic resonance of Mn spin decoherence in quantum-confined geometries.

Sections 9.2 and 9.3 of this chapter overview the magneto-optical Faraday effect and the details of time-resolved Faraday rotation experiments, respectively. Spin relaxation measurements in DMS quantum wells in zero magnetic field and in longitudinal magnetic fields are covered in Sects. 9.4 and 9.5. Section 9.6 details the electron and hole spin relaxation and spin coherence that is observed in transverse magnetic fields, and the spin coherence of the embedded Mn ions is treated in Sect. 9.7.

9.2 Origins of the Magneto-Optical Faraday Effect in Diluted Magnetic Semiconductors

In this chapter, measurements of dynamic spin coherence and spin relaxation in diluted magnetic semiconductors (DMSs) are principally investigated using ultrafast optical techniques based on time-resolved Faraday rotation (TRFR). This Section briefly reviews the origins of the Faraday effect in DMS systems.

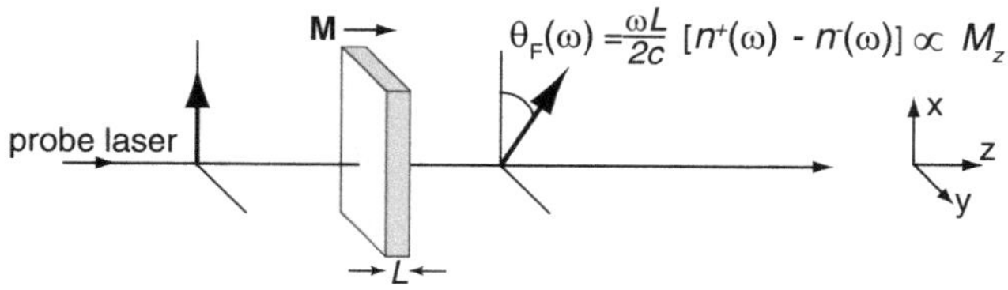

Fig. 9.1 The magneto-optical Faraday effect: Rotation of the optical polarization plane of linearly polarized light upon transmission through a magnetized material. A Faraday rotation can also occur when carriers (electrons and holes) in the material are spin-polarized

The "magneto-optical Faraday effect" refers to the rotation of the plane of polarization of linearly polarized light upon transmission through a magnetized material, as depicted in Fig. 9.1 (a close relative, the magneto-optical Kerr effect, refers to the polarization rotation of reflected light). Optical Faraday rotation $\theta_F(\omega)$ results from unequal indices of refraction for right- and left-circularly polarized light, $n^{\pm}(\omega)$:

$$\theta_F(\omega) = \frac{\omega L}{2c}[n^+(\omega) - n^-(\omega)], \tag{9.1}$$

where ω is the angular frequency of the light ($\hbar\omega$ is the photon energy), c is the speed of light, and L is the effective thickness of the sample (note that L may be considerably less than the actual sample thickness, for example in the case of a DMS quantum well embedded in an otherwise nonmagnetic structure). A difference between the indices of refraction n^+ and n^- can arise from the Zeeman effect in an applied magnetic field, or may result from an intrinsic magnetization **M**. In DMS, it is usually the case that $n^+ - n^-$ (and therefore θ_F) is directly proportional to the sample magnetization **M** along the light propagation direction.[1] Further, a difference between n^+ and n^- can also arise when spin-polarized *carriers* (electrons and holes) are introduced into the conduction and valence bands: As will be discussed in detail shortly, this *electronic* contribution to θ_F can be exploited to measure the spin dynamics of carriers in semiconductors.

Static Faraday rotation studies have been used for decades as a versatile, noninvasive, and sensitive measure of magnetization in diluted magnetic semiconductors. In DMS materials exhibiting paramagnetic behavior (e.g., those containing low concentrations of noninteracting magnetic ions), measured Faraday rotations exhibit the characteristic Brillouin-function dependence on temperature and applied magnetic field [2]. Alternatively, DMS systems exhibiting ordered (ferro- or antiferromagnetic) or frustrated (glassy) magnetism exhibit history-dependent Faraday rotations as a function of temperature or applied field [3].

[1] A notable exception being when the Zeeman splitting exceeds the absorption width, for photon energies near the absorption resonance; see [1].

High-sensitivity Faraday rotation techniques are ideally suited to time-resolved optical pump-probe studies, wherein pump-induced perturbations to the sample magnetization (and/or to the carrier spin polarization) are measured by the Faraday rotation imparted to a time-delayed optical probe pulse. In contrast with ultrafast optical spectroscopies that explicitly require the presence of photoexcited electrons and holes such as time-resolved photoluminescence, time-resolved Faraday rotation (TRFR) is sensitive to changes in magnetization, which in DMS materials may persist long after all photoexcited carriers have recombined. Coupled with femtosecond time resolution, TRFR measurements can reveal the spin coherence and spin relaxation of photoexcited carriers on short timescales, as well as the much longer-lived precession and relaxation of the perturbed magnetic ions.

Methods for time-resolved Faraday rotation were first applied to DMS materials in 1985 by Awschalom and co-workers [4], who studied the dynamics of magnetic polaron formation in bulk $Cd_{1-x}Mn_xTe$ at low temperatures. Subsequent investigations of DMS systems have used TRFR to reveal spin relaxation and spin scattering of optically injected excitons [5, 6], optically induced magnetization [1, 7, 8], optical coherence between Zeeman-split exciton states [7, 9], and the spin coherence of electrons, holes, and local magnetic ions [10–23].

Time-resolved Faraday rotation signals in II–VI and III–V semiconductors are typically largest in the spectral vicinity of band-to-band optical transitions, and can arise from both *electronic* and *magnetic* contributions. Figure 9.2 shows schematically how pump-induced Faraday rotation signals arise in both cases. For simplicity, the diagrams depict the fundamental band-edge optical absorption transitions as having highly idealized Lorentzian lineshapes. These absorption transitions – $\alpha^+(\omega)$ (solid line) and $\alpha^-(\omega)$ (dotted line) – correspond to excitation of spin-down and spin-up excitons that couple preferentially to right/left-circularly polarized light, in accordance with the standard angular momentum optical selection rules. We adopt in this chapter the usual convention whereby "σ^+" and "σ^-" circular polarizations correspond to excitation of spin-down and spin-up states, respectively. Drawn below the absorption resonances are the corresponding indices of refraction $n^+(\omega)$ and $n^-(\omega)$, and the resulting Faraday rotation, $\theta_F(\omega) \propto [n^+(\omega) - n^-(\omega)]$.

Figure 9.2a illustrates a purely electronic contribution to the measured Faraday rotation in a TRFR experiment. On short timescales, following selective photoexcitation of, for example, the $\alpha^-(\omega)$ transition by a circularly polarized σ^- pump pulse, the presence of spin-polarized electrons and holes reduces the oscillator strength of the α^- transition due to phase-space filling, as drawn. Note there may also be small energy shifts and/or broadening of both $\alpha^+(\omega)$ and $\alpha^-(\omega)$ due to Pauli blocking, screening and bandgap renormalization [11, 24, 25]; for simplicity, these additional effects are not considered here. The corresponding index of refraction, $n^-(\omega)$, is therefore also reduced in amplitude, leading directly to the induced Faraday rotation $\theta_F(\omega)$ shown in the lowest row. Note that the pump-induced Faraday rotation is asymmetric with respect to the resonance energy and can be either positive or negative (or zero!) depending on the photon energy of the probe light. As the electrons and holes spin-relax and occupy both spin states equally, this electronic contribution to $\theta_F(\omega)$ decreases to zero. Thus, TRFR studies on short timescales can

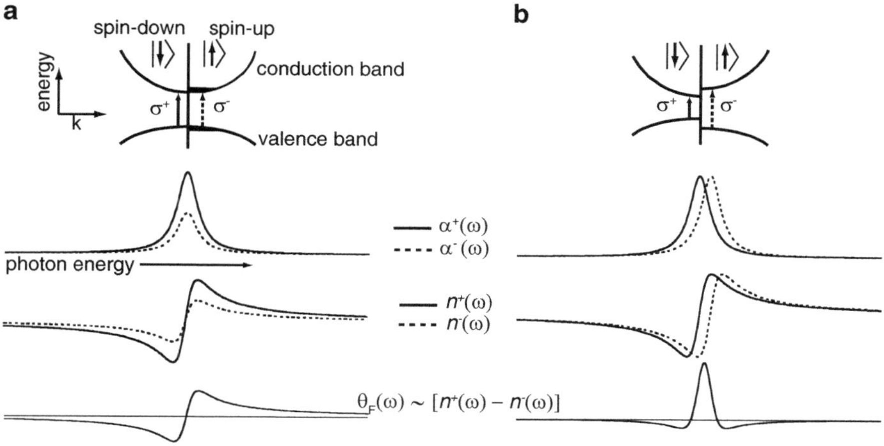

Fig. 9.2 Diagrams illustrating how pump-induced Faraday rotation signals can arise from either (**a**) electronic contributions, such as photo-bleaching of a particular spin state by electrons and holes, or (**b**) magnetic contributions, such as an induced Zeeman splitting due to a change in magnetization. The idealized Lorentzian absorption resonances $\alpha^+(\omega)$ (*solid line*) and $\alpha^-(\omega)$ (*dotted line*) couple to right/left circularly polarized light, $\sigma^\pm$. Their associated indices of refraction $n^\pm(\omega)$ are shown below, followed by the resulting Faraday rotation, $\theta_F(\omega)$

provide a measure of carrier spin relaxation and spin coherence in both magnetic and nonmagnetic semiconductors.

Alternatively, Fig. 9.2b shows a purely magnetic contribution to the measured Faraday rotation, as might be observed in DMS materials. If the optical pump pulse perturbs the equilibrium magnetization of the magnetic ions in the sample (for example, by heating or by direct angular-momentum transfer from spin-polarized photocarriers), then the Zeeman energy splitting between the $\alpha^\pm(\omega)$ optical transitions is perturbed. The corresponding indices of refraction therefore also shift in energy (as drawn), leading to an induced Faraday rotation $\theta_F(\omega)$ that is symmetric with respect to the unperturbed absorption resonance. Pump-induced changes to the magnetization typically relax on the characteristic timescale of the magnetic ions, which can greatly exceed the electronic carrier lifetimes at low temperatures [6,26].

In practice, both electronic and magnetic processes occur in TRFR experiments on DMS materials, although usually the electronic contributions dominate on short timescales, while purely magnetic effects dominate the dynamics on long timescales. The spectral shapes of the induced Faraday rotation resonances $\theta_F(\omega)$ are rarely as idealized and symmetric as depicted in Fig. 9.2, but rather depend explicitly on the actual lineshape of the $\alpha^\pm(\omega)$ transitions in the sample. Finally, the reader should note that in DMS quantum wells containing excess electrons or holes, the effect of an applied magnetic field alone can also alter the $\alpha^\pm(\omega)$ oscillator strengths as shown in [27].

9.3 Time-Resolved Faraday Rotation

Using a balanced polarization detection scheme and lock-in detection of the pump-induced signal, the sensitivity of TRFR measurements can often be limited primarily by the fundamental photon counting statistics of the probe light (that is, "photon shot noise"). Pump-induced Faraday rotations in the nanoradian range are measurable (in a 1 Hz detection bandwidth) using only microwatts of probe laser power. Together with the greatly enhanced magneto-optical properties in DMS materials due to the $sp-d$ exchange interaction, TRFR is thus an extremely sensitive probe of magnetization, even in systems as thin as a single magnetic monolayer.

Figure 9.3 shows a schematic of a typical TRFR experiment that is used to probe spin coherence and spin relaxation in DMS materials [1, 11]. The samples are mounted on a variable-temperature insert within a magneto-optical cryostat. Pump and probe pulses are derived from one or more ultrafast lasers that are usually tuned in photon energy near the band-edge absorption resonance of the sample. A circularly polarized pump pulse excites spin-polarized electrons and holes (oriented along $\pm\hat{z}$, the sample normal) in accordance with the angular momentum selection rules. Subsequent changes to the carrier spin polarization and to the sample magnetization along the $\hat{z}$ direction are then measured by the additional pump-induced Faraday rotation, $\Delta\theta_F(t)$, imparted on a weaker, time-delayed, linearly polarized probe pulse.

In longitudinal magnetic fields (B_z, the Faraday geometry), the electron, hole, and Mn spin eigenstates are most naturally oriented and quantized along the $\hat{z}$ direction. This is the direction along which carrier spins are initially injected by circularly polarized pump pulses, as well as the direction along which spin and magnetization are measured by the Faraday effect. Alternatively, in *transverse* magnetic fields (B_x, the Voigt geometry), electron and Mn spin eigenstates have a natural

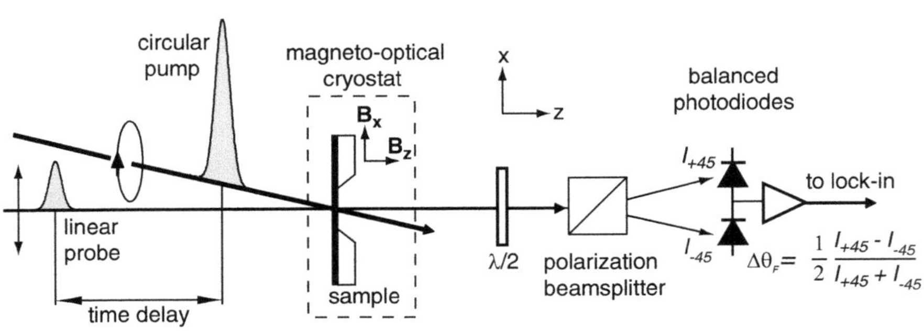

Fig. 9.3 Schematic of a typical time-resolved Faraday rotation (TRFR) experiment. Spin-polarized electrons and holes are optically injected in the sample by a circularly polarized ultrafast pump pulse. Subsequent changes in the $\hat{z}$-component of the electron, hole, and Mn spin polarization are measured by the pump-induced Faraday rotation imparted on a time-delayed, linearly polarized probe pulse. Small induced rotations are revealed using a combination of a polarization rotator (half-wave plate, $\lambda/2$), polarization beamsplitter, and balanced photodiodes

basis in the $\hat{x}$ direction, which is orthogonal to the injection and observation direction. Circularly polarized pump pulses whose spectral bandwidth overlaps multiple Zeeman levels can therefore excite coherent superpositions of these spin eigenstates, and probe pulses measure the $\hat{z}$ projection of these superpositions as they evolve in time. Semiclassically, this corresponds to the measurement of spin precession in the time domain, and the measured free-induction decay directly reveals the effective transverse spin relaxation time of the ensemble, T_2^*.

The pump beam is typically intensity-modulated (chopped) or polarization-modulated (from right- to left-circular) to facilitate detection of $\Delta\theta_F$ using lock-in amplifiers. In addition, the probe beam may also be modulated to mitigate unwanted background signals from scattered pump light. If the timescale of the measured spin dynamics exceeds the repetition period of the pulsed laser (typically $\sim$10 ns for unamplified Ti:sapphire oscillators), then the induced magnetization will not have relaxed back to equilibrium by the time the next pump pulse arrives, leading to the buildup of a background signal. To avoid these effects, an acousto-optic pulse-picker can be used to reduce the laser repetition rate [1], typically by a factor of 10–100.

Measuring $\Delta\theta_F$ with nanoradian resolution usually requires cancelation of extrinsic noise sources, including laser intensity and polarization noise. A photodiode "polarization bridge" is therefore used, which in principle can detect signals down to the fundamental shot-noise limit determined by photon counting statistics. The design and operating principle of the polarization bridge are straightforward: After transmission through the sample, the linearly polarized probe beam is split into two orthogonally polarized linear components at $\pm 45°$. The power in both "arms" of the bridge is measured by two matched photodiodes arranged head-to-tail as shown, giving photocurrents I_{+45} and I_{-45}. When the polarization rotator (half-wave plate, $\lambda/2$) is adjusted so that $I_{\pm 45}$ are equal, the photocurrent difference is zero (except for small, fluctuating currents due to shot noise) and the polarization bridge is "balanced." Any pump-induced Faraday rotation $\Delta\theta_F$ away from this balance point generates a finite difference between the I_{+45} and I_{-45} photocurrents, which is amplified in a low-noise, high gain current-to-voltage amplifier and then detected using lock-in techniques. Simple trigonometry gives $\Delta\theta_F$ in terms of the normalized difference of the photocurrents: $(I_{+45}-I_{-45})/(I_{+45}+I_{-45}) = \sin(2\Delta\theta_F) \simeq 2\Delta\theta_F$ for small rotations. Note further that detecting the pump-induced changes to the *sum* of the photocurrents $(I_{+45} + I_{-45})$ measures the net pump-induced transmission, from which the total population of carriers (independent of spin) can be inferred.

The utility of this scheme lies in the fact that intensity fluctuations from the laser appear *equally* in both arms of the bridge and are therefore canceled, providing considerable common-mode noise rejection. Laser polarization fluctuations are also effectively nulled by using a high-quality linear polarizer in the probe beam before the sample, to convert polarization fluctuations into intensity fluctuations. In this way, shot-noise limited measurements are possible even in the presence of much larger laser noise, and microwatts of probe laser power can yield detection sensitivities in the nanoradian range. Consider this typical example: A total probe power of $P_0 = 10\,\mu$W at 825 nm is incident on the polarization bridge. When balanced, the photocurrent in each arm of the bridge is $I_{\pm 45} = R(P_0/2)$, where R is the photodiode responsivity ($R \approx 0.5$ amperes/watt for silicon photodiodes at this wavelength). Root-mean-square fluctuations of the photocurrents give shot noise currents $i_{\rm rms} = \sqrt{2qI_{\pm 45}}$ amps/$\sqrt{\rm Hz}$. Summing the (uncorrelated) shot noise from the two photodiodes in quadrature gives a Faraday rotation noise floor $2\Delta\theta_{\rm F}^{\rm min} = (\sqrt{2}\, i_{\rm rms})/(I_{+45} + I_{-45}) = \sqrt{2q/RP_0}$. In this example, $\Delta\theta_{\rm F}^{\rm min} \simeq 130$ nrad/$\sqrt{\rm Hz}$ due to shot noise, which typically exceeds by about an order of magnitude other sources of detector and amplifier noise (e.g., Johnson noise) when using unbiased photodetectors and high-gain ($\geq 10^6$ V/A), low-bandwidth (≤ 100 kHz) amplifiers.

9.4 Spin Relaxation in Zero Magnetic Field

Before addressing the coherence properties of spins precessing in transverse magnetic fields, it is instructive to first briefly discuss the spin relaxation of carriers and magnetic ions in zero magnetic field (this section), and in longitudinal magnetic fields (next section). Figure 9.4 shows a typical example of carrier spin relaxation in zero magnetic field, where spin-up and spin-down exciton states[2] are degenerate in energy and the equilibrium sample magnetization **M** is zero. These TRFR data were taken on a single 30 Å wide ZnSe/$Zn_{0.80}Cd_{0.20}Se$ quantum well into which magnetic Mn atoms were introduced "digitally" during epitaxial growth [5]. Specifically, three quarter-monolayers of MnSe are spaced evenly (every 2.75 monolayers) in this quantum well (hence this sample is called the "$3 \times \frac{1}{4}$ ml MnSe" well). Digital introduction of the magnetic atoms allows certain freedom to engineer the magnetic properties of the quantum well independent from the electronic properties [5, 28]. For the very thin quarter-monolayer MnSe layers in this sample, and due to the effects of Mn diffusion and segregation during growth (~1 monolayer), the Mn atoms in this quantum well behave largely as independent paramagnets. For the purposes of this chapter, these digital magnetic quantum wells containing $\frac{1}{4}$-monolayers of MnSe can effectively be regarded as conventional ZnCdMnSe DMS quantum wells having a uniform magnetic concentration ($x_{\rm Mn} \simeq 8\%$) unless otherwise stated.

The data points show the pump-induced Faraday rotation following the photoexcitation of spin-up and spin-down excitons ($\mathbf{S}_{\rm ex} \parallel \pm\hat{z}$) using σ^- and σ^+ circularly polarized pump pulses. Both pump and probe pulses are derived from the same ultrafast Ti:sapphire laser, which is frequency-doubled in a nonlinear crystal and tuned to the peak of the induced Faraday resonance at ~450 nm (that is, this quantum well's band-edge). As discussed in Sect. 9.2, TRFR signals on short timescales arise from the spin-selective phase-space filling of electron and hole spin states, thus providing a measure of the net spin of the photoexcited carriers in the quantum well.

[2] The notion of "spin-up" and "spin-down" excitons (having higher and lower Zeeman energy) is not strictly defined in zero field, however for continuity with the following sections we use this terminology here. Spin-up and spin-down excitons may be regarded as having total spin parallel and antiparallel to the $\hat{z}$ axis.

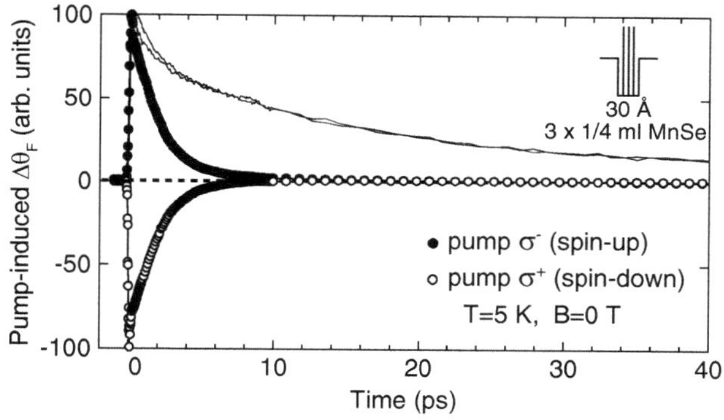

Fig. 9.4 Pump-induced Faraday rotation measured in zero magnetic field in a single ZnSe/$Zn_{0.80}Cd_{0.20}Se$ quantum well containing three equally spaced quarter-monolayer MnSe planes ("3 × $\frac{1}{4}$ ml MnSe"; see text). *Solid* (*open*) data points show the rapid spin relaxation of spin-up (spin-down) excitons following photoexcitation with σ^- (σ^+) circularly polarized light. Spin relaxation is symmetric at zero field, as expected. Also shown (*thin lines*) are the time-resolved transmission decays, as simultaneously measured by the *sum* of the photodiode currents. These slower decays are independent of pump polarization, and provide a measure of the exciton recombination time. Reprinted with permission from [11]

As expected, the pump-induced Faraday rotation inverts sign for opposite circular polarizations of the pump beam, and it is verified that no TRFR signal exists when equally populating both exciton spin states with linearly polarized pump pulses.

The rapid decay of the TRFR signal within ~5 ps reflects the spin relaxation of the electrons and holes as well as their recombination. As mentioned in the previous Section, distinguishing between the two effects is accomplished by simultaneously measuring pump-induced changes to the *sum* of the photodiode currents, $I_{+45} + I_{-45}$. This is effectively a time-resolved transmission measurement that is sensitive only to the net *number* of photoexcited electrons and holes (independent of spin), giving an accurate measure of the exciton recombination time. The thin solid lines in Fig. 9.4 show the decay of the pump-induced sum current for both circular pump polarizations. As expected, these signals and their much longer decays (~22 ps) are independent of the photo-injected carrier spin orientation.

In this magnetic quantum well, it is therefore clear that complete spin relaxation of electrons and holes is achieved on very short timescales of a few picoseconds, which is much faster than the ~22 ps exciton recombination lifetime. This behavior is typical of DMS materials having relatively high Mn concentrations of order 10%. However, the reader should note that carrier spin relaxation times are generally found to be over an order of magnitude *longer* in similar but *non*-magnetic ZnSe/$Zn_{0.80}Cd_{0.20}Se$ quantum wells. This striking difference highlights the important role of the local Mn atoms in dominating the rapid spin relaxation of electrons and holes in DMS systems. Similar trends have been observed in TRFR studies of other DMS materials such as CdMnTe [22], and these effects will be discussed at length in Sect. 9.6.4, after the coherence properties of electron and hole spins are introduced.

We note finally that at zero magnetic field, in this DMS quantum well (and others), no long-lived perturbations to the underlying Mn ions are observed. Such a perturbation would indicate the presence of optically induced magnetization and should manifest as a long-lived nonzero TRFR. However, in applied longitudinal or transverse magnetic fields, pump-induced changes to the Mn magnetization *do* occur: The case of longitudinal magnetic fields is shown in the next section, and the case of transverse fields is covered in Sect. 9.6.

9.5 Spin Dynamics in Longitudinal Magnetic Fields

Applying longitudinal magnetic fields (B_z) lifts the energy degeneracy between spin-up and spin-down excitons, so that the spin-down exciton state (lower-energy, σ^+) becomes energetically favorable. As shown in Fig. 9.5, TRFR data now develop pronounced asymmetries resulting from the preferential spin relaxation of electrons and holes to this lower-energy spin state. Figure 9.5 shows TRFR studies of the same 30 Å wide $3 \times \frac{1}{4}$ ml "digital" magnetic quantum well, but now in the presence of a small longitudinal field ($B_z = 0.25\,\mathrm{T}$) that splits the spin-up and spin-down exciton states by 4.9 meV. Spin relaxation of spin-up excitons to the spin-down state is very fast: By 2 ps, the majority of the injected spin-up excitons have relaxed to the spin-down state, as evidenced by the negative sign of $\Delta\theta_F(t)$ for $t > 2\,\mathrm{ps}$. Conversely, the spin relaxation of injected spin-down excitons is suppressed: $\Delta\theta_F(t)$ remains negative and decays to zero more slowly.

Regardless of the pump helicity (σ^+ or σ^-), the electron and hole spins in this sample achieve an equilibrium spin polarization after ∼15 ps. This quasi-equilibrium condition corresponds to a majority of the photoinjected excitons in the

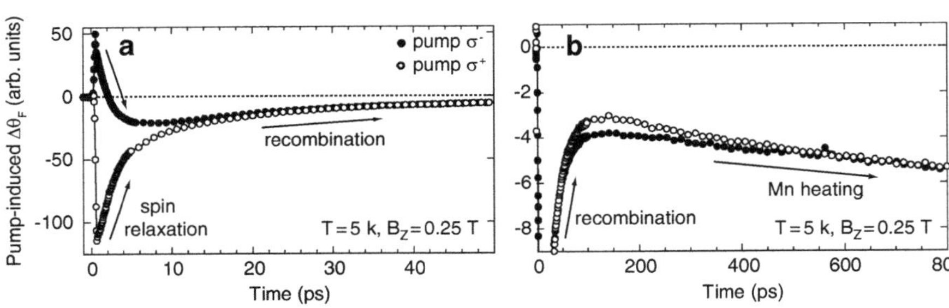

Fig. 9.5 $\Delta\theta_F(t)$ in the same "$3\times\frac{1}{4}$ ml" digital magnetic quantum well in a small *longitudinal* magnetic field (Faraday geometry; $B_z = 0.25\,\mathrm{T}$). (**a**) On short timescales, the injected spin-polarized excitons preferentially spin-relax to the lower-energy σ^+ state, and achieve thermalization within ∼15 ps. (**b**) On longer timescales (<100 ps), these thermalized excitons recombine completely, after which the embedded Mn moments dynamically warm up over hundreds of picoseconds. Reprinted with permission from [11]

low energy spin-down state, giving a negative pump-induced Faraday rotation. Subsequently, as this partially polarized exciton population recombines, $\Delta\theta_F(t)$ decays toward zero with the characteristic exciton recombination time of about 22 ps, as expected and as observed in Fig. 9.4.

However, following the complete recombination of excitons by ~100 ps, the TRFR signal is seen to *increase* again in the negative direction (Fig. 9.5b). Since no excitons remain, this purely magnetic effect arises from a pump-induced perturbation to the equilibrium magnetization M_z of the Mn ions. These data provide direct evidence for a long-lived dynamic heating (reduced magnetization) of the Mn ions in the quantum well. The most notable feature of this induced magnetization is that the magnitude *and sign* of the long-lived signal are independent of the photoinjected spin polarization (σ^+ or σ^-). Both pump helicities generate a decrease of the net Faraday rotation (i.e., a decrease of M_z), strongly suggesting a heating of the Mn ions by nonequilibrium phonons that are generated by the photoinjected carriers. The magnitude of this longlived heating signal is strongly dependent on temperature and magnetic field, and its complete recovery back to equilibrium occurs on the microsecond timescales characteristic of spin-lattice relaxation, therefore requiring the use of a pulse-picker to reduce the repetition rate of the Ti:sapphire laser so that the decay can be resolved [5, 6]. For a comprehensive discussion of spin and energy transfer between carriers and the embedded Mn ions in DMS systems, the reader is referred to [29, 30] and to Chap. 8 by Yakovlev and Merkulov.

In this and in other "digital" magnetic quantum wells in longitudinal fields, no evidence of a long-lived *spin-dependent* induced magnetization was observed [6]. That is, there is no evidence that spin-polarized electrons and holes directly impart any persistent or long-lasting angular momentum to the Mn ions through the carrier-ion exchange interaction, $J_{sp-d}\mathbf{S}\cdot\mathbf{s} = J_{sp-d}(S_z s_z + S_- s_+ + S_+ s_-)$. Instances of spin-dependent optically induced magnetization in DMS systems are not uncommon, however: The diagonal "mean-field" part of this interaction ($S_z s_z$) has been identified in polaron formation [4] and in photoinduced magnetization studies [31–34]. Similarly, the off-diagonal "flip-flop" terms ($S_- s_+ + S_+ s_-$) are thought to be at least partially responsible for the long-lived magnetization observed in earlier TRFR studies [1, 7]. It will be important to distinguish these mechanisms from those which induce a long-lived *oscillatory* Mn spin precession in the presence of transverse applied fields, to be discussed shortly in Sect. 9.7.

9.6 Electron and Hole Dynamics and Spin Coherence in Transverse Magnetic Fields

Having discussed in the preceding sections the monotonic spin relaxation that occurs in zero magnetic field and in longitudinal fields B_z, we turn now to the *oscillatory* signals that arise in TRFR experiments when electron spins precess coherently in *transverse* magnetic fields ($B_x > 0$; the Voigt geometry). In transverse fields, the two eigenstates of spin-$\frac{1}{2}$ electrons are quantized along the transverse field direction (that is, $|\pm\hat{x}\rangle$), which is now orthogonal to the pump and probe direction ($\hat{z}$). Circularly polarized pump photons having spin angular momentum $\pm\hbar\hat{z}$ therefore no longer couple to specific electron spin eigenstates, as was the case in longitudinal fields. Rather, these pump pulses, which photoexcite electrons having initial spin $\mathbf{s}_e \| \pm\hat{z}$, excite a *coherent superposition* of the two electron spin eigenstates:

$$\mathbf{s}_e(t=0) = |\pm z\rangle = \frac{1}{\sqrt{2}}(|+x\rangle \pm |-x\rangle) \tag{9.2}$$

Ignoring spin relaxation for the moment, the time evolution of this coherent superposition can be written as

$$\mathbf{s}_e(t) = \frac{1}{\sqrt{2}}(|+x\rangle \pm \mathrm{e}^{\mathrm{i}\omega_L t}|-x\rangle), \tag{9.3}$$

where $\hbar\omega_L = g_e^* \mu_B B_x$ is the Zeeman energy splitting between the electron spin eigenstates and g_e^* is the effective electron g-factor. In DMS materials, the Zeeman splitting and associated Larmor precession frequency ω_L can be quite large, of order several terahertz ($\omega_L/2\pi \simeq 10^{12}$ Hz) in modest magnetic fields, due to the $s{-}d$ exchange-amplified spin splitting in the conduction band. The probe laser measures the $\hat{z}$-component of this coherent spin superposition – that is, $\langle \mathbf{s}_e(t)| + \hat{z}\rangle$ – which oscillates in time. Semiclassically, this may be regarded as simply a precession of the electron ensemble spin at the Larmor frequency ω_L, whose decay provides a measure of the ensemble's transverse spin relaxation time, T_2^*.

The decay and decoherence of this precessing ensemble may result from a variety of sources. In nonmagnetic II–VI and III–V compound semiconductors, electron spin relaxation and decoherence can result from momentum- and energy-dependent spin-orbit effects, from scattering with holes, impurities and/or phonons, and from finite lifetime effects (e.g., recombination). All these mechanisms also exist in DMS materials, along with an additional spin relaxation channel due to direct spin scattering with the embedded magnetic ions [34, 35]. In addition, spatial and temporal variations in the local magnetic environment in DMS systems can lead to a spread of spin precession frequencies and relaxation rates within the electron ensemble, and therefore an associated rapid "dephasing" of the observed precession signal. Recent experiments, to be discussed in Sect. 9.6.4, point to the importance of this latter effect in determining the decoherence of electron spin ensembles in CdMnTe.

And what of the photoexcited spin-polarized heavy-holes? In most quantum wells, the effects of quantum confinement and *hh-lh* splitting on the valence band mandate that the heavy-hole spins are pinned along the growth direction and do not precess, but rather relax monotonically (and often very quickly, <1 ps). This convenient fact allows separate identification of electron and hole spin relaxation times, as discussed in detail in Sect. 9.6.3.

9.6.1 Terahertz Electron Spin Precession

In close analogy with pulsed nuclear magnetic resonance, the TRFR experiment measures the free-induction decay (precession and dephasing/decoherence) of an electron ensemble whose spin orientation is initially prepared in a direction orthogonal to the applied transverse field. Measurement of the electron spin free-induction decay directly in the time domain provides a very accurate measure of g_e^*, and also provides the electron spin ensemble's decoherence time T_2^* as a function of temperature, applied field, and magnetic doping. Typically, $T_2^* < 10$ ps at low temperatures in DMS materials having $x_{Mn} > 1\%$.

Figure 9.6a shows the pump-induced Faraday rotation measured in a single DMS quantum well in zero magnetic field (monotonic decays) and also in a transverse magnetic field of $B_x = 1$ T. This 120 Å wide $Zn_{.77}Cd_{.23}Se/ZnSe$ well contains a four monolayer barrier of $Zn_{.90}Mn_{.10}Se$. Fast oscillations ($\omega_L/2\pi \simeq 160$ GHz) are clearly observed, and $\Delta\theta_F$ inverts sign when pumping with opposite circular polarization, as expected. The spin splitting associated with these oscillations, $\hbar\omega_L \simeq 660\,\mu$eV, is only about one-sixth of the *total* Zeeman splitting between spin-up and spin-down heavy-hole excitons (electron + hole) measured separately in 1 T longitudinal magnetic fields. This ratio is almost exactly the ratio given by the exchange parameters $N_0\alpha$ and $N_0\beta$ that, respectively, describe the $s-d$ (electron–Mn) and $p-d$ (hole-Mn) exchange interaction in $Zn_{1-x}Mn_xSe$ [36] (specifically, $N_0\alpha \simeq 0.29$ eV and $N_0\beta \simeq -1.4$ eV, so that $\alpha/(\alpha-\beta) = 1/5.8$). For these reasons, the observed oscillations are ascribed to the spin precession of electrons alone.

A population of coherently precessing electron spins also appears as oscillations in time- and polarization-resolved photoluminescence measurements, where these spin precession effects were first observed in GaAs by Heberle and co-workers [37]

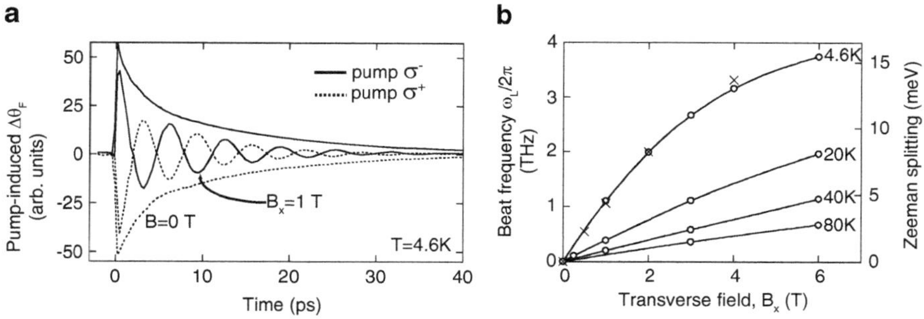

Fig. 9.6 (**a**) $\Delta\theta_F(t)$ measured in a 120 Å wide $Zn_{.77}Cd_{.23}Se/ZnSe$ QW containing four monolayers of $Zn_{.90}Mn_{.10}Se$, in zero field and in a transverse magnetic field (Voigt geometry; $B_x = 1$ T), in response to pumping with σ^- and σ^+ circularly polarized pulses (*solid* and *dashed lines*). $\Delta\theta_F(t)$ oscillates about zero as electron spins, initially oriented along $\pm\hat{z}$, precess about B_x. (**b**) The electron precession frequency $\omega_L/2\pi$ and corresponding conduction-band Zeeman splitting in a 120 Å wide "24 × $\frac{1}{8}$ ml" digital magnetic QW vs. B_x, at 4.6, 20, 40, and 80 K. *Lines* are Brillouin-function fits. *Crosses* indicate the *total* exciton Zeeman splitting (electron + hole), scaled down by a factor of 5.7. Reprinted with permission from [10, 11]

in 1994. Subsequent studies of spin precession based on time-resolved Faraday- and Kerr-rotation have since been used to measure electron g-factors in a variety of different II–VI and III–V semiconductor compounds [38–45]. These electron "spin beats" are an example of an *intra*-band spin coherence between Zeeman-split spin levels in the conduction band, much in the same way that nuclear free-induction decays in pulsed NMR studies reflect a spin coherence between nuclear spin levels. Importantly, this intra-band electron spin coherence requires no memory of the optical phase of the pump laser. As such, these oscillatory phenomena are distinct from the *inter*-band ("optical" or "orbital") coherences between *exciton* states that have been investigated in many ultrafast studies of semiconductors [7,46–49].

In accord with the enhanced Zeeman splittings realized in DMS materials, the Larmor frequency of coherent electron spin precession can be quite fast even in modest magnetic fields, as shown in Fig. 9.6b. Precession frequencies approaching 4 THz ($\omega_L/2\pi = 4 \times 10^{12}$ Hz, or $\hbar\omega_L \simeq 16$ meV) are possible in 6 T transverse fields in those DMS structures that exhibit strong J_{sp-d} coupling (that is, wavefunction overlap) between the electronic carriers and the local Mn ions. Figure 9.6b shows that the electron spin precession frequency tracks the paramagnetic magnetization of the Mn ions in the quantum well, scaling with temperature and applied transverse field as a $B_{5/2}$ Brillouin function. The data points represented by crosses ($\times$) in Fig. 9.6b show the total Zeeman splitting of the heavy-hole exciton (electron + hole) in this quantum well at 4.6 K (as measured by the exciton absorption splitting in longitudinal fields), scaled down by a factor of 5.7. The good agreement supports the interpretation that these oscillations arise from electrons alone: As mentioned above, earlier reflectivity measurements in bulk $Zn_{1-x}Mn_xSe$ indicate nearly the same ratio (5.8) between exciton and electron Zeeman splitting [36].

9.6.2 Tuning Electron Spin Precession via Wavefunction Control

The measured precession frequency ω_L enables a very accurate measurement of the effective electron g-factor, g_e^*. In DMS quantum wells, g_e^* is enhanced by the s–d exchange interaction: $g_e^* = g_e + N_0\alpha\langle S_z\rangle f(\psi)/\mu_B B$, where g_e is the intrinsic electron g-factor (or order unity in wide-gap semiconductors), $N_0\alpha$ is the usual s–d exchange parameter for electrons, $\langle S_z\rangle$ is the effective average spin per Mn ion (which follows a modified $B_{5/2}$ Brillouin function), and $f(\psi)$ is the overlap of the square of the electron wavefunction with the embedded Mn ions. In bulk DMS, $f(\psi)$ is equivalent to x_{Mn}, the magnetic concentration. This wavefunction overlap can be tuned from sample to sample by controlling either the concentration of Mn ions or, in epitaxially-grown structures, the quantum well dimensions and the placement of Mn ions within the quantum well [11]. Figure 9.7 shows the electron precession frequency measured by TRFR in four $ZnSe/Zn_{0.80}Cd_{0.20}Se$ "digital magnetic" quantum wells having different widths (30, 60, 120, and 240 Å), but

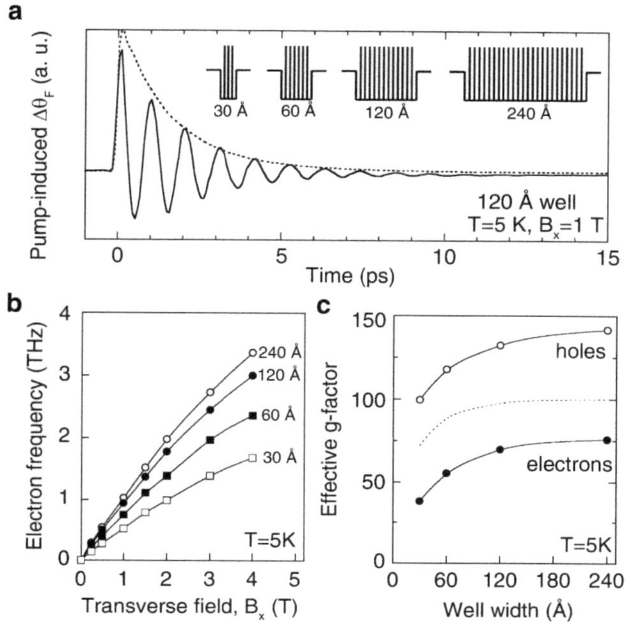

Fig. 9.7 (**a**) $\Delta\theta_F(t)$ in the 120 Å wide "12 × $\frac{1}{4}$ ml MnSe" digital magnetic QW in a transverse field $B_x = 1$ T and in zero field (*dotted line*). (**b**) Electron precession frequency vs. B_x measured in digital magnetic QWs of different width but similar magnetic environment ($\frac{1}{4}$ ml of MnSe spaced every $2\frac{3}{4}$ monolayers; schematics shown in (**a**)). (**c**) The electron and hole effective g-factors vs. QW width. Reprinted with permission from [11]

nominally identical magnetic environments ($\frac{1}{4}$-monolayers of MnSe spaced every $2\frac{3}{4}$ monolayers; $\langle x_{\mathrm{Mn}}\rangle \simeq 8\%$). The effects of quantum confinement change the degree of overlap between the electron/hole wavefunctions and the Mn ions – for example, in thin wells, more of the wavefunction "leaks out" into the barriers, leading to reduced overlap with the Mn in the well. Again, the electron spin precession frequencies follow a modified $B_{5/2}$ Brillouin function (see Fig. 9.7b), and the corresponding spin splittings at 5 K and low field indicate that $g_e^* = 38.3, 55.2, 69.6$, and 75.8, in order of increasing well width (Fig. 9.7c). The increase of g_e^* results from increased overlap of the electron wavefunction with the MnSe planes in wider wells (the widest well approaches bulk-like conditions), a trend that can be qualitatively reproduced with a 1-D Schrödinger solution for each well (dotted line).

Comparison of g_e^* with the heavy-hole *exciton* (electron + hole) Zeeman splitting measured in longitudinal magnetic fields allows to characterize the J_{p-d} enhanced heavy-hole g-factor, defined here through the relation $g_{\mathrm{ex}}^* \simeq g_e^* + 3g_{\mathrm{hh}}^*$. Both electron and hole g-factors depend strongly (and qualitatively similarly) on the confinement potential (Fig. 9.7c).

The ability to tune the wavefunction overlap between carriers and Mn ions *within* a single structure was recently shown by Myers and co-workers [19], who

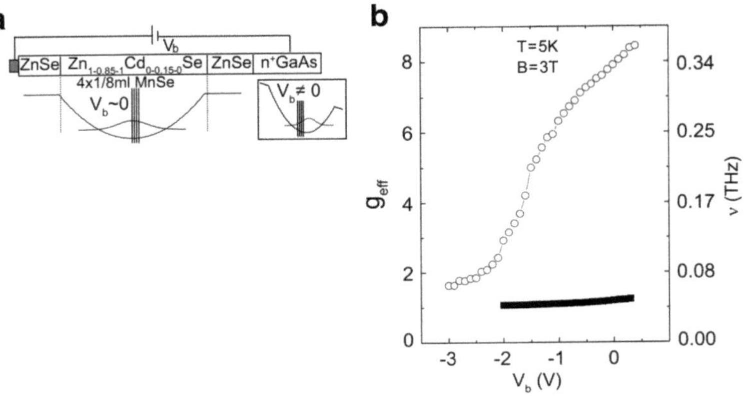

Fig. 9.8 Electrical tuning of the carrier-Mn wavefunction overlap in a gated parabolic QW containing a narrow region of paramagnetic Mn ions. The electron spin precession frequency is tuned by over a factor of four with gate bias (V_b) in this structure (*open circles*). Little change is observed in a similar but nonmagnetic structure (*solid*). Reprinted with permission from [19]

demonstrated in-situ electrical control over the carrier wavefunctions in a gated parabolic DMS quantum well. In doing so, the effective strength of the J_{sp-d} exchange interaction – and therefore the electron spin precession frequency – are tuned using external voltage. Figure 9.8a shows a schematic of the ZnCdSe parabolic quantum well used in this study. Four $\frac{1}{8}$-monolayer paramagnetic MnSe planes were incorporated into the well, and front and back gates permit vertical electrical biasing of the structure, which spatially shifts the well's minimum and therefore the carrier wavefunctions with respect to the fixed Mn ions. Figure 9.8b shows the measured electron g-factor and precession frequency as a function of applied gate voltage in this parabolic DMS well, and also in a nonmagnetic but otherwise identical well. The electron spin precession frequency varies by over a factor of four as the wavefunction overlap between carriers and local Mn spins is tuned with external voltage.

9.6.3 Separating Electron and Hole Spin Dynamics

In addition to the coherently precessing electron spins measured in these ultrafast studies, there are, of course, an equal number of optically excited hole spins also present in the quantum well. This raises the question of whether the presence of hole spins appears in the time-domain data. Ultrafast studies of spin relaxation in zero magnetic field and in longitudinal magnetic fields generally reveal monotonic, multi-exponential decays. Separate contributions from electrons, holes, and excitons are often inferred from the shape and timescale of the various decay components [50, 51]. Alternatively, studies in *n*- or *p*-type material can help identify the spin relaxation of holes and/or electrons [52].

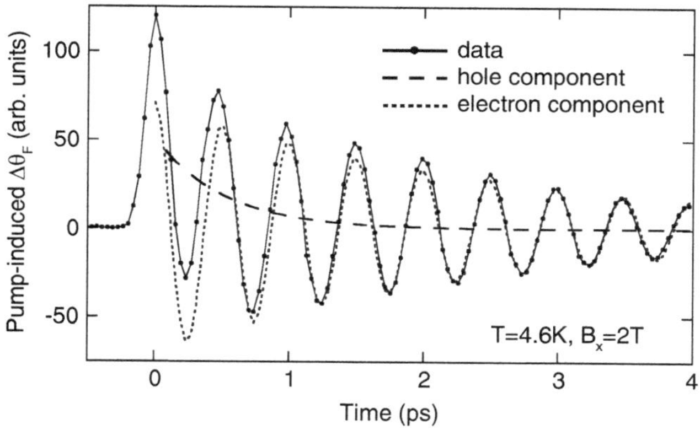

Fig. 9.9 $\Delta\theta_F(t)$ in a 120 Å wide $Zn_{.77}Cd_{.23}Se$ quantum well containing 24 equally spaced eighth-monolayers of MnSe ("24 × $\frac{1}{8}$ ml"). The data show the rapid monotonic spin relaxation of heavy holes superimposed on the oscillatory signal from precessing electrons. *Dotted* and *dashed lines* show fits to the electron and hole components of the data, respectively. Reprinted with permission from [10]

In transverse magnetic fields, however, the photoexcited electron spins generate a unique oscillatory signal, allowing contributions from other sources (holes and Mn ions) to be readily distinguished. The data in Figs. 9.6 and 9.7 show that, after the first few picoseconds following excitation, the TRFR signal oscillates about zero at a single frequency corresponding to electron spin precession. Closer inspection of the data on short timescales, however, shows that the electron oscillations are offset vertically by a nonzero quantity. As shown in the characteristic data of Fig. 9.9, the TRFR signal is best fit by the sum of an exponentially-decaying cosine (the electron spins) *and* a fast monotonic exponential decay that arises from the rapid spin relaxation of the photoinjected hole population:

$$\Delta\theta_F(t) = A_e \exp(-t/\tau_e)\cos(\omega_L t) + A_h \exp(-t/\tau_h), \tag{9.4}$$

where $\tau_{e,h}$ are the spin decoherence (spin relaxation) times of the electron and hole ensembles. Heavy-hole spins in these quantum wells are not expected to precess: Rather, as discussed in [53], the $j_z = \pm\frac{3}{2}$ heavy-hole spins are constrained to lie along the growth axis of the quantum well ($\hat{z}$) by the effects of quantum confinement and strain. These effects originate in the spin-orbit coupling of the p-like valence band, leading to a splitting between light-hole and heavy-hole states (the conduction band, being s-like, experiences no such spin-orbit effects and, thus, the electron spin is nominally isotropic). Spin relaxation of holes is presumed to occur through mixing of valence band states away from $\mathbf{k} = 0$. The ability to distinguish between the temporal evolution of the electron and hole spin populations (oscillatory vs. monotonic decays) enables one to study in detail the role of applied fields, temperature, and magnetic doping upon both hole spin relaxation as well as electron spin decoherence.

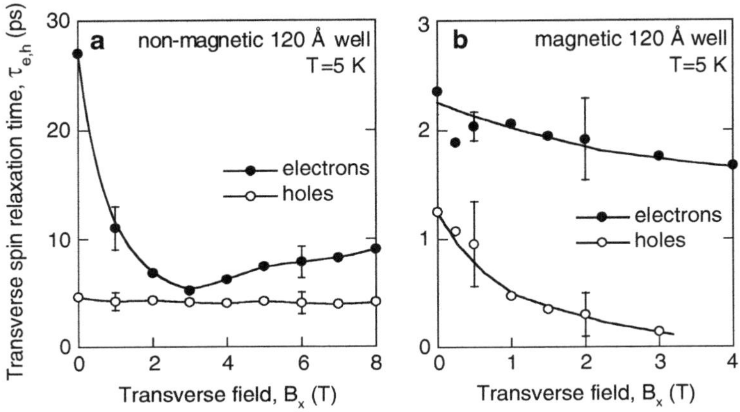

Fig. 9.10 Data comparing the electron and hole transverse spin relaxation times in a (**a**) nonmagnetic and (**b**) magnetic 120 Å "digital magnetic" quantum well. *Lines* are guides to the eye. Note the different vertical axis scales. Reprinted with permission from [11]

9.6.4 Electron and Hole Spin Relaxation and Dephasing in DMS

Figure 9.10 compares the measured electron and hole transverse spin relaxation times vs. B_x in a nonmagnetic ZnSe/$Zn_{.80}Cd_{.20}Se$ 120 Å single quantum well, and in an identical magnetic well containing twelve quarter-monolayer planes of MnSe ("$12 \times \frac{1}{4}$ ml"; $\langle x_{\mathrm{Mn}} \rangle \sim 8\%$). Clearly, the introduction of Mn ions into an otherwise nonmagnetic II–VI quantum well markedly accelerates the spin relaxation of both electrons and holes, even at zero field. In DMS materials, these [10, 11] and subsequent TRFR studies [13, 17, 22, 54] have revealed a strong correlation between the magnetic concentration x_{Mn} and the measured spin relaxation *rates* ($\tau_{\mathrm{e,h}}^{-1}$). For electrons, these trends appear to suggest that spin-flip scattering between electrons and Mn ions (i.e., $s-d$ exchange scattering [34, 35]) dominates spin decoherence processes.

However, recent TRFR studies of bulk $Cd_{1-x}Mn_xTe$ as a function of temperature, magnetic field, and x_{Mn} by Rönnburg and co-workers [22] make a compelling case for a motional-narrowing type of electron ensemble spin decoherence due to *dephasing* by the inhomogeneous magnetization fluctuations that are "seen" by the mobile photoinjected electrons. Figure 9.11 summarizes these results. The measured transverse spin relaxation (ensemble dephasing) times, τ_{e}, in these bulk samples are of the same order of magnitude as those measured in DMS quantum wells, in contrast to the expectation that direct spin-flip scattering should be greatly reduced in the bulk. Further, τ_{e} increases with temperature (for a fixed concentration x_{Mn}) and decreases with x_{Mn} (for a fixed temperature), in remarkably good agreement with a model of precessional dephasing wherein the diffusing electron spins experience fluctuating local magnetic fields due to the embedded Mn ions.

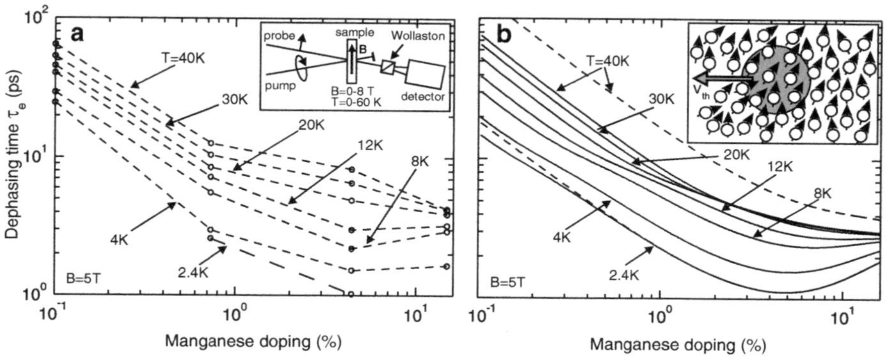

Fig. 9.11 (**a**) Evidence for a motional-narrowing component to the electron spin dephasing time in bulk $Cd_{1-x}Mn_xTe$ epilayers as a function of x_{Mn} and temperature at $B_x = 5\,T$. (**b**) A motional-narrowing model of electron spin dephasing due to fluctuating magnetic fields from the Mn ions. Reprinted with permission from [22]

The measured hole spin lifetimes decrease quickly in DMS quantum wells with applied transverse field (see Fig. 9.10b). Even though the *hh*-*lh* splitting may be quite large in these quantum wells (tens of meV), the magnitude of the J_{p-d} exchange enhanced Zeeman splitting in the valence band can be comparable even in modest transverse fields of order 1 T, strongly mixing heavy and light holes and giving very fast hole spin relaxation [55]. It should also be noted that in *bulk* zincblende DMS, heavy and light hole bands are degenerate at $\mathbf{k} = 0$ and hole spins are not constrained to lie along any particular direction. Under these conditions, a very fast spin *precession* of holes has been recently identified [22] in TRFR studies of bulk $Cd_{1-x}Mn_xTe$ epilayers.

9.6.5 Spin Precession Overtones and Electron Entanglement

A recent and exciting development is the observation by Bao and co-workers [15, 16] of time-domain electron spin precession at not only the fundamental Larmor frequency $\omega_L = g_e^* \mu_B B_x/\hbar$, but also at *twice* and at *three* times ω_L. It can be argued that these overtones necessarily imply the existence of true quantum-mechanical spin entanglement between multiple electrons (up to three, in this case).

An example of their data, acquired by time-resolved Kerr rotation studies in a $Cd_{1-x}Mn_xTe$ quantum well, is shown in Fig. 9.12. Entanglement and correlation are due to an impulsive stimulated Raman scattering process, driven by the exchange interaction between the excitons that are optically injected at the quantum well resonance and the paramagnetic impurities residing in the well (i.e., the donor-bound electrons and/or embedded Mn ions). These studies suggest possible routes toward ultrafast creation and control of spin-entangled particles for quantum information processing and computing schemes.

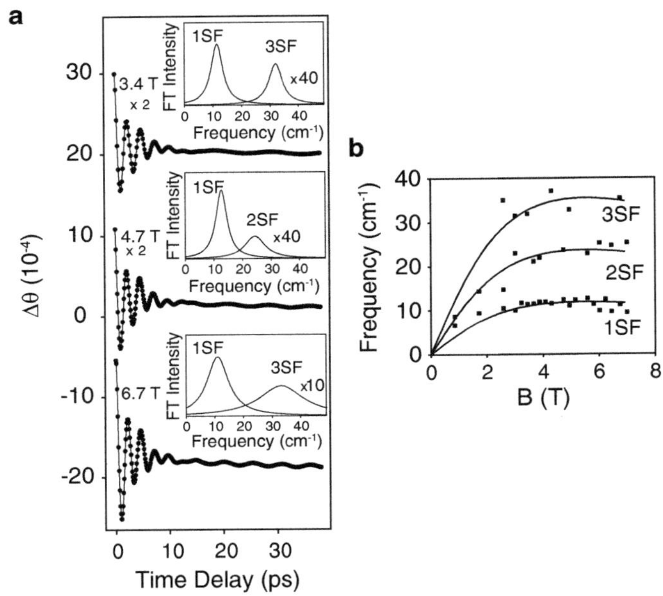

Fig. 9.12 Differential Kerr rotation in a CdMnTe QW at 2 K. Only donor-related transitions are shown in the Fourier spectra. The first (second) overtone of the electron spin-flip is denoted by 2SF (3SF). (**b**) Frequency vs. magnetic field for the donor spin-flip fundamental, 1SF, and its overtones. Reprinted with permission from [16]

9.7 Coherent Spin Precession of the Embedded Mn Ions

In the previous sections, TRFR was used to determine the influence of the J_{sp-d} interaction on the dynamics of electron and hole spins. The local Mn spins in DMS quantum wells and epilayers generated extremely rapid THz electron spin precession as well as fast electron and hole spin relaxation. In this Section, we now consider the converse effect – the influence of the photoexcited, spin-polarized electrons and holes on the embedded Mn spins.

In marked contrast to the TRFR experiments in longitudinal magnetic fields discussed in Sect. 9.5, where no *spin-dependent* longlived perturbation to the Mn ions was observed (only spin-independent Mn heating), measurements in transverse magnetic fields reveal a strong transient coupling between the spin-polarized holes and the embedded Mn, mediated by the J_{p-d} exchange interaction. This exchange coupling "tips" a macroscopic ensemble of Mn spins away from the axis of the applied field, leading to a subsequent free-induction decay of the Mn ensemble at gigahertz frequencies that can be directly measured in the time domain. These measurements enable all-optical time-domain electron paramagnetic resonance in single DMS quantum wells.

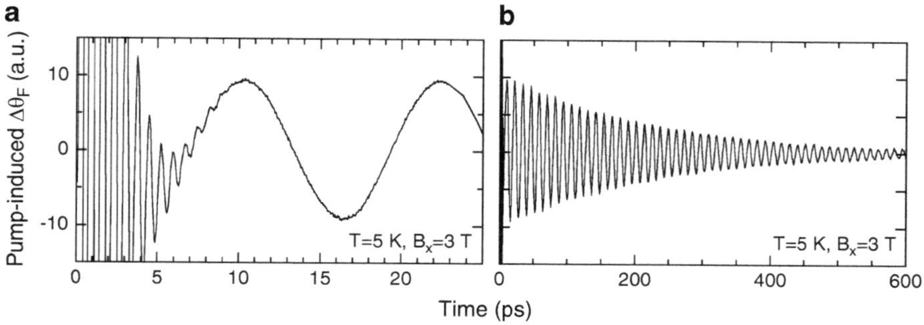

Fig. 9.13 (**a**) $\Delta\theta_F(t)$ measured in the "3×$\frac{1}{4}$ ml MnSe" digital magnetic QW in the Voigt geometry, showing the last few fast electron beats giving way to much slower oscillations due to precessing Mn spins. (**b**) A complete view of this Mn free-induction decay, persisting for hundreds of picoseconds. Reprinted with permission from [11]

9.7.1 Long-Lived Oscillations from Mn Spin Precession

Following the complete spin relaxation of holes and precessing electrons in transverse magnetic fields B_x (as indicated by the disappearance of the fast electron beats), TRFR measurements in DMS quantum wells reveal an additional oscillatory signal having a much slower frequency and longer decay time. A clear example is shown in Fig. 9.13a, acquired in the 30 Å wide "3×$\frac{1}{4}$ ml MnSe" digital magnetic quantum well. This oscillatory signal persists for hundreds of picoseconds (as shown in Fig. 9.13b), and exhibits a g-factor and decay envelope consistent with the free-induction decay of a coherently perturbed ensemble of Mn spins precessing in synchrony about B_x. This long-lived TRFR signal is purely magnetic in origin, since all the photoexcited carriers have recombined in this quantum well by 100 ps.

The oscillation frequency increases linearly with B_x and corresponds to the known Mn g-factor $g_{\rm Mn} = 2.01$. The signal is absent in all nonmagnetic samples, and is present (with the same $g_{\rm Mn}$) in every magnetic sample. The induced Faraday rotation can be fit almost perfectly with an exponentially decaying sinusoid,

$$\Delta\theta_F(t) = A_{\rm Mn} \exp(-t/\tau_{\rm Mn}) \sin(\omega_{\rm Mn} t + \phi), \tag{9.5}$$

where $\tau_{\rm Mn}$ is the transverse spin relaxation time of the Mn ensemble, $\omega_{\rm Mn} = g_{\rm Mn}\mu_B B_x/\hbar \simeq 2\pi \times 28$ GHz/T is the Mn Larmor spin precession frequency, and the phase ϕ is an additional fitting parameter whose significance will be discussed below. In keeping with the temperature-independent g-factor of Mn in DMS systems, the measured Mn precession frequency shows no temperature dependence (this is in marked contrast to the electron precession frequency, which scales with temperature as a modified $B_{5/2}$ Brillouin function).

9.7.2 A Model for Coherent "Tipping" of the Mn Ensemble

The observation of coherent Mn precession following photoexcitation of spin-polarized electrons and holes implies a mechanism involving the simultaneous and rapid "tipping" of a large number of Mn spins all in the same direction. If the Mn spins were perturbed in random directions, or if the tipping process occurred over timescales comparable to the Mn precession period, then no long-lived signal would be observed due to the averaging of the individual Mn-spin precession phases. There is no "dc" or offset component to the signal – $\Delta\theta_F(t)$ oscillates about zero within the noise limits of the experiment, and the oscillation amplitude scales linearly with pump laser fluence. Moreover, an oppositely oriented circular pump results in a 180° phase shift of the Mn precession signal, and no long-lived signal at all is observed when pumping unpolarized excitons with linearly polarized light.

The tipping mechanism originates in the coherent rotation of the net Mn magnetization about the transient p–d exchange field of the photoexcited hole spins. Figure 9.14 schematically illustrates this model, which proceeds as follows: (a) Before the pump pulse arrives ($t < 0$), the Mn spins are partially aligned along the applied transverse field B_x, giving a net Mn magnetization $\mathbf{M} = M_x$. (b) Immediately following photoexcitation, the electrons and holes are spin-polarized along $\hat{z}$. The electron spins begin precessing, while the polarized hole spins generate an exchange field along $\hat{z}$ that decays exponentially with the hole spin lifetime, τ_h. This exchange field exerts a transient torque on the Mn spins, rotating the net magnetization M_x (that is, *many* Mn spins) initially into the $\hat{y}$ axis. (c) At long times, after the holes and electrons have recombined, the perturbed Mn magnetization continues to precess about the applied field B_x, leading to small oscillations of the measured component of magnetization, M_z. The signal decays away as the perturbed Mn ensemble decoheres, in analogy with nuclear spin free-induction decays in pulsed NMR.

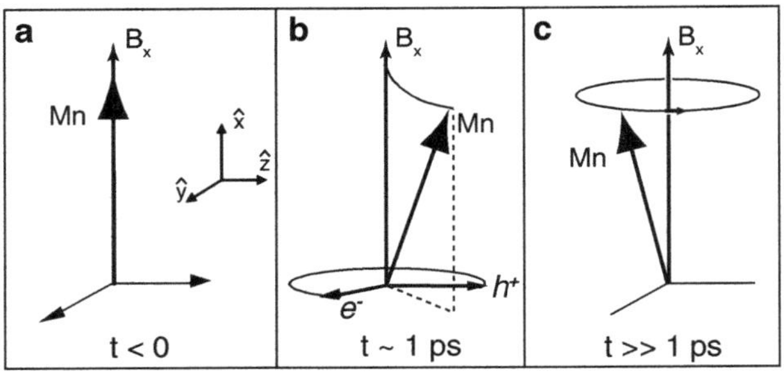

Fig. 9.14 A model for coherent spin rotation of the embedded Mn. The net Mn magnetization M_x is (**a**) oriented initially along the applied transverse field B_x (Voigt geometry), and then is (**b**) tipped into the $\hat{y}$ direction by the torque from the transient exchange field of the photoexcited hole spins ($\mathbf{j}_h \parallel \pm\hat{z}$). Finally, (**c**) after the carriers have recombined, the perturbed Mn spins continue to precess about B_x

Prior to these TRFR studies, a collective Mn tipping mechanism of precisely this type was postulated in 1995 by Stühler and co-workers to explain the remarkable appearance of up to *fifteen* Mn spin-flip lines in low-temperature resonant Raman scattering of $Cd_{1-x}Mn_xTe$ quantum wells in transverse magnetic fields [56,57]. (An example of their Raman data can be found in Chap. 3 of this book, Fig. 3.10). The Mn tipping mechanism is effectively a single-hole phenomenon: the *p-d* exchange field from a single spin-polarized heavy hole tips the large number of Mn spins that lie within its wavefunction, $\Psi_{\rm hh}(\mathbf{r})$. The number of Mn ions within the hole wavefunction is estimated to be of order 10–100 in these digital magnetic quantum wells, where the planar Mn density is of order 10^{13}–10^{14} cm^{-2}. A complete theory describing the quantum dynamics of this classically large angular momenta transfer was subsequently formulated by Kavokin and Merkulov [58], and the interested reader is referred to this work and also to Chap. 3 of this book. The essential elements of the model may be described briefly as follows: The exchange Hamiltonian describing the hole-Mn interaction is

$$H_{p-d} = \frac{1}{3}\beta \mathbf{S}_i \cdot \mathbf{J}|\Psi_{\rm hh}(\mathbf{R}_i)|^2, \tag{9.6}$$

which corresponds to an exchange magnetic field oriented along $\hat{z}$,

$$B_z^{\rm exch}(\mathbf{R}_i) = \frac{1}{3g_{\rm Mn}\mu_{\rm B}}\beta J_z|\Psi_{\rm hh}(\mathbf{R}_i)|^2. \tag{9.7}$$

Here, β is the usual $p-d$ exchange parameter, and $\mathbf{S}_i$ is the spin of the ith Mn ion at position $\mathbf{R}_i$, and $\mathbf{J} = J_z = \pm\frac{3}{2}$ is the spin of the heavy hole. This exchange field acts on the local magnetic ions for a time $\sim\tau_{\rm h}$, the hole spin lifetime. Thus, immediately following photoexcitation, the Mn spins near a heavy hole are subject to a total magnetic field

$$\mathbf{B}_T(t) = \hat{x}B_x + \hat{z}B_z^{\rm exch}\exp(-t/\tau_{\rm h}), \tag{9.8}$$

which tips the Mn spins into the $\hat{y}$-axis. The time evolution of the Mn magnetization $\mathbf{M}$ can be derived from the Bloch equation (ignoring damping terms),

$$\frac{\mathrm{d}\mathbf{M}}{\mathrm{d}t} = -\frac{g_{\rm Mn}\mu_{\rm B}}{\hbar}\mathbf{M}\times\mathbf{B}_T(t). \tag{9.9}$$

In the limit that $\tau_{\rm h}$ is very short (much shorter than the Mn precession period), the equilibrium magnetization $\mathbf{M}(0) = M_x$ is impulsively tipped into the $\hat{y}$ axis, and the measured magnetization, M_z, commences as a sinusoid.

The $s-d$ exchange field of the electron spins is ignored because of its much ($\sim$5×) weaker coupling to the Mn spins, and also due to the fact that the rapid THz precession of the electron spins averages to nearly zero over a single Mn precession period.

9.7.3 Amplitude and Phase of the Mn Free-Induction Decay

Both the amplitude and the phase of the measured Mn-spin precession support the model of coherent tipping by a transient hole exchange field. The amplitude of the Mn beats is expected to scale with the initial Mn magnetization M_x, and therefore the beat amplitude should increase linearly from zero with increasing applied field B_x. This trend is indeed observed, as shown in the raw data of Fig. 9.15a, where the amplitude of the Mn free-induction decay is large at $B_x = 1$ T, much smaller at 0.25 T, and identically zero at 0 T (i.e., $M_x = 0$ at zero field, and therefore no Mn precession signal exists). More quantitatively, Fig. 9.15b shows the amplitude of the Mn beats as a function of B_x in the series of digital magnetic quantum wells having different well widths but similar magnetic environments (quarter-monolayer MnSe planes; see drawings in Fig. 9.7). For any given well, the Mn beat amplitude increases linearly from zero with applied field B_x from 0–1 T, in support of the "tipping" scenario. The observed rolloff at higher B_x is due to the decreasing hole spin relaxation time, and also because the samples' Verdet constant (Faraday rotation per unit field along $\hat{z}$) begins to decrease in large B_x [11].

The initial *phase* of the Mn free-induction decay also favors a model of a transient "tipping" torque due to the hole spins. Because a coherent, impulsive rotation about the hole exchange field B_z^{exch} initially tips the Mn spins into the $\hat{y}$ axis, the Mn precession signal ($\propto M_z$) should commence as a *sinusoid* and not a co-sinusoid; that is, the Mn beat signal should have near-zero amplitude when extrapolated back to zero time delay. This behavior is indeed approximately observed, and can be seen in the raw data of both Figs. 9.13a and 9.15a.

The reader should note, however, that the transient tipping torque due to the hole exchange field persists for a *finite* amount of time – τ_h, the hole spin lifetime. The

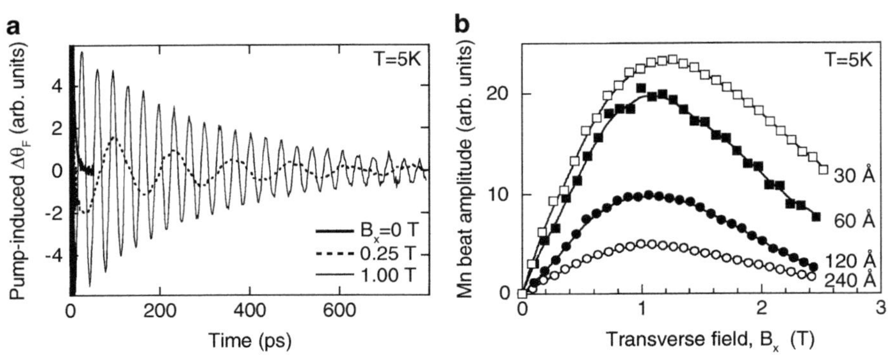

Fig. 9.15 (**a**) $\Delta\theta_F(t)$ showing Mn spin precession in a DMS quantum well for $B_x = 1.0, 0.25$, and 0 T. The amplitude of the Mn beats decreases with field and is identically zero at B=0, in support of a model for coherent "tipping" of the net Mn moment. (**b**) Amplitude of the Mn beats vs. B_x for the four digital magnetic QWs with similar magnetic environment but different well widths. At low fields, the Mn beat amplitude increases linearly with field. Reprinted with permission from [11]

torque felt by the Mn spins is therefore not strictly a δ-function-like impulse. Consequently, the sinusoidal Mn precession signal may exhibit an overall phase shift (that is, $\Delta\theta_F \propto \sin(\omega_{Mn}t + \phi)$) that is due to the finite time over which tipping occurs. Several ultrafast studies have extracted a measure of τ_h based on the observed phase shift ϕ of the Mn free-induction decay in CdMnTe quantum wells [12, 13]. On the other hand, direct time-domain data in ZnMnCdSe "digital" magnetic wells clearly show that a nonzero phase ϕ of the Mn beats accumulates dynamically over the 10–50 ps timescale of the carrier recombination time, which is much longer than either τ_h or τ_e. For an example of this data, see Fig. 14 of [11]: The Mn spins begin precessing slowly, then accelerate over ~50 ps timescales to a higher frequency commensurate with $g_{Mn} = 2.01$. Though not completely understood, this phase shift may accrue because, during this time, the Mn spins experience the applied field B_x reduced by the demagnetization field of the spin-relaxed electrons and holes. This effect may be related to the renormalization (reduction) of the Mn Larmor frequency that is theoretically expected when Mn ions interact with, e.g., a 2D gas of heavy holes [59].

9.7.4 Exchange Fields, Tipping Angles, and Mn-Spin Manipulation

From static and time-resolved Faraday rotation data, it is possible to estimate [12,60] the average angle φ_{tip} through which the Mn spins are coherently tipped by the transient hole exchange field. It is necessary to know only the ratio of the perturbed magnetization δM_z to the total magnetization M_x, so that $\varphi_{tip} \simeq \delta M_z/M_x$. The measured Mn beat amplitude is directly proportional to δM_z, and the total magnetization M_x is derived from static Faraday rotation studies that "calibrate" the low-field response $\theta_F(B_z)$ of a given sample in longitudinal magnetic fields. From this simple analysis, average tipping angles in the range $\varphi_{tip} \approx 200$–400 microradians can be inferred. Using a hole spin lifetime $\tau_h \approx 1$ ps, the Bloch equations yield an average exchange field of order 2 mT. Of course, these values represent an average over all the Mn in the quantum well, only some of which lie near a photoexcited hole. Locally (within a hole's wavefunction), φ_{tip} and the exchange field may be much higher. Given the photoinjected exciton density (10^{10}–10^{11} cm^{-2}) and a typical hole wavefunction radius of 25 Å, the local exchange field and tipping angles are likely at least two orders of magnitude larger than these average values.

The average Mn tipping angle can be directly manipulated in real-time using multiple pump pulses. This ultrafast optical control of the Mn-spin precession was demonstrated by Akimoto and co-workers [61] in time-resolved Kerr rotation studies of $Cd_{1-x}Mn_xTe$ quantum wells using *two* pump pulses. An example of their data is shown in Fig. 9.16, where the measured Mn-spin precession signal can be either amplified or suppressed by circularly polarized pump pulses arriving at the appropriate times. Using a transverse field $B_x = 3$ T, the Mn precession period is 12 ps. Co-circularly polarized pump pulses arriving 12 ps apart amplify the Mn

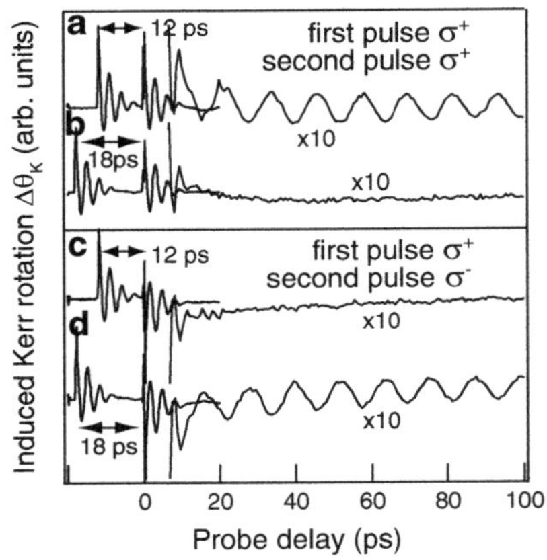

Fig. 9.16 Time-resolved Kerr rotation signals at 5 K and 3 T in a 40 Å wide $Cd_{1-x}Mn_xTe$ quantum well that is excited by double pump pulses. The Mn precession period is 12 ps. (**a**, **b**) Co-circular pump pulses (σ^+, σ^+) arrive 12 ps and 18 ps apart, respectively, amplifying and suppressing the Mn beats. (**c**, **d**) Cross-circular pump pulses (σ^+, σ^-) arrive 12 ps and 18 ps apart. Reprinted with permission from [61]

beat signal, while a separation of 18 ps suppresses the beat signal because the second pump arrives exactly out of phase with the Mn precession. Cross-circular pump pulses, which tip the Mn spins in opposite directions, have exactly the opposite effect, as expected.

9.7.5 *All-Optical Time-Domain Paramagnetic Resonance of Submonolayer Magnetic Planes*

The envelope of the Mn free-induction decay provides a measure of the Mn transverse spin-relaxation time, τ_{Mn}. This allows for the exciting possibility of performing all-optical paramagnetic spin-resonance studies of the small numbers of Mn atoms present in single quantum wells, where the reduced dimensionality from 3-D to 2-D directly influences the formation of frustrated or ordered magnetic phases [62]. Figure 9.17a shows τ_{Mn} vs. temperature as measured by TRFR in three 120 Å wide Zn(Cd,Mn)Se quantum wells: One well contains three equally spaced full MnSe monolayers ("3 × 1 ml"), another contains a four monolayer barrier of $Zn_{.90}Mn_{.10}Se$ ("4 ml 10%"), and the last contains 24 eighth-monolayers of MnSe ("24 × $\frac{1}{8}$ ml"). Due to ~1 monolayer of segregation and interdiffusion during epitaxial growth, the maximum local Mn densities in these wells are ≈50%, 10%, and 8%.

Reassuringly, the temperature dependence of τ_{Mn} measured by TRFR in the "4 ml 10%" quantum well agrees with that measured in a *bulk* $Zn_{.90}Mn_{.10}Se$ crystal using

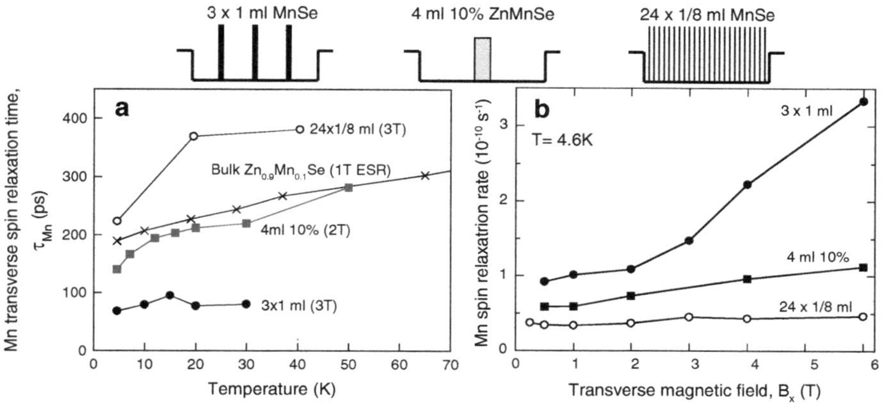

Fig. 9.17 Paramagnetic spin resonance of Mn moments in three quantum wells having similar width, but containing very different local Mn densities (~50%, 10%, 8%). (**a**) τ_{Mn} vs. temperature. (**b**) τ_{Mn} vs. B_x. Mn–Mn interactions are manifest as a strong field dependence of the dephasing rate when the local Mn density is large. Reprinted with permission from [10]

a traditional EPR spectrometer (crosses, ×), confirming the realization of an all-optical time-domain spin resonance method to probe small numbers of Mn spins. Notably, τ_{Mn} decreases dramatically as the local Mn concentration increases from 8 to 50%, indicating strong short-range spin–spin interactions (and its associated spin decoherence) when the local Mn density is large. Thus, although the $24 \times \frac{1}{8}$ ml and the 3×1 ml wells contain an equal number of Mn spins, their very different local spin densities (8 and 50%) are clearly evidenced in their disparate dephasing times (230 ps and 70 ps), pointing to the crucial role of nearest-neighbor spin-spin interactions. Overall, τ_{Mn} increases at elevated temperatures, in agreement with traditional EPR studies of bulk DMS. These variations of τ_{Mn} with temperature and local spin density are in qualitative agreement with exchange-narrowing models [63] that relate τ_{Mn} to anisotropic and isotropic spin–spin interactions, as well as to static and dynamic spin–spin correlations.

The sub-picosecond time resolution of TRFR is well-suited to study the very fast Mn dephasing times that occur at low temperatures. Moreover, TRFR works equally well over a wide range of applied magnetic fields (0.25–6 T in these studies), permitting flexible frequency-dependent spin-resonance from 7 to 170 GHz, in contrast to fixed-frequency conventional EPR spectrometers based on microwave cavities. Figure 9.17b shows the Mn spin dephasing *rate* τ_{Mn}^{-1} in these quantum wells as a function of magnetic field (frequency) at low temperature. τ_{Mn}^{-1} varies only weakly with field at low local Mn density (in the $24 \times \frac{1}{8}$ ml well), but is strongly field dependent in the case of high local Mn density (the 3×1 well), highlighting the role of applied field in determining local spin-spin interactions of Mn clusters.

9.8 Conclusions

The measurements of coherent electron, hole, and Mn spin dynamics discussed in this chapter do not by any means represent a complete accounting of all such studies in DMS materials to date. Indeed, exciting recent developments in this field also include time-resolved Faraday/Kerr rotation studies of exciton–polariton spin coherence in DMS microcavities [21], spin precession of carriers, and Mn ions in self-assembled CdMnSe/ZnSe quantum dots [23], and the "softening" of certain spin precession resonances near a ferromagnetic transition in p-doped CdMnTe quantum wells [18], to name a few. These studies, as well as the coherent spin precession experiments described in the preceding sections of this chapter, rely on the remarkable ability to separately distinguish electron, hole, and Mn spin dynamics due to their very different temporal evolutions in applied transverse magnetic fields. Using these techniques, ultrafast studies have accurately measured the J_{sp-d} exchange-enhanced carrier g-factors in DMS materials, and have identified the role of the Mn spins in causing rapid spin dephasing and spin relaxation of the electrons and holes. Conversely, the effect of a transient hole exchange field leads to a coherent "tipping" and subsequent free-induction decay of the embedded Mn spins, permitting all-optical Mn spin resonance of truly two-dimensional spin distributions in quantum structures. We anticipate that ultrafast optical studies of electron, hole, and Mn spin coherence will continue to provide critical interaction parameters in next-generation DMS structures and devices.

Acknowledgements

Many individuals were closely involved in this work, foremost among them being David Awschalom and Nitin Samarth. It is my distinct pleasure to thank them both for their enthusiastic support and guidance. I am also indebted to Jeremy Baumberg and Frank Flack for their important contributions to this work.

References

1. J.J. Baumberg, S.A. Crooker, D.D. Awschalom, N. Samarth, H. Luo, J.K. Furdyna, Phys. Rev. B **50**, 7689 (1994)
2. J.K. Furdyna, J. Kossut, in *Diluted Magnetic Semiconductors*. Semiconductors and Semimetals, vol. 25 (Academic, San Diego, 1988)
3. H. Kett, W. Gebhardt, U. Krey, J.K. Furdyna, J. Magn. Magn. Mater. **25**, 215 (1981)
4. D.D. Awschalom, J.M. Halbout, S. von Molnár, T. Siegrist, F. Holtzberg, Phys. Rev. Lett. **55**, 1128 (1985)
5. S.A. Crooker, D.A. Tulchinsky, J. Levy, D.D. Awschalom, R. Garcia, N. Samarth, Phys. Rev. Lett. **75**, 505 (1995)
6. S.A. Crooker, D.D. Awschalom, N. Samarth, IEEE J. Sel. Topics Quant. Electron. **1**, 1082 (1995)

7. J.J. Baumberg, D.D. Awschalom, N. Samarth, H. Luo, J.K. Furdyna, Phys. Rev. Lett. **72**, 717 (1994)
8. R. Pankoke et al., Appl. Phys. Lett. **68**, 2615 (1995); C. Buss et al., Phys. Rev. Lett. **78**, 4123 (1997)
9. Th. Östreich, K. Schönhammer, L.J. Sham, Phys. Rev. Lett. **75**, 2554 (1995)
10. S.A. Crooker, J.J. Baumberg, F. Flack, N. Samarth, D.D. Awschalom, Phys. Rev. Lett. **77**, 2814 (1996)
11. S.A. Crooker, D.D. Awschalom, J.J. Baumberg, F. Flack, N. Samarth, Phys. Rev. B **56**, 7574 (1997)
12. R. Akimoto, K. Ando, F. Sasaki, S. Kobayashi, T. Tani, Phys. Rev. B **57**, 7208 (1998)
13. C. Camilleri et al., Phys. Rev. B **64**, 085331 (2001)
14. F. Teppe, M. Vladimirova, D. Scalbert, T. Wojtowicz, J. Kossut, Phys. Rev. B **67**, 033304 (2003)
15. J.M. Bao, A.V. Bragas, J.K. Furdyna, R. Merlin, Nat. Mat. **2**, 175 (2003)
16. J.M. Bao, A.V. Bragas, J.K. Furdyna, R. Merlin, Phys. Rev. B **71**, 045314 (2005)
17. C.J.P. Smits et al., Phys. Rev. B **70**, 115307 (2004)
18. D. Scalbert, F. Teppe, M. Vladimirova, S. Tatarenko, J. Cibert, M. Nawrocki, Phys. Rev. B **70**, 245304 (2004)
19. R.C. Myers, K.C. Ku, X. Li, N. Samarth, D.D. Awschalom, Phys. Rev. B **72**, 041302(R) (2005)
20. R.C. Myers, M. Poggio, N.P. Stern, A.C. Gossard, D.D. Awschalom, Phys. Rev. Lett. **95**, 017204 (2005)
21. A. Brunetti et al., Phys. Rev. B **73**, 205337 (2006)
22. K.E. Rönnburg, E. Mohler, H.G. Roskos, K. Ortner, C.R. Becker, L.W. Molenkamp, Phys. Rev. Lett. **96**, 117203 (2006)
23. M. Scheibner, T.A. Kennedy, L. Worschech, A. Forchel, T. Slobodskyy, G. Schmidt, L.W. Molenkamp, Phys. Rev. B **73**, 081308(R) (2006)
24. J.B. Stark, W.H. Knox, D.S. Chemla, Phys. Rev. B **46**, 7919 (1992)
25. P. Płochocka, P. Kossacki, W. Maślana, J. Cibert, S. Tatarenko, C. Radzewicz, J.A. Gaj, Phys. Rev. Lett. **92**, 177402 (2004)
26. T. Strutz, A.M. Witowski, P. Wyder, Phys. Rev. Lett. **68**, 3912 (1992)
27. W. Maślana et al., Phys. Rev. B **63**, 165318 (2001)
28. M.K. Kneip, D.R. Yakovlev, M. Bayer, G. Karczewski, T. Wojtowicz, J. Kossut, Appl. Phys. Lett. **88**, 152105 (2006)
29. M.K. Kneip et al., Phys. Rev. B **73**, 035306 (2006)
30. M.K. Kneip et al., Phys. Rev. B **73**, 045305 (2006)
31. H. Krenn, W. Zawadzki, G. Bauer, Phys. Rev. Lett. **55**, 1510 (1985)
32. D.D. Awschalom, J. Warnock, S. von Molnár, Phys. Rev. Lett. **58**, 812 (1987)
33. H. Krenn, K. Kaltenegger, T. Dietl, J. Spałek, G. Bauer, Phys. Rev. B **39**, 10918 (1989)
34. G. Bastard, L.L. Chang, Phys. Rev. B **41**, 7899 (1990)
35. Y.G. Semenov, Phys. Rev. B **67**, 115319 (2003)
36. A. Twardowski, T. Dietl, M. Demianiuk, Solid State Comm. **48**, 845 (1983)
37. A.P. Heberle, W.W. Rühle, K. Ploog, Phys. Rev. Lett. **72**, 3887 (1994)
38. M. Oestreich, S. Hallstein, A.P. Heberle, K. Eberl, E. Bauser, W.W. Rühle, Phys. Rev. B **53**, 7911 (1996)
39. J.M. Kikkawa, D.D. Awschalom, Phys. Rev. Lett. **80**, 4313 (1998)
40. J.A. Gupta, D.D. Awschalom, X. Peng, A.P. Alivisatos, Phys. Rev. B **59**, 10421 (1999)
41. B. Beschoten et al., Phys. Rev. B **63**, 121202 (2001)
42. P. Palinginis, H. Wang, Phys. Rev. Lett. **92**, 037402 (2004); Y. Shen, A.M. Goebel, G. Khitrova, H.M. Gibbs, H. Wang, Phys. Rev. B **72**, 233307 (2005)
43. S. Ghosh et al., Appl. Phys. Lett. **86**, 232507 (2005)
44. E.A. Zhukov, D.R. Yakovlev, M. Bayer, G. Karczewski, T. Wojtowicz, J. Kossut, Phys. Stat. Sol. (b) **243**, 878 (2006); arXiv:0707.4294
45. Z.G. Chen et al., Phys. Rev. B **75**, 115320 (2007)
46. S. Bar-Ad, I. Bar-Joseph, Phys. Rev. Lett. **66**, 2491 (1991)
47. A.P. Heberle, J.J. Baumberg, K. Kohler, Phys. Rev. Lett. **75**, 2598 (1995)

48. S.T. Cundiff, R. Hellmann, M. Koch, G. Mackh, A. Waag, G. Landwehr, W.H. Knox, E.O. Göbel, J. Opt. Soc. Am. B **13**, 1263 (1996)
49. T. Amand et al., Phys. Rev. Lett. **78**, 1355 (1997)
50. M.Z. Maialle, E.A.D.A. Silva, L.J. Sham, Phys. Rev. B **47**, 15776 (1993)
51. A. Vinattieri, J. Shah, T.C. Damen, D.S. Kim, L.N. Pfeiffer, Phys. Rev. B **50**, 10868 (1994)
52. T.C. Damen, L. Vina, J.E. Cunningham, J. Shah, L.J. Sham, Phys. Rev. Lett. **67**, 3432 (1991)
53. R.W. Martin et al., Phys. Rev. B **42**, 9237 (1990)
54. R. Akimoto, K. Ando, S. Kobayashi, F. Sasaki, T. Tani, Phys. Rev. B **56**, 9726 (1997)
55. E. Tsitsishvili, H. Kalt, Phys. Rev. B **73**, 195402 (2006)
56. J. Stühler, G. Schaack, M. Dahl, A. Waag, G. Landwehr, K.V. Kavokin, I.A. Merkulov, Phys. Rev. Lett. **74**, 2567 (1995)
57. J. Stühler, G. Schaack, M. Dahl, A. Waag, G. Landwehr, K.V. Kavokin, I.A. Merkulov, J. Cryst. Growth **159**, 1001 (1996)
58. K.V. Kavokin, I.A. Merkulov, Phys. Rev. B **55**, R7371 (1997)
59. K.V. Kavokin, Phys. Rev. B **59**, 9822 (1999)
60. S.A. Crooker, Ultrafast optical probes of spin dynamics in magnetic semiconductor heterostructures, Ph.D. thesis, UC Santa Barbara (1997)
61. R. Akimoto, K. Ando, F. Sasaki, S. Kobayashi, T. Tani, J. Appl. Phys. **84**, 6318 (1998)
62. N. Samarth et al., Phys. Rev. B **44**, 4701 (1991); D.D. Awschalom, G. Grinstein, J.M. Hong, L.L. Chang, Phys. Rev. Lett. **59**, 1733 (1987)
63. N. Samarth, J.K. Furdyna, Phys. Rev. B **37**, 9227 (1998); D.L. Huber, Phys. Rev. B **31**, 4420 (1985)

Chapter 10
Spectroscopy of Spin-Polarized 2D Carrier Gas, Spin-Resolved Interactions

F. Perez and P. Kossacki

Abstract In this chapter, we describe a novel understanding allowed by use of diluted magnetic semiconductor as a material in quantum wells with carrier gas. We discuss how the possibility of spin polarization of the 2D carrier gas gave new insight into both interband and intraband excitations allowing for spin resolution in the interactions. Two principal experimental tools will be used: interband spectroscopy (photoluminescence and transmission or reflectivity) and electronic resonant Raman scattering, for probing, respectively, interband and intraband excitations in systems with 2D carrier gas. Effects of carrier–carrier interactions are analyzed, e.g., charged exciton formation and many-body enhancement of Zeeman splitting.

10.1 Introduction

Structures containing quasi–two-dimensional carrier (electron or hole) gases were first obtained in silicon metal oxide semiconductor field effect transistors and in doped heterostructures of III–V compounds such as GaAs/(Ga,Al)As heterojunctions. These systems have been extensively studied by magneto-transport and various optical spectroscopies [1]. The disorder due to ionized impurities and possible potential fluctuations have been reduced to such a low level that the behavior of the electron system was dominated by Coulomb interactions between carriers at a quantum level. As such heterostructures behave like model 2D systems, they shed considerable light on the quantum many-body aspects of such systems.

F. Perez (✉)
Institute des NanoSciences de Paris, CNRS Université Paris 6, France
e-mail: florent.perez@insp.jussieu.fr

P. Kossacki
Institute of Experimental Physics, Faculty of Physics, University of Warsaw, Hoża 69 00-681 Warsaw, Poland
e-mail: Piotr.Kossacki@fuw.edu.pl

J. Kossut and J.A. Gaj (eds.), *Introduction to the Physics of Diluted Magnetic Semiconductors*, Springer Series in Materials Science 144,
DOI 10.1007/978-3-642-15856-8_10,

Charge–charge interactions and/or their interplay with magnetic orbital quantization have first to be investigated in depth. Specifically, because in those materials a static magnetic field B_0 leads to the cyclotron energy $\hbar\omega_c$ greater than the intrinsic spin splitting energy $g\mu_B B_0$. Among many remarkable observations, we recall striking effects discovered in the last three decades such as observation of charged excitons [2–4], integer [5] and fractional quantum Hall [6] effects, composite fermions [7], electron wave focusing [8]. Also, in plane dispersion of well-defined low-lying collective excitations, such as plasmons in the two-dimensional electron gas, has been investigated at very low temperatures by intraband spectroscopies [9].

High-quality II–VI doped heterostructures appeared later [10]. Their specific properties added new insights stemming from a smaller Bohr radius a_B^* (or equivalently a larger exciton binding energy) than in common GaAs-based heterostructures. Indeed, the unambiguous identification of the charged excitons in quantum wells (QWs) was possible [11]. For this reason also, all studies of excitonic complexes were much easier. The intensive development of the II–VI technology was also motivated by efforts to fabricate opto-electronic devices working in the range of green and blue light [12].

A second striking opportunity offered by II–VI materials is the giant Zeeman effect induced by insertion of magnetic impurities such as Mn atoms (see Chap. 1) to form diluted magnetic semiconductors (DMSs) [13]. The use of DMS heterostructures gave the unique possibility to reach high-spin polarization of carriers at low magnetic fields, sufficiently small to keep Landau orbital quantization negligible compared to the spin quantization. This opened the experimental access to a situation exactly reversed to that of GaAs quantum wells [123]. In this chapter, we will describe a novel understanding allowed by such situation for both interband and intraband excitations of a two-dimensional spin-polarized carrier system.

10.2 Preliminaries

10.2.1 Typical Samples with Spin-Polarized Carriers

Studies of spin-polarized carriers gained momentum by introduction of diluted magnetic semiconductors (DMSs) in semiconductor heterostructures. The spin polarization may be obtained also in nonmagnetic materials, however, usually a very strong magnetic field is required. Then the Landau quantization plays an important role. In the case of DMS materials, the giant Zeeman effect allows one to obtain spin polarization by applying relatively small magnetic fields. Although there is a large variety of different DMS compounds, including both II–VI and III–V materials, it is not easy to achieve sufficient degree of control over their properties, particularly in the case of III–V DMS's. High-quality 2D carrier gas is so far very difficult to obtain in III–V DMS heterostructures. This is related to difficulties in incorporation of magnetic ions as neutral centers. Moreover, even neutral magnetic ions often form deep centers acting as traps for carriers. Therefore, good quality 2D carrier gas in DMS heterostructures is usually obtained in II–VI materials, such as tellurides or selenides.

A sufficiently high carrier gas concentration in quantum wells is obtained using several methods common for different groups of materials, such as modulation doping or optical excitation [14]. In II–VI semiconductors, Indium and Iodine are commonly used as donors, and Nitrogen – as an acceptor. The control of the carrier gas density may be achieved either by variation of the doping density or of the distance between doped layer and the quantum well. A particularly useful design was developed by Wojtowicz et al. [15] with spatial in-plane profiling giving variable carrier density in a single quantum well. Moreover, an illumination of the sample with photons of energy greater than the band gap of the barrier material usually allows one to tune gradually the carrier density. In addition to standard doping, in CdTe -based heterostructures one can use (Cd,Mg)Te surface states as acceptor centers [16]. This gives a unique possibility of very simple design of samples and an efficient way of obtaining hole gas in quantum wells.

Two specific examples of the design are presented below. They were used for observation of positively and negatively charged excitons, respectively. In the first case, the barriers were made of $Cd_{1-y-z}Mg_yZn_zTe$, in which the Mg content ($y = 0.25 - 0.28$) determined the valence band offset, while the presence of Zn ($z = 0.08 - 0.07$) ensured a good lattice match to the $Cd_{0.88}Zn_{0.12}Te$ substrate. The barriers were doped with nitrogen at a distance of 20 nm from the QW on the surface side. Such a design makes possible to tune the hole concentration isothermally by varying the intensity of illumination with photons of energy greater than the band gap of the barrier material [17]. It was possible to tune the hole concentration in those samples in the range from $2 \times 10^{10}\,cm^{-2}$ to $3 \times 10^{11}\,cm^{-2}$.

On the other hand, the samples used for observation of X^- and intra-band excitations were one-side modulation-doped $(Cd,Mn)Te/Cd_{0.75}Mg_{0.25}Te$ heterostructures. The remote layer of iodine donors was located 10 nm from the QW. Its width was changed in steps along the sample. This made possible probing different densities of the electron gas in steps by selecting a spot on the sample surface [15]. Additionally, for a given spot, the electron density could be increased further by illuminating the sample with light of energy higher than the energy gap of the barrier. The mechanism of this effect is based on the competition between the QW and surface acceptor states [16] which both can trap carriers, and is similar to one used in p-type samples. The maximum attainable electron density in those "wedge"-structures was $2 \times 10^{11}\,cm^{-2}$.

The introduction of magnetic ions allows one to tune spin polarization of the carriers by applying a small magnetic field. The magnetic ions may be placed in the barrier or in the quantum well. The first solution has an advantage of low composition disorder of binary QW material combined with significant Zeeman splitting. This was studied extensively in works related to spin tracing (see Chap. 4). However, the doping of (Cd,Mn)Te barrier material is very difficult. Therefore in studies of higher carrier gas densities other barrier materials are more convenient and the diluted magnetic material usually forms a quantum well. This design offers a wide range of possibilities from a complete spin polarization of the carrier gas, through carrier-mediated ferromagnetism [18–22] to spin splitting engineering [23].

In the latter case, the giant Zeeman splitting of the states in the quantum well is well described by the same formulae as those developed for the bulk material (see Chap. 1). Small discrepancies between bulk Zeeman effect and the effect observed in quantum wells were reported [24–26], and were attributed to k-vector dependence of the exchange interaction. They result in a slight decrease of the Zeeman effect.

A particular introduction of Mn ions in the quantum wells opens a possibility of tailoring the potential profile. This was employed by varying the Mn concentration across the well. Such design was used for growing parabolic and half parabolic quantum wells and applied, for example, for precise determination of relative valence band offsets [27].

10.2.2 Modeling of Spin Polarized 2D Carrier Gas in $II_{1-x}Mn_xVI$ Quantum Wells

We consider in this section a diluted magnetic modulation-doped quantum well with a fraction x of magnetic Mn impurities substituting atoms on cation sites, e.g., $Cd_{1-x}Mn_xTe$. In such quantum wells, two subsystems have to be considered: the first one is composed of electrons or holes populating the first conduction (respectively, valence) subband in the well. They originate from n-type (respectively, p-type) dopant impurities located in the barrier. This itinerant spin subsystem is coupled to the second subsystem formed by spins of electrons localized on the manganese impurities introduced in the well. These electrons, occupying the d-shell of each Mn atom, behave like a rigid localized 5/2 spin. For the sake of simplicity, in the following section we will discuss the formalism for the electrons. We will limit the discussion to the cases when the Landau quantization is negligible for the carriers. In particular, this condition is met for the magnetic field direction parallel to the quantum well plane and the quantum well width much smaller than the magnetic length. Adaptation to p-type quantum well and other directions of the magnetic field is possible. However, one has to keep in mind that the spin of the electron is isotropic while that of heavy holes is strongly anisotropic. Therefore, the discussed formalism might be applied directly to p-type case only for the configuration with small magnetic field perpendicular to the quantum well plane.

10.2.2.1 Two Coupled Spin SubSystems

The coupling between two spin subsystems is described by $s-d$ exchange interaction [28, 29]. We write the Hamiltonian in terms of spin densities (in virtual crystal approximation):

$$\hat{H}_{\mathrm{sd}} = -\alpha \int\int \phi^2(y)\hat{s}\left(\boldsymbol{r}_{\|}\right) \cdot \hat{\boldsymbol{S}}\left(\boldsymbol{r}_{\|}, y\right) \mathrm{d}^2 r_{\|} \mathrm{d}y, \qquad (10.1)$$

where three-dimensional spatial coordinates are split into $(\boldsymbol{r}_{\|}, y)$, y is the direction of the growth and $\boldsymbol{r}_{\|}$ is the $x-z$ plane projection parallel to the well. $\phi(y)$ is the single electron envelope wave function of the first quantized level in the well. We assume here that the splitting between the lowest levels is sufficient to neglect modification of the electron wave function due to interaction with other carriers and Mn spins. The $\hat{\boldsymbol{s}}(\boldsymbol{r}_{\|})$ and $\hat{\boldsymbol{S}}(\boldsymbol{r}_{\|}, y)$ denote spin density operators of the 2D conduction electrons and 3D Mn-electrons, respectively. The Mn spin density operator is given by $\hat{\boldsymbol{S}}(\boldsymbol{r}_{\|}, y) = \sum_i \hat{\boldsymbol{S}}_{\rm i}\delta(\boldsymbol{r} - \boldsymbol{R}_{\rm i})$, where $\boldsymbol{R}_{\rm i}$ runs over Mn sites. The constant $\alpha(>0)$ is the $s-p$ exchange integral between s-like conduction electrons and d electrons of Mn ion. A static magnetic field $\boldsymbol{B}_0 = B_0 z$ is applied in-plane and we will choose z as being the spin quantization axis. The standard DMS Hamiltonian of the two coupled subsystems in the presence of the field is then defined by [30]:

$$\hat{H}_{\rm DMS} = \hat{H}_{\rm Gas} + \hat{H}_{\rm sd} + \hat{H}_{\rm Mn} \tag{10.2}$$

$\hat{H}_{\rm Mn}$ is the Mn-spin Hamiltonian:

$$\hat{H}_{\rm Mn} = g_{\rm Mn}\mu_{\rm B} \iint \boldsymbol{B}_0 \cdot \hat{\boldsymbol{S}} {\rm d}^2 r_{\|} {\rm d}y, \tag{10.3}$$

where we neglect the direct antiferromagnetic coupling between Mn spins. This coupling results in pairing of Mn spins [28], which reduces the average amount of spin per cation site x to $x_{\rm eff}$ (denoted by x_0 in Chap. 1).

$\hat{H}_{\rm Gas}$ is the two-dimensional electron gas (2DEG) Hamiltonian in the presence of a static in-plane magnetic field $\boldsymbol{B}_0$ sufficiently low in magnitude to neglect any orbital quantization. In practice, this approximation is valid until the magnetic length $l_{\rm m} = (\hbar/eB_0)^{1/2}$ remains larger than the width w of the quantum well. Typical widths of quantum wells are in the range 100–200 Å, where this approximation is valid for magnetic fields as high as 5 T. The validity of the approximation can be easily checked experimentally by measuring the 2DEG luminescence line. Indeed, in the presence of Landau quantization, the luminescence line acquires structures periodically spaced with $\hbar\omega_{\rm c}$. Since we neglect the orbital quantization, the 2DEG Hamiltonian reads:

$$\hat{H}_{\rm Gas} = \sum_i \frac{\hat{\boldsymbol{p}}_{\rm i}^2}{2m_{\rm e}^*} + \frac{1}{2}\frac{1}{4\pi\varepsilon_{\rm s}} \sum_{i \neq j} \frac{e^2}{\left|\boldsymbol{r}_{{\rm i}\|} - \boldsymbol{r}_{{\rm j}\|}\right|} + g_{\rm e}\mu_{\rm B} \int \boldsymbol{B}_0 \cdot \hat{\boldsymbol{s}}(\boldsymbol{r}_{\|})\, {\rm d}^2 r_{\|}, \tag{10.4}$$

where the kinetic energy of electrons remains undisturbed by the magnetic field, $\mu_{\rm B}$ is the electron Bohr magneton ($\mu_{\rm B} > 0$); $g_{\rm e}$ and $g_{\rm Mn}$ are conduction and manganese g-factors. $\varepsilon_{\rm s}$ is the semiconductor static dielectric constant. At sufficiently low temperature, the system described by $\hat{H}_{\rm DMS}$ can undergo a ferromagnetic transition (see Chap. 1). In the ferromagnetic state, the two spin subsystems are strongly coupled so that they have a common dynamics dominated by mixed excitations involving both the conduction electrons and Mn-electrons spins [31]. Unfortunately, the ferromagnetic state in 2D electron gas in $Cd_{1-x}Mn_xTe$ quantum wells has never been observed as its Curie temperature is estimated to be in the range of 10 mK [30]. However, two-dimensional hole systems in $Cd_{1-x}Mn_xTe$ have been shown to become ferromagnetic even at 3 K for $x \sim 4\%$ [18–20] and up to 10 K in disordered system [32]. Nevertheless, investigation of spin-dependent interactions by either inter- or intraband spectroscopy is more convenient in the paramagnetic phase as it provides an additional tuning parameter: the spin-polarization degree which is a function of an external magnetic field. For this reason, we first restrict the discussion to the paramagnetic state of $\hat{H}_{\mathrm{DMS}}$.

10.2.2.2 Spin-Polarized 2DEG in a Paramagnetic Phase

To start, let us assume that in the paramagnetic state the two spin subsystems are weakly coupled. Therefore, their spin dynamics is independent (limits of this assumption will be discussed in Sect. 10.4.3). Further, we assume that the conduction spin density couples to the Mn-spin density only in the low frequency and long wavelength limit of the Mn-spin fluctuations. This is the core of the mean-field approximation already developed in Chap. 1. It means that each Mn spin is considered as frozen in the same statistic thermal average state $\langle \boldsymbol{S}\,(B_0, T)\rangle$ determined by the presence of $\boldsymbol{B}_0$ and $d{-}d$ exchange interactions. This average $\langle \boldsymbol{S}\,(B_0, T)\rangle$ is modeled by the modified Brillouin function [28]:

$$\langle \boldsymbol{S}(B_0, T)\rangle = \langle S_z(B_0, T)\rangle\, \boldsymbol{z} = S_0 B_{5/2}(B_0, T + T_0)\, \boldsymbol{z}, \qquad (10.5)$$

where S_0 and T_0 are empirical parameters appearing due to $d{-}d$ interactions (see Chap. 1). Reciprocally, Mn spins are coupled to a frozen nonfluctuating conduction spin density. The coupling Hamiltonian $\hat{H}_{\mathrm{sd}}$ can be replaced by two mean-field exchange terms[1]:

$$\hat{H}_{\mathrm{sd}} \approx -x_{\mathrm{eff}}\bar{N}_0\alpha \int \hat{\boldsymbol{s}}\left(\boldsymbol{r}_{\|}\right) \cdot \langle \boldsymbol{S}\,(B_0, T)\rangle\, \mathrm{d}^2 r_{\|}$$
$$-n_{2\mathrm{D}}\alpha \int\!\!\int \phi^2\,(y)\, \hat{\boldsymbol{S}}\left(\boldsymbol{r}_{\|}, y\right) \cdot \langle \boldsymbol{s}\rangle \mathrm{d}^2 r_{\|} \mathrm{d}y \qquad (10.6)$$

$n_{2\mathrm{D}}$ is the conduction electrons equilibrium density. In the first term, the y integral domain has been cut by the homogeneous Mn-distribution in the well of width w, hence, $\bar{N}_0 = N_0 \int_0^w \phi^2\,(y)\,\mathrm{d}y$ where N_0 is the number of cation sites per unit volume. The first (second) term in (10.6) leads to the Overhauser (Knight) shift.

---

[1] Some authors prefer using x_{eff} ($x_{\mathrm{eff}} \cdot S = x \cdot S_0$) when defining the magnetization. In (10.6), the value S_0 should thus be understood as equal to 5/2.

These terms, respectively, shift the intrisic Zeeman energies $g_e\,\mu_B\,B_0$ and $g_{Mn}\,\mu_B\,B_0$ of conduction and Mn- electrons. In the following, as we are interested in describing the spin-polarized 2D electron gas (SP2DEG), we will concentrate on conduction spin degrees of freedom. Thus, we keep in $\hat{H}_{DMS}$ only $\hat{H}_{Gas}$ and the first mean-field term of (10.6). In this, we obtain the $\hat{H}_{SP2DEG}$ Hamiltonian, which can be written in the second quantization formalism on the basis of plane wave one electron states as:

$$\hat{H}_{SP2DEG} = \hat{H}_0 + \hat{H}_{Coul} = \sum_{\boldsymbol{k},\sigma} E^0_{\boldsymbol{k},\sigma} c^+_{\boldsymbol{k},\sigma} c_{\boldsymbol{k},\sigma} + \frac{1}{2} \sum_{\boldsymbol{q},\boldsymbol{k},\boldsymbol{k}',\sigma,\sigma'} V(\boldsymbol{q})\, c^+_{\boldsymbol{k}+\boldsymbol{q},\sigma} c^+_{\boldsymbol{k}'-\boldsymbol{q}',\sigma'} c_{\boldsymbol{k}',\sigma'} c_{\boldsymbol{k},\sigma}, \tag{10.7}$$

where $E^0_{\boldsymbol{k},\sigma} = \hbar^2 k^2/2m^*_e + sgn(\sigma)\, Z(B_0)/2$ is the energy of single electron state with the spin $\sigma = \uparrow, \downarrow$ ($sgn(\uparrow) = +1$) and the in-plane wavevector $\boldsymbol{k}$, $V(\boldsymbol{q}) = F(q)\, e^2/2\varepsilon_s q L^2$ is the spatial Fourier transform of the bare Coulomb potential, expressed as product of the 2D Coulomb interaction with a form factor $F(q)$ depending on $\phi(y)$ [33], L^2 is the sample area. In the second summation in (10.7), the $q = 0$ term has to be omitted because it is canceled by the positive charge background of the ionized donors, which suppresses the $V(\boldsymbol{q})$ divergence. $Z(B_0)$ is the total bare Zeeman energy of the conduction electrons, given by the sum of the Overhauser shift and the intrinsic Zeeman term:

$$Z(B_0) = -x_{eff}\bar{N}_0\alpha\,\langle S_z(B_0, T)\rangle - |g_e|\,\mu_B B_0. \tag{10.8}$$

We have explicitly expressed the opposite signs of the two Zeeman contributions: the Overhauser shift is positive because the CdTe s–d exchange integral [28] $N_0\alpha = 0.22\,\text{eV}$ and the Mn g-factor $g_{Mn} = 2.007$, which means that Mn spins are antiparallel to the magnetic field, while the intrinsic Zeeman splitting is negative because of the electron g-factor [34] $g_e = -1.64$. In practice, for well width as large as 150 Å, the electrons are strongly confined in the well, so $\bar{N}_0 \approx N_0$. As we neglect the dynamical coupling with Mn-spins, the energy $Z(B_0)$ represents the total static external magnetic influence on the electrons in the well. Even for low Mn concentration such as $x = 1\%$ ($x_{eff} \approx 0.86\%$) [35], $Z(B_0)$ has a maximum value of $Z_{max} \approx 4.75\,\text{meV}$, which is comparable with usual value of the Fermi energy $E_F \approx 10\,\text{meV}$ of the unpolarized 2DEG. This will induce high equilibrium spin polarization degree $\zeta = (n_\uparrow - n_\downarrow)/(n_\uparrow + n_\downarrow)$. Without the Coulomb interaction between the electrons, the "bare" spin polarization degree $\zeta_0$would be fully determined by Z:

$$\zeta_0 = -Z/2E_F = -m^*_e Z/\hbar^2 k_F^2 \tag{10.9}$$

Here, we have assumed zero temperature and parabolic bands; $k_F = \sqrt{2\pi n_{2D}}$ is the Fermi wavevector. When the Coulomb interaction is retained, we will see in the following that the bare Zeeman energy is enhanced [36] to Z^*, hence the net spin polarization ζ is also greater than ζ_0:

$$\zeta = -Z^*/2E_F \qquad (10.10)$$

Equivalent formulae might be obtained for the heavy hole gas. Differences are that $N_0\alpha$ has to be replaced by $N_0\beta$ which is usually larger (four times larger in (Cd,Mn)Te), the effective mass and, strong spin anisotropy of hole. Particularly, the anisotropy appearing when the direction of magnetic field is rotated in the plane, has to be taken into account.

It is remarkable that, thanks to the DMS giant Zeeman effect, we deal here with an original and model situation: a highly polarized two-dimensional paramagnetic conducting system embedded in a semiconductor heterostructure. Indeed the Hamiltonian (10.7) is similar to that of a paramagnetic metal whose magnetic excitations have been described in the past in the 3D case [37] and are commonly measured by electron spin resonance [38]. The difference between our case and the paramagnetic metal is that the spin polarization degree is here much greater and comparable to that of a ferromagnetic metal. In the latter, magnetic excitations are zero-sound spin waves [39] and Stoner excitations [40] centered at a very high energy $2E_F\zeta$. They can be probed by spin-polarized electron energy loss spectroscopy [41]. We will see in the following that resonant Raman scattering experiments on two-dimensional semiconductor structures give access to the dispersion of magnetic excitations. Figure 10.2 shows the model band structure of the system with SP2DEG. Another important difference between metals and our system is related to relatively low carrier density in semiconductors. Typically in p or n-type systems $k_F^{-1} \sim 200$ Å, which has to be compared with the typical Mn–Mn average distance ~ 5 Å. This justifies the mean-field approach [42].

10.3 Properties of Quantum Well with Spin-Polarized Carrier Gas: Interband Spectroscopy

The interband spectroscopy is a powerful tool of investigation of the 2D systems. This is a routine method of the quantum well characterization used in semiconductor technology. It gives fast and easy access to parameters of a quantum well such as its width, homogeneity, and strain (Chap. 4). There are also techniques particular to DMS materials such as spin tracing (Chap. 4) developed and used for studies of interface mixing [43]. All of them are based on relatively simple measurements of the absorption or photoluminescence spectra or equivalent experiments. The interband spectra are usually dominated by exciton transitions. This is particularly true at low- or zero-magnetic field. However, the introduction of carriers leads to important changes in the spectra. The neutral exciton line is quenched, and simultaneously a charged exciton line appears. As the carrier concentration increases, an evolution toward Fermi Edge Singularity (FES) and band-to band transitions is observed.

Studies of the impact of carriers on the exciton transitions were performed for many nonmagnetic systems. The carrier gas was obtained either by doping or by optical excitation. Several important questions related to properties of charged excitons have been addressed in studies of II–VI quantum wells. For example, detailed studies were performed of: (1) the selection rules [44–46], (2) the dynamics [47,48], (3) the role of the carrier gas localization [49, 50], (4) the localization of charged exciton in QW [50–52], (5) the stability of the complex in high magnetic field and (6) the interaction between neutral and charged exciton states [52, 53]. The properties of charged excitons in quantum Hall regime were also studied [44, 54]. In particular, several optical transitions, such as exciton cyclotron resonance and resonance with charged exciton, were identified [23, 55].

Some results were obtained for spin-polarized carrier gas created by strong optical excitation using polarized pulses of light [56]. However, such studies were limited to relatively high temperatures of the carrier system. It was only owing to the use of the DMS that a freedom of independent tuning of the densities of carriers in different spin subbands appeared.

10.3.1 Low Carrier Density: Charged Excitons

At low carrier densities, the fingerprint of the presence of carriers in the QW is usually an observation of the charged exciton. In the simplest picture the charged exciton is a three-particle complex: a trion. The fundamental state of the trion is spin singlet and its direct creation by light requires presence of a preexisting carrier. During formation process, the electron-hole pair is bound to a majority carrier. Spins of the two majority carriers in given trion are opposite. In the simplest case of the absorption experiment, the electron–hole pair is photocreated, and the spins of photocreated carriers are defined by the circular polarization of the light. Thus, the polarization of the trion absorption is sensitive to the spin polarization of the preexisting carrier gas.

We will begin the discussion with the simplest case of the charged exciton observation. Transitions related to charged excitons are observed both in absorption and in photoluminescence (PL) experiments. They are seen as narrow lines below the energy of the neutral excitons (see Fig. 10.1). The distance in energy between the neutral and charged exciton (dissociation energy) depends on the semiconductor system. For bulk crystals, it is typically only a few percent of the neutral exciton binding energy [57]. It is usually too small to make charged exciton line sufficiently resolved for experimental observation. Confinement in a quantum well enhances the neutral exciton (X) binding energy as well as the charged exciton dissociation energy. Similarly as X-binding energy, the dissociation energy varies with QW width and increases significantly with confinement of the carriers. In different samples the dissociation energy varies from <1 meV to >10 meV but always remains smaller than the binding energy of the neutral exciton to neutral donor (D^0X) or acceptor (A^0X) in the same quantum well. For example, in an 80 Å-wide

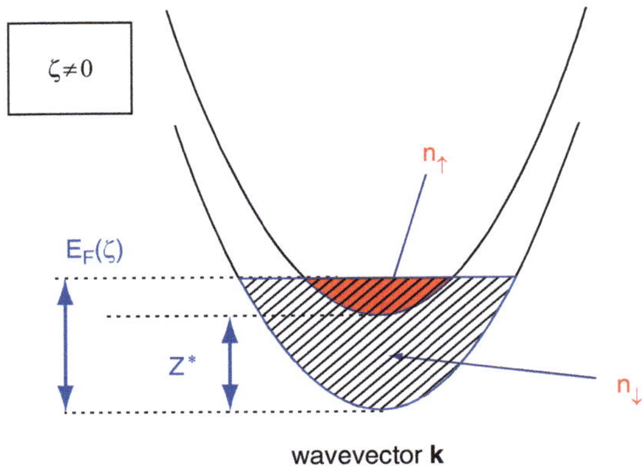

Fig. 10.1 Schematics of the model spin-polarized two-dimensional carrier gas with equilibrium spin polarization $\zeta = (n_\uparrow - n_\downarrow)/(n_\uparrow + n_\downarrow)$. $E_F(\zeta)$ is the Fermi energy of the spin-polarized ground state; $E_F(0) = E_F$. Z^* is the renormalized Zeeman energy

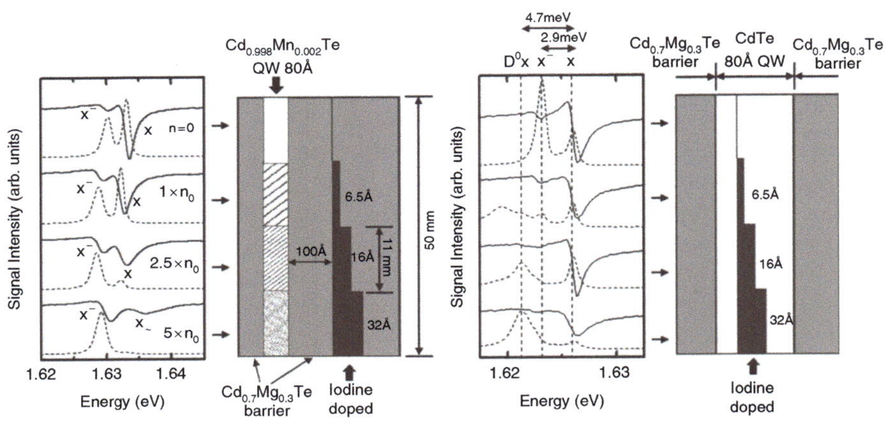

Fig. 10.2 Photoluminescence (*dashed lines*) and reflectivity (*solid lines*) of four regions of "wedge"-doped structures with iodine donors introduced either 100 Å away from $Cd_{0.998}Mn_{0.002}Te$ QW (*left panel*) or inside the CdTe QW (*right panel*). The width of QW was 80 Å in both structures. Details of the structure designs are shown by the schemes on the right-hand side of each panel. Different shading of the QW region in the *left panel* represents varying 2DEG concentration ($n_0 = 3 \times 10^{10}\,\mathrm{cm}^{-2}$). From [24]

CdTe quantum well with (Cd,Mg,Zn)Te barriers, these energies are approximately 2 meV, 2 meV, and 4 meV for X^-, X^+, and D^0X respectively [11,24,58]. The example dependence of X^- dissociation energy versus QW width in CdTe/(CdMg)Te quantum wells is presented on Fig. 10.3a. Similar dependences were obtained for different materials such as ZnSe, and for both kinds of trions: positively and negatively charged [59–61]. The experimental values usually agree well with theoretical predictions which were done by many authors [60, 62–65]. However, comparing

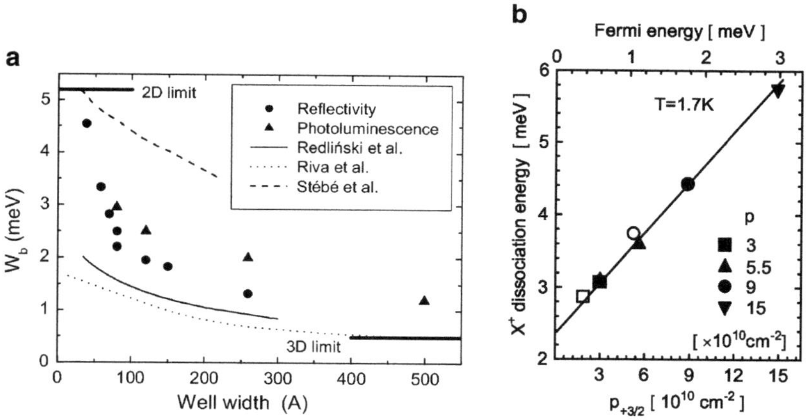

Fig. 10.3 (**a**) Dissociation energy of negatively charged exciton as a function of the CdTe/(CdMg)Te quantum well width. *Points* represent experimental data, *curves* were taken from numerical simulation by Riva [63], Stebe [62], and Redliński [60]. (graph from Kutrowski [61]). (**b**) Dissociation energy measured as a distance between X and X^+ absorption lines as a function of the hole gas concentration in the spin subband promoting the X^+ formation (*full symbols*: complete spin polarization of holes, *open symbols*: no applied field). Data obtained for 8 nm wide $Cd_{0.998}Mn_{0.002}$.Te quantum well with $Cd_{0.66}Mg_{0.27}Zn_{0.07}$Te barriers, (graph from [70])

experimental and theoretical data has to be done with some caution. Most theoretical models consider charged exciton as a three-particle complex (two holes and one electron – X^+ or two electrons and one hole – X^-). In real experiment we deal with nonzero carrier concentration. Actually it was shown that the distance between charged and neutral excitons seen in the absorption increases with carrier gas concentration. This leads to scatter of experimental data. To be correct, one should compare the experimental value extrapolated to the zero carrier concentration with a theoretical one.

The dependence of the dissociation energy on the background carrier concentration was observed and studied in many absorption experiments without and with magnetic field. The increase was found to be linear with the Fermi energy of the degenerate carrier gas. This effect is common to positively [17] and negatively charged excitons [66], and occurs in different materials [59, 67]. This observation might be explained in a simple intuitive way: both the neutral exciton and the charged exciton are due to the existence of a bound level, which appears in the 2D gas in the presence of a carrier of opposite sign (i.e., of a hole in the case of an electron gas, or of an electron in the case of a hole gas). The neutral exciton then corresponds to a single occupancy of this level. The charged exciton involves, in addition to the creation of the exciton, the transfer of a carrier of opposite spin from the Fermi level down to the bound level. If we assume that the binding energy of both states is the same, we obtain that the X-X^+ (X-X^-) splitting contains, in addition to the binding energy, a contribution equal to the Fermi energy. A similar effect was predicted by Hawrylak [68] for high concentration of electron gas and for dispersionless holes. A more accurate calculation still remains a challenge [69]. Nevertheless, several aspects of the experimental findings can be explained. For example, it was shown that the dissociation energy is a linear function of the concentration of carriers in one hole spin subband only (the one with the spin opposite to that of the photocreated hole) (see Fig. 10.3b).

The slope of the linear dependence is very close to the predicted value of 1. The scatter of values reported in the literature results most probably from the accuracy of the carrier density determination. For example, a value of 1.7 for X^+ in [17] is due to underestimation of the hole gas density by a factor 1.5 [71].

10.3.1.1 Intensity of Neutral and Charged Absorption Lines

The simplest property, which provides information about the density of the carrier gas, is the intensity of the absorption or reflectivity lines related to the charged exciton. An example of spectra for different carrier gas concentrations is presented in Fig. 10.1 [17, 24]. The relative intensity of charged and neutral excitonic lines depends strongly on carrier gas density and the temperature. Neutral exciton dominates both kinds of spectra (PL and reflectivity) for the lowest densities. When the carrier density increases, the low-energy (trion) line becomes more intense while the other one weakens and finally disappears. The limit for neutral exciton observation is different for absorption (reflectivity) and PL experiments. For holes, it is about $6 \times 10^{10}\,\mathrm{cm}^{-2}$ and $2 \times 10^{10}\,\mathrm{cm}^{-2}$, respectively. In absorption, the presence of carriers attenuates the neutral exciton line. Simultaneously, the presence of carriers, necessary for creation of charged excitons, enhances the X^+ or X^- line with respect to the X line. For low-carrier density, the intensity of X^+ or X^- line is proportional to the density [11, 45]. At higher carrier gas densities, the intensity starts to saturate. This initial proportionality may be used to identify charged exciton transition. However, the precise value of carrier density is not always easy to determine.

A proper identification of the trion line might be based on selection rules of absorption in a magnetic field. The fundamental state of the charged exciton (in zero- and low-magnetic fields) is a singlet. Therefore, the photo-created carrier has to be of opposite spin than the preexisting one. Defined spin of the preexisting carrier determines circular polarization of absorbed photon. Particularly, this is the case of experiments in a magnetic field when the carrier gas is spin polarized. Then the absorption line can be observed in only one circular polarization. The necessary condition for such simple selection rule is sufficiently large value of the g-factor (spin splitting larger than the Landau splitting). This condition is particularly easy to fulfill in semimagnetic semiconductors where the g-factor may be of the order of thousands [28].

Figure 10.4a shows an example of transmission spectra at a low hole density, taken at several magnetic fields in both circular polarizations. We observe a characteristic population behavior of the low-energy line: its intensity increases with the field in σ^- polarization, while it decreases in σ^+. At the field of about 0.3 T, the σ^+

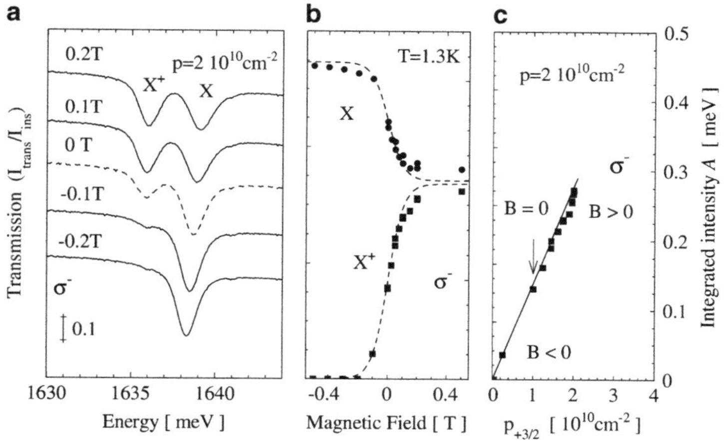

Fig. 10.4 Transmission of 8 nm wide $Cd_{0.998}Mn_{0.002}Te$ quantum well with $Cd_{0.66}Mg_{0.27}Zn_{0.07}Te$ barriers in a magnetic field, obtained for low hole gas concentration ($p = 2 \times 10^{10}\,cm^2$) at the temperature 1.3 K: spectra (**a**), integrated line intensities as a function of a magnetic field (**b**), and of the hole gas concentration in the spin subband occupied by holes with proper spin orientation allowing X^+ formation (**c**) [70]

component disappears completely. In σ^+ polarization, the creation of X^+ exciton in its singlet state involves a photocreated spin-up hole and a preexisting spin-down hole. Thus, this observation reflects the absence of such spin-down holes. Note, however, that in a $Cd_{1-x}Mn_xTe$ QW, contrary to CdTe QWs, the giant Zeeman effect in the conduction and valence bands leads to the same sign of the circular dichroism for charged excitons of both signs.

Let us note that g-factors of electron and hole are different in most cases. In the case of (CdMn)Te, the giant Zeeman splitting is four times larger in the valence band than in the conduction band. Knowing the Zeeman splitting and the total carrier density, it is possible to calculate carrier densities in both spin subbands separately. An example of a quantitative test of a proportionality between the carrier density and absorption intensity was done for the X^+ transition [17]. First, the measured exciton splitting was fitted with phenomenological expressions given in [43] for the bulk. Then, the intensities of both transitions were analyzed by fitting two Gaussian functions to the absorption spectra. The integrated intensity of the respective lines is presented versus magnetic field in Fig. 10.4b. At a constant total hole concentration p, we expect the intensity of the X^+ line to be proportional to the population of holes with the appropriate spin, i.e., $S_z = 3/2$ in the σ^+ polarization. Using the Maxwell-Boltzmann distribution between the Zeeman split subbands to describe the hole concentration, one obtains for the intensity,

$$A_{\pm}(Z) = A(\infty)\frac{1}{1+\exp(\mp Z/k_{\mathrm{B}}T)}, \tag{10.11}$$

where Z denotes the valence band Zeeman splitting and is taken to be positive for one spin direction and negative for the other one, T is the carrier gas temperature, and k_B is the Boltzmann constant. $A(\infty)$ is the maximum intensity, achieved at complete hole spin polarization. Indices + and − of the intensity denote the two circular polarizations of light. Results shown in Fig. 10.4c confirm the proportionality of the X^+ intensity to the population of holes with the relevant spin. The Maxwell-Boltzmann distribution was used to describe the population ratio at low hole density where the holes are likely to be localized [2,52,72]. The same assumption was done in [11] for a CdTe QW. At higher densities, the Fermi–Dirac distribution should be used (see (10.9)). In spite of the fact that carriers with higher k-vectors give smaller contribution to the absorption oscillator strength, it was shown that even for relatively high hole gas concentrations the variation of the X^+ or X^- oscillator strength in magnetic field might be described as proportional to the carrier population in one spin subband [16, 17].

This proportionality was tested for different systems. Quantum wells with extremely low Mn concentration are of particular interest. Their giant Zeeman splitting may be comparable to the splitting related to intrinsic g-factors. The giant Zeeman contribution saturates in a magnetic field of a few tesla, while the intrinsic component is linear with the field. If the sign of both components is opposite, it is possible to find a value of magnetic field for which the total Zeeman splitting vanishes. It was demonstrated for (Cd,Mn)Te QW with electrons [55]. The electron splitting changes its sign at 10 T for 0.2% Mn. The intensity of the charged exciton absorption was following the electron gas density in a proper spin subband.

The change of the trion oscillator strength is accompanied by a significant change of the neutral exciton line intensity. The two changes are proportional to each other if the total carrier density is kept constant (see Fig. 10.4b). To describe this opposite behavior, the idea of "intensity stealing" has been suggested [11, 73], which could be justified by the existence of a sum rule in a closed system formed by the neutral exciton and the charged exciton. However, the total intensity (in a given polarization) is generally observed to vary as the carrier population changes, which suggests that some intensity is transferred, e.g., to band-to-band transitions. The use of DMS quantum wells allowed to change the population of each spin subband independently and to understand the meaning of the different components of the line intensities.

It was established that to a reasonable approximation the two oscillator strengths, measured at constant total hole concentration p, are linear functions of densities of holes with the appropriate spin ($p_\pm$). Moreover, it is well known that the exciton intensity is reduced in the presence of carriers, due to various effects, such as screening and phase space filling (PSF). This reduction has been widely studied in nonmagnetic quantum wells, as a function of the total carrier density [73–75]. Therefore, the total intensities of X and X^+ lines might be described by phenomenological expressions:

$$A_\pm^X(p, p_+, p_-) = A_0^X \eta_X(p)(1 - \sigma\, p_\mp) \tag{10.12}$$

and

$$A_{\pm}^{CX}(p, p_+, p_-) = A_0^{CX}\sigma\, p_{\mp}\eta_{CX}(p), \tag{10.13}$$

where the reduction factor $\eta_{CX}(p)$ depends only on the total population, and not on its distribution between the two spin states. The same is true for the neutral exciton reduction factor $\eta_X(p)$. The "cross section for intensity stealing" was introduced as σ. The same factor σ was chosen to scale the trion absorption, so that A_0^X and A_0^{CX} can be compared directly: Their ratio depends on the extension of the wave functions in the two forms of the exciton.

The example parameters were determined for X^+ in 8 nm quantum well [17]. The cross-section for intensity transfer corresponds to $\sigma^{-1} = 7 \times 10^{10}\,\text{cm}^{-2}$. This describes the interaction between an exciton and a free hole with antiparallel spin, which gives rise to the formation of X^+. It corresponds to the radius 200 Å, much larger than Bohr radius of exciton. Such large cross-section describing the possibility to form a charged exciton is reminiscent of the giant oscillator strength of excitons bound to donors in bulk semiconductors. The characteristic spin independent reduction factors were described by phenomenological functions $\eta_X = \exp(-p/p_X)$ and $\eta_{CX} = \exp(-p/p_{CX})$ with a characteristic density $p_x = 5.8 \times 10^{10}\,\text{cm}^{-2}$, and $p_{cx} = 7.0 \times 10^{10}\,\text{cm}^{-2}$. The extrapolated optical densities of neutral exciton and charged exciton were $A_0^X = 0.9\,\text{meV}$ and $A_0^{CX} = 1.3\,\text{meV}$, respectively (Fig. 10.5).

The proper understanding of the obtained dependences was achieved as a result of time-resolved pump–probe experiments [76, 77]. Ultra short light pulses were used to generate a significant population of neutral or charged excitons. The controlled occupation of the selected states allowed us to distinguish between phase space filling and other mechanisms, contributing to the empirical formula (10.13). To excite only one transition (charged or neutral exciton), the pump pulse was shaped to a spectral width <1 nm and a duration of about 2 ps, thus creating either X or X^+, depending on the photon energy. The spin of photocreated carriers was

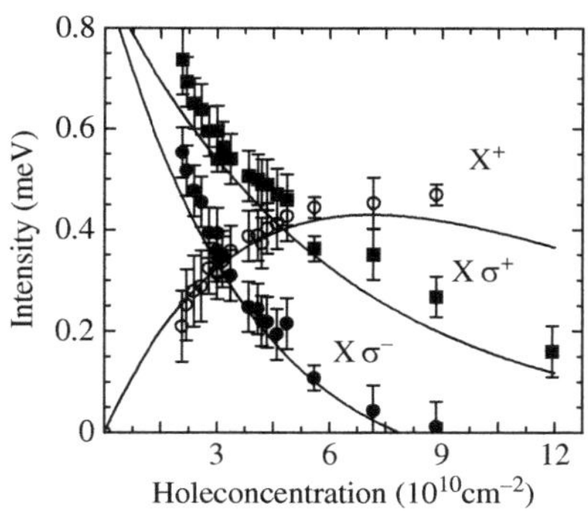

Fig. 10.5 Integrated intensity as a function of the hole gas concentration obtained for holes fully polarized by the magnetic field (*squares* – σ^+ polarization, *circles* – σ^- polarization). *Full symbols* represent the X line and *open symbols* the X^+ line [17]

controlled by circular polarization of the pump beam. The pump and probe pulses were focused to a common spot on the sample with diameter smaller than 100 μm, and the spectrum of the probe pulse transmitted through the sample was recorded as a function of the pump–probe delay. The duration of the probe laser pulse was below 100 fs and the spectral width was sufficient to use laser light for obtaining whole transmission spectra.

The average power of the pump beam was typically 300 μW, which results in the creation of a few times 10^{10} cm^{-2} excitons by each pulse. The probe beam was detected behind the sample in both circular polarizations, measuring the absorption associated with the creation of either X or X^+ having electron–hole pairs of the same spin as the pump (σ^+, co-polarized) or opposite to it (σ^-, cross-polarized). Therefore, one could analyze the influence of the charged exciton on the neutral exciton and separate spin-dependent contribution by separation of the light polarization. Additionally, by applying magnetic field we were able to resolve all four polarization configurations: pump with σ^- and probe with σ^+ $(-/+)$ or σ^- $(-/-)$ polarization, and pump with σ^+ and probe with σ^+ $(+/+)$ or σ^- $(+/-)$ polarization. To analyze the neutral and charged exciton line intensities, two Gaussian functions were fitted to the cw or time-resolved spectra. In some experiments, difference spectra were analyzed to expose the change in the optical absorption under the influence of the pump pulse.

Figure 10.6a shows optical density spectra of the quantum well containing 2.5×10^{10} holes per cm^2, taken 9 ps before and 0.5 ps after the pump pulse. The pump beam was tuned to the charged exciton energy and both beams had the same circular polarization. A decrease of the X^+ line intensity is visible, accompanied with an increase of the X line intensity. The bleaching of the charged exciton line is

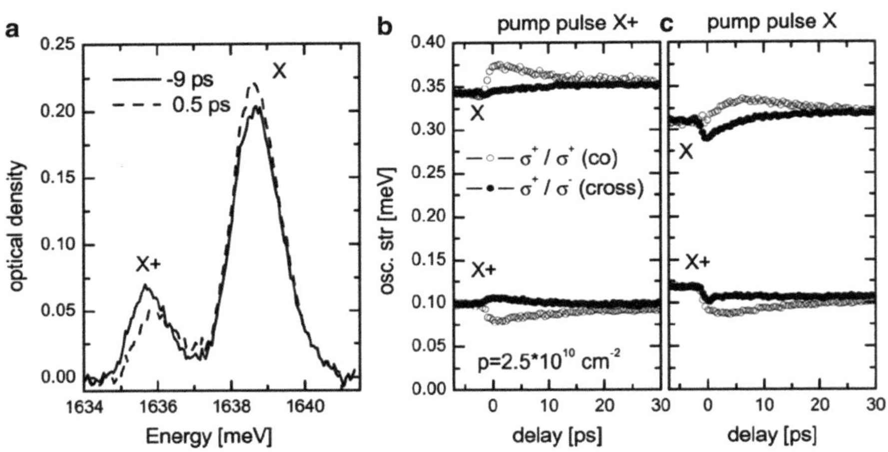

Fig. 10.6 (**a**) Optical density for a QW with hole density $p \approx 2.5 \times 10^{10}$ cm^{-2}, at negative delay (-9 ps, *solid line*) and short positive delay (0.5 ps, *dashed line*), for a co-polarized pump pulse tuned to X^+. (**b**) Evolution of the X and X^+ oscillator strength with pump pulse tuned to X^+. (**c**) Evolution of X and X^+ with pump pulse tuned to X [76]

easy to explain: photocreation of a charged exciton requires the presence of a free hole; to a first approximation, the charged exciton line intensity is proportional to the free-hole density. The strong pump pulse binds a significant fraction of the holes in photocreated charged excitons and thus creation of new charged excitons is less probable since less holes are available. The most important result is the increase of the neutral exciton line intensity. It demonstrates that free holes screen electron–hole interaction in the exciton, reducing its oscillator strength. The pump pulse, creating the charged excitons, binds the free holes, excludes them from the screening, enhancing thus the neutral exciton line intensity. This observation proves that the effect of "oscillator strength stealing" is not caused by PSF. The holes bound in trions are still present in the proper spin subband and occupy phase space in which photocreated carriers are formed. Figure 10.6b shows the evolution of both line intensities as a function of delay of the probe after the pump pulse tuned to the X^+ energy. Significant intensity changes occur only at the same polarization of both pulses, whereas at cross-polarization no significant effect is visible. This shows clearly that the interactions involved are spin dependent since each light polarization selects a spin orientation of excitons. The absence of X^+ bleaching in crossed polarizations stems from the fact that holes of only one spin orientation (opposite to that of the photocreated ones) can participate in the charged exciton creation. This restriction is imposed by Pauli exclusion principle, which requires that the two-hole spins in the charged exciton are opposite. The same Pauli principle prevents the free holes from approaching neutral excitons with the same hole spin to produce an efficient screening. Both X^+ bleaching and X intensity enhancement are thus caused by free holes of the same spin orientation, and observed in the same polarization configuration.

The detailed analysis of the data in the magnetic field provided also information on the interaction between neutral excitons created by the same (as well as opposite) circular polarizations of light. In both cases, the optical density of the neutral exciton line was reduced. The reduction was described by an effective cross-section. The corresponding radius for the interaction with a neutral exciton was determined to be 80 Å in both polarizations [77]. In the case of co-polarized spectra, it has been described quantitatively as a result of phase space filling [78]. This mechanism, however, does not apply to cross-polarized spectra. In the latter case, the main effect to consider is the possibility to form – or not – a biexciton. However, the cross-section which describes the formation of the biexciton from a first exciton created by the light in the cross-polarized spectra (and the simultaneous decrease of the neutral exciton) has a value close to that representing PSF. Actually, it is not unexpected that the excitonic Bohr radius give the order of magnitude of both effects, PSF and biexciton formation. In short, within a cross-section which is essentially determined by the excitonic Bohr radius, the absorption of a photon is forbidden for one polarization due to PSF, while it gives rise to the formation of a biexciton in the other polarization.

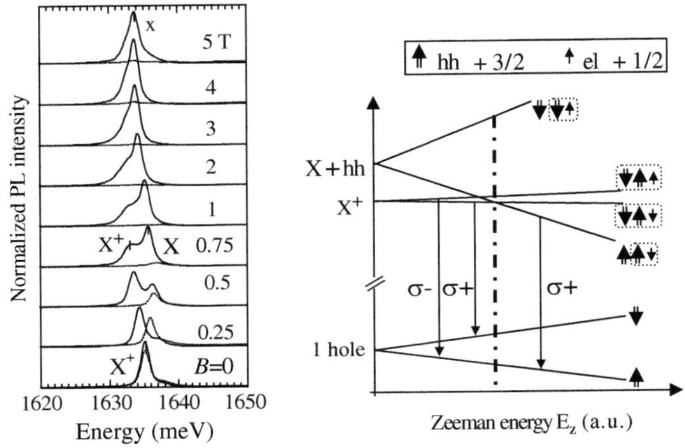

Fig. 10.7 *Left panel*: PL spectra, at 1.7 K, at various values of the applied field, from a $Cd_{0.9963}Mn_{0.0037}Te$ QW. *Solid lines* are for σ^+ spectra and *dotted line* for σ^-. *Right panel*: Three-carrier diagram of optical transitions, *single arrows* show the electron spin and *double arrows* the hole spin; the *dash-dotted line* marks the level crossing [71]

10.3.2 Photoluminescence

The giant Zeeman splitting of the hole state can be large enough to destabilize the singlet state of the positively charged exciton. A more favorable situation corresponds then to a neutral exciton (emitting in σ^+ polarization) and a free hole with the same spin direction as the hole forming the exciton (triplet configuration).

Figure 10.7 shows example PL spectra at different values of the magnetic field, in both circular polarizations, observed in sample with 0.37% Mn in the QW. The carrier density was estimated to be below $2 \times 10^{10}\,\mathrm{cm}^{-2}$. At zero field, the spectrum is dominated by a single line related to the charged exciton. A much weaker line due to the neutral exciton appears as a shoulder on the high energy side. On application of the magnetic field, the relative intensity of both lines changes. In σ^- polarization, the neutral exciton line (the shoulder) disappears completely above 0.3 T, and the intensity of the charged exciton line progressively decreases; it disappears at 1 T. In σ^+ polarization, the PL intensity of the neutral exciton increases with respect to the X^+ one. The intensities of both lines become comparable at 0.6 T, and at higher fields the neutral exciton line dominates the spectra. This transition from the charged to the neutral exciton is governed by the value of the giant Zeeman splitting. It results from different Zeeman splittings of the three-particle system (two holes and an electron) either arranged as a charged exciton, or containing a neutral exciton. A schematic diagram of the three-particle levels is shown in the right panel of Fig. 10.7. On application of a magnetic field, the two states of the three-particle system experience different giant Zeeman shifts. In the fundamental state of the charged exciton, the two holes are arranged in a singlet state, therefore the Zeeman shift of the three-particle system is equal to the shift of the electron only, which is rather small: in (Cd,Mn)Te, the giant Zeeman shift of the heavy hole is four times larger than the electron shift. The bright state of the neutral exciton is composed of a hole and an electron with opposite spin directions. The remaining hole is free, therefore the component with two holes in the majority band (which emits in σ^+ polarization) shows a strong redshift under applied field, equal to the sum of the individual shifts of the electron and the two holes. At zero field, the charged exciton state is lower in energy since it is a bound state, and this remains true at low field. When the heavy-hole Zeeman splitting is equal to the X^+ dissociation energy (Fig. 10.7), the neutral exciton level crosses the level of the charged exciton emitting in the σ^+ polarization, so that the neutral exciton level becomes lower in energy. The mechanism is similar to the crossing induced by the giant Zeeman splitting between the A^0X bound excitons and free excitons in bulk diluted magnetic semiconductors [79].

The ratio of the X and X^+ intensities experimentally observed varies smoothly, making it difficult to point precisely in the crossing. Actually, the intensity ratio results from the dynamics between the X and X^+ states, and, in addition, disorder introduces a broadening of the densities of states.

It was recognized [47] that the dynamical nonequilibrium distribution between the X and X^+ populations leads to the observation of some PL intensity from the upper level. In particular, at zero field, the neutral exciton luminescence is observed due to a formation time of the charged exciton comparable with the decay time of the neutral exciton. This formation time is strongly sensitive to the carrier density [70]. As the carrier density is increased, it becomes shorter so that a more abrupt transition is expected.

10.3.3 Spectroscopy of Spin-Polarized Carrier Gas with High Densities of Carriers

An increase of the carrier gas density results in an evolution of the shape of the photoluminescence spectra. For very low densities, the emission spectrum consists of two sharp peaks corresponding to recombination of the trion and of the neutral exciton. As the density increases, the X peak weakens very fast. Initially, the trion peak remains quite sharp and intense. For larger densities, the emission spectrum broadens drastically and a redshift occurs. The main intensity is now at the lower energy edge, and the line width between the main peak and a weak, high-energy shoulder is approximately equal to the Fermi energy [80, 81]. Such spectra, characteristic for degenerate carrier gas, were observed for n-type samples and densities above 10^{11} cm^{-2} (see Fig. 10.8). In p-type samples, the main PL line keeps its charged exciton character also at high carrier densities (at least up to 5×10^{11} cm^{-2}). This is probably due to the larger effective mass of heavy holes. The Fermi energy is smaller for the same carrier density and, therefore, the binding energy of the charged exciton is relatively larger in the heavy-hole system. The band-to-band character of the photoluminescence was observed only for spin polarized hole gas.

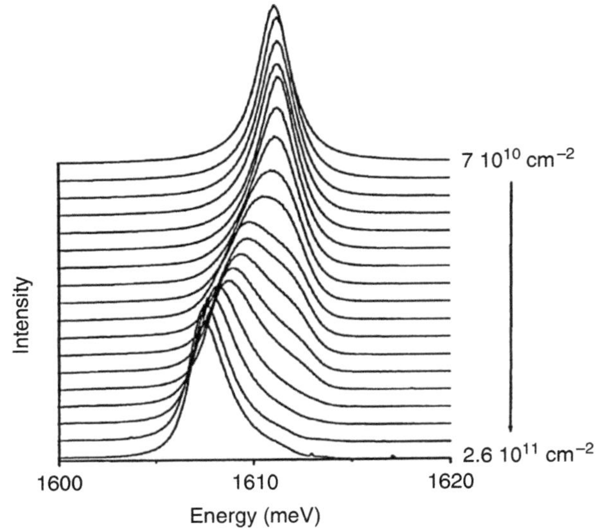

Fig. 10.8 Emission spectra at 2 K for a 100 Å-wide CdMnTe (0.2% Mn) quantum well as a function of electron density varied by illuminating the cap layer with blue laser light (from [81])

The presence of the giant Zeeman splitting in DMS QWs additionally suppresses the charged exciton also at higher carrier densities as we will see. Similarly to the case of very low hole densities, the giant Zeeman splitting of the hole can be large enough to destabilize the singlet state of the charged exciton. We will discuss this effect in a greater detail in the case of *p*-type (Cd,Mn)Te/(Cd,Mg)Te quantum wells.

Figure 10.9a shows PL spectra, at different values of the applied magnetic field in both circular polarizations, observed in the sample with the carrier density estimated to be $p = 4.2 \times 10^{11}\,\mathrm{cm}^{-2}$. Similarly to the low-density case, a single line is observed at zero field. At low field (see the spectra at 0.5 T), the giant Zeeman effect is already sizable; the high energy line is σ^--polarized and is reasonably sharp, while the low energy line, σ^+-polarized, is broader, with a weak tail on the low energy side. The Zeeman splitting observed here is the same as that measured for the same sample with the carrier gas depleted. At higher fields, the PL is fully σ^+ polarized, with a typical double structure containing a high energy line labeled D_{high}, and a low energy line labeled D_{low}. The transition from the single to the double line in σ^+ polarization occurs at about 0.6 T, close to the value causing the X/X^+ crossing observed for the same sample but at low carrier density (marked by the dashed–dotted lines in Fig. 10.9b, c). The shape of the double line remains constant up to 3 T in Fig. 10.9a. Above this value of the field, a progressive change of the shape occurs, which is attributed to emergence of the Landau levels.

The transition between the two kinds of PL spectra (excitonic line and the double-line) is related to the level crossing of the initial states, with the first state involving a pair of holes forming a singlet (charged exciton recombination at low field), the other one with all hole spins having the same orientation. Here, the field at which

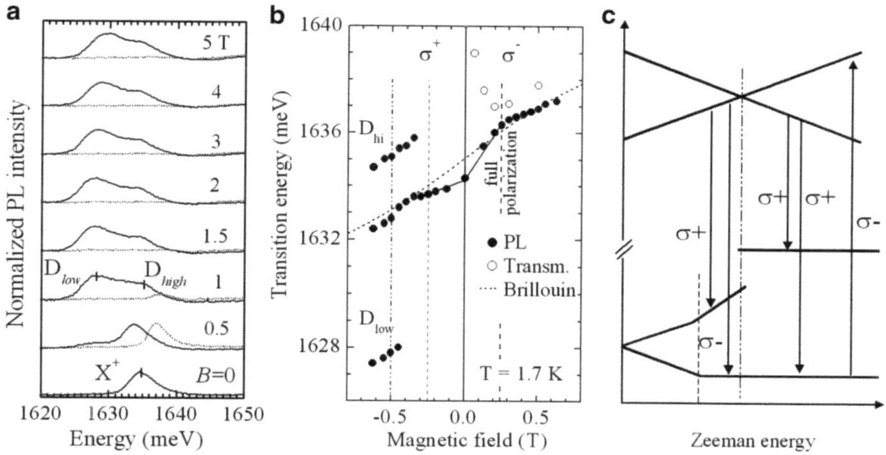

Fig. 10.9 (**a**) PL spectra of a $Cd_{0.9963}Mn_{0.0037}Te$ QW, at 1.7 K, at various values of the applied field. *Solid lines* are for σ^+-polarized spectra and *dotted line* for σ^-, (**b**) Positions of the transmission lines (*open circles*) and PL lines (*full circles*), vs. the applied magnetic field. The σ^- data are plotted at negative fields. The *dotted line* represents the Brillouin function. (**c**) schematic diagram of the initial and final states and of the optical transitions, In panels (**b**) and (**c**), *dashed vertical lines* indicate the threshold field values for complete hole polarization, and the *dotted vertical line* indicates the destabilization of X^+ [71]

the transition takes place is defined by a value of the valence band Zeeman splitting, and the transition is more or less abrupt, depending on the relaxation processes between the two levels. It is worth mentioning that, as expected, the transition is more abrupt in stationary experiments with very weak cw excitation, than in time-resolved experiments for which the system is excited further from quasi–equilibrium conditions.

The determination of the magnetic field at which the change of the nature of lines is observed allows one to estimate the charged exciton dissociation energy. This energy is equal to the valence band Zeeman splitting at the jump from the single X^+ line to the D_{high}–D_{low} doublet. The dissociation energy in this sense was found to be carrier density independent and was between 2 and 3 meV for all samples measured in [71].

The crossing of the singlet and the triplet levels plays also an important role in the polarization of the total PL signal. The two-hole singlet states, which are the initial states of PL transitions in both circular polarizations differ only by the spin of the electron. Thus, the splitting of this level is only 1/5 of the exciton Zeeman splitting (in (Cd,Mn)Te, where $\alpha/\beta = -1/4$). Such a small conduction electron splitting leads to comparable occupations of both spin states and hence to comparable PL intensities in both polarizations. Additionally, the spin flip time of the electron is relatively long and even comparable to the X^+ lifetime [47, 70]. This prevents fast spin relaxation between the charged exciton singlet states. A completely different situation takes place after the singlet-triplet crossing. Then, only the σ^+ transition is possible from the lowest initial state. The opposite triplet state (both hole spins down, not shown in Fig. 10.9c) has an energy higher by the sum of the values of the exciton and hole splitting. This is in agreement with the PL polarization experimentally observed: The σ^- line intensity remains almost the same as that in σ^+ below the jump, and it vanishes completely after the jump.

Several characteristics of the double line observed in σ^+ polarization suggest that this transition is quite similar to the band-to-band transition. In particular, it was pointed out that the line shape, its dynamics after pulsed excitation (including fast rise time) and the coincidence in energy of the lower component with the energy between first Landau levels extrapolated to $B = 0$, might be easily explained in terms of band-to band transition. In such a picture, the upper component of the double line involves the transition to the fundamental state of the hole gas. Such a transition is an indirect one since the electron recombines with a hole at the Fermi level. The lower component of the double line is related to a transition to an excited state of the hole gas. The excitation involves no spin flip (all holes are in the majority spin subband). In a band-to-band picture, it is the state after recombination at $k = 0$, in which one electron is left at the top of the valence band (Fig. 10.10). The

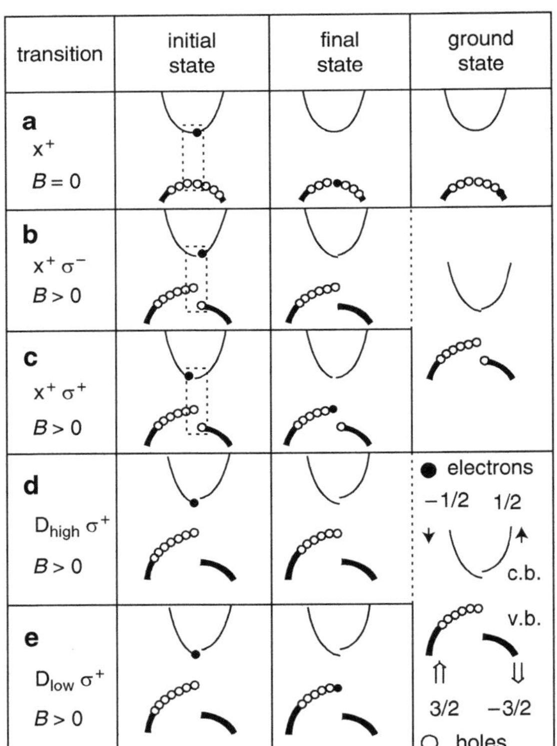

Fig. 10.10 Schematic diagram of the initial and final states in optical transitions [71]

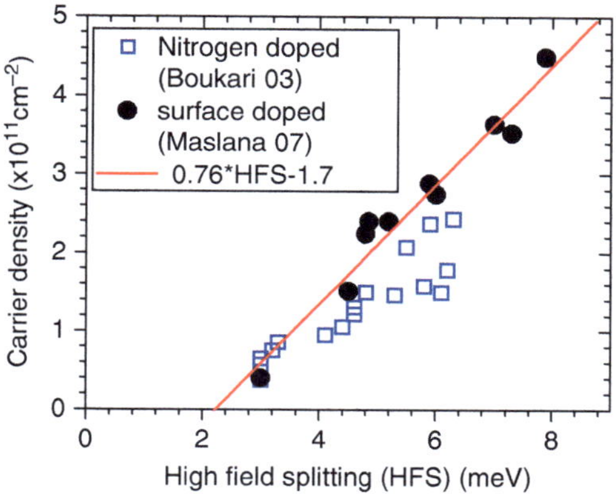

Fig. 10.11 Calibration of $D_{\text{high}} - D_{\text{low}}$ splitting (HFS) by the carrier density [82]

energy of the carrier gas excited state increases with the carrier density, as shown in Fig. 10.11. At low carrier density, the energy of this excitation as seen in the splitting of the double line is larger than the Fermi energy of the hole gas. It appears also to deviate from the linear dependence on the carrier density. This suggests that the triplet state of the exciton, which is thought to have in some cases a small but finite binding energy with respect to the free carrier continuum. The energy in the final state should be discussed in terms of excitations of the hole gas (plasmon, combination of single particle excitations, many body excitations, etc.) [83] with a total wavevector equal to k_{F}. At large carrier density, the $D_{\text{high}} - D_{\text{low}}$ splitting tends to match the Fermi energy, as expected for the simple band-to-band transitions. Therefore, the splitting of the double line might be used as a convenient tool to measure the hole gas density [84].

Another important issue in these measurements is the actual value of the valence band Zeeman splitting at which the hole gas becomes completely polarized. It is determined by the analysis of the evolution of the position of the PL and absorption lines (see Fig. 10.9b). When the optical transition occurs in the spin subband with the degenerate carrier gas, the absorption appears at a higher energy than the emission. This is due to the fact that in both transitions the final states might be excited ones. In the absorption, it results in an increase of the photon energy, by the energy of the excitation, while in PL, the photon energy is decreased. For the particular field, where the absorption and transmission line match, the proper spin subband becomes empty and the hole gas is fully polarized. We can determine the corresponding Zeeman energy Z as being 4/5 of the splitting of the σ^+/σ^- PL lines in Fig. 10.9a [85]. At full polarization, the Zeeman energy equals twice the Fermi energy E_{F} at zero field. Figure 10.12 gives the ratio of $Z^* = 2E_{\text{F}}$ (where E_{F} has been determined independently, as explained above) to the Zeeman energy Z. It was found that these two quantities are not equal but the system behaves as if the Zeeman energy driving the spin polarization degree was enhanced compared to the bare Zeeman energy Z imposed by Mn system upon the hole gas [85]. Data

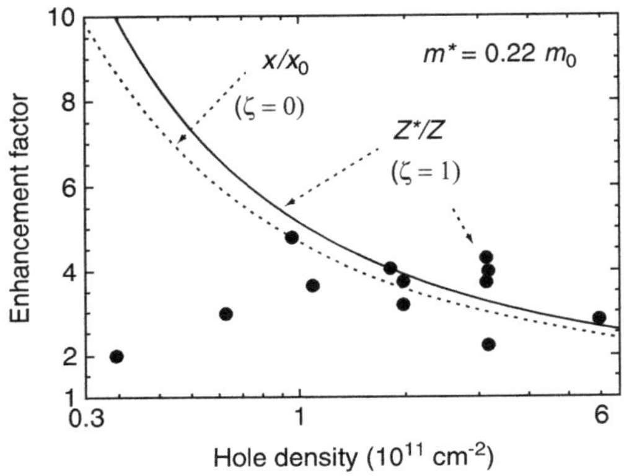

Fig. 10.12 Enhancement factor of the spin splitting, as a function of the carrier density. Using $m^* = 0.22m_0$, the lines give the result of the calculation at complete polarization (*solid line*) and vanishing polarization (*dashed line*); *symbols* are experimental data. After [85]

points are compared with a theoretical calculation of the Zeeman enhancement (see below) using the heavy-hole mass $m^* = 0.22m_0$ obtained from the single particle valence band calculation (as described by Fishman [86]). Such an enhancement of the Zeeman energy has been evidenced by Raman scattering performed on n-type quantum wells in similar samples [36] in an experiment, which will be detailed in the following section. This enhancement is linked with the spin susceptibility enhancement, which results from many-body interactions as predicted earlier for metallic systems [87] and extensively investigated by magneto-transport measurements on 2DEG systems in GaAs heterojunction [88] and in SI–MOSFET [89]. Indeed, as given by (10.9) and (10.10), the ratio Z^*/Z is exactly:

$$Z^*/Z = \zeta/\zeta_0, \tag{10.14}$$

where ζ is the spin polarization degree and ζ_0 is the same quantity defined for noninteracting carriers. Since the spin susceptibility is defined as $\chi = \partial m_z/\partial b_z$, where the magnetization $m_z \propto n_{2D}\zeta$ and b_z is the magnetic field acting upon the carriers, the spin susceptibility enhancement is related to the Zeeman energy enhancement [36],

One may ask why the splitting of the PL line observed in Fig. 10.9a leads to the bare Zeeman energy rather than to the quantity Z^*, which represents the Zeeman energy of interacting carrier gas per single particle. This will be explained in the next section. Indeed, the PL line of a carrier gas might be seen as a single particle band-to-band recombination [68] as it does in Fig. 10.9a for magnetic fields above 1T. It is natural to expect that the splitting of the σ^+/σ^- PL lines corresponds to the single particle Zeeman energy Z^*. One key observation is that this splitting does not depend on the carrier density. Therefore, it involves no Coulomb contribution. It has been shown to be a consequence [85] of the fact that the final state has to be the same after the recombination process of both σ^+ and $+\sigma^-$ photons [71].

Furthermore, recombination of a σ^+ (resp. σ^-) photon destroys a majority (resp. minority) spin carrier (see Fig. 10.10b, c). Then to reach the same final state, I–e the ground state (see Fig. 10.10), the carrier system has to deexcite a $k = 0$ spin-flip excitation when a minority spin carrier recombines [85]. As the energy difference between this two recombination process does not depend on the density, the spin-flip excitation is consequently collective and its zone center energy is the bare Zeeman energy Z, as stated by the Larmor theorem (see Sect. 10.4.1.1). This property has been demonstrated beyond any doubt by comparison of Raman and PL measurements carried out on n-type samples [80]. Figure 10.13 shows magneto-PL spectra obtained for 2DEG confined in (Cd,Mn)Te quantum wells. In this case, where the electron density is $\sim 10^{11}$ cm^{-2} the charged exciton is already destabilized and the PL exhibits single particle band-to-band recombination with a characteristic asymmetric shape. The splitting between the peaks of the σ/π lines is compared with the Z and Z^* determined by Raman scattering on the same sample.

The spin susceptibility enhancement directly measured in the hole system [85] plays a key role in carrier-induced ferromagnetism. So far, we have neglected the influence of the carrier spin polarization on the magnetization of the Mn system.

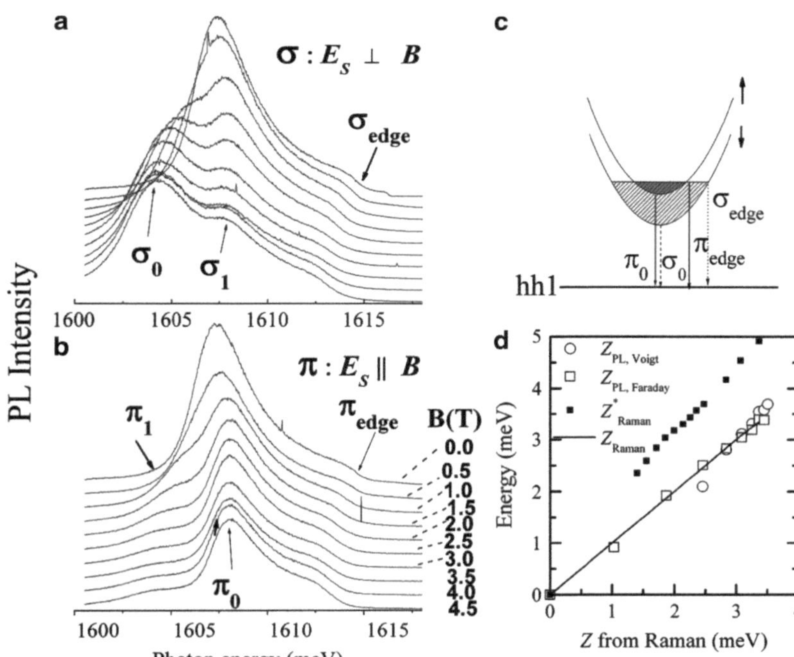

Fig. 10.13 (**a**) and (**b**): PL spectra in the Voigt configuration obtained for various in-plane magnetic field for both polarizations, respectively, perpendicular (σ) and parallel (π) to the field. (**c**) Schematic diagram showing the transitions involved in the PL process. (**d**) Splitting of the σ_0 and π_0 peaks ($Z_{PL,Voigt}$) compared with Z and Z* determined by Raman scattering (From [80])

However, in the quantum wells with a few percent of Mn ions and carrier density above 10^{10} cm^{-2} such assumption is no longer valid. As it was first predicted theoretically [90, 91] and demonstrated experimentally [18, 21], the interaction between both systems leads to the appearance of a ferromagnetic phase. The transition is governed by the two susceptibilities: the magnetic susceptibility of Mn ions in "empty" DMS material, and spin susceptibility of the carrier gas. The first one is influenced by the antiferromagnetic interaction between Mn magnetic moments and takes the maximum value for about 10% of Mn ions [16, 43]. The second one is enhanced by carrier–carrier interactions. In fact, this spin-susceptibility enhancement is necessary to get positive Curie temperature in such systems as (Cd,Mn)Te quantum wells. The experimentally observed transition temperatures (usually below 10 K) can be explained with values of the enhancement factor Z^*/Z between 2 and 3.

10.4 Intraband Excitations: Raman Scattering

In this section, we will focus on intraband excitations of a spin-polarized two-dimensional carrier gas probed by resonant Raman scattering. In usual p-type quantum wells such as (Cd,Mn)Te/(Cd,Mg)Te, due to the strain and confinement, the heavy and light holes are strongly decoupled so that the first populated heavy hole level is purely of spin $\pm\frac{3}{2}$. Spin flip processes of such heavy holes require a change of the total momentum from $\pm\frac{3}{2}$ to $\mp\frac{3}{2}$ and are of higher order in the Raman two photon process. This renders their observation very difficult. Therefore, in the following, we deal with spin polarized electron gas for which the spin-flip processes are accessible to light.

As mentioned, 2D interacting electron systems were first obtained in GaAs/(Ga,Al)As doped heterostructures. Well-defined excitations of the Fermi disk of the two-dimensional electron gas have been investigated at very low temperatures by intraband spectroscopies. As these excitations are comparable with the Fermi energy (a few meV), far infrared transmission and electronic resonant Raman scattering (ERRS) [92] in the visible range are the most powerful methods for such purpose. But in the former, the microwave electromagnetic field couples directly to the electrons and probes only zone center excitations. Moreover, when a static magnetic field is applied, the Kohn theorem [93] prevents the 2DEG properties at microwave frequencies to be sensitive to Coulomb interactions unless the in-plane translational invariance is broken [94]. On the contrary, ERRS was able to probe excitations with nonzero in-plane momentum where many-body interactions manifest themselves. Dispersions of intrasubband and intersubband excitations have been thus determined. Intrasubband plasmons, single particle excitations (SPEs) [9], inter-subband plasmons and spin density excitations [95] have been observed in the absence of external magnetic field in a standard modulation doped GaAs quantum well. In the integer quantum Hall regime, spin-waves, inter-Landau level magneto-plasmons and spin flip waves (SFWs) have been evidenced [96]. Spin excitations of fractional states have also been investigated [97].

Recently, the spin-polarized 2D electron gas (SP2DEG) in $Cd_{1-x}Mn_xTe/Cd_{1-y}Mg_yTe$ n-type modulation doped quantum wells [98] (see in Sect. 10.2.2.2) has been successfully introduced as a model paramagnetic system to investigate spin excitations by ERRS [123]. A novel understanding [99] of spin-resolved exchange-correlations interaction and dynamical spin-susceptibilities has been developed on the basis of these ERRS results. Moreover, such paramagnetic n-type (Cd,Mn)Te quantum wells are a high quality test-bed for the study of low-lying magnetic excitations, which are involved [31] in the carrier mediated ferromagnetism [100] occurring in DMSs.

10.4.1 Probing Spin-Flip Excitations

Spin-flip excitations are transverse (with respect to the quantization axis defined by the field) magnetic excitations where the spin of electrons is flipped collectively or individually. Two techniques are usually employed to measure spin flip excitations energies: electron spin resonance (ESR) and electronic resonant Raman scattering (ERRS). In the first one, a rotating standing microwave magnetic field $\boldsymbol{b}_1(\boldsymbol{r},t) = \widetilde{b}_+(\boldsymbol{q}=0,\omega)(\cos\omega t\cdot\boldsymbol{x}+\sin\omega t\cdot\boldsymbol{y})$ is applied perpendicularly to the equilibrium spin direction (direction of $\boldsymbol{B}_0$) and couples directly to the spin through the Hamiltonian $\hat{H}_{\text{pert}}$:

$$\begin{aligned}\hat{H}_{\text{pert}} &= g_{\text{e}}\mu_{\text{B}}\int \mathrm{d}^2 r_{\|}\boldsymbol{b}_1\left(\boldsymbol{r}_{\|},t\right)\cdot\int \hat{\boldsymbol{s}}\left(-\boldsymbol{q}\right)\mathrm{e}^{\mathrm{i}\boldsymbol{q}\cdot\boldsymbol{r}_{\|}}\mathrm{d}^2 q \\ &= g_{\text{e}}\mu_{\text{B}}\widetilde{b}_+\left(\boldsymbol{q}=0,\omega\right)\hat{s}_-\left(0\right)\mathrm{e}^{+\mathrm{i}\omega t}+c.c.\end{aligned} \tag{10.15}$$

where we have introduced Fourier components of the transverse spin densities operators:

$$\hat{s}_\pm(\boldsymbol{q}) = \int\left[\hat{s}_x\left(\boldsymbol{r}_{\|}\right)\pm i\hat{s}_y\left(\boldsymbol{r}_{\|}\right)\right]\mathrm{e}^{\mathrm{i}\boldsymbol{q}\cdot\boldsymbol{r}_{\|}}\mathrm{d}^2 r_{\|}. \tag{10.16}$$

The perturbation (10.15) will induce oscillation of the 2DEG transverse magnetization given by the expectation value of the operator: $\hat{\boldsymbol{m}}_\pm(\boldsymbol{q}) = -g_{\text{e}}\mu_{\text{B}}\hat{\boldsymbol{s}}_\pm(\boldsymbol{q})$ at $\boldsymbol{q}=0$. ESR probes the $q=0$ total magnetization precession. But heating of conduction electrons and the small number of spins prevent ESR to be sensitive to carrier spin flip excitations in quantum wells [101]. In practice, in DMS, ESR reveals spin-flip excitations associated with Mn-spin degrees of freedom [102], with such high accuracy, that the Knight shift due to exchange coupling with the 2DEG has been observed [103].

The ERRS process is a second-order perturbation involving two photons with wavevectors $\boldsymbol{k}_{i(s)}$, frequencies $\omega_{i(s)}$ and polarizations $\boldsymbol{e}_{i(s)}$. The two photons couple to two real or virtual interband electron-hole pairs. Spin-flip processes are allowed by the Raman mechanism, if the incoming and scattered photons have orthogonal polarizations (this will be referred to later as the cross-polarized case) [104]. The fact that one can see Raman processes involving a spin flip of electrons does not stem from a two-photon nature of the process but is, rather, related to spin-mixed nature of the intermediate valence hole state $|v\rangle$ involved in the process. Indeed, when calculating the Raman cross-section for spin-flip excitations we have [104]:

$$\frac{\mathrm{d}^2\sigma}{\mathrm{d}\omega\mathrm{d}\Omega} \propto \sum_M \left| \sum_{c,v,k} \frac{\langle c\uparrow| e_\mathrm{i}\cdot p\,|v\rangle\,\langle v|\, e_\mathrm{s}\cdot p\,|c\downarrow\rangle - \langle c\uparrow| e_\mathrm{s}\cdot p\,|v\rangle\,\langle v|\, e_\mathrm{i}\cdot p\,|c\downarrow\rangle}{(E_{\mathrm{c}k}-E_{\mathrm{v}k})^2-(\hbar\omega_\mathrm{i})^2} \right.$$
$$\left. \times \langle M|\, c^+_{k+q\uparrow} c_{k\downarrow}\,|0\rangle_0 \right|^2 \delta\left(E_\mathrm{M}-E_0-\hbar\omega\right). \quad (10.17)$$

In (10.17), $|c\sigma\rangle$ are conduction band states of spin σ, $|0\rangle$ and $|M\rangle$ are many-body ground and excited states (here, the summation reduces to spin-flip excitations) and $\langle\rangle_0$ denotes ground state averaging over the thermal equilibrium ensemble. Photons wavevector and the difference between the two photon frequencies have been neglected. The k-dependence in optical matrix elements has also been neglected. The Raman process is governed by two conservation laws: the translational invariance along in plane directions implies conservation of the wavevectors parallel components and the energy is also conserved. This defines the Raman transferred in-plane wavevector $\boldsymbol{q}$ and the Raman shift frequency ω

$$\begin{cases} q = (k_\mathrm{i}-k_\mathrm{s})_\parallel \\ \omega = \omega_\mathrm{i}-\omega_\mathrm{s} \end{cases} \quad (10.18)$$

The transferred wavevector can be tuned in amplitude (by changing the projection of $\boldsymbol{k}_{i(s)}$ with respect to the 2DEG plane) or in direction (by turning the sample around the growth axis). In quasi-back-scattering geometry, the $\boldsymbol{q}$ amplitude can be maximized as $q = (4\pi/\lambda)\sin\theta$, where θ is the angle of incidence, λ the incoming photon wavelength.

One immediately sees in (10.17) that the two electron-hole pairs share the same hole level which has to couple through the light with both spin-up and spin-down electron states. This is possible only if one can find hole states which are not pure in spin (because of spin orbit coupling). The resonant denominator in the Raman cross-section is essential for observation of the scattered light with presently available detector sensitivity. This restricts the choice of λ to values close to the quantum well absorption edge. In $\mathrm{Cd_{1-x}Mn_xTe}$, the maximum transferred wavevector is then lower than $16\,\mu\mathrm{m}^{-1}$.

To model the Raman response, one usually assumes, first, that the factor $\gamma_{\uparrow\downarrow}(c,v,\boldsymbol{k})$ multiplying $\langle M|\, c^+_{k+q\uparrow} c_{k\downarrow}\,|0\rangle_0$ in the sum $\sum_{c,v,k}$ of (10.17) is a constant and equals to $\gamma_{\uparrow\downarrow}$. This is, of course, reasonable only when the incoming photon is far from any optical transition of the host crystal (i.e., the denominator is far from its zeros). In this "out-of-resonance" situation, the Raman cross-section simplifies [105] and it reads:

$$\frac{\mathrm{d}^2\sigma}{\mathrm{d}\omega\mathrm{d}\Omega} \propto \left|\gamma_{\uparrow\downarrow}\right|^2 \sum_M \left|\langle M|\, s^+ (q)\, |0\rangle_0\right|^2 \delta\left(E_M - E_0 - \hbar\omega\right)$$
$$= \frac{\hbar}{\pi}\frac{1}{1-e^{-\beta\hbar\omega}} \left|\gamma_{\uparrow\downarrow}\right|^2 \mathrm{Im}\chi_+ (\boldsymbol{q}, \omega) . \tag{10.19}$$

Here, we have made use of expressions of the transverse spin density operators in second quantization:

$$\widehat{s}_+(\boldsymbol{q}) = \sum_k c^+_{k+q,\uparrow} c_{k,\downarrow} \quad \text{and} \quad \widehat{s}_-(\boldsymbol{q}) = \sum_k c^+_{k+q,\downarrow} c_{k,\uparrow} \tag{10.20}$$

and introduced $\chi_+ (\boldsymbol{q}, \omega)$, the dynamical transverse spin susceptibility defined by:

$$\widetilde{m}_+ (\boldsymbol{q}, \omega) = -\left(g_e\mu_B\right)^2 \chi_+ (\boldsymbol{q}, \omega)\, \widetilde{b}_+ (\boldsymbol{q}, \omega) , \tag{10.21}$$

where $\widetilde{m}_+ (\boldsymbol{q}, \omega)$ is the Fourier transform of the transverse magnetization induced by a periodic rotating field $\boldsymbol{b}_1(\boldsymbol{r}, t) = \widetilde{b}_+(\boldsymbol{q}, \omega)[\cos(\omega t - \boldsymbol{q}\cdot\boldsymbol{r}_\parallel)\cdot\boldsymbol{x} + \sin(\omega t - \boldsymbol{q}\cdot\boldsymbol{r}_\parallel)\cdot\boldsymbol{y}]$.

The second equality in (10.19) is the fundamental *fluctuation-dissipation theorem* [106], which links the dissipation ($\mathrm{Im}\chi_+$) to the dynamics of transverse spin fluctuations given by the dynamical spin structure factor:

$$S_{s_+s_-} (\boldsymbol{q}, \omega) = \frac{1}{2\pi} \int_{-\infty}^{\infty} \langle \hat{s}_+ (\boldsymbol{q}, t)\, \hat{s}_- (-\boldsymbol{q})\rangle_0\, \mathrm{e}^{-\mathrm{i}\omega t} \mathrm{d}t$$
$$= \sum_F \left|\langle F|\, s^+ (q)\, |I\rangle_0\right|^2 \delta\left(E_F - E_I - \hbar\omega\right), \tag{10.22}$$

where $\hat{s}_+ (\boldsymbol{q}, t) = \mathrm{e}^{it/\hbar \hat{H}_{\mathrm{SP2DEG}}} \hat{s}_+ (\boldsymbol{q})\, \mathrm{e}^{-it/\hbar \hat{H}_{\mathrm{SP2DEG}}}$ is the Heisenberg time dependent operator of $\hat{s}_+ (\boldsymbol{q})$ and $\langle\rangle_0$ denotes the average over the thermal equilibrium ensemble.

However, in strong resonance with a particular electron–hole transition, one has to take into account variation of the $\gamma_{\uparrow\downarrow} (c, v, \boldsymbol{k})$ factor with $\boldsymbol{k}$. Then the Raman response can be expressed by:

$$\frac{\mathrm{d}^2\sigma}{\mathrm{d}\omega\mathrm{d}\Omega} \propto \sum_M \left|\sum_k \gamma_{\uparrow\downarrow} (\boldsymbol{k})\, \langle M|\, c^+_{k+q\uparrow} c_{k\downarrow}\, |0\rangle_0\right|^2 \delta\left(E_{\mathrm{M}} - E_0 - \hbar\omega\right) \tag{10.23}$$

In the latter, only the resonant term has been retained. Hence, the Raman response is no longer trivially proportional to the imaginary part of the transverse susceptibility. But, since the variation of the $\gamma_{\uparrow\downarrow} (c, v, \boldsymbol{k})$ factor scales with the sharpness of the resonance (linewidth of the corresponding absorption peak), if the latter is smooth compared to peaks in the $\mathrm{Im}\chi_+$spectrum, then, one expects only small modifications of the nonresonant response. Features observed in the resonant situation will be discussed in Sects. 10.4.1.1 and 10.4.2.3. For the sake of simplicity, here, we will just remark on two aspects of the response given in (10.23). First, due to the resonance condition, the transverse response is now "dressed" by the coupling with photons. Second, particular $\boldsymbol{k}$ states located on a circle defined by the zero of the denominator have an enhanced weight in the sum. This will raise their oscillator strength in the Raman spectrum.

In conclusion, the cross-polarized ERRS probes fluctuations of the system at any $\boldsymbol{q}$, while ESR probes dissipation at $q = 0$ only. When out of resonance and $q = 0$, both responses are strictly equivalent. In the following paragraphs, we will see how the resonance introduces difference between the two responses.

10.4.1.1 Larmor Theorem in (Cd,Mn)Te Quantum Wells

Under the nonresonant condition, the $q = 0$ Raman response is equivalent to the ESR response and we expect to see peaks at Raman shift values corresponding to the oscillation eigenfrequencies of the magnetization. In Fig. 10.14a, we depicted

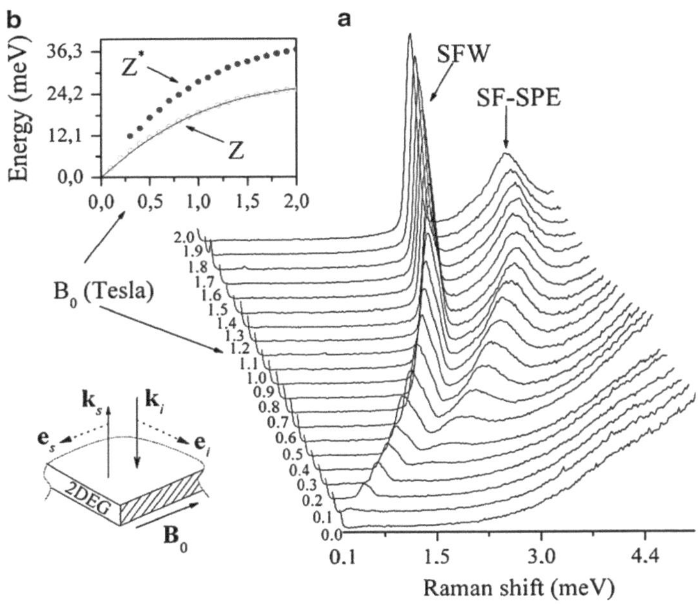

Fig. 10.14 (**a**) Cross-polarized Raman spectra with incident and scattered beam along the growth axis of the quantum well (see inset), taken for various values of in-plane magnetic field B_0. Spin-flip excitations are probed at $q = 0$. The low energy line is the collective one (SFW) and the other is the individual one (SF-SPE) – see text. (**b**) Peak position of the two precedent lines reproduced as a function of the magnetic field. The SFW peak energy has been fitted with (10.8) to obtain the manganese concentration $x = 0.75\%$ and the electron temperature $T = 1.5$ K. Taken from [36]

the Raman spectra obtained at 1.5 K in the Voigt configuration with in-plane magnetic field using a sample with a 2DEG embedded in a $Cd_{0.9925}Mn_{0.0075}Te/Cd_{0.8}Mg_{0.2}Te$ quantum well. The illumination and collection were along the growth axis with crossed polarizations, so that the spin-flip excitations with vanishing momentum ($q = 0$) were probed. Each spectrum shows two coexisting Raman lines. One of them is narrow and lower in energy, the other is broader and higher in energy. Both lines shift with the applied magnetic field. Thus, there are two eigenfrequencies for each magnetic field. The system has the following symmetries: in-plane translation and the rotation along the spin quantization axis. It is natural to think that the two modes are associated with internal degrees of freedom of the collective conduction electron spin system. One frequency might be a collective excitation while the second – an individual spin-flip process. Coexistence of both individual and collective excitation is the core of the Landau theory of Fermi liquids which has been successful since more than 20 years in describing the 2DEG behavior at high and intermediate electron densities. Excitations of the noninteracting system naturally evolve, in the presence of Coulomb interaction, into the individual excitations. Coulomb interaction leads additionally to collective excitations. We are left to associate a line with its corresponding excitation. At $q = 0$, flipping the spin of a single electron requires an energy Z^* (see Fig. 10.2), and all such single particle excitations (SF-SPE) are degenerate. In a spin-flip process, there is no net charge perturbation induced, such that direct Coulomb terms do not contribute. Higher order terms such as exchange and correlations [107] will together with the Zeeman splitting $Z(\mathrm{B}_0)$ determine the spin-flip excitation energy. Exchange is a ferromagnetic-type interaction, which leads to self-alignment of electron spins. It dominates over the correlations for the intermediate 2DEG densities considered here ($n_{2D} = 2.8 \times 10^{11}\,\mathrm{cm}^{-2}$). Hence, flipping the spin of a single electron without disturbing the other spins opposes the exchange interaction. This suggests that degenerate SF-SPEs are related to the broad high energy line, while the collective motion is related to the narrow low energy line. In the long-wavelength collective motion of conduction electron spins, all spins rotate in phase, thus, they can be considered as a single macroscopic spin. Its precession is governed only by external static magnetic fields (including the exchange effective magnetic fields). By "external", we mean influences that are not due to interactions within the conduction electron system itself. This property of the zone center collective spin motion is also known as the *Larmor theorem* [108]. It is a consequence of rotational invariance of the spin degrees of freedom similar to the *Kohn theorem* [93], which applies to the orbital degrees of freedom in a system with translational invariance. As mentioned in (Cd,Mn)Te quantum wells, the external magnetic influence upon electrons is the sum of the direct coupling with B_0 and the exchange-coupling with Mn spins, as stated by (10.8). As a consequence, the collective spin-flip energy (narrow low energy line) is exactly the bare Zeeman energy $Z(B_0)$. In other words, no manifestations of spin-dependent Coulomb interactions, such as exchange and/or correlations [109], are present at the long wavelength, paramagnetic collective resonance, a behavior already observed for orbital degrees of freedom probed by cyclotron resonance [110]. In a microscopic point of view, this fact is explained by the following: flipping the ensemble of spins induces no change of their relative orientations, hence, no contribution of spin-dependent Coulomb interactions is involved. This is true at $q = 0$ only as will be shown in Sect. 10.4.2. This assignment of the collective mode is confirmed in Fig. 10.14b, where the energy dependence on B_0 is reproduced. After proper adjustment of the Mn concentration x and the Mn-spin temperature T, the low energy line follows the bare Zeeman energy given by (10.8).

It is remarkable that in the Raman spectra of Fig. 10.14, uncorrelated individual spin-flip modes coexist with the collective mode of the magnetization oscillation. Such observation was never made in ESR measurements. This coexistence is indeed a consequence [111] of the resonant condition needed for the observation of electronic Raman effect as explained in the discussion of (10.3)–(10.9). We pointed out that the resonant coupling with the photons enhanced the oscillator strength of some particular $\boldsymbol{k}$ states. This means that particular $c^{+}_{k+q\uparrow}c_{k\downarrow}|0\rangle$ excitations will have an enhanced weight in the spectrum. For $q = 0$, they correspond to zone center SF-SPE. They are degenerate with the energy Z^*. The ratio Z^*/Z is linked with the spin-susceptibility enhancement as shown in 10.13). Contrary to common intuition, the Coulomb interaction between conduction electrons manifests itself in a single-particle spin-flip transition and does not show up in the collective one. Such a behavior, being a consequence of the spin-rotational invariance, was already known for orbital degrees of freedom in 2DEGs. Indeed, the cyclotron resonance, which is the excitation of the collective "Kohn" orbital-mode shows no influence of the Coulomb interaction when the 2DEG translational invariance is not broken [112]. Finally, we show how SF-SPE coexist with the collective mode in the spin fluctuations spectrum probed by ERRS, while their weight in the dissipation spectrum is strongly weakened (screened) by the presence of the collective mode. For zone center excitations, signatures of the Coulomb interaction are present in the individual modes only.

10.4.2 Dispersion of Spin-Flip Excitations of the Spin-Polarized Two-Dimensional Electron Gas

In this subsection, we will turn to the theoretical definition of the individual and collective excitations of the SP2DEG and investigate their wavevector dependence both theoretically and experimentally. According to the fluctuation-dissipation theorem, the spin-flip excitations are related to peaks in the dissipation spectrum. It is then natural to evaluate the response of the SP2DEG Hamiltonian of (10.7) to a perturbation by a rotating magnetic field introduced in (10.21). The latter couples to the transverse spin density operators, given by (10.16) via the Hamiltonian (10.15).

The perturbation induces oscillations of the transverse magnetization. We intend to determine the time evolution of $\langle \hat{m}_{\pm}(\boldsymbol{q},t)\rangle_0 = -g_e\mu_B \langle \hat{s}_{\pm}(\boldsymbol{q},t)\rangle_0$, where $\langle\rangle_0$ represents the ground state expectation value and the time dependence of operators is determined in the Heisenberg picture in presence of the perturbation b_1. From (10.19), we see that the magnetization is a coherent superposition of transverse spin fluctuations $\langle c^+_{k+q,\uparrow} c_{k,\downarrow}(t)\rangle_0$ resulting from the excitation of a single electron from the state $|\boldsymbol{k}, \downarrow\rangle$ to the state $|\boldsymbol{k}+\boldsymbol{q}, \uparrow\rangle$. This suggests considering the equation of motion of the operator $\hat{v}_{k,q}(t) = c_{k+q,\uparrow} c_{k,\downarrow}(t)$:

$$i\hbar \frac{\mathrm{d}}{\mathrm{d}t} \hat{v}_{k}(t) = \left[\hat{v}_{k,q}(t), \hat{H}_{\mathrm{SP2DEG}} + \hat{H}_{\mathrm{pert}}\right]$$
$$= \left[\hat{v}_{k,q}, \hat{H}_0 + \hat{H}_{\mathrm{Coul}} + g_{\mathrm{e}}\mu_{\mathrm{B}}\widetilde{b}_+(\boldsymbol{q},\omega)\hat{s}_-(-\boldsymbol{q})\mathrm{e}^{+\mathrm{i}\omega t} + \mathrm{c.c.}\right]^{\hat{H}_b}, \tag{10.24}$$

where $\hat{v}_{k,q} = c^+_{k+q,\uparrow} c_{k,\downarrow} = \hat{v}_{k,q}(0)$ and the commutator has to be evaluated at the end in the Heisenberg picture. The first and third commutators in (10.24) are straightforward:

$$\left[\hat{v}_{k,q}, \hat{H}_0\right] = \left(E^0_{k,\downarrow} - E^0_{k+q,\uparrow}\right)\hat{v}_{k,q} \tag{10.25}$$

$$\left[\hat{v}_{k,q}, \hat{s}_-(-\boldsymbol{q})\right] = n_{k+q,\uparrow} - n_{k,\downarrow}. \tag{10.26}$$

The second commutator involves $\hat{H}_{\mathrm{Coul}}$ and consists of so many complex terms that it is impossible to handle without further approximations. We shall reproduce here the commonly used approximations. First, we keep only those contributions, which may be written as products of occupation number operators $n_{k,\sigma} = c^+_{k,\sigma} c_{k,\sigma}$ with $\hat{v}_{k,q}$. This corresponds to the random-phase approximation (RPA) [107]. Such operator products are multiplied by prefactors representing contributions of the Coulomb interaction involved in spin-flip processes. These prefactors can be found in the frame of the local spin density approximation (LSDA) [113] valid for q lower than the Fermi wavevector, and are given by [113]:

$$G^T_{\mathrm{xc}} = -\frac{2}{L^2}\frac{1}{n^2_{\mathrm{2D}}}\frac{1}{\zeta}\frac{\partial E_{\mathrm{xc}}}{\partial \zeta}, \tag{10.27}$$

where E_{xc} is the exchange-correlation part of the ground state energy per unit surface. The form of this prefactor can be understood as follows: the correction we are looking for is the change in the Coulomb energy resulting from the interaction of a given spin fluctuation $\langle \hat{v}_{k,q}(t)\rangle_0$ with other. Such a spin fluctuation, destroying a spin down electron to populate a spin-up state, disturbs the spin polarization degree ζ. As these excitations are close to the ground state, the energy change is expressed as a derivative of the ground state energy with respect to the disturbed quantity, here the spin polarization degree. Of course, there is no contribution of the Hartree term as there is no disturbance of the charge density. The local approximation makes this interaction to be the same for all $v_{k,q}$. This yields [114]:

$$\left[\hat{v}_{k,q}, \hat{H}_{\text{Coul}}\right] \cong G_{\text{xc}}^{T}\left\{\hat{v}_{k,q}\sum_{k'}\left(n_{k'+q,\uparrow}-n_{k',\downarrow}\right)+\left(-n_{k+q,\uparrow}+n_{k,\downarrow}\right)\sum_{k'}\hat{v}_{k',q}\right\} \tag{10.28}$$

After inserting all these commutators in (10.24) and taking the expectation value in the ground state, we find the equation of motion for the transverse spin fluctuation:

$$\begin{aligned}&\left\{i\hbar\frac{\mathrm{d}}{\mathrm{d}t}+E^{0}_{k+q,\uparrow}-E^{0}_{k,\downarrow}-G_{\text{xc}}^{T}L^{2}\left(n_{\uparrow}-n_{\downarrow}\right)\right\}\left\langle\hat{v}_{k,q}(t)\right\rangle_{0}\\&=\left(f_{k+q,\uparrow}-f_{k,\downarrow}\right)\left\{-G_{\text{xc}}^{T}\sum_{k'}\left\langle\hat{v}_{k',q}(t)\right\rangle_{0}+g_{\mathrm{e}}\mu_{\mathrm{B}}\widetilde{b}_{+}(\boldsymbol{q},\omega)\,\mathrm{e}^{\mathrm{i}\omega t}\right\}\end{aligned} \tag{10.29}$$

To obtain (10.29), we have neglected the time dependence of $n_{k,\sigma}$ to keep only terms which are first order in the perturbation. This leads to the introduction of equilibrium occupation numbers $f_{k,\sigma} = \langle n_{k,\sigma}\rangle_0$ and spin densities $n_\sigma = L^{-2}\sum_k f_{k,\sigma}$. The term with $\widetilde{b}_{+}(\boldsymbol{q},\omega)$ on the right-hand side of (10.29) acts as a source term. We will now analyze this equation of motion.

10.4.2.1 Spin-Flip Single Particle Excitations

The first term on the right-hand side of (10.29) couples the transverse spin fluctuation $\langle\hat{v}_{k,q}(t)\rangle_0$ with others having a different initial state $\langle\hat{v}_{k',q}(t)\rangle_0$. Let us switch off this coupling and the source terms in (10.29). From this equation, it results that $\langle\hat{v}_{k,q}(t)\rangle_0$ oscillates with a well-defined frequency, which we relate to an excitation energy: $\widetilde{E}_{k+q,\uparrow}-\widetilde{E}_{k,\downarrow}$, with $\widetilde{E}_{k,\sigma}$ defined as:

$$\widetilde{E}_{k,\sigma} = E^{0}_{k,\sigma} - G_{\text{xc}}^{T}L^{2}n_\sigma = E^{0}_{k,\sigma} + \frac{2}{n_{2\text{D}}^{2}}\frac{n_\sigma}{\zeta}\frac{\partial E_{\text{xc}}}{\partial\zeta} \tag{10.30}$$

The correction to $E^{0}_{k,\sigma}$ in (10.30) is the electron self-energy, which accounts for the Coulomb interaction with other electrons. Hence, $\widetilde{E}_{k,\sigma}$ is the energy of the single electron state$|\boldsymbol{k},\sigma\rangle$. Within the local assumption (self energy independent of $\boldsymbol{k}$), the electron mass remains unchanged by the Coulomb interaction, but the bare Zeeman splitting is renormalized to Z^*:

$$Z^{*} = Z + 2n_{2\text{D}}^{-1}\frac{\partial E_{\text{xc}}}{\partial\zeta} \tag{10.31}$$

Z^* is the quantity driving the equilibrium spin polarization degree as shown in (10.10). Finding the equilibrium state requires solving (10.10) and (10.30) self-consistently, once we know the exchange-correlation energy functional [115].

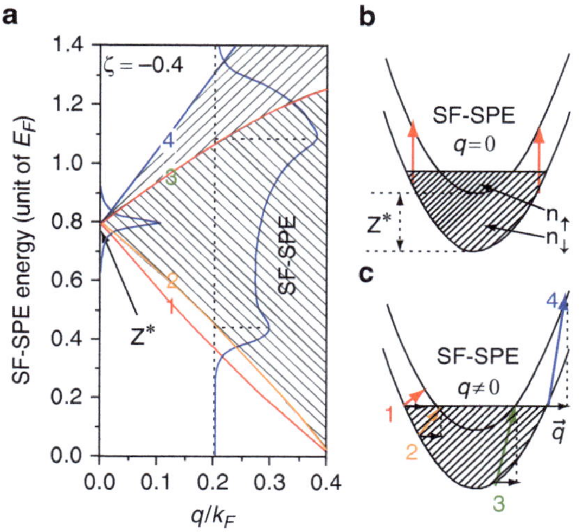

Fig. 10.15 (**a**) Typical theoretical dispersions of spin-flip single particle excitations (SF-SPE) of the SP2DEG calculated for $\zeta = -0.4$. The hatched area is the SF-SPE continuum. Lines numbered 1, 2, 3, 4 correspond, respectively, to the excitations 1, 2, 3, 4 in (**c**): Line 4 (1) is the excitation of a spin down electron with initial wavevector $k = k_{F\downarrow}$ ($k = -k_{F\downarrow}$). Lines 2 and 3 are limits, where the number of excitations is restricted due to filling of the spin-up subband. Overlaid are the calculated spectra, given by imaginary part of $\Pi_{\uparrow\downarrow}$ (see text), for $q/k_F = 0$ and 0.2 including finite temperature and disorder. The zero of the amplitude axis has been shifted correspondingly. (**b**) and (**c**) Schematic of spin-split subbands indicating representative SF-SPE excitations for (**b**) $q = 0$ and (**c**) nonzero q

Quantities such as $\widetilde{E}_{k+q,\uparrow} - \widetilde{E}_{k,\downarrow}$define the *spin flip single particle excitation* (SF-SPE) continuous spectrum. Its boundaries are defined by the occupancy conditions: $n_{k+q,\uparrow} < 1$ and $n_{k,\downarrow} > 0$. At zero temperature, the latter condition leads to the external boundaries of the continuum (see Fig. 10.15):

$$E_{4/1}(q) = Z^* \pm \hbar v_{F,\downarrow} q + \hbar^2 q^2/2m_e^* \qquad (10.32)$$

while the former defines the boundaries of the exclusion zone:

$$E_{3/2}(q) = Z^* \pm \hbar v_{F,\uparrow} q - \hbar^2 q^2/2m_e^*. \qquad (10.33)$$

We have introduced the 0K spin-resolved Fermi velocities and wavevectors:

$$v_{F,\sigma} = \hbar k_{F,\sigma}/m_e^* k_{F,\sigma} = k_F\sqrt{1 + sgn(\sigma)\,\zeta}. \qquad (10.34)$$

The low energy boundary reaches 0 for $q = q_l = k_{F,\downarrow} - k_{F,\uparrow}$. This corresponds to spin-flip processes that remain on the Fermi circles. For q above this limit, the SF-SPE continuum extends from 0 to $E_4(q)$.

We now plug the source term in (10.28). The magnetic field will drive coherently every spin fluctuation. It is natural to assume that they all have the same time dependence of the form $e^{-\eta t} e^{i\omega t}$ with a decay rate η, which accounts for coupling to any disorder source not included in (10.28). By taking the Fourier transform of (10.28) and summing over $\boldsymbol{k}$, we find the magnetization amplitude:

$$\widetilde{m}_{+}(\boldsymbol{q},\omega) = -(g_e\mu_B)^2\, \Pi_{\downarrow\uparrow}(\boldsymbol{q},\omega)\, \widetilde{b}_{+}(\boldsymbol{q},\omega), \tag{10.35}$$

where $\Pi_{\downarrow\uparrow}(\boldsymbol{q},\omega) = \sum_{\boldsymbol{k}} \frac{f_{\boldsymbol{k}+\boldsymbol{q},\uparrow} - f_{\boldsymbol{k},\downarrow}}{\widetilde{E}_{\boldsymbol{k}+\boldsymbol{q},\uparrow} - \widetilde{E}_{\boldsymbol{k},\downarrow} - \hbar\omega - i\hbar\eta}$ is the noninteracting spin-flip response function.

SF-SPE appear as poles of the real part of $\Pi_{\downarrow\uparrow}(\boldsymbol{q},\omega)$ and lead to delta-function singularities in the imaginary part. In Fig. 10.15, typical $\mathrm{Im}\Pi_{\downarrow\uparrow}(\boldsymbol{q},\omega)$ have been plotted for intermediate spin polarization degree and nonvanishing q. The SF-SPE spectrum has a characteristic asymmetric double peak structure. The fact that the values between the peaks are lower is due to a restriction of the number of excitations due to the filling of the up-spin subband. In itinerant ferromagnetic metals, SF-SPE are called *Stoner excitations*.

We now switch on the coupling between different transverse spin fluctuations in the equation of motion (10.29). This has two effects: first, SF-SPEs become short lived and second, their coherent superposition develops collective spin flip modes which produce oscillations of the magnetization.

10.4.2.2 Spin-Flip Waves

We consider now solutions of (10.28) for the case when the coupling between transverse spin fluctuations is present. We follow the same procedure as above and find the magnetization amplitude which gives the transverse spin susceptibility:

$$\chi_{+}(\boldsymbol{q},\omega) = \frac{\Pi_{\downarrow\uparrow}(\boldsymbol{q},\omega)}{1 + G_{\mathrm{xc}}^{T}\Pi_{\downarrow\uparrow}(\boldsymbol{q},\omega)}. \tag{10.36}$$

According to the *fluctuation-dissipation* theorem [106], singularities of the imaginary part of $\chi_{+}(\boldsymbol{q},\omega)$ define the excitation spectrum of the system. Two types of singularities are possible: the first type arises from singularities of $\Pi_{\downarrow\uparrow}(\boldsymbol{q},\omega)$ itself, which determined the previously defined SF-SPE; the second type arises when:

$$1 + G_{\mathrm{xc}}^{T}\Pi_{\downarrow\uparrow}(\boldsymbol{q},\omega) = 0. \tag{10.37}$$

This leads, for each value of q, to a pole which corresponds to a collective mode called the *spin-flip wave* (SFW). A small q expansion of (10.37) leads to the dispersion of the SFW [99]:

$$\hbar\omega_{\rm SFW} = Z - \frac{1}{|\zeta|}\frac{Z}{Z^* - Z}\frac{\hbar^2}{2m_{\rm e}^*}q^2 \tag{10.38}$$

We see that at $q = 0$, $\hbar\omega_{\rm SFW} = Z$, in agreement with the Larmor theorem.

10.4.2.3 Interplay Between Individual and Collective Excitations

Similarly to the situation for $q = 0$, also for nonvanishing q's, we expect to see signatures of both the collective and the individual excitations in the Raman spectra. An example of those, obtained in a quantum well containing $n_{\rm 2D} = 3.8 \times 10^{11}\,{\rm cm}^{-2}$ electrons and with $x = 0.75\%$ of Manganese are plotted in Fig. 10.16a for various scattering angles. A magnetic field $B_0 = 2.0\,{\rm T}$ was applied while the sample was immersed in Helium bath at the temperature T = 1.5 K. One clearly sees the negative dispersion of the narrow line (SFW), which evolves to lower energies when the transferred wavevector is increased. At higher energy, the SF-SPE peak,

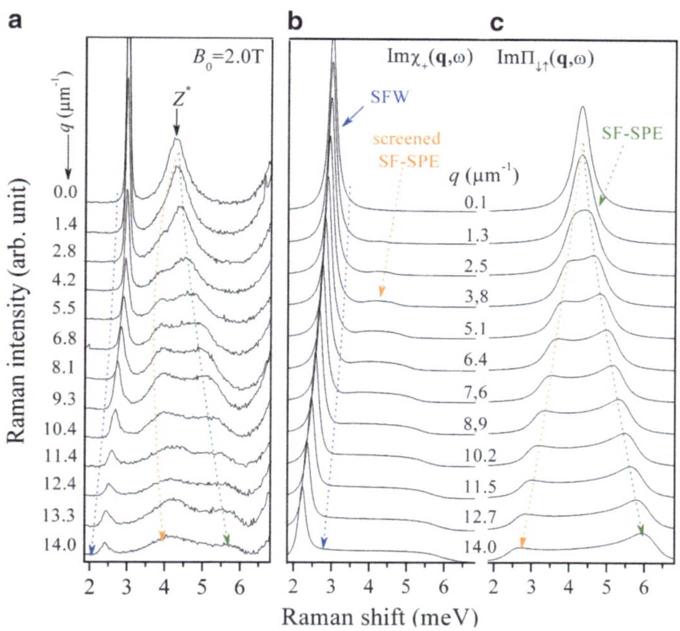

Fig. 10.16 (**a**) Cross-polarized Raman spectra obtained for $B_0 = 2.0\,{\rm T}$ on a quantum well with $n_{2D} = 3.8 \times 10^{11}\,{\rm cm}^{-2}$ and Manganese concentration $x = 0.75\%$ at Helium bath temperature $T = 1.5\,{\rm K}$. The scattering angle has been varied to change the transferred wavevector $\boldsymbol{q}$ as indicated. (**b**) and (**c**) Calculation of ${\rm Im}\chi_+(\boldsymbol{q}, \omega)$ and ${\rm Im}\Pi_{\downarrow\uparrow}(\boldsymbol{q}, \omega)$ with density $n_{\rm 2D} = 4.9 \times 10^{11}\,{\rm cm}^{-2}$, spin polarization degree $\zeta = 0.18\%$ and temperature $T = 1.5\,{\rm K}$. The homogeneous broadening $\hbar\eta$ was set to 0.12 meV and 0.25 meV, respectively. ${\rm Im}\Pi_{\downarrow\uparrow}(\boldsymbol{q}, \omega)$ reveals SF-SPE while the collective response ${\rm Im}\chi_+(\boldsymbol{q}, \omega)$ shows the SFW peak coexisting with screened SF-SPE. Taken from [124]

degenerate at $q = 0$, evolves with increasing q into a continuum with a characteristic double peak structure as discussed in previous section. As shown in Fig. 10.16b, SF-SPE are strongly screened by the collective mode in the calculated dissipation spectrum, although they remain visible as a high energy tail close to the collective peak. It is clear that $\mathrm{Im}\chi_+ (\boldsymbol{q}, \omega)$ reproduces only partially the experimental spectra. This puzzling issue, previously observed for unpolarized 2DEG and discussed in Sects. 10.4.1 and 10.4.1.1, is a consequence of the strong resonance condition needed for the observation of electronic Raman effect. A calculation of the resonant Raman response is available only for the unpolarized 2DEG [111]. For the spin-polarized 2DEG, we expect that Raman spectra are also a combination of $\mathrm{Im}\chi_+ (\boldsymbol{q}, \omega)$ and the individual response $\mathrm{Im}\Pi_{\downarrow\uparrow} (\boldsymbol{q}, \omega)$. This combination reveals the competition between two types of coupling of the individual excitations: with Coulomb interaction or with the electromagnetic field. Looking at Fig. 10.16a, one understands that for SF-SPE far from the SFW peak, the photon coupling is able to enhance their contribution contrary to the case of SF-SPEs close to the SFW peak, which remain screened by the Coulomb coupling. As seen in Fig. 10.17a, this manifests itself in an apparent blue shift of the low energy peak SF-SPE continuum.

10.4.2.4 Dispersion of Spin-Flip Excitations

In Fig. 10.17a, we show the dispersion of the SFW and characteristic features of the SF-SPE continuum reconstructed from the above Raman spectra. The SFW dispersion is compared with the solution of (10.37) calculated with $n_{2D} = 4.9 \times 10^{11}\,\mathrm{cm}^{-2}$ and $\zeta = 18\%$, while the SF-SPE peaks are compared with peaks of $\mathrm{Im}\Pi_{\downarrow\uparrow} (\boldsymbol{q}, \omega)$ (see caption for details). The coupling between the SFW and the low-energy branch of SF-SPE is attested by the gradual increase of the SFW line width suggesting an increasing damping when the SFW approaches the SF-SPE continuum. This will be discussed in the next Sect. 10.4.2.5. Meanwhile, the SFW dispersion remains insensitive to the coupling and is perfectly reproduced by the solution of (10.37). One may ask why the local approximation introduced in (10.27) is so good? Let us recall that the exchange-correlation Coulomb term given in (10.27) is responsible for the existence of the SFW and its dispersion law. This interaction is short range with k_{F}^{-1} being the characteristic length [116], so that for the range of probed wavevectors for which $q \ll k_{\mathrm{F}}$, this approximation is perfectly valid. Departure from the local approximation is expected when q becomes closer to k_{F}.

A surprising fact is the negative slope of the SFW dispersion, which always lies below the low energy boundary of the SF-SPE continuum. We understand the SFW energy shift as a function of wavevector $\Delta = \hbar\omega_{\mathrm{SFW}} (q) - Z$ as a consequence of the Pauli repulsion, which is stronger for parallel spins than for antiparallel spins [116]. A simultaneous flip of all the spins ($q = 0$) induces no change of the Pauli repulsion, contrary to the $q \neq 0$ case where the spins are periodically antiparallel for each $\lambda = 2\pi/q$. Then a reduction of the repulsion occurs, being stronger for shorter λ. According to (10.38), for small q, Δ is linear in q^2, with a slope γ depending on the

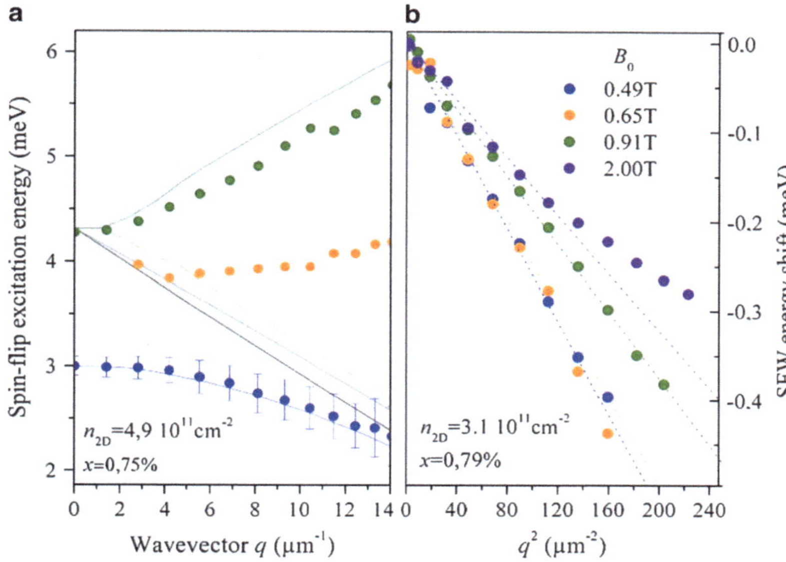

Fig. 10.17 (**a**) Spin-flip excitations dispersions reconstructed from Fig. 10.16a compared with theoretical ones. The SF-SPE continuum is plotted as a *dashed* domain at zero temperature. Overlaid are the peak positions of the SF-SPE continuum, calculated for zero disorder, zero temperature (*solid lines*) and from $\mathrm{Im}\Pi_{\downarrow\uparrow}(\boldsymbol{q},\omega)$ with finite temperature 1.5 K and broadening $\hbar\eta = 0.25$ meV (*dashed lines*). *Symbols* are experimental peaks extracted from Fig. 10.16a. Below the SF-SPE continuum, the SFW is shown (*circles*). The vertical error bars assigned to each experimental point represent the width of the SFW line. The SFW dispersion and SF-SPE continuum have been calculated with $\mathrm{n}_{2\mathrm{D}} = 4.9 \times 10^{11}\,\mathrm{cm}^{-2}$, and $\zeta = 0.18\%$. (**b**) Shift of the SFW energy (Δ) with the wavevector q plotted as a function of q^2 for various magnetic fields obtained on a sample with $n_{2\mathrm{D}} = 3.1 \times 10^{11}\,\mathrm{cm}^{-2}$, and $\mathrm{x} = 0.79\%$

density and the spin-polarization degree:

$$\frac{\hbar^2}{2m_e^*}\gamma^{-1} = |\zeta|\left(\frac{Z^*}{Z} - 1\right). \tag{10.39}$$

When $\zeta \to 0$, Z^*/Z tends to the static spin susceptibility enhancement [36] χ/χ_0, which remains greater than unity. When $|\zeta|$ increases, Z^*/Z also slightly increases, thus, the slope γ is monotonously decreasing with the spin polarization degree, as qualitatively reproduced in Fig. 10.17b.

10.4.2.5 Damping of the Spin-Flip Waves

As shown in the discussion of Fig. 10.17a, the damping of the SFW increases continuously with the wavevector q. The Random Phase Approximation used in the equation of motion (10.29) is unable to reproduce this behavior. Indeed, in the RPA, the damping of SFW is independent of q, except when the SFW enter the SF-SPE continuum, where the SFW modes are completely damped. Thus, to reproduce the continuous variation of the damping, we are left to explore more accurate dynamical equations of both the individual spin-flip excitations $\hat{v}_{k,q} = c^{+}_{k+q,\uparrow} c_{k,\downarrow}$ and the transverse collective modes $\hat{s}_{+}(\boldsymbol{q})$. Taking the exact commutator of the latter with the full Hamiltonian $\hat{H}_{\mathrm{SP2DEG}}$ leads to:

$$\begin{aligned} i\hbar \frac{\mathrm{d}}{\mathrm{d}t} \hat{s}_{+}(\boldsymbol{q}, t) &= \left[\hat{s}_{+}(\boldsymbol{q}), \hat{H}_{\mathrm{SP2DEG}} \right]^{\hat{H}_b} = \left[\hat{s}_{+}(\boldsymbol{q}), \hat{H}_0 + \hat{H}_{\mathrm{Coul}} \right]^{\hat{H}_b} \\ &= Z(B_0)\, \hat{s}_{+}(\boldsymbol{q}, t) + \sum_{\boldsymbol{k}} \left(E_{\boldsymbol{k}+\boldsymbol{q}} - E_{\boldsymbol{k}} \right) \hat{v}_{\boldsymbol{k},\boldsymbol{q}}(t), \end{aligned} \qquad (10.40)$$

where $E_{\boldsymbol{k}} = \hbar^2 k^2 / 2m_{\mathrm{e}}^{*}$ is the bare kinetic energy of a single electron state (see (10.7)). Equation (10.40) is a consequence of the spin rotational invariance of $\hat{H}_{\mathrm{Coul}}$, which conserves the macroscopic spin ($\hat{H}_{\mathrm{Coul}}$ conserves also the macroscopic momentum). Hence, the commutator $[\hat{s}_{+}(\boldsymbol{q}), \hat{H}_{\mathrm{Coul}}] = 0$. On the contrary, the kinetic Hamiltonian does not conserve the spin density operator $\hat{s}_{+}(\boldsymbol{q})$ and, thus, couples the collective motion to the individual modes $\hat{v}_{\boldsymbol{k},\boldsymbol{q}}$ as found in the second term on the right-hand side of (10.40). This is definitely true, except for the zone center collective mode ($q = 0$), where this coupling vanishes. The zone-center motion equation reads:

$$\begin{aligned} i\hbar \frac{\mathrm{d}}{\mathrm{d}t} \hat{s}_{+}(\boldsymbol{q} = 0, t) &= \left[\hat{s}_{+}(\boldsymbol{q} = 0), \hat{H}_{\mathrm{SP2DEG}} \right]^{\hat{H}_b} \\ &= Z(B_0)\, \hat{s}_{+}(\boldsymbol{q} = 0, t), \end{aligned} \qquad (10.41)$$

which shows that the homogenous ($q = 0$) magnetization mode is a true eigenstate of $\hat{H}_{\mathrm{SP2DEG}}$ and that its precession frequency is given by the external magnetic influence (*Larmor's theorem*). Of course, this mode should be damped by any other sources of magnetic disorder and/or fluctuations, which are external to the electron degrees of freedom and, thus, not present in $\hat{H}_{\mathrm{SP2DEG}}$. Fluctuations of the Mn spins are largely the dominant relaxation source of this mode (see Chap. 9). On the contrary, nonzero q modes experience a strong intrinsic damping, inherent to the motion of carriers supporting the transverse spin modes. This is a fundamental difference between our case and a ferro- or paramagnetic insulator having a localized spin system that develops magnon modes. For the latter, the kinetic damping of (10.40) does not exist. Further examination of the coupling term shows that, for symmetry reasons, the damping is proportional to q^2. Indeed, as $E_{\boldsymbol{k}+\boldsymbol{q}} - E_{\boldsymbol{k}} = \hbar^2 \boldsymbol{k} \cdot \boldsymbol{q} / m_{\mathrm{e}}^{*} + \hbar^2 q^2 / 2m_{\mathrm{e}}^{*}$, the sum over $\boldsymbol{k}$ states averages out the $\boldsymbol{k} \cdot \boldsymbol{q}$ contribution, but the q^2 contribution remains. This q^2 damping of inhomogenous magnetization modes has recently been investigated theoretically in [117] and experimentally in DMS quantum wells [118] by Raman scattering.

10.4.3 Beyond the Decoupled Model

In the preceding sections, we assumed that the two-spin subsystems formed, respectively, by the conduction and Mn electrons are dynamically decoupled. Only the static action of the Mn spins was kept in the description. This description was in very good agreement with the experiment when the spin-flip dynamics of both systems are very different. However, this assumption does not hold for magnetic fields and wave-vectors q, such that the SFW energy becomes comparable with the spin flip mode of Mn-electrons. As the direct coupling between neighboring Mn spins is very weak in diluted systems and as the Mn-spins are randomly distributed, the dispersion of the Mn-spin-flip mode is flat for wavevectors $q\bar{d}_{\mathrm{Mn}} \ll 1$, where $\bar{d}_{\mathrm{Mn}}$ is the average distance between Mn atoms. This spin flip mode energy is then equal to $g_{\mathrm{Mn}}\mu_{\mathrm{B}}B_0$. According to (10.8) and (10.38), the resonance between electronic and Mn spin flips at $q = 0$ at the magnetic field B_{R} is given by the equation:

$$x_{\mathrm{eff}}\bar{N}_0\alpha\,|\langle S_z\,(B_{\mathrm{R}},T)\rangle| = (|g_{\mathrm{e}}| + g_{\mathrm{Mn}})\,\mu_{\mathrm{B}}B_{\mathrm{R}}, \tag{10.42}$$

which has always a solution. A coupling of the two spin-flip modes has been evidenced in Raman spectra [103], by measuring the anticrossing occurring at the resonance field.

Figure 10.18 shows the Raman spectra obtained on a 10-nm thick modulation-doped quantum well of $Cd_{0.998}Mn_{0.002}Te$. The anticrossing of precession frequencies and the crossing of lifetimes were also observed in time-domain experiments in [119]. For further information about time-domain experiments probing the transverse spin dynamics, we refer to Chap. 9.

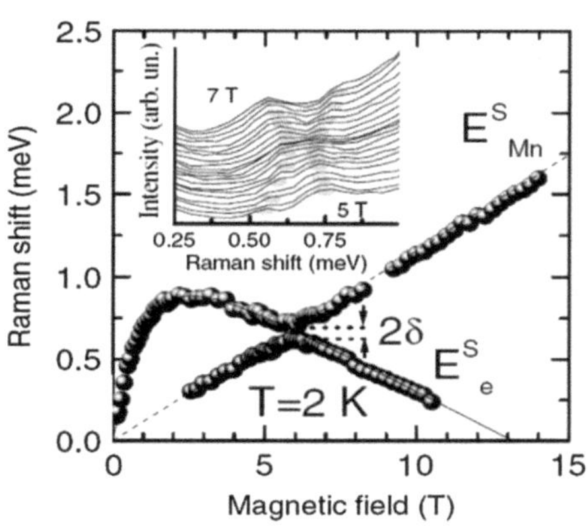

Fig. 10.18 Raman shift of the spin-flip wave and spin-flip Mn^{2+}-mode obtained on a 10 nm-thick CdMnTe/CdMgTe single quantum well with estimated electron density $n_{2D} = 1.0 \times 10^{11}\,\mathrm{cm}^{-2}$ and an effective Mn concentration $x_{\mathrm{eff}} = 0.2\%$. The expected behavior for decoupled modes is indicated by the *solid* and *dashed lines*. The inset shows the Raman spectra (every 0.1 T) in the region of the avoided crossing. The *shaded* regions are a guide to the eye. From [103]

The anticrossing gap 2δ extracted from the measurements shown in Fig. 10.18 is a consequence of the *Knight shift* K of the Zeeman energy of Mn spins [30]. The *Knight shift* is the Mn-equivalent of the exchange term in (10.8) occurring for the conduction electrons [120]. The presence of this coupling reveals the mixed nature of both spin-flip modes. The electron mode is dominated by the conduction electron spin-flip and possesses a small spin-flip contribution from the Mn electrons, while the reverse occurs for the Mn mode. A full theoretical description of collective mixed modes in doped DMS materials was first carried out by Mauger et al. [121]. It has been more recently rewritten for p-type bulk DMS [31] and applied to n-type DMS quantum wells in [122] In this full dynamical description, carrier spins and Mn spins modes develop two branches: an "acoustic" branch corresponding to the "in phase" precession of both spin ensembles and an "optic" branch, where the carrier and Mn spins ensemble precess "out of phase". In the former, the carrier contribution dominates, contrary to the latter, where the Mn contribution is the most important. They, respectively, correspond to the high- and low-energy modes in Fig. 10.18, for fields below the resonance field (6 T). As both spin systems have different g-factor, the spin-rotational invariance of the full dynamical problem is broken. The zone center precession frequency of each of these mixed modes is no longer given by the external magnetic influence. The model of [31] gives the precession frequencies for the zone center mixed modes of 2D electrons and Mn:

$$\hbar\omega = g_{\mathrm{Mn}}\mu_{\mathrm{B}}B_0 + K + K\Delta\frac{2\pi\hbar^2}{m_{\mathrm{e}}^* Z^*} \times \Pi_{\downarrow\uparrow}(\boldsymbol{q} = 0, \omega), \tag{10.43}$$

where

$$\Delta = -x_{\mathrm{eff}}\bar{N}_0\alpha \left\langle S_z\left(B_0, T\right)\right\rangle \tag{10.44}$$

is the *Overhauser shift* already found in (10.8). Equation (10.43) describes the Mn point of view, and can be explained as follows. The magnetic influence on the Mn has three contributions corresponding, respectively, to the three terms in (10.43): (1) the external magnetic field; (2) the static exchange field of 2D electrons given by the *Knight shift* $K = -\alpha n_{2\mathrm{D}}\zeta/w$, and (3) the dynamical contribution of the exchange field modulated by the precession of the electron spins. The latter is itself an electron response (through $\Pi_{\downarrow\uparrow}(\boldsymbol{q} = 0, \omega)$) to the exchange field modulation due to the Mn spins precession. This feedback loop explains why, in the third term of (10.43), α enters in the second power. Depending on the respective orientation of spins, the dynamical correction can be positive or negative. Indeed for p-type material, Mn and hole spins are antiparallel, the correction is negative while for n-type material, both spins are parallel leading to a positive correction. For the optical mode, this correction, relative to Δ, is of the order of $c = |n_{2\mathrm{D}}\zeta/2w| \left(x_{\mathrm{eff}}\bar{N}_0 S\right)^{-1}$, which represents the ratio of the spin concentration present in both systems (S is the Mn spin 5/2). For usual gas densities, this gives a very small relative correction $c \approx 10^{-4}$. However, in [122], the Coulomb interaction between electrons has been omitted, so that, the positive dynamical correction, no matter how small it is, leads to an optic mode situated above the SF-SPE continuum. Coulomb-exchange between electrons shifts this mode below the SF-SPE continuum, in agreement with experiment (see Fig. 10.17a).

10.5 Conclusions and Outlook

In this chapter, we discussed interband and intraband spectroscopic properties of the carrier gas in quantum wells with diluted magnetic semiconductor. The introduction of magnetic ions in quantum well material allowed spin polarization of carrier gas without application of strong magnetic field. This brought a novel understanding of several phenomena.

In particular, in the interband spectroscopy, the transitions related to charged exciton formation were analyzed in a broad range of carrier densities. The different mechanisms determining the exciton oscillator strength were discussed. It was shown that the leading mechanism, which suppress the absorption due to neutral exciton formation in presence of free carriers is related to the spin dependent interactions which are not a simple phase space filing. The possibility of the spin polarization of hole gas allowed also to interpret photoluminescence from the quantum well with high carrier density. The transition between charged exciton and band-to band recombination was demonstrated.

The giant Zeeman effect was also used as a tool in studies of the enhancement of the spin susceptibility of carrier gas due to many-body interactions.

In the second part of the chapter, we discussed the intraband excitations. They can be studied by Raman spectroscopy. It was shown that the cross-polarized Raman response is the signature of the spin-flip excitation spectrum, where individual and collective excitations of the electron gas coexist. Because of the giant Zeeman effect, the energy of transverse spin excitations and their dispersions are measurable with the help of standard ERRS setups. An approach assuming that the spins of the carriers and Mn spin degrees of freedom are dynamically decoupled, reproduced the experimental results with very good accuracy as long as the magnetic field is below the resonance. At the zone center, in the collective mode, electrons precess with the Larmor frequency, where all external magnetic influences contribute: the magnetic field and the static exchange field due to Mn polarization. No contribution of Coulomb-exchange between carriers contribute to the mode at $q = 0$. However, Coulomb interaction between carriers is essential to reproduce the wavevector variation of the collective mode energy. It also shifts the energy of individual modes toward higher energies. Apart from the homogenous magnetization mode, which is a true eigenstate of the decoupled picture, inhomogenous modes are intrinsically progressively damped by the motion of the carriers. This strong damping cannot be avoided since the system is conducting one. The dynamical coupling between the carriers and Mn spins leads to spin modes, which are mixed in nature. Very little is actually known about their dispersion. Let us stress, however, that these issues are of great importance for a better understanding of ferromagnetism and spin transport in DMS structures.

References

1. T. Ando, A. Fowler, F. Stern, Rev. Mod. Phys. **54**, 437 (1982)
2. G. Finkelstein, H. Shtrikman, I. Bar-Joseph, Phys. Rev. Lett. **74**, 976 (1995)
3. H. Buhmann, L. Mansouri, J. Wang, P.H. Beton, N. Mori, L. Eaves, M. Henini, M. Potemski, Phys. Rev. B **51**, 7969 (1995)
4. A.J. Shields, J.L. Osborne, M.Y. Simmons, M. Pepper, D.A. Ritchie, Phys. Rev. B **52**, R5523 (1995)
5. K. von Klitzing, G. Dorda, M. Pepper, Phys. Rev. Lett. **45**, 494 (1980)
6. For a review, see H.L Störmer, D.C. Tsui, A.C. Gossard, Rev. Mod. Phys. **71**, S298 (1999)
7. J.K. Jain, Phys. Rev. Lett. **63**, 199 (1989)
8. H. van Houten et al., Phys. Rev. B **39**, 8556 (1989)
9. G. Fasol, N. Mestres, H.P. Hughes, A. Fischer, K. Ploog, Phys. Rev. Lett. **56**, 2517 (1986); B. Jusserand, D.R. Richards, G. Fasol, G. Weimann, W. Schlapp, Surf. Science **229**, 394 (1990)
10. N.B. Brandt, V.V. Moshchalkov, Adv. Phys. **33**, 193 (1984)
11. K. Kheng, R.T. Cox, Y. Merle d'Aubigné, Franck Bassani, K. Saminadayar, S. Tatarenko, Phys. Rev. Lett. **71**, 1752 (1993)
12. M. Klude, T. Passow, R. Kroeger, D Hommel, Electron. Lett. 37, 1119 (2001); M. Klude, G. Alexe, C. Kruse, T. Passow, H. Heinke, D. Hommel, Phys. State Sol. (B) **229**, 497 (2002)
13. J.K. Furdyna, J. Appl. Phys. **64**, 4 (1988)
14. S. Lovisa, R.T. Cox, T. Baron, M. Keim, A. Waag, G. Landwehr, Appl. Phys. Lett. **73**, 656 (1998)
15. T. Wojtowicz, M. Kutrowski, G. Karczewski, J. Kossut, Appl. Phys. Lett. **73**, 1379 (1998)
16. W. Maślana, M. Bertolini, H.Boukari, P. Kossacki, D. Ferrand, J.A. Gaj, S. Tatarenko, J. Cibert, Appl. Phys. Lett. **82**, 1875 (2003)
17. P. Kossacki, J. Cibert, D. Ferrand, Y. Merle d'Aubigné, A. Arnoult, A. Wasiela, S. Tatarenko, J.A. Gaj, Phys. Rev. B **60**, 16018 (1999)
18. A. Haury, A. Wasiela, A. Arnoult, J. Cibert, S. Tatarenko, T. Dietl, Y.M. d'Aubigne, Phys. Rev. Lett. **79**, 511 (1997)
19. P. Kossacki et al., Physica E **6**, 709 (2000)
20. H. Boukari, P. Kossacki, M. Bertolini, D. Ferrand, J. Cibert, S. Tatarenko, A. Wasiela, J.A. Gaj, T. Dietl, Phys. Rev. Lett. **88**, 207204 (2002)
21. P. Kossacki, A. Kudelski, J.A. Gaj, J. Cibert, S. Tatarenko, D. Ferrand, A. Wasiela, B. Deveaud, T. Dietl, Physica E **12**, 344 (2002)
22. D. Ferrand, J. Cibert, A. Wasiela, C. Bourgognon, S. Tatarenko, G. Fishman, T. Andrearczyk, J. Jaroszyński, S. Koleśnik, T. Dietl, B. Barbara, D. Dufeu, Phys. Rev. B **63**, 085201 (2001)
23. D.R. Yakovlev et al., Phys. Rev. Lett. **79**, 3974 (1997)
24. T. Wojtowicz, M. Kutrowski, G. Karczewski, J. Kossut, B.Koenig, A. Keller, D.R. Yakovlev, A. Waag, J. Geurts, W. Ossau, G. Landwehr, I.A. Merkulov, F.J. Teran, M. Potemski, J. Cryst. Growth **214–215**, 378 (2000)
25. U. Zehner, B. Kuhn-Heindrich, W. Ossau, A. Waag, G. Lanwehr, H.H. Cheng, R.J. Nicholas, Acta Phys. Pol. **90**, 989 (1996)
26. W. Ossau, B. Kuhn-Heindrich, A. Waag, G. Landwehr, J. Cryst. Growth **159**, 1046 (1996)
27. T. Wojtowicz, M. Kutrowski, M. Surma, K. Kopalko, G. Karczewski, J. Kossut, M. Godlewski, P. Kossacki, Nguyen The Khoi, Appl. Phys. Lett. **68**, 3326 (1996)
28. J.A. Gaj, R. Planel, G. Fishman, Solid State Comm. **29**, 435 (1979)
29. J.A. Gaj, *Semiconductors and Semimetals*, vol. 25, chap. 7 (Academic Press, NY, 1988)

30. J. König, A.H. MacDonald, Phys. Rev. Lett. **91**, 77202 (2003)
31. J. König, H.H. Lin, A.H. MacDonald, Phys. Rev. Lett. **84**, 5628 (2000)
32. W. Maślana et al., *Physics of Semiconductors*, ed. by A.R. Long, J.H. Davies. Inst. of Physics, Conf. Series 171, p. 294 (2002)
33. A. Gold, Phys. Rev. B **35**, 723 (1987)
34. A.A. Sirenko et al., Phys. Rev. B **56**, 2114 (1997)
35. W. Grieshaber, A. Haury, J. Cibert, Y. Merle d'Aubigné, A. Wasiela, J.A. Gaj, Phys. Rev. B **53**, 4891 (1996)
36. F. Perez, C. Aku-leh, D. Richards, B. Jusserand, L. Smith, D. Wolverson, G. Karczewski, Phys. Rev. Lett. **99**, 26403 (2007)
37. L.L. van Zandt, Phys. Rev. **162**, 399 (1967); P.M. Platzman, P.A. Wolff, Phys. Rev. Lett. **18**, 280 (1967)
38. J.R. Sambles, D.A. Mace, G.L. Dunifer, Phys. Rev. B **31**, 7408 (1985)
39. T. Izuyama, D.J. Kim, R. Kubo, J. Phys. Soc. Jpn. **18**, 1025 (1963); G. Vignale, K.S. Singwi, Phys. Rev. B **32**, 2824 (1985)
40. E.C. Stoner, Acta Metall. 2, 260 (1954)
41. R. Vollmer, M. Etzkorn, P.S. Anil Kumar, H. Ibach, J. Kirschner, Phys. Rev. Lett. **91**, 147201 (2003)
42. Y. Merle d'Aubigne et al., Physica E **3**, 169 (1998)
43. J.A. Gaj, W. Grieshaber, C. Bodin-Deshayes, J. Cibert, G. Feuillet, Y. Merle d'Aubigné, A. Wasiela, Phys. Rev. B **50**, 5512 (1994)
44. S. Lovisa, R.T. Cox, N. Magnea, K. Saminadayar, Phys. Rev. B **56**, R12787 (1997)
45. K. Kheng, Ann. Phys. **20**, 229 (1995)
46. G.V. Astakhov, V.P. Kochereshko, D.R. Yakovlev, W.Ossau, J. Nurnberger, W. Faschinger, G. Landwehr, T. Wojtowicz, G. Karczewski, J. Kossut, Phys. Rev. B **65**, 115310 (2002)
47. E. Vanelle, M. Paillard, X. Marie, T. Amand, P. Gilliot, D. Brinkmann, R. Lévy, J. Cibert, S. Tatarenko, Phys. Rev. B **62**, 2696 (2000)
48. V. Ciulin, P. Kossacki, S. Haacke, J.-D. Ganière, B. Deveaud, A. Esser, M. Kutrowski, T. Wojtowicz, Phys. Rev. B **62**, R16310 (2000)
49. K. Kheng, R.T. Cox, T. Baron, K. Saminadayar, S. Tatarenko, J. Cryst. Growth **159**, 443 (1996)
50. D. Brinkmann J. Kudrna, P. Gilliot, B. Hönerlage, A. Arnoult, J. Cibert, S. Tatarenko, Phys. Rev. B **60**, 4474 (1999)
51. H.P. Wagner, H.-P. Tranitz, R. Schuster, Phys. Rev. B **60**, 15542 (1999)
52. P. Gilliot, D. Brinkmann, J. Kudrna, O. Crégut, R. Lévy, A. Arnoult, J. Cibert, S. Tatarenko, Phys. Rev. B **60**, 5797 (1999)
53. T. Brunhes, R. André, A. Arnoult, J. Cibert, A. Wasiela, Phys. Rev. B **60**, 11568 (1999)
54. H. Yokoi, Y. Kakudate, S. Fujiwara, Y. Kim, S. Takeyama, G. Karczewski, T. Wojtowicz, J. Kossut, Phys. Status Solidi B **229**, 681 (2002)
55. T. Wojtowicz, M. Kutrowski, G. Karczewski, J. Kossut, F.J. Teran, M. Potemski, Phys. Rev. B **59**, R10437 (1999)
56. L. Vina, J. Phys. Cond. Matt. **11**, 5929 (1999) and references therein
57. M.A. Lampert, Phys. Rev. Lett. **1**, 450 (1958)
58. T. Wojtowicz, M. Kutrowski, G. Karczewski, J. Kossut, Acta Phys. Pol. A **94**, 199 (1998)
59. G.V. Astakhov, D.R. Yakovlev, V.P. Kochereshko, W.Ossau, W. Faschinger, J. Puls, F. Hanneberger, S.A.Crooker, Q.McCulloch, D.Wolverson, N.A.Gippius, A. Waag, Phys. Rev. B **65**, 165335 (2002)
60. P. Redliński, J. Kossut, Solid State Comm. **118**, 295 (2001)
61. M. Kutrowski, PhD thesis, Institute of Physics Polish Academy of Sciences (2001)
62. B. Stébé, E. Feddi, A. Ainane, F. Dujardin, Phys. Rev. B **58**, 9926 (1998) and references therein
63. C. Riva, F.M. Peeters, K. Varga, Phys. Rev. B **61**, 13873 (2000)
64. B. Stébé, G. Munschy, L. Stauffer, F. Dujardin, J. Murat, Phys. Rev. B **56**, 12454 (1997)
65. A. Esser, E. Runge, R. Zimmermann, W. Langbein, Phys. Rev. B **62**, 8232 (2000)

66. V. Huard, R. Cox, K. Saminadayar, A. Arnoult, S. Tatarenko, Phys. Rev. Lett. **84**, 187 (2000)
67. B.E. Cole, T. Takamasu, K. Takehana, R. Goldhahn, D. Schulze, G. Kido, J.M. Chamberlain, G. Gobsch, M. Henini, Physica B **249–251**, 607 (1998)
68. P. Hawrylak, Phys. Rev. B **44**, 3821 (1991); S.A. Brown, J.F. Young, J.A. Brum, P. Hawrylak, Z. Wasilewski, Phys. Rev. B **54**, R11082 (1996)
69. R.A. Suris, V.P. Kochereshko, G.V. Astakhov, D.R. Yakovlev, W. Ossau, J. Nurnberger, W. Faschinger, G. Landwehr, T. Wojtowicz, G. Karczewski, J. Kossut, Phys. Stat. Solidi B **227**, 343 (2001)
70. P. Kossacki, J. Physics C **15**, R471 (2003) and references therein
71. P. Kossacki, H. Boukari, M. Bertolini, D. Ferrand, J. Cibert, S. Tatarenko, J.A. Gaj, B. Deveaud, V. Ciulin, M. Potemski, Phys. Rev. B **70**, 195337 (2004)
72. G. Eytan, Y. Yayon, M. Rappaport, H. Shtrikman, I. Bar-Joseph, Phys. Rev. Lett. **81**, 1666 (1998)
73. R.B. Miller, T. Baron, R.T. Cox, K. Saminadayar, J. Cryst. Growth **184–185**, 822 (1998)
74. S. Schmitt-Rink, D.S. Chemla, D.A.B. Miller, Adv. Phys. **38**, 89 (1989)
75. M. Combescot, P. Nozieres, J. Phys. **32**, 913 (1971)
76. P. Plochocka, P. Kossacki, W. Maslana, J. Cibert, S. Tatarenko, C. Radzewicz, J.A. Gaj, Phys. Rev. Lett. **92**, 177402 (2004)
77. P. Kossacki, P. Plochocka, B. Piechal, W. Maslana, A. Golnik, J. Cibert, S. Tatarenko, J.A. Gaj, Phys. Rev. B **72**, 035340 (2005)
78. S. Schmitt-Rink, D.S. Chemla, D.A.B. Miller, Phys. Rev. B **32**, 6601 (1985)
79. R. Planel, J. Gaj, C. Benoit, J. Phys. C **41**(C5), C5–39 (1980)
80. C. Aku-Leh, F. Perez, B. Jusserand, D. Richards, W. Pacuski, P. Kossacki, M. Menant, G. Karczewski, Phys. Rev. B **76**, 155416 (2007)
81. V. Huard, R.T. Cox, K. Saminadayar, C. Bourgognon, A. Arnoult, J. Cibert, S. Tatarenko, Physica E **6**, 161 (2000)
82. W. Maślana, Ph.D. thesis, University of Warsaw (2007)
83. B. Jusserand, G. Karczewski, G. Cywiński, T. Wojtowicz, A. Lemaître, C. Testelin, C. Rigaux, Phys. Rev. B **63**, 161302 (2001)
84. W. Maślana, P. Kossacki, P. Płochocka, A. Golnik, J.A. Gaj, D. Ferrand, M. Bertolini, S. Tatarenko, J. Cibert, Appl. Phys. Lett. **89**, 052104 (2006)
85. H. Boukari, F. Perez, D. Ferrand, P. Kossacki, B. Jusserand, J. Cibert, Phys. Rev. B **73**, 115320 (2006)
86. G. Fishman, Phys. Rev. B **52**, 11132 (1995)
87. D.R. Hamman, A.W. Overhauser, Phys. Rev. **143**, 183 (1966); S.H. Vosko, J.P. Perdew, Can. J. Phys. **53**, 1385 (1975)
88. J. Zhu, H.L. Stormer, L.N. Pfeiffer, K.W. Baldwin, K.W. West, Phys. Rev. Lett. **90**, 056805 (2003)
89. V.M. Pudalov, M.E. Gershenson, H. Kojima, N. Butch, E.M. Dizhur, G. Brunthaler, A. Prinz, G. Bauer, Phys. Rev. Lett. **88**, 196404 (2002)
90. T. Dietl, A. Haury, Y. Merle d'Aubigné, Phys. Rev. B **55**, 3347 (1997)
91. T. Junwirgh, B. Lee, A.H. MacDonald, Phys. E. **10**, 153 (2001)
92. For a review, see G. Abstreiter, M. Cardona, A. Pinczuk, in *Light Scattering in Solids IV*, ed. by M. Cardona, G. Güntherodt (Springer, Heidelberg, 1984), p. 5
93. W. Kohn, Phys. Rev. **123**, 1242 (1961)
94. K. Ensslin, D. Heitmann, H. Sigg, K. Ploog, Phys. Rev. B **36**, 8177 (1987)
95. A. Pinczuk, S. Schmitt-Rink, G. Danan, J.P. Valladares, L.N. Pfeiffer, K.W. West, Phys. Rev. Lett. **63**, 1633–1636 (1989)
96. A. Pinczuk, B.S. Dennis, D. Heiman, C. Kallin, L. Brey, C. Tejedor, S. Schmitt-rink, L.N. Pfeiffer, K.W. West, Phys. Rev. Lett. **68**, 3623 (1992)
97. I. Dujovne, A. Pinczuk, M. Kang, B.S. Dennis, L.N. Pfeiffer, K.W. West, Phys. Rev. Lett. **95**, 056808 (2005)
98. G. Karczewski, J. Jaroszynski, A. Barcz, M. Kutrowski, T. Wojtowicz, J. Kossut, J. Cryst. Growth **184–185**, 814 (1998)

99. F. Perez, Phys. Rev. B **79**, 045306 (2009)
100. T. Dietl, H. Ohno, F. Matsukara, J. Cibert, D. Ferrand, Science **287**, 1019 (2000)
101. L.C. Smith, S.J. Bingham, J.J. Davies, D. Wolverson, Appl. Phys. Lett. 87, 202101 (2005)
102. N. Samarth, J.K. Furdyna, Phys. Rev. B **37**, 9227 (1988)
103. F.J. Teran, M. Potemski, D.K. Maude, D. Plantier, A.K. Hassan, A. Sachrajda, Z. Wilamowski, J. Jaroszynski, T. Wojtowicz, G. Karczewski, Phys. Rev. Lett. **91**, 077201 (2003)
104. D.C. Hamilton, A.L. MacWorther, in *Light Scattering Spectra of Solids*, ed. by G.B. Wright (Springer, Heidelberg, 1969), p. 309
105. M.V. Klein, in *Light Scattering in Solids*, ed. by M. Cardona (Springer, Heidelberg, 1975), p. 150
106. G.F. Giuliani, G. Vignale, *Quantum Theory of the Electron Liquid* (Cambridge University Press, Cambridge, 2005)
107. D. Pines, *Elementary Excitations in Solids* (W.A. Benjamin, New York, 1964)
108. L.D. Landau, E.M. Lifshitz, *The Classical Theory of Fields*, chap. 5, 2nd edn. (Pergamon Press, Oxford, 1962), p. 120
109. A.K. Rajagopal, J.C. Kimball, Phys. Rev. B **15**, 2819 (1977)
110. For a recent discussion see, Y.A. Bychkov, G. Martinez, Phys. Rev. B **72**, 195328 (2005)
111. S. Das Sarma, D.W. Wang, Phys. Rev. Lett. **83**, 816 (1999)
112. C. Faugeras, G. Martinez, A. Riedel, R. Hey, K.J. Friedland, Yu. Bychkov, Phys. Rev. B **75**, 035334 (2007)
113. A.K. Rajagopal, Phys. Rev. B **17**, 2980 (1978)
114. T. Izuyama, D.J. Kim, R. Kubo, J. Phys. Soc. Jpn. **18**, 1025 (1963)
115. C. Attacalite, S. Moroni, P. Gori-Giorgi, G.B. Bachelet, Phys. Rev. Lett. **88**, 256601 (2002)
116. P. Gori-giorgi, S. Moroni, G. Bachelet, Phys. Rev. B **70**, 115102 (2004)
117. E. Hankiewicz, G. Vignale, Y. Tserkovnyak, Phys. Rev. B **78**, R020404 (2008)
118. J. Gomez, F. Perez, E.M. Hankiewicz, B. Jusserand, G. Karczewski, T. Wojtowicz, Phys. Rev. B **81**, 100403(R) (2010)
119. M. Vladimirova, S. Cronenberger, P. Barate, D. Scalbert, F.J. Teran, A.P. Dmitriev, Phys. Rev. B **78**, 081305 (2008)
120. T. Story et al., Phys. Rev. Lett. **77**, 2802 (1996)
121. A. Mauger, D. Mills, Phys. Rev. B **28**, 6553 (1983)
122. D. Frustiglia, J. König, A.H. MacDonald, Phys. Rev. B **70**, 45205 (2004)
123. B. Jusserand, F. Perez, D. Richards, G. Karczewski, T. Wojtowicz, C. Testelin, D. Wolverson, J.J. Davies, Phys. Rev. Lett. **91**, 086802 (2003)
124. C. Aku-Leh, F. Perez, B. Jusserand, D. Richards, G. Karczewski, to appear in Phys. Rev. B (2010)

Chapter 11
Quantum Transport in Diluted Magnetic Semiconductors

Jan Jaroszynski

Abstract The chapter highlights selected electric charge transport phenomena studied recently in low-dimensional structures of DMSs. The first part describes transport phenomena related to the quantum interference of scattered electron waves in diffusive transport regime at the boundary of metal-insulator transition. The second part is devoted to Landau quantization of electronic states, as quantum Hall effect and related phenomena. The focus is put on dramatic influence of exchange interaction between magnetic ions and charge carriers on transport.

11.1 Magnetically Doped Low-Dimensional Semiconductor Structures

We start from a short preview of available low-dimensional devices for transport studies in DMSs. There are two types of semiconductor devices containing two-dimensional electronic systems (2DESs). In the field effect transistor (FET) also known as metal-insulator-semiconductor (MIS) structure, 2D conducting layer is formed at the interface between a semiconductor and an insulator, as shown in Fig. 11.1a, b. The insulator is equipped with electric gate. The electric field perpendicular to the interface attracts electrons from the semiconductor. As a result, a few nanometer wide potential dip is formed at the interface. Since its width is comparable with electronic wavelength λ_e, the motion perpendicular to the interface is quantized into discrete energy levels corresponding to the electronic standing waves. However, in such quantum well (QW), the carriers can move in the plane parallel to the interface, forming a two-dimensional system of electrons (2DES). In particular, the most popular FET has Si for the semiconductor and SiO_2 for the insulator.

J. Jaroszynski
National High Magnetic Field Laboratory, 1800 East Paul Dirac Drive, Tallahassee, FL 32310, USA
e-mail: jaroszy@magnet.fsu.edu

J. Kossut and J.A. Gaj (eds.), *Introduction to the Physics of Diluted Magnetic Semiconductors*, Springer Series in Materials Science 144,
DOI 10.1007/978-3-642-15856-8_11,

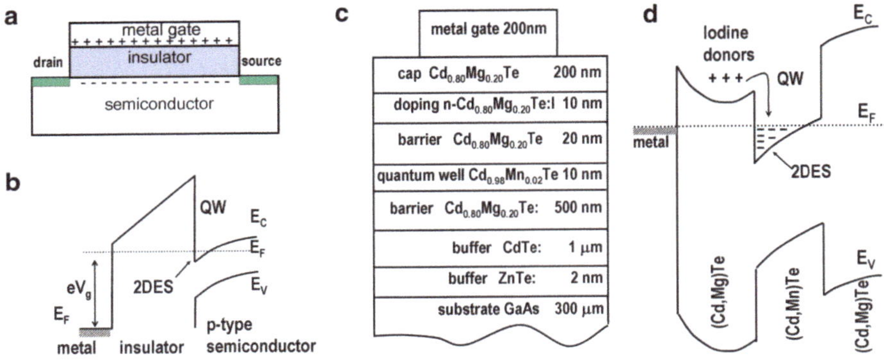

Fig. 11.1 (**a**) Metal-insulator-semiconductor (MIS) device and (**b**) its band diagram showing conduction band E_c, valence band E_v, and Fermi level E_F. Conducting 2DES is formed at the interface between the insulator and the p-type semiconductor as a consequence of the positive voltage V on the metal gate electrode. (**c**) Modulation doped (Cd,Mn)Te/(Cd,Mg)Te heterostructure and its (**d**) band diagram. A 2DES is formed in the undoped (Cd,Mn)Te when electrons from donors located in (Cd,Mg)Te barrier move into the quantum well. The top electric gate makes additional changes of the electron density possible

The earliest transport investigation of 2DES involving DMSs was carried out on metal-insulator-semiconductor structures fabricated on the surface of (Hg,Mn)Te [1]. It was followed by a study of 2DES formed at naturally grown inversion layers in the grain boundaries of the same material [2].

Further progress occurred owing to advances of the molecular beam epitaxy (MBE). This technique allows fabrication of thin films of semiconductor with atomic resolution. Moreover, it is possible to grow two (or more) semiconducting layers alternately. Another type of two-dimensional electron system is formed in the *heterostructures* of two semiconductors, where quantum well is formed due to differences of their energy gaps as shown in Fig. 11.1c, d.

The first MBE grown DMSs [3–5] were used mainly for spectroscopic, structural, and magnetic studies. However, the progress in doping techniques [6–12] made it possible to grow samples also for transport studies.

The very important advantage of MBE is the possibility of modulation doping. In this method, dopants are introduced selectively only to some layers of the structure, far from the conducting channel. Thus, although they deliver conducting carriers, their contribution to the scattering potential is very limited. This is crucial for obtaining 2DES with high carrier mobility μ. For example, in bulk doped thin films of (Cd,Mn)Te μ is typically lower than 10^3 cm^2/Vs. In modulation doped (Cd,Mn)Te/(Cd,Mg)Te heterostructures μ reaches 10^5 cm^2/Vs, which is the highest mobility observed in wide-gap DMSs [11]. However, introduction of magnetic ions into host material usually decreases its quality. To avoid this, magnetic ions are often put into barriers [13] or inserted digitally [9] only to selected layers of the QW. This allows one to obtain high electron mobilities and still to observe effects of magnetic ions on transport phenomena.

At the same time owing to low temperature MBE technique, it was possible to grow III–V semiconductors containing magnetic ions, e.g., (In,Mn)As and (Ga,Mn)As [14–17], avoiding notorious problems with clustering of transition metal atoms in III–V semiconductors.

Further reduction of sample dimensionality for transport studies can be achieved with the help of lithographic process (which usually involves electron beam lithography followed by dry etching) or using split electrical gates, which confine 2DES into one-dimensional wires or zero-dimensional dots.

However, to observe effects of low dimensionality on electronic transport, not necessarily samples with sizes of electron wavelength λ_e are needed. As it is shown below, striking quantum transport phenomena show up when the sample has reduced dimensionality with respect to some other, usually longer, length scales, such as mean free path ℓ or phase coherence length L_ϕ.

11.2 Quantum Phenomena in Diffusive Transport Regime

At low temperatures doped DMSs usually find themselves in the diffusive transport regime, where quantum interference between the electronic wave functions plays a crucial role in the charge transport. Electronic waves scatter from randomly distributed impurities and other static defects *elastically*, i.e., without changing their energy, thus preserving the phase coherence. The elastic mean free path ℓ_e is typically of the order of tens of nanometers. On the other hand, collisions with other electrons, phonons, and magnetic impurities are *nonelastic* and destroy the electronic phase. At low temperatures, the coherence length L_ϕ, on which electrons maintain their phase, can reach hundreds of micrometers. Thus, L_ϕ can be much longer than ℓ_e. As a result, quantum interference of scattered electron waves leads to an important modification of the charge transport in comparison with classical Drude formula even in 3D samples.

The conductance G of the sample is related to the transmission probability of partial electron waves along all the possible paths between points M and N in Fig. 11.2a:

$$G \propto \left| \sum_p a_p \exp(i\varphi_p) \right|^2 = \sum_p |a_p|^2 + 2 \sum_{pp'} |a_p||a_{p'}| \cos(\varphi_p - \varphi_{p'}). \quad (11.1)$$

The quantum corrections to G originate from the second sum on the right side of (11.1), which contains interference terms between different paths p and p', where:

$$\varphi_p - \varphi_{p'} = \frac{2\pi}{\lambda_e}(L_p - L_{p'}) + 2\pi \frac{BS_{pp'}}{h/e} \quad (11.2)$$

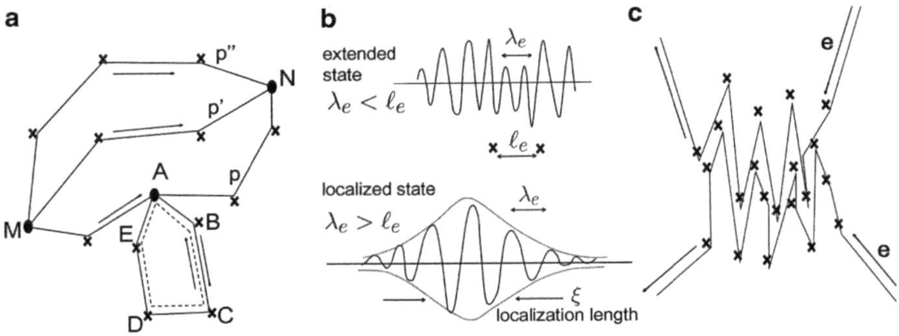

Fig. 11.2 (**a**) Diffusive (where diffusion means scattering) transport regime: running electron waves $\psi = a\exp[i(\boldsymbol{kr} - \omega t)]$ scatter from impurities (*crosses*) elastically, i.e., changing only direction of momentum $\boldsymbol{k}$ while conserving kinetic energy $E = \hbar\omega$. Thus, the phase of the electron wave $\phi = \boldsymbol{kr} - \omega t$ does not change. The impurities are spaced randomly in the host crystal, so their presence is often referred to as disorder. (**b**) Interference of running electron waves can produce a standing wave. This leads to localized state when $\lambda_e > \ell_e$, i.e., when the wavelength of electrons is longer than their elastic mean free path. Typical wave functions of extended state (i.e., waves running through the whole crystal) with mean free path ℓ_e (*top*) and localized state (*bottom*) with localization length ξ. They correspond to metallic and insulating phases, respectively. (**c**) Schematic illustration of a modification of electron–electron interactions by disorder (scattering from impurities). In the presence of disorder, two electrons can be localized in a given region of space for a long time, in contrast to clean conductors. This strongly enhances interactions between electrons

is a phase difference acquired along two paths due to their different lengths L_p and $L_{p'}$, and magnetic flux $BS_{pp'}$ between them, where h/e is the magnetic flux quantum. The phase difference depends on λ_e.

Generally, the contribution of interference terms in macroscopic sample disappears after averaging, except for self-crossing trajectories as A-B-C-D-E-A in Fig. 11.2a and its time reversed path A-E-D-C-B-A, which have identical phases in the absence of B, hence interference in A is always constructive. Thus, the probability of return to the point of departure $|a^+ + a^-|^2 = 4|a|^2$ is then twice the classical result, a phenomenon often called coherent backscattering, quantum localization, or weak localization. The backscattering reduces the diffusion constant and, hence, the conductivity. In relatively clean solids (when $2\pi\ell_e/\lambda_e = k_F\ell_e \gg 1$), this results in small negative correction to the conductivity and is referred to as weak localization. However, when $\ell_e \sim \lambda_e$, the quantum interference leads to a complete localization, and metal-to-insulator transition takes place when running waves corresponding to extended states convert to standing waves localized in space, as shown in Fig. 11.2b.

Importantly, quantum localization strongly enhances electron–electron interactions, since localized particles interact several times before they leave the given region of space, as depicted in Fig. 11.2c.

Quantum interference is extremely sensitive to symmetry breaking factors. Magnetic field, which breaks the time reversal symmetry, decreases the localization, since self-crossing paths in Fig. 11.2a are no longer equivalent. The decrease of localization drives resistance R back toward Drude value and results in negative magnetoresistance, often called weak localization magnetoresistance.

In turn, when inelastic spin-flip scattering of electrons from localized spins is substantial, the quantum coherence is lost completely. This dephasing could be strong around $B = 0$ in the paramagnetic phase.

Another important symmetry breaking mechanism in DMSs is the giant spin splitting, since it suppresses interactions between two electrons with the same kinetic energy but opposite spins.

The role of quantum interference is further enhanced in small samples, when L_ϕ becomes comparable or larger than the sample dimensions. In this *mezoscopic* transport regime, such phenomena as Aharonov-Bohm (AB) effect and universal conductance fluctuations (UCFs) are observed. AB effect occurs in, e.g., small rings, when a sweeping magnetic field changes the phase difference in two narrow arms of the ring, and constructive/destructive interference results in conductance oscillations periodic in B. In turn, in small samples with an arbitrary shape, sample-specific universal conductance fluctuations manifest themselves as reproducible, but aperiodic, changes in the conductance as a function of an external control parameter, such as magnetic field, gate voltage, which changes k_F, or between samples with identical number of impurities but their different distribution. When the scattering potential evolves slowly with the time, a time-dependent UCFs resulting in conductance noise are observed. At sufficiently low temperature, the UCFs have an amplitude of the order of e^2/h that is nearly independent of the sample size and shape, hence the name "universal."

11.3 Magnetoresistance

11.3.1 Paramagnetic Phase

DMSs usually show complicated, nonmonotonic, and strongly temperature dependent magnetoresistance (MR), i.e., resistance changes as a function of external magnetic field. It is much stronger than in nonmagnetic, doped semiconductors. As a very typical example, Fig. 11.3 shows magnetotransport in thin films of n-$Zn_{1-x}Mn_xO$. The same qualitatively behavior was found in other bulk (Cd,Mn)Se [18], (Cd,Mn)Te [19, 20], (Zn,Mn)Te [21], (Zn,Co)O [22, 23], (Zn,Fe)O [24], (Sn,Mn)O_2 [25], (Ti,Co)O_2 [26] as well as in low-dimensional (Zn,Cd,Mn)Se [27,28], (Cd,Mn)Te [13,29,30] doped DMSs. In nonmagnetic ZnO, a weak ($\sim$0.1%) and negative MR is observed. As in other doped semiconductors, it originates from destructive influence of the magnetic field on the backscattering, often referred to as weak localization MR. This is an orbital effect, since it stems just from an influence of the external magnetic field on the phase of electronic waves and no spin degrees of freedom are involved. Quite generally, negative MR is a characteristic feature of nonmagnetic semiconductors at low temperatures.

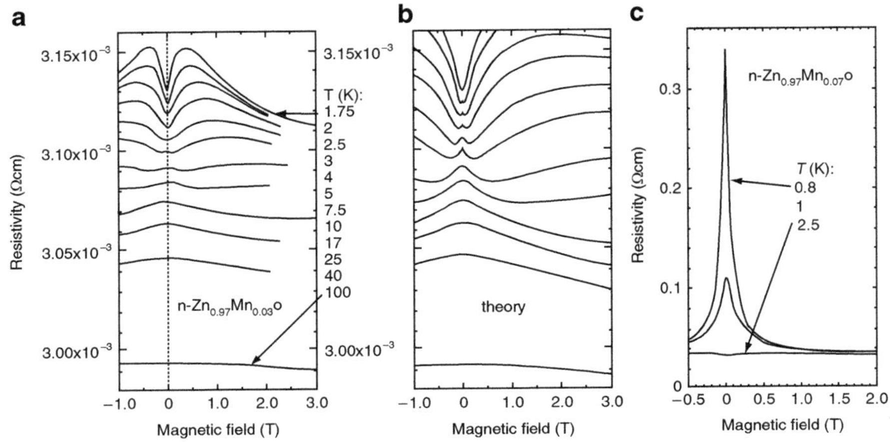

Fig. 11.3 (**a**) Measured and (**b**) calculated within no adjustable parameter resistivity changes in the magnetic field for n-$Zn_{0.97}Mn_{0.03}O$. (**c**) ρ vs. B at various T for $Zn_{0.93}Mn_{0.07}O$. The dramatic increase of resistance at low temperatures is destroyed by the magnetic field. Reprinted from [35]

When Mn ions are introduced into n-ZnO, effects of much stronger magnitude are clearly visible in Fig. 11.3a. Moreover, in (Zn,Mn)O magnetoresistance is characterized by a competition between positive and negative contributions. The large magnitude of MR as well as its dependence on the field and temperature indicates that phenomena brought about by the presence of the Mn spins dominate over the weak-localization effects specific to nonmagnetic n-ZnO. It has been shown [18, 27, 30] that the giant spin splitting of band states in DMSs affects considerably quantum correction to the conductivity associated with the disorder modified electron–electron interactions. This results in a positive MR provided that the Mn ions are not already spin-polarized in the absence of the magnetic field. Furthermore, the spin-splitting leads to a redistribution of the electrons between spin subbands. This diminishes localization by increasing the carrier kinetic energy [31] in one of the spin subbands c.f. Fig. 11.9a. Fig. 11.3b shows MR calculations for (Zn,Mn)O taking into account both single-electron and many-body quantum effects in the weakly localized regime. These calculations reproduce well the interplay between negative and positive MR in weak magnetic fields.

However, at lower temperature, an abrupt increase of resistivity is observed in $Zn_{0.93}Mn_{0.07}O$ sample, as depicted in Fig. 11.3c. It is accompanied by a strong negative contribution to MR, which is too large to be explained in the framework of the weak localization theory. No such effect has been observed in nonmagnetic CdSe, ZnO, CdTe, and ZnP. In contrast, it has been found in (Cd,Mn)Se [18,32,33], (Cd,Mn)Te [20], (Zn,Mn)P [34] and linked with the bound magnetic polaron (BMP) formation. BMP is a cloud of Mn ions mutually polarized by an electron bound to impurity (see Chap. 7). A strong increase of ρ originates from spin-disorder scattering from BMPs, which is in turn suppressed by external magnetic field destroying BMPs. However, no quantitative model has been proposed so far. In addition, what is puzzling, these phenomena are observed deep into the metallic phase, where all the donors are ionized and cannot contribute to the BMP formation.

11.3.2 Magnetic Polarons or Nanoclustering

A new light on the role of bound magnetic polarons was shed recently by experimental study of magnetotransport in modulation-doped n-(Cd,Mn)Te quantum wells [36]. Similarly to bulk DMSs, a dramatic upturn of the resistivity at some low temperature T^* was found, where T^* strongly depends on electron density n_s. This is T-dependent metal-to-insulator transition, since above the T^* the resistance barely depends on T, while below T^* strongly increases with decreasing T. The resistance upturn is again accompanied by a strong negative MR as it is shown in Fig. 11.4 and as it is observed in bulk DMSs. However, in contrast to bulk DMSs, in modulation-doped quantum well donors are set back away from the conducting channel, thus no formation of bound magnetic polarons is expected. Instead, it is postulated that the observed behavior stems from the competition between the antiferromagnetic (AF) exchange characterizing the insulating phase of DMSs and the ferromagnetic (FM) correlations induced by conducting electrons. This results in the formation of FM metallic bubbles embedded within a carrier-poor, magnetically disordered matrix and promotes an electronic nanoscale phase separation. This, in turn, results in magnetotransport behavior known as colossal magnetoresistance observed in several oxides containing magnetic ions [37, 38].

According to the Zener theory of carrier-mediated FM, a ferromagnetic transition is expected to occur at very low $T_C \ll 1$ K in bulk n-type zinc-blende DMS [39], such as n-(Cd,Mn)Te. However, recent Monte–Carlo simulations [40–43] indicate a formation of isolated ferromagnetic bubbles and the colossal magnetoresistance-like behavior below $T^* \gg T_C$. At $B = 0$, the FM bubbles are oriented randomly, which enhances resistance since the electrons have to change their spin orientation to flow from one bubble to another. This is reminiscent to the celebrated phenomenon

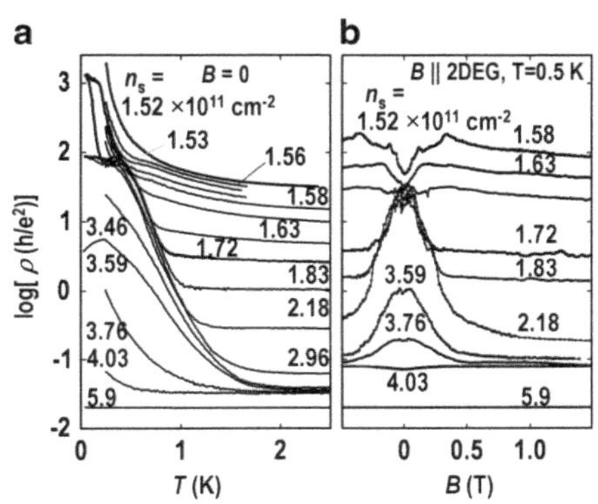

Fig. 11.4 (**a**) Resistivity $\rho(T)$ for different electron densities n_s measured at $B = 0$ in (Cd,Mn)Te modulation-doped heterostructure. (**b**) $\rho(B)$ at $T = 0.5$ K for different n_s. B is in the sample plane. Reprinted from [36]

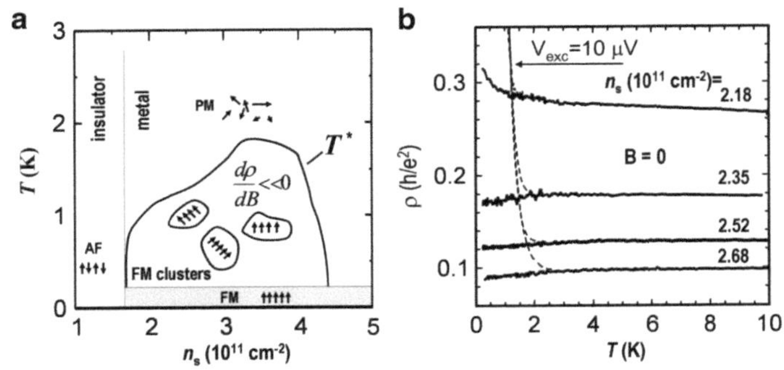

Fig. 11.5 (**a**) Different transport regimes in the $n_s - T$ plane. AF and FM states are sketched tentatively. (**b**) Metal-insulator transition in modulation doped gated n-(Cd,Mn)Te quantum well. Thick and dotted lines show $\rho(T)$ at high- and low-excitation voltages, $V_{\text{exc}} = 500$ and $10\,\mu$V, respectively, for different electron densities n_s. Reprinted from [36]

known as giant magnetoresistance (GMR) [44, 45], when the resistance between two macroscopic ferromagnetic films separated by a nonmagnetic spacer strongly depends on whether they have the same or opposite direction of magnetization. Thus, when magnetic field aligns the bubbles, a strong negative MR follows. These effects are expected to exist in both 3D and 2D cases but should be stronger in 2D than in 3D, in agreement with the experimental results as seen in comparison with T^* observed in (Cd,Mn)Se [18].

Another striking observation is the lack of ρ dependence on n_s. The $\rho(T, B = 0)$ (Fig. 11.4a) and $\rho(T = 0.5$ K, $B)$ (Fig. 11.4b) traces collapse for a wide range of densities $1.6 \leq n_s \leq 3.3 \times 10^{11}$ cm^{-2}. This is consistent with the Zener theory of ferromagnetism [39, 43], which predicts that FM ordering temperature depends merely on the carrier density of states (DOS) and magnetic susceptibility of the Mn ions. However, DOS does not depend on n_s in 2D, resulting in this striking behavior. At the lowest n_s, where the carriers become localized, the local FM order is destroyed, and the intrinsic AF interactions between the Mn ions dominate [43], thus the anomalous resistance behavior vanishes.

Figure 11.5a shows a phase diagram at the n_s-T plane, as determined experimentally. It indicates that the "dome" of anomalous behavior is, in fact, located just on the metallic side of the MIT. It is characterized by abrupt T-dependent localization, strong negative MR, and the lack of $\rho(n_s)$ dependence.

11.3.3 Metal-Insulator Transition in Magnetic 2D Systems

Doped DMSs usually are close to localization boundary and are characterized by extremely strong magnetoresistance. This often leads to magnetic field-induced metal-insulator transition (MIT). The MIT was studied previously in bulk (Hg,Mn)Te and (Cd,Mn)Se [46]. In these materials increasing magnetic field induces a transition from insulating phase to the metallic phase, in contrast to what is usually observed in nonmagnetic semiconductors. The MIT in bulk materials is now well understood. However, recent observation of MIT in 2D electron system in Si-MOSFET [47, 48] and other 2D nonmagnetic systems was somewhat unexpected, since scaling theory of localization [49] predicts that metallic phase does not exist in 2D. While this 2D MIT is still under strong debate, it was shown [36] that 2D MIT occurs also in magnetically doped 2DES. Figure 11.5b shows apparent metallic behavior (i.e., $d\rho/dT > 0$) in modulation doped (Cd,Mn)Te quantum well. Metallic phase is clearly seen at elevated temperatures above $T^* \approx 2$ K, while at lower T it is destroyed by anomalous T-dependent localization described above. However, when transport is measured at higher voltage excitations ($V_{\rm exc} > 100\,\mu$V, which still do not cause Joule selfheating), the enormous increase of ρ is suppressed and the metallic phase continues down to the lowest T.

This is yet another argument against BMP mechanism in this system since BMP with the binding energy of a few meV should not be that sensitive to the excitation voltages $V \sim 100\,\mu$V, i.e., 0.1 meV. At the same time, such a low electric field could drag electrons between adjacent ferromagnetic bubbles.

The 2D MIT in (Cd,Mn)Te heterostructures occurs at $n_s = 2.2 \times 10^{11}\,\rm cm^{-2}$, if $d\rho/dT = 0$ is regarded as a criterion for the transition. It corresponds to $\rho_c \approx 0.25h/e^2$, which is the lowest critical resistivity among all the systems where 2D MIT was observed so far. It resembles 2D MIT in nonmagnetic GaAs samples where $\rho(T)$ dependence in the metallic phase is also weak [50] in contrast to that observed in Si-MOSFETs [48].

11.3.4 Magnetoresistance in Ferromagnetic Semiconductors

The presence of a giant spin-splitting, specific to DMS in a paramagnetic phase, gives rise to large positive MR. This is in contrast to ferromagnetic semiconductors, e.g., (Ga,Mn)As, in which *negative* MR points to the presence of a coupling between localized spins and charge carriers. This is because in ferromagnetic semiconductors localized spins are already aligned at $B = 0$, so mechanism of MR must be different.

Figure 11.6 shows magnetoresistance in (Ga,Mn)As, which is typical for ferromagnetic semiconductors. Again, a competition between positive and negative contributions to MR is clearly visible like in a typical paramagnetic II–VI DMSs. However, the origin of these contributions is quite different. The magnitude of the observed negative magnetoresistance in Fig. 11.6a and the temperature dependence of ρ strongly suggest that the spin disorder scattering by thermodynamic fluctuations of magnetization is involved. A maximum of ρ around T_C can be interpreted as a result of critical scattering by packets of the Mn spins with a ferromagnetic short-range order characterized by a correlation length comparable to the wavelength of the carriers. Thus, the negative MR can be understood as the reduction of scattering by alignment of spins by B. Well above T_C, where there is only small

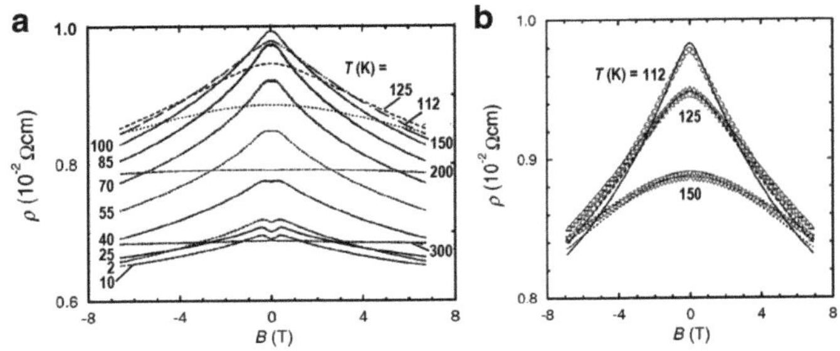

Fig. 11.6 (**a**) Resistivity ρ of $Ga_{0.947}Mn_{0.053}As$ at various temperatures. The sample is ferromagnetic below $T_C = 110$ K. (**b**) Negative magnetoresistance at three different temperatures above T_C. The *solid lines* show fits using (11.3). Reprinted with permission from [16]

spin correlation among Mn ions, one can use the spin disorder scattering formula to describe the B dependence of the resistance [51]:

$$\rho = 2\pi^2 \frac{k_F}{pe^2} \frac{m^2 J_{pd}^2}{h^3} n_{Mn} [S(S+1) - \langle \mathbf{S} \rangle^2], \tag{11.3}$$

where k_F is Fermi wave number, e the elementary charge, m the effective mass of carrier, h Planck constant, n_{Mn} the concentration of Mn ions, p concentration of holes, and $\langle \mathbf{S} \rangle$ the thermal average of S. As Fig. 11.6b shows, (11.3) describes well the experimental data. However, the negative MR hardly saturates even in rather strong magnetic field and occurs also at low T, where the Mn spins are fully ferromagnetically ordered therefore a giant spin-splitting makes spin-disorder scattering inefficient. Moreover, the absence of spin-disorder scattering makes the coherence length L_ϕ much longer. It is possible that under such conditions the negative magnetoresistance is due to suppression of localization resulting from quantum interference, i.e., weak localization MR [52].

The small positive magnetoresistance observed at low T and weak B is usually attributed to the tilt of the magnetization from its original in-plane direction at $B = 0$. MR in this region strongly depends on the field direction with respect to both current direction and crystallographic orientation as shown in Fig. 11.7. This is a signature of anisotropic magnetoresistance (AMR), which is characteristic for ferromagnetic semiconductors [52, 53], while in paramagnetic II–VI DMSs low-field spin-related MR is isotropic [30]. However, generally the temperature and magnetic field dependence of resistivity in ferromagnetic semiconductors is still poorly understood. Recent studies of the AMR [55] down to 30 mK around MIT in (Ga,Mn)As samples show that in the lowest temperature regime the transport can be well described by the 3D scaling theory of Anderson's localization in the presence of spin scattering [56]. Another study [54] points out striking analogies between (Ga,Mn)As and some manganites, such as nonmonotonic $\rho(T)$ dependence, and quantitatively describes transport in (Ga,Mn)As, as shown in Fig. 11.7c, using a scaling theory of localization for strongly disordered ferromagnets [57].

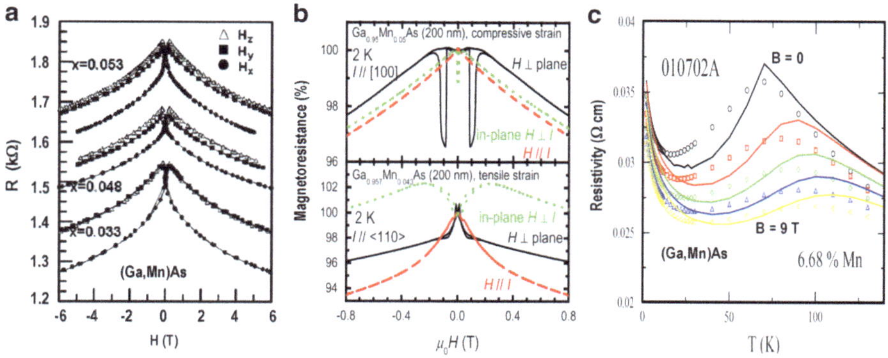

Fig. 11.7 (**a**) Low-field magnetoresistance in (Ga,Mn)As films reveals strong anisotropy and depends on field, current and sample orientation [53] as well as (**b**) on strain [52]. (**c**) Resistance of (Ga,Mn)As film as a function of T at magnetic fields $B = 0, 1, 3, 6$ and 9 T. *Dots* represent experimental data, *solid lines* are theoretical fits. Reprinted with permission from (**a**) Elsevier [7], (**b**) APS [6, 54]

11.4 Universal Conductance Fluctuations in Diluted Magnetic Semiconductors

11.4.1 Spin-Splitting Driven Conductance Fluctuations

An influence of magnetic ions in DMSs on coherent transport phenomena was studied in free standing wires of (Cd,Mn)Te made from MBE grown thin films [29]. The wires dimensions were $5\times0.3\times0.3\,\mu$m. Figure 11.8a, b show the magnetoresistance traces in CdTe and $Cd_{0.99}Mn_{0.01}Te$ wires, respectively. Weak field magnetoresistance and aperiodic but reproducible resistance fluctuations are clearly seen in both materials. However, the observed UCFs are quite different in the case of magnetic wire. As shown by dotted line in Fig. 11.8b, the characteristic features on the UCF pattern shift with either T or B, a behavior not observed in nonmagnetic wire. UCF pattern in (Cd,Mn)Te wire scales with the temperature similarly to the low-B positive MR. This strongly suggests an involvement of s–d exchange Zeeman splitting. Figure 11.8c where traces from Fig. 11.8b are plotted as a function of magnetization, shows that both positive MR and UCF scale with the magnetization M to which the spin-splitting is proportional.

According to a model [29] proposed to explain this novel mechanism of UCF generation, it originates from a giant spin-splitting, which induces redistribution of the carriers between spin subbands substantially changing electronic wavelengths λ_F at Fermi level, as it is shown in Fig. 11.9a. This according to (11.2) in Sect. 11.2 alters quantum interference and generates conductance fluctuations when spin splitting changes.

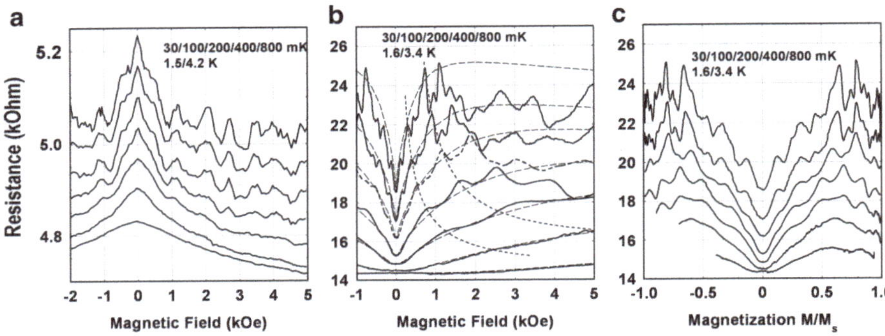

Fig. 11.8 ρ vs. B at various T for the submicron wires of CdTe (**a**) and (Cd,Mn)Te (**b**). *Dashed lines* in (**b**) are guides for the eye and visualize a strong dependence of ρ features in (Cd,Mn)Te. (**c**) Data of (**b**) plotted as a function of magnetization normalized to its saturation value M_S. Reprinted from [58]

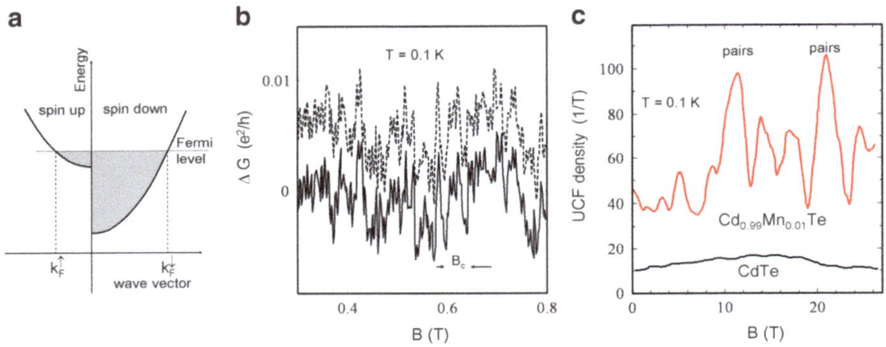

Fig. 11.9 (**a**) Illustration how strong spin splitting alters wave vector at Fermi level. (**b**) UCF measured in a wire made of (Cd,Mn)Te heterostructure containing 2DES. (**c**) Density of UCF measured in wires made of CdTe and (Cd,Mn)Te. In the latter, density of UCF strongly increases in the regions of magnetization steps. Reprinted with permission from [60]

11.4.2 Conductance Fluctuations in Modulation Doped Wires

Figure 11.9b shows magnetoresistance data measured at $T = 0.1$ K in small samples made of (Cd,Mn)Te heterostructures containing 2DES. Data clearly show random but reproducible UCF. However, in contrast to the case of DMS wires patterned from uniformly doped thin films described previously, their correlation field B_{corr}, which is a distance between characteristic features of fluctuation pattern, does not change with external B. In addition, characteristic points of the pattern do not shift with T. The temperature affects only the UCF magnitude. This is because in modulation doped heterostructures with donors set-back far from the conduction channel, elastic mean free path becomes very long in comparison with electron wavelength and $\ell_e \gg \lambda_F$. In other words, distances between scattering centers are substantially larger as well as typical area of the loops of self-crossing trajectories as A-B-C-D-E-A in Fig. 11.2a. Thus, small change of B produces substantial phase shift in (11.2). This makes B_{corr} much smaller and density of the fluctuation per field unit $dUCF/dB \propto 1/B_{corr}$ much higher in comparison with these for UCF generated by Zeeman splitting. As a result, only UCF generated by orbital effect of magnetic field upon phase of electronic waves are observed in modulation doped wires.

11.4.3 Magnetization Steps Observed by Means of Conductance Fluctuations

The new mechanism of spin splitting induced UCF described above produces fluctuations as long as the magnetization M changes with external magnetic field, i.e., when magnetic susceptibility $\chi \neq 0$. In DMSs, M usually saturates at relatively low-field, $B < 1$ T at low T. However, it is well known that some of the randomly distributed magnetic ions constitute clusters, which do not contribute to the total M because of intrinsic short range AF interaction. When B is strong enough to overcome internal AF interactions, the ions from the cluster align their spins with the external B and start contributing to M. This results in sudden increase of M, called a magnetization step. In particular, nearest neighbor pairs of Mn in (Cd,Mn)Te contribute to M steps at $B \approx 10.4$ and 19.4 T [59].

According to the above-mentioned model of spin-splitting driven UCF, the density of the fluctuations, i.e., average number of fluctuations per magnetic field unit, is proportional to the magnetic susceptibility of the localized spins. Thus, UCF measurements make it possible to study quantitatively the subsystem of magnetic ions.

In $Cd_{0.99}Mn_{0.01}Te$ wire, a strong enhancement of the UCF density is observed [60] around 11 and 20 T (Fig. 11.9c). These anomalies are assigned to a step-like increase of the magnetization, which results from the field-induced change in the ground state of the nearest neighbor pair of Mn ions. Such a conclusion is supported by a previous observation of the magnetization steps in $Cd_{1-x}Mn_xTe$ [59] for the same values of the magnetic field. Moreover, at appropriately low temperatures, additional features can be seen. They are attributed to distant-neighbor exchange interactions. These observations strongly support the model of UCF driven by spin-splitting.

11.4.4 Time-Dependent Conductance Fluctuations in the Spin Glass Phase

Diluted magnetic semiconductors, which contain randomly distributed localized spins, are usually paramagnets if temperature is high enough. In the absence of an externally applied magnetic field, the spins are oriented randomly because of thermal motion. These thermal fluctuations are quite fast of the order of picoseconds. However, II–VI DMSs have also intrinsic antiferromagnetic superexchange coupling between the moments. Randomness and AF coupling are the key ingredients for the magnetic phase known as spin glass. In this phase, a given localized spin (here Mn ion) tries to configure itself in direction opposite to its neighbor. This is easy when there is only one another spin around. But, if there are three of them and distance between them is identical (i.e., they constitute a triangle), it becomes impossible. If two equilateral are antiparallel, then the third is *frustrated*, i.e., cannot find antiparallel configuration to each of its two neighbors. Situation complicates further in the case of many such spins in the sample. Spin glasses have many ground states, i.e., spin configurations with the lowest total energy. Thus, the system slowly wanders between different ground states and never explores all of them on experimental time scales, in analogy to structural glasses, as window glass, which slowly flows even at room temperature.

In [61], universal conductance fluctuations are used to extract information about this slow spin dynamics in small mezoscopic $Cd_{1-x}Mn_xTe$ samples with high Mn content, up to $x = 0.2$. According to magnetization measurements [62], $Cd_{1-x}Mn_xTe$ freezes into spin glass phase at, e.g., $T_f \approx 0.3$ K and 2 K for $x = 0.07$ and 0.20, respectively.

Figure 11.10b shows amplitudes δG of the field-dependent UCF in $Cd_{1-x}Mn_xTe$ wires for $x = 0$, 0.01 and 0.07. At higher T, the amplitude is substantially reduced in $x = 0.07$ sample by inelastic spin disorder scattering when electrons scatter from Mn ions changing their spin direction, and thus their energy, *which destroys their phase*. A substantial increase of the fluctuation amplitude occurs at $T_g \approx 0.3$ K, the temperature corresponding to the freezing of the Mn subsystem into spin-glass phase. In this phase, the localized Mn spins are frozen, i.e., they cannot thermally fluctuate as fast as in the paramagnetic phase. This leads to the suppression of the spin disorder scattering. On the other hand, mesoscopic conductance is sensitive

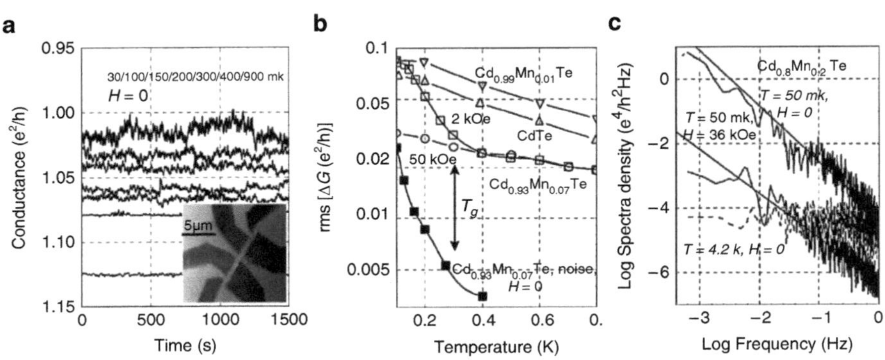

Fig. 11.10 (**a**) Conductance G as a function of time at $B = 0$ in the wire of n-$Cd_{0.93}Mn_{0.07}Te$ at selected T down to 30 mK. (**b**) Amplitude of UCF as a function of T in n-$Cd_{1-x}Mn_xTe$ wires with different Mn concentrations (*open symbols*) and noise amplitude in n-$Cd_{0.93}Mn_{0.07}Te$ wire. The *arrow* marks the bulk value of the spin-glass freezing temperature for $x = 0.07$. (**c**) Noise power spectra in $Cd_{0.8}Mn_{0.2}Te$ wire at selected temperatures and magnetic fields. *Straight lines* show $1/f^{\alpha}$ dependence. Reprinted from [61]

to the specific scattering potential imbedded in the particular sample. Indeed, as it is clearly seen from Fig. 11.2, it is enough to change (e.g., move) just one scattering center to modify resulting amplitude of interfering electronic waves, i.e., conductance.

In the presence of the frozen spins, the time-reversal symmetry is broken. Thus, closed clock- and counterclockwise trajectories (as path ABCDEA in Fig. 11.2a) are not equivalent any more, while magnetic flux (or more precisely, vector potential) between different paths p, p', p'' etc. is nonzero even in the absence of the external magnetic field. This strongly influences interference of electronic waves. Thus, if spins slowly fluctuate with time, as in spin-glass phase, they will cause time-dependent conductance fluctuations, i.e., conductance noise. These phenomena provide real-time probe of spin dynamics. Indeed, below $T_f \approx 0.3\,\mathrm{K}$ an immense electrical noise appears at the onset of the freezing point, as shown in Fig. 11.10a, b. Its amplitude decreases with increasing temperature and magnetic field and it is absent in both diamagnetic CdTe and low-composition paramagnetic $Cd_{1-x}Mn_xTe$ samples. These facts strongly suggest that slow dynamics of localized spins is observed by means of coherent transport.

Figure 11.10c shows the resistance noise power spectra S_R measured in $x = 0.20$ wire. While a white (i.e., frequency independent) noise is observed at $T = 4.2\,\mathrm{K}$, at low temperatures S_R becomes f dependent, $S_R \propto 1/f$. After applying magnetic field, the noise power is substantially reduced, since under B slow evolution of spin glass phase is suppressed. A higher-order statistics of the noise power made it possible to determine that droplet model of the spin glass phase better describes the observed slow dynamics than the hierarchical model. It was an important experimental observation for a still strongly debated nature of spin-glasses.

11.4.5 Mesoscopic Transport in III-Mn-V Semiconductors

Recently, mesoscopic transport was measured also in III-Mn-V nanowires and rings [63–65]. Figure 11.11a shows magnetoresistance traces measured in ferromagnetic ($T_C = 55\,\mathrm{K}$) nanowires of (Ga,Mn)As. There are strong UCFs clearly seen. Their amplitude increases with lowering T and decreasing sample length. At the lowest T, it reaches universal value $\delta G = e^2/h$. The analysis of the length and temperature dependence of the observed UCF enabled the authors of [63] for the first time to determine the coherence length in ferromagnetic semiconductor $L_\phi = 100\,\mathrm{nm}$ at $T = 10\,\mathrm{mK}$ and $L_\phi \propto T^{-1/2}$. The same dependence was also found in [64]. This dependence suggests electron–electron interaction as a dominant dephasing mechanism. Figure 11.11b shows Aharonov-Bohm oscillations observed in a (Ga,Mn)As ring sample. They are superimposed on UCF fluctuation, thus difficult to resolve. It is also possible that, in fact, the observed oscillations originate from the same spin-splitting driven mechanism like in paramagnetic wires described previously in Sect. 11.4.1. Indeed, they are accompanied by positive MR, which strongly suggest that not all Mn are spin polarized and that polarization increases at the small B range, i.e., not all of the Mn ions contribute to ferromagnetic ordering at $B = 0$.

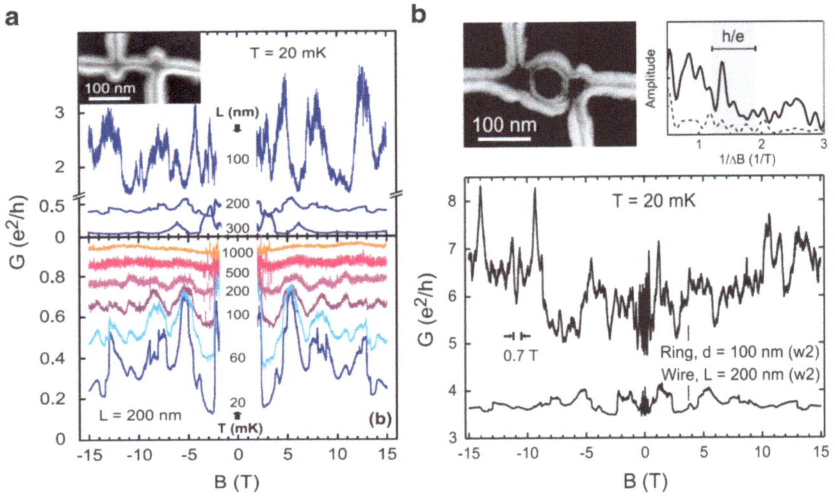

Fig. 11.11 (**a**) Conductance G vs. B of the 200 nm wide wire for different lengths (*top*) and temperatures (*bottom*) between 20 mK and 1 K. (**b**) Comparison of the magnetoconductance trace of the (Ga,Mn)As ring sample (*top*) with the conductance of a wire of similar length and 20 nm width (*bottom*). Corresponding Fourier transform taken from the conductance of ring and wire (*top right*). The region where AB oscillations are expected is highlighted. Reprinted with permission from [63]

Results [65] of mesoscopic MR in ferromagnetic (In,Mn)As wires ($T_C = 27{,}47$ K) provide another evidence for this suggestion. The observed low frequency noise is strongly suppressed by B at low T. In a ferromagnetic system with the time-reversal symmetry already broken at $B = 0$, no decrease in noise power is expected since the system is already fully spin polarized at $B = 0$. However, the observed suppression of the noise strongly implies that the dominant source of time-dependent UCF is the spin disorder scattering of the carriers from fluctuating magnetic disorder. One possibility is that the noise is associated with Mn spins, perhaps at the edges or in some isolated regions of the sample, not fully participating in the bulk FM order of the system. Spin disorder scattering of slowly fluctuating local magnetization could cause time-dependent UCF, as in the aforementioned (Cd,Mn)Te spin glass case. At sufficiently large B, those moments would be saturated, removing the above source of fluctuations.

11.5 Anomalous Hall Effect

In ferromagnetic materials (and paramagnetic materials in a magnetic field), the Hall resistivity includes an additional contribution, known as the anomalous Hall effect (AHE). The AHE is also called extraordinary, spontaneous, or spin Hall effect;

however, the latter is recently referred to as the transverse spin imbalance rather than electric charge imbalance [66–68]. The AHE depends directly on the magnetization of the material, and is often much larger than the ordinary Hall effect. Empirically:

$$\rho_{xy} = R_{\mathrm{H}}B + R_{\mathrm{S}}M, \tag{11.4}$$

where $R_{\mathrm{H}} = 1/ne$ is known as Hall coefficient, while the constant R_{S} is called the anomalous Hall coefficient. The anomalous contribution is often seen as proportional to the magnetization of the sample and becomes constant once the magnetization has reached its saturation value.

The ordinary contribution originates from the external magnetic field perpendicular to the sample when Lorentz force acting on the current carriers gives rise to a transverse voltage, which balances the force, because electric current cannot be really deflected in the finite sample. Although the AHE was discovered by E.H. Hall in ferromagnets almost simultaneously with the ordinary effect, and despite many years of experimental studies in different ferromagnets, spinels, type-II superconductors, Kondo-lattice materials, and magnetically doped metals, the origin of the AHE is still vigorously debated. Notably, this effect is not due just to the contribution of the magnetization to the total magnetic field. On the one hand, there are predictions [69, 70] that AHE is an effect of the spin–orbit interaction on spin-polarized conduction electrons. In other words, it arises from a general property of how electrons move in a periodic lattice and how what is now called Berry geometrical curvature gives rise to additional term in transverse carrier velocity. Reference [71] pedagogically explains the issue. This mechanism is now referred to as intrinsic AHE since it originates from symmetry considerations and should occur in perfectly periodic lattice without any impurities. This model predicts $R_{\mathrm{S}} \propto \rho^2$. On the other hand, extrinsic (i.e., related to impurities) mechanisms called skew scattering [72] and side jumps [73] were proposed later. In both models, the AHE arises from spin–orbit scattering from impurities and vanishes in purely periodic lattices. These models give $R_{\mathrm{S}} \propto \rho$ and $R_{\mathrm{S}} \propto \rho^2$, respectively.

Figure 11.12 shows that in the presence of spin-dependent scattering a transverse spin imbalance builds up and spin Hall effect follows. When the beam of electrons entering the sample is additionally spin polarized, also the charge imbalance occurs and the anomalous Hall effect is observed.

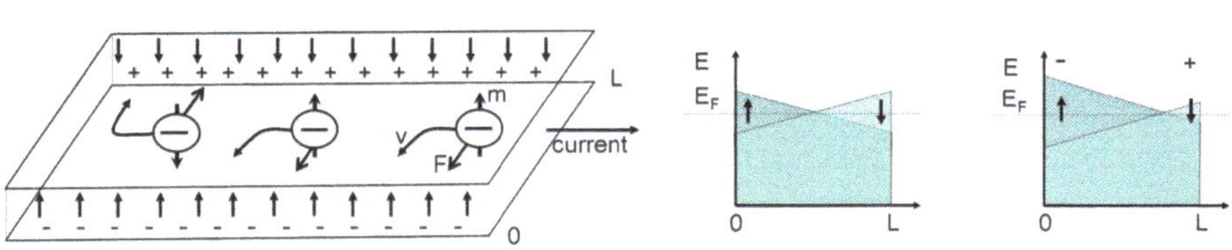

Fig. 11.12 In the spin Hall effect, spin-dependent scattering of the moving electrons causes spin imbalance, in a direction perpendicular to the current flow. If additional carriers are spin polarized (e.g., when there are more spin-up then spin-down carriers), also charge imbalance results and anomalous Hall effect occurs (after [66])

The AHE plays a very important role in DMSs, especially ferromagnetic. It serves as an easy transport measure of the magnetization; however, it notoriously obscures normal Hall effect, making it difficult to determine the carrier concentration. DMSs offer a worthwhile opportunity to study AHE. Unlike in most of the systems studied before where it is difficult to change parameters such as carrier concentration, spin polarization, and diagonal resistivity to quantitatively test existing models of the AHE, usually it is much easier in semiconductors. This promises an opportunity to clarify the mechanisms of the AHE.

Early reports of the AHE in ferromagnetic p-type III–V (In,Mn)As [14] (see Fig. 11.13a), (Ga,Mn)As [16] as well as in p-type II–VI (Zn,Mn)Te [21] (see Fig. 11.13b) reported a linear dependence of $R_S \propto \rho$ upon resistivity, strongly suggesting that skew scattering is responsible for its appearance. However, later theoretical studies [74] successfully interpreted the AHE in (III,Mn)V as a result of intrinsic mechanism, with a quantitative agreement with the experimental data in DMSs. Moreover, this theoretical model was strongly supported by the high temperature AHE measurements in paramagnetic phase of (Ga,Mn)As [75].

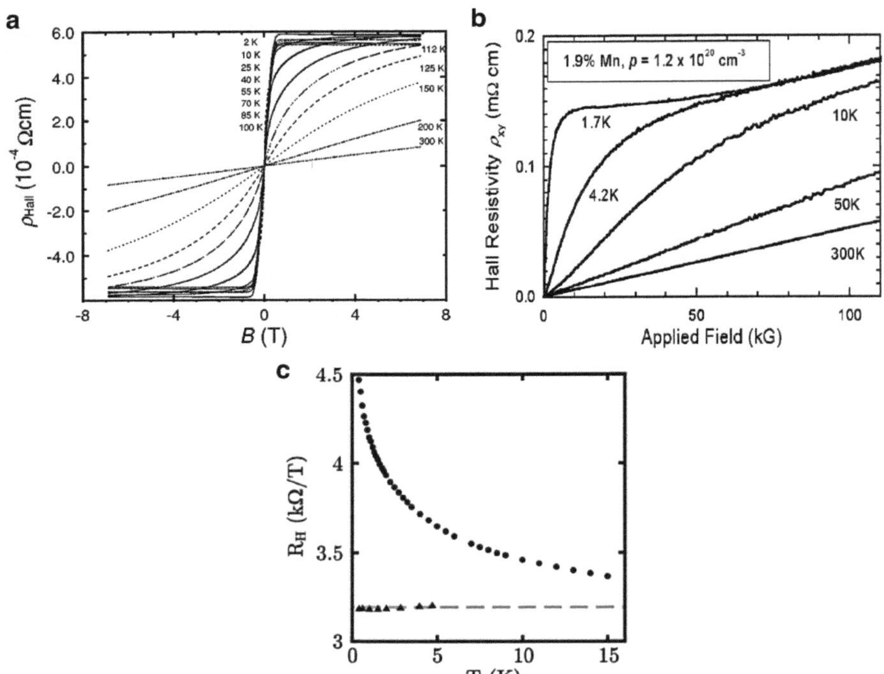

Fig. 11.13 (**a**) The observed Hall resistivity ρ_{xy} in thin films of p-(Ga,Mn)As and (**b**) p-(Zn,Mn)Te. (**c**) Hall coefficient vs. T in n-(Zn,Cd,Mn)Se quantum well (*top*) compared to R_H obtained from Shubnikov–de Haas oscillations (*bottom*). Reprinted with permission from [16] (**a**), [21] (**b**), [76] (**c**)

The AHE study in $Sb_{2-x}Cr_xTe_3$ [77] revealed again $R_S \propto \rho$ in disagreement with the clean-limit theory [74], despite that the level of impurities in $Sb_{2-x}Cr_xTe_3$ is similar to III–V DMSs. In turn, the AHE studied in n-(Zn,Co)O thin films across metal insulator transition [23, 78] showed $R_S \propto \rho^{1.4}$ dependence and was interpreted as a result from different contributions. Measurements of the AHE in the hopping regime [79] in (Ga,Mn)As evidenced a sublinear dependence of R_S on ρ with qualitative agreement with the corresponding theory [80]. The observed behavior is inconsistent with theories of the AHE in good metals, and also disagrees with predictions for a hopping AHE in manganites [81]. The study of the AHE in compensated insulating (Ga,Mn)As [82] shows that a strong AHE can exist also in the spin-glass phase.

Recently, AHE was also observed in n-type magnetically doped 2DES n-(Zn,Cd, Mn)Se modulation doped quantum wells [76]. In these structures, the normal Hall contribution can be easily extracted from the carrier density measured independently from SdH oscillations (Fig. 11.13c). Moreover, the magnetization can be measured by magneto-luminescence while carrier density and hence disorder could be tuned by electric gate. The AHE temperature dependence was found to follow paramagnetic Brillouin-like magnetization of Mn ions. The results show clearly linear AHE dependence on resistance and are interpreted as a result of skew scattering mechanism. At the same time, theoretical studies of the AHE in paramagnetic 2DES [83] carried out within intrinsic mechanism framework cannot explain the experimental data.

The above examples show that although a lot has been done about the nature of the anomalous contribution to the Hall resistivity in diluted magnetic semiconductors, the situation is still unclear and calls for further experimental and theoretical efforts.

11.6 Quantum Hall Effect in Diluted Magnetic Semiconductors

11.6.1 Introduction to the Integer Quantum Hall Effect

The quantum Hall effect (QHE) is one of the most fascinating phenomena discovered in condensed matter physics during last decades. It was first observed in silicon MOSFETs and then intensively studied in GaAs/(Ga,Al)As as well as in other III–V two-dimensional structures. QHE originates from two ingredients, Landau quantization of electronic energy levels and disorder-induced localization. In the 2DES, the density of states at zero-magnetic field is $\rho(E) = g_s g_v(m^*/2\pi\hbar^2)$, where g_s and g_v are spin and valley degeneracy, respectively, i.e., $\rho(E)$ does not depend on energy. In the presence of a strong magnetic field, the energy states contract into Landau levels separated by cyclotron energy $\hbar\omega_c = \hbar eB/m^*$. Each LL is split into two spin subbands as shown in Fig. 11.14a. In typical nonmagnetic 2DES the energy

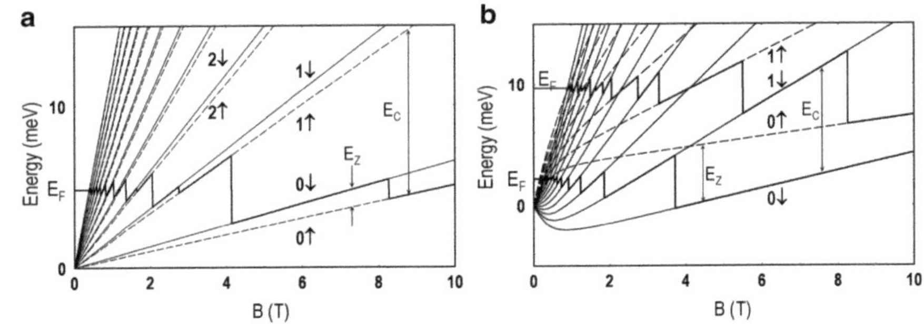

Fig. 11.14 Landau level fan diagrams in nonmagnetic (**a**) and magnetically doped (**b**) quantum wells. *Dashed* (*thin solid*) *lines* belong to Landau levels with spin-up (spin down). These energy diagrams were constructed using CdTe conduction band parameters for 10 nm wide quantum wells. In particular: effective mass for electrons is $m^* = 0.10m_0$, i.e., the cyclotron (Landau) energy $\hbar eB/m^*$ is $\approx$1 meV/T, electron Landé factor $g_e = -1.67$ so spin splitting in CdTe is $\approx$1/8 meV/T, saturation value of the exchange part of the spin splitting in $Cd_{0.985}Mn_{0.015}Te$ is roughly 5 meV. Fermi energy $E_F(B=0) = 4.8$ meV for CdTe corresponds to the electron density $n_s = 2 \times 10^{11}$ cm^{-2}, while for $Cd_{0.985}Mn_{0.015}Te$ $E_F(B=0) = 9.7$ meV and $E_F(B=0) = 2.4$ meV correspond to $n_s = 4 \times 10^{11}$ cm^{-2} and $n_s = 1 \times 10^{11}$ cm^{-2}, respectively. Importantly, in the latter case the electronic transport occurs in fully spin-down polarized Landau levels

of LLs linearly depends on magnetic field: $E_{N,\uparrow(\downarrow)} = (N + 1/2)\hbar\omega_c \pm \frac{1}{2}g^*\mu_B B$. where $N = 0, 1, 2 \ldots$ is Landau level index, $g^*\mu_B B$ is a spin splitting (Zeeman energy), and g^* is Landé factor. Usually, the ratio of Zeeman to cyclotron energy is small, e.g., ~1/20 in GaAs and ~1/8 in CdTe. Each Landau level is strongly degenerated and contains $eB/\hbar$ electronic states per area unit.

Due to disorder (caused by impurities), sharp Landau levels evolve to broader energy bands. Energy states in the center of the band are extended, and localized elsewhere. Only the extended states can carry current at zero temperature. The experiments show that if Fermi energy lies within localized states, the Hall resistance has fixed values (plateaux) $\rho_{xy} = h/e^2/i$, where i is a number of LLs below E_F. At the same time, the longitudinal resistance ρ_{xx} vanishes. Indeed, when E_F is not at LL center and electron density increases (or the magnetic field is decreased) the localized states gradually fill up without any change in occupation of the extended states below E_F, thus without any change in the Hall resistance. It is only as the Fermi level passes through the center of LL that the longitudinal resistance becomes appreciable $\rho_{xx} > 0$ and the Hall resistance ρ_{xy} makes its transition from one plateau step to the next.

11.6.2 Dramatic Modification of Energy Diagram by a Giant s–d Exchange

Incorporation of magnetic ions into quantum well modifies LLs diagram dramatically. A strong s–d exchange coupling between band and localized d-electrons leads to a giant spin splitting $E_z = |E_{N,\uparrow} - E_{N,\downarrow}|$, which is proportional to magnetization M described by modified Brillouin function [84]:

$$E_{N,\uparrow(\downarrow)} = (N+\frac{1}{2})\frac{\hbar eB}{m^*} \pm \frac{1}{2}\left[g^*\mu_B B + \alpha N_0 x_{\text{eff}} S \mathrm{B}_S\left(\frac{Sg\mu_B B}{k_B[T+T_{\text{AF}}]}\right)\right]. \quad (11.5)$$

For (Cd,Mn)Te $m^* = 0.10m_0$ and $g_e = -1.67$ [85] are the effective mass and Landé factor of the electrons in CdTe; $\alpha N_0 = 0.22\,\text{eV}$ is the s–d exchange energy [84, 86, 87], and B_S is the Brillouin function, in which $S = 5/2$ and $g = 2.0$. The functions $x_{\text{eff}}(x) < x$ and $T_{\text{AF}}(x) > 0$ [84] describe the reduction of magnetization $M(T, B) = g\mu_B x_{\text{eff}} N_0 S \mathrm{B}_S(T, B)$ by antiferromagnetic interactions. For $x \sim 0.01$, E_z reaches value 5 meV. Thus, in the low-B range, it substantially exceeds the cyclotron energy $\hbar\omega_c \approx 1$ meV per Tesla. E_z is also comparable with Fermi energy, as shown in Fig. 11.14b.

Several important consequences follow (11.5). (1) spin subbands are well-resolved; (2) positions of electronic levels strongly depend on the temperature; (3) for low electron density only spin-down LLs are occupied, thus electron gas is fully spin-polarized; (4) since M rises rapidly with B and then saturates, also E_z dependence on the magnetic field B is strongly nonlinear. This results in many crossings of Landau spin sublevels, so that in an independent-electron picture, energies come into "coincidence" for particular values of B_c. According to (11.5) LL crossing occurs when E_z is an integer multiple of $\hbar\omega_c$. (5) Moreover, s–d induced spin splitting and that one related to intrinsic g^* have opposite signs. Thus, LLs with the same orbital index N but opposite spins cross at high magnetic fields.

As a consequence, many striking spin-dependent transport phenomena were observed in magnetic 2D systems.

11.6.3 Early Observations of Landau Quantization in Diluted Magnetic Semiconductors

The earliest study of quantum transport in 2DES involving magnetic ions was carried out on MIS structures prepared on the surface of p-type (Hg,Mn)Te [1]. Under magnetic field, pronounced SdH oscillations were observed in the inversion layer as a function of gate voltage, similar to these observed in Si-MOSFETs and other nonmagnetic metal-insulator-semiconductor structures. However, in the presence of magnetic ions, the positions of SdH oscillations showed strong dependence on the T, providing evidence of an influence of sp–d exchange on Landau level energies.

In turn, the earliest observation of QHE in DMS was actually performed not on MBE grown structures, but on grain boundaries in bulk (Hg,Cd,Mn)Te [88]. Bulk ingots of HgTe and its alloys with CdTe and MnTe, grown either by the Bridgman or the solid state recrystallization methods, consist usually of differently oriented single-crystalline grains of diameter ~5–10 mm. As grown HgTe alloys are p-type.

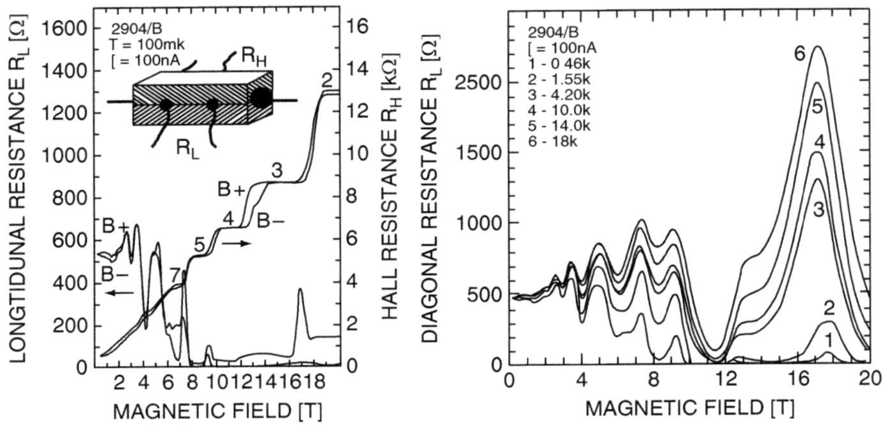

Fig. 11.15 Longitudinal and Hall resistances at 100 mK of 2DES with electron density $n_s = 1.1 \times 10^{12}\,\mathrm{cm}^{-2}$ naturally occurring in $Hg_{0.79}Cd_{0.19}Mn_{0.02}Te$ for two directions of the magnetic field. Sample dimensions: distance between lateral probes $L = 0.7$ mm, width $W = 0.5$ mm. Inset: schematic view of the sample with a single grain boundary. (**b**) Longitudinal resistance measured at various temperatures. Reprinted with permission from [88]

However, at the grain boundaries there are n-type two-dimensional (2D) inversion layers with surprisingly high electron mobility $\mu \sim 5 \times 10^4\,\mathrm{cm}^2/\mathrm{Vs}$. This, after tedious isolating and contacting to a single grain boundary, made it possible to study the 2D transport. Figure 11.15 shows results of transport measurements performed on such grain boundary in $Hg_{0.79}Cd_{0.19}Mn_{0.02}Te$ host crystal. Well-developed quantum Hall plateaux are clearly seen at $\rho_{xy} = h/e^2/i$ for $i = 2, 3, 4, 5, 7$. The absence of the 6th plateau signals overlap of the corresponding LLs. The data demonstrate the Hall resistance quantization with better than 0.1% precision.

11.6.4 Quantum Hall Effect Scaling

According to theory [89], at $T = 0$ in the regime of integer QHE, the Hall resistivity $\rho_{xy} = h/e^2/i$, longitudinal resistivity $\rho_{xx} = 0$, and the states at the Fermi level are localized. The exceptions are regions of measure zero in between Hall plateaux, where E_F coincides with a singular delocalized state in the center of the LL. As E_F approaches the center that has energy E^* at the magnetic field B^*, corresponding localization–delocalization transition can be described as a divergence of the localization length $\xi \propto |E_\mathrm{F} - E^*|^\alpha \propto |B - B^*|^\alpha$ with an universal exponent $\alpha = 7/3$.

However, at finite $T > 0$ the divergence of the ξ is removed and delocalized states at the centers of LLs become broadened on the energy scale. This results in transitional region between QHE states, where ρ_{xx} is finite and ρ_{xy} gradually changes between adjacent plateaux. In this regime, transport coefficients obey temperature scaling, in particular the half-width of SdH maxima $\Delta B \propto T^{\kappa}$, while the maximum derivative of the Hall resistivity $\mathrm{d}\rho_{xy}/\mathrm{d}B|_{\max} \propto T^{\kappa}$ [89]. The universal scaling exponent κ is related to the localization exponent α through the relation $\kappa = p/\alpha \approx 0.4$, where the exponent $p \approx 1$ describes divergence of the phase coherence length $L_{\varphi} \propto T^{-p/2}$ as $T \to 0$. If, however, the phase coherence length becomes greater than a sample size $L_{\phi} \geq L$, the scaling is suppressed.

11.6.5 Temperature Scaling

Due to a large ratio of the Zeeman to Landau splittings, it is possible to examine the QHE scaling in DMSs at high LL, since in contrast to nonmagnetic QHE systems, the dependence of the localization length on the distance to the center of the Landau level is not obscured by the two overlapping densities of states originating from the adjacent spin subbands. This possibility was explored in the study of temperature and size QHE scaling in (Cd,Mn)Te nanostructures [90] containing a substantial concentration of localized spins.

It was found that for large 2D samples the inverse width of the resistance peaks and the slope of the Hall resistance in between the plateaux, $1/\Delta B$, $\mathrm{d}\rho_{xy}/\mathrm{d}B$ (as shown in Fig. 11.16a), respectively, obey the characteristic power law $T^{-\kappa}$. In $Cd_{0.997}Mn_{0.003}Te$ sample $\kappa = 0.42 \pm 0.05$ for $1.5 < \nu < 4.5$ in the temperature range $4.2\,\mathrm{K} > T > 50\,\mathrm{mK}$. For either higher ν or for the higher Mn concentration, κ becomes smaller.

11.6.6 QHE Scaling in Small Samples: Dimensional Effects

In narrow (widths $\leq 10\,\mu\mathrm{m}$) Hall bars of $Cd_{1-x}Mn_xTe$ [90], the width of the resistance maxima becomes independent of the temperature below a characteristic temperature T_{C}, as shown in Fig. 11.16b. At the same time, magnetoresistance reveals the presence of the UCF. Their amplitude increases with decreasing temperature down to 50 mK without any indication of a saturation. This demonstrates that the apparent saturation of $1/\Delta B$ below T_{C} is caused by a size effect, not by heating of the electron gas, since in the latter case UCF amplitude would also saturate below T_{C}. The data also make it possible to evaluate the temperature dependence of the coherence length to be $L_{\phi} \propto T^{-0.55}$.

11.6.7 Quantum Hall-Insulator Transition

Quantum Hall liquid to insulator transition, which is closely related to QHE scaling, was studied in a strongly localized modulation doped (Zn,Cd,Mn)Te heterostructures [28] on both sides of QHE state with $\nu = 1$, as shown in Fig. 11.16c. In

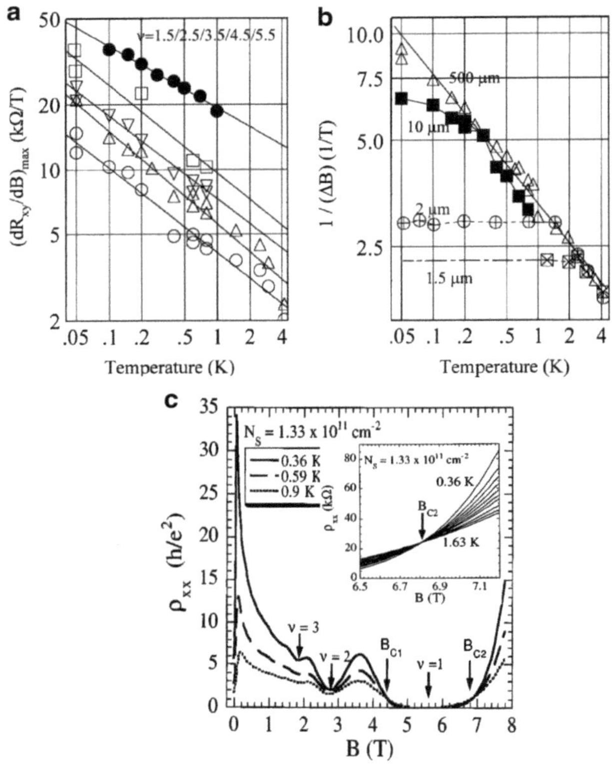

Fig. 11.16 (**a**) Temperature scaling of the slope of the Hall resistance in between the plateaux in $Cd_{0.997}Mn_{0.003}Te$ (*open symbols*) and $Cd_{0.98}Mn_{0.02}Te$ (*full symbols*) quantum wells. (**b**) Temperature scaling of inverse peak widths in the longitudinal resistivity for different widths of wires made of (Cd,Mn)Te QWs [90] (**c**) The longitudinal resistivity in (Zn,Cd,Mn)Se QW as a function B. The critical fields B_{C1} and B_{C2} demarcate transitions between insulating and QH states. The inset shows ρ_{xx} around the upper critical field B_{C2} at various temperatures. Reprinted with permission from [28]

the vicinity of that transition, it is expected that resistivity scales according to $\rho_{xx} = f(|B - B_{cn}|/T^{\kappa})$, where f is some function and B_{cn} ($n = 1, 2$) are (lower, higher) critical fields at which ρ_{xx} does not depend on T, as shown in the inset of Fig. 11.16c. At both critical field, ρ_{xx} scales according to the above formula. The critical exponent κ describing the scaling at higher critical field $B_c \approx 6.8\,T$ was found to be $\kappa \approx 0.52$, i.e., slightly higher than that in nonmagnetic QHE systems, while κ at lower critical field $B_c \approx 4.4\,T$ was much larger and its values scattered for different electron densities. It is possible that T was too high to explore the critical region, in which scaling should be universal and density independent. Anyway, the obtained scaling calls for further research for possible new universality in this fully polarized electron liquid.

11.6.8 Dramatic Modification of Shubnikov–de Haas Oscillations

Reference [13] reports on the first clear observation on how magnetic ions affect electronic transport in 2DES MBE manufactured structure. In this case, five 12 nm nonmagnetic CdTe quantum wells were separated by magnetic 50 nm wide (Cd,Mn)Te barriers. Modulation doping was obtained by inserting bromine into (Cd,Mn)Te. Transport measurements showed clearly SdH oscillations, while measurements in tilted B ultimately indicated 2D character of transport. From strong temperature dependence of SdH oscillations, it was clear that $s-d$ exchange contributed to LL energy.

Figure 11.17a shows (Zn,Cd,Mn)Se [27] structure used for the extensive transport studies of MBE grown 2DES in DMSs. The structure consists of 10.5 nm wide (Zn,Cd,Mn)Se quantum well. The well is modulation doped by the symmetrically placed ZnSe:Cl doping layers, which are separated from the QW by undoped ZnSe barriers. The electrons from doping layers migrate into QW and constitute 2DES. Because a positive charge left on the donor impurities in the barriers is set far from the conducting 2D channel, the scattering from the ionized impurities is strongly suppressed. Thus, owing to modulation doping, the mobility is substantially enhanced in comparison with uniform doping in bulk crystals.

The quantum well is a digital alloy, where MnSe layers are inserted into $Zn_{0.8}Cd_{0.2}Se$ nonmagnetic host. Generally, QWs with magnetic ions introduced

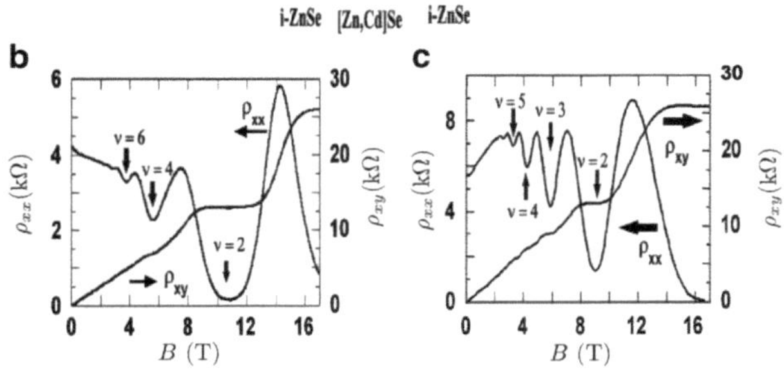

Fig. 11.17 (**a**) Structure of modulation doped n-(Zn,Cd,Mn)Se/(Zn,Cd)Se:Cl heterostructure. Mn ions are inserted digitally into the quantum well. Longitudinal ρ_{xx} and transverse ρ_{xy} sheet resistances at 4.2 K in nonmagnetic (**b**) and magnetic (**c**) structures, demonstrating the observation of an IQHE in each case. The figures also indicate the filling factors ν. Reprinted with permission from [27]

digitally have higher mobilities, since electrons could move in structurally better nonmagnetic (Zn,Cd)Se, still having strong interaction with magnetic ions.

Figure 11.17b, c compare magnetotransport results in nonmagnetic and magnetic samples. Despite similar electron densities and mobilities, the results are quite different. In low-magnetic fields, a weak, negative MR is observed in nonmagnetic sample, while much stronger positive MR dominates in magnetic ones. These effects originate from suppression of quantum interference by magnetic field, and by modifications of quantum corrections to the conductivity arising from the giant spin splitting, respectively, as described previously in Sect. 11.3. Generally, the magnetoresistance is stronger in 2D than in 3D samples.

In both samples, the quantum Hall effect and SdH oscillations are clearly seen up to filling factor $\nu = 6$. However, only even plateaux ($\nu = 2, 4, 6\ldots$, but also $\nu = 1$) are observed in nonmagnetic sample, while both even and odd are well resolved in magnetic one. This is a manifestation of s–d exchange-enhanced spin splitting, resulting in a high spin polarization of the electron gas beginning to be observable already at large Landau level filling factors, i.e., at small fields.

Another systematic study [91] on n-(Zn,Cd,Mn)Se/ZnSe heterostructures revealed further striking effects resulting from the strong s–d modification of energy levels in DMS. Figure 11.18a shows longitudinal magnetoresistance measured at four different temperatures. As it is clearly seen, some SdH minima shift toward lower B when T increases. Moreover, the observed SdH oscillations are not periodic when plotted as a function of the inverse magnetic field, in contrast to what is usually observed in nonmagnetic QHE systems. In particular, the minima of SdH oscillations do not occur at integer filling factors when LL are supposed to be full while Fermi level should lie in between two LLs. These effects stem from a giant and temperature dependent s-d spin splitting, which dramatically modifies

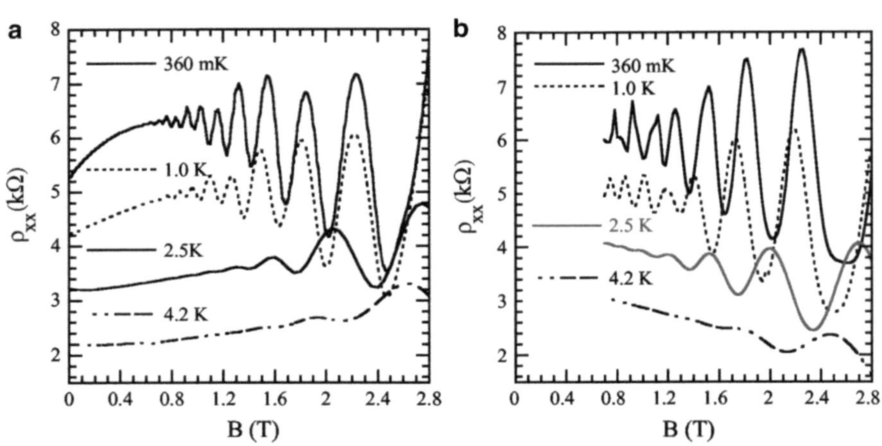

Fig. 11.18 (**a**) The measured ρ_{xx} vs. B at different temperatures in n-(Zn,Cd,Mn)Se. Traces are vertically shifted for clarity. (**b**) Simulations of these magnetoresistance data as a function of temperature and magnetic field. Reprinted with permission from [91]

energy of LLs, resulting in striking features as Landau levels of opposite spin cross, as described previously and shown in Fig. 11.14. Thus, large spin-splitting leads to well-resolved spin subbands at some fields, while in the others results in LL overlapping, which in turn strongly modifies density of states. In particular, when two LL cross and Fermi level lies at their centers, as shown in Fig. 11.20a, ρ_{xx} maximum, instead of minimum is observed, despite that this situation corresponds to an integer filling factor. At the same time, the maximum is enhanced due to doubled density of states at LL crossing.

Figure 11.18b shows simulations of these transport data made within a model involving extended states centered at each Landau level and two-channel conduction for spin-up and spin-down electrons. The simulations reproduce transport data with high accuracy while energy levels calculated using (11.5) and corresponding density of states agree well with these found from independent magnetization measurements [92] on the same sample.

Figure 11.19a, b present results of magnetotransport studies in (Cd,Mn)Te heterostructure [93] with high electron mobility $\mu \approx 6\times10^4$ cm^2/Vs and relatively low Mn contents $x = 0.003$, which corresponds to exchange enhanced spin splitting at saturation $\Delta_{\text{sat}} \approx 1.8$ meV. Thus, since $\hbar\omega_c \approx 1$ meV/T in (Cd,Mn)Te, LLs do not cross above $B_c \approx 1.8$ T, where $\Delta_{\text{sat}} = \hbar\omega_c$. Indeed, Fig. 11.19a shows precise QHE quantization and well resolved SdH minima exactly at integer filling factors.

A closer inspection of the low-magnetic field region in Fig. 11.19b reveals well-pronounced SdH oscillations with minima corresponding to filling factors as high

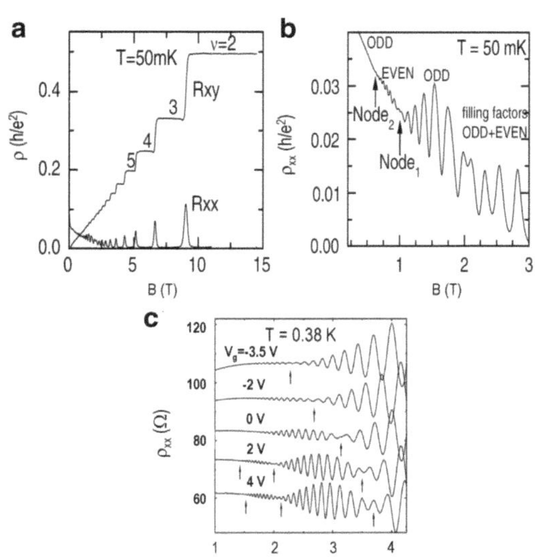

Fig. 11.19 (**a**) ρ_{xx} and ρ_{xy} measured at 50 mK in (Cd,Mn)Te QW with the record high electron mobility $\mu \approx 60{,}000$ cm^2/Vs. (**b**) Zoom in the low-field ρ_{xx} data showing the beating pattern of SdH oscillations. (**c**) ρ_{xx} vs. B measured at different gate voltages in n-(Hg,Mn)Te quantum well. Reprinted with permission from [93] (**a**, **b**) and [94] (**c**)

as $\nu = 53$. Moreover, the minima are seen clearly for both even and odd fillings (although not simultaneously) at low fields down to $B \simeq 0.5\,\mathrm{T}$ for odd filling factors. In nonmagnetic CdTe quantum wells with similar mobility minima at odd ν are usually resolved only for $B > 3\,\mathrm{T}$. At the same time, the oscillation pattern dramatically depends on temperature.

At low-field region, SdH oscillations show a beating pattern with clearly seen nodes, where their amplitude is suppressed while the amplitude reaches maximum when $E_Z = N\hbar\omega_c$ with $N = 1, 2, 3, \ldots$ This condition actually corresponds to the crossings of LLs, where doubled density of states enhances ρ_{xx} while the gap between overlapping LL reaches maximum. In turn, at the nodes, where $E_Z = (N + 1/2)\hbar\omega_c$ LL lie equidistantly, which minimizes the gap between LLs. Observation of odd or even minima depends on the number of the lowest, spin-down LLs, which do not cross. As in the case of (Zn,Cd,Mn)Se structures, a simple model based on (11.5) and overlapping LLs describes well quantum transport observed in this high mobility (Cd,Mn)Te sample.

In addition, resistively detected electron spin resonance experiment [95] was performed on the same high mobility (Cd,Mn)Te heterostructures. It revealed an anomalously large Knight shift, observed for magnetic fields for which the energies for the excitation of free carriers and Mn spins are almost identical. These findings suggest the existence of a magnetic-field-induced ferromagnetic order at low temperatures.

Results of magnetotransport studies [94, 96] in gated n-(Hg,Mn)Te modulation doped magnetic quantum wells also reveal strongly T-dependent beating pattern of SdH oscillation at low B. A systematic measurement of the node positions made it possible to separate the gate-voltage-dependent Rashba spin–orbit splitting (which arises from the vertical, i.e., perpendicular to the heterostructure, electric field gradient) from the T-dependent Zeeman splitting [94]. It was found that Rashba splitting is larger than or comparable with the s–d exchange energy in the narrow gap magnetic 2DES even at moderately high magnetic fields. Further studies and analysis of SdH in these n-(Hg,Mn)Te QWs [96] allowed the authors to determine exchange constants $N_0\alpha$, $N_0\beta$, the antiferromagnetic temperature T_0, and the effective spin of S_0 of Mn subsystem. The latter was found to have different T dependence in comparison with the bulk (Hg,Mn)Te.

11.6.9 Quantum Hall Ferromagnetism in Diluted Magnetic Semiconductors

Landau level crossing can have much more profound consequences than just anomalous SdH pattern due to modified density of states described above. If LL corresponding to the opposite spin orientations overlap, the spin degree of freedom is *not* frozen by the field. In other words, although $B \neq 0$, electrons with opposite spins have the same energy, i.e., effectively Zeeman splitting is zero. For instance, spin subband $0\uparrow$ coincides with $1\downarrow$ at field $B_c \approx 4\,\mathrm{T}$, as shown in Fig. 11.14b.

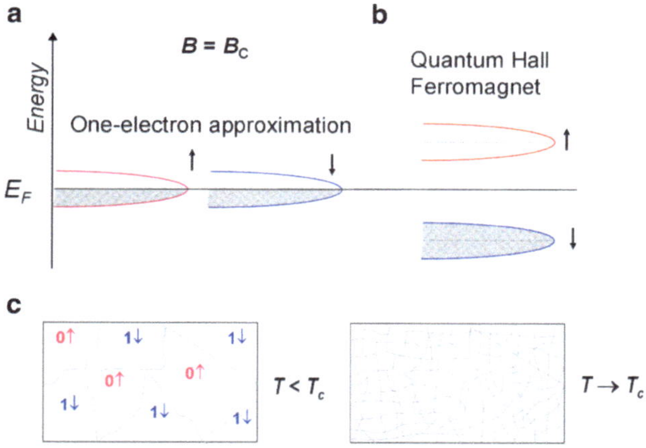

Fig. 11.20 Schematic view of QHFM mechanism. (**a**) Within the one electron approximation, when two LLs coincide at Fermi level SdH maximum is observed at integer filling factor. (**b**) LLs split into one fully occupied and one empty when it is energetically favorable. (**c**) In different regions of the sample either spin-up or spin-down polarization of the electron liquid prevails. Domain walls scatter electrons and result in additional resistance spikes at LLs coincidence. Domains become smaller with increasing T and they vanish above T_C

Under such circumstances, a spontaneous spin order may appear at low temperatures [97], leading to the state being known as the quantum Hall ferromagnet (QHFM). Reference [98] describes the QHFM in popular way. It should be stressed that QHFM refers to magnetic order of electrons, not localized spins, which remain in paramagnetic phase.

Figure 11.20a shows two half-filled overlapping LL. In one electron approximation, the density of states doubles and, since E_F lies on the centers of LLs, enhanced SdH maximum is observed, despite integer filling factor ($\nu = 1/2 + 1/2+$ number of fully occupied LLs at lower energy). However, when these LLs split into one fully occupied and one empty, final situation (Fig. 11.20a) is energetically favorable. This is because electrons with the same spins have to avoid each other in space due to Pauli exclusion principle. This minimizes the total Coulomb energy. It happens when energy gain J is larger than LL width. Since the width of LL decreases with decreasing T, at some critical temperature T_C the transition takes place. Now, E_F lies in the gap between LL, and the quantum Hall state with $\rho_{xx} = 0$ should be recovered. However, on either side of the crossing, the spin polarization of the electron system has opposite direction, e.g., $0 \uparrow$ for $B < B_c$ and $1 \downarrow$ for $B > B_c$. Thus, due to local differences in Mn concentration, which changes exchange part of spin splitting, domains of different spin directions coexist as shown in Fig. 11.20c. Domain walls introduce scattering, thus longitudinal resistance $\rho_{xx} \neq 0$, while it should be zero in the QHE regime. Instead, characteristic ρ_{xx} spikes appear at $B = B_c$. These new peaks are distinct from the usual SdH maxima between QHE minima [99].

The LL arrangement corresponding to such Ising QHFM has been realized in various III–V 2DES [100–102], however to bring LL into coincidence in these nonmagnetic structures either tilting the sample to reduce cyclotron energy (which is proportional to magnetic field perpendicular to 2DES plane) or crossing LL from different electric levels in wide QWs, or double QW is necessary. Magnetic 2DES offers an opportunity to investigate QHFM when $B \perp$2DES. This makes it possible to avoid complications from in-plane field effects. Moreover, only moderate B values are needed in contrast to tilted field experiments, where often very strong total B is needed to maintain its perpendicular component $B_{\perp}$ strong enough. Such a systematic study [103] was performed in modulation doped n-$Cd_{1-x}Mn_xTe$ quantum wells.

Figure 11.21a presents Hall ρ_{xy} and longitudinal ρ_{xx} resistivities, the latter revealing the presence of a strong resistance spike at the magnetic field $B_c \approx 5.8$ T. This B_c corresponds to the LL filling factor $\nu \equiv n_s h/eB \approx 2$ at the crossing of $0\uparrow$ and $1\downarrow$ spin subbands. Figure 11.21b indicates a broad SdH maximum at $T = 8$ K in this region, resulting from two overlapped LLs (as in Fig. 11.20a. When lowering T a sharp QHFM spike appears instead, while SdH maximum shifts toward lower B. According to Fig. 11.21b, c, the spike has a maximum value at $T_C \approx 1.3$ K, when the number of domains is large, as shown in Fig. 11.20c, and scattering from their walls is the most efficient.

Figure 11.22a, b show ρ_{xx} measured at various electron densities n_s. QHFM spikes are clearly seen not only at $B_c \approx 5.8$ T, but also around $B_c \approx 3$ and 2 T. Spikes around 3 T correspond to crossing of $0\uparrow$ and $2\downarrow$ LLs for lower n_s and to $1\uparrow$ and $3\downarrow$ for higher n_s, when more LLs are populated, while at 2 T to $2\uparrow$ and $5\downarrow$. Generally, positions of the spikes agree with fields where LLs crossing are predicted by (11.5). However, it is clearly seen that spikes corresponding to, e.g., ($0\uparrow$, $2\downarrow$) are substantially shifted toward higher field with respect to these at ($1\uparrow$,$3\downarrow$). The observed shift stems from the exchange interactions between electrons in LLs at the coincidence with these in the fully occupied LLs$\downarrow$ lying deep ("frozen") below Fermi energy. This contribution again reflects Pauli exclusion principle. For instance, lowering LL$\downarrow$ at the crossing increases the number of the majority spin-down electrons, which have to avoid each other due to Pauli principle, and thus reducing Coulomb repulsion energy. At the same time, when LL$\downarrow$ is lowered, the crossing point shifts toward higher field, as seen in Fig. 11.14b. This effect is smaller for higher LLs since their energetic distance to the lowest LL is larger. Actually, in magnetic 2DES we deal with a unique situation when a number of frozen LL$\downarrow$ *increases* as *B decreases*. To obtain such a situation in nonmagnetic 2DES, a very strong, tilted field is necessary.

Figures 11.21c and 11.22b provide an experimental evidence that the QHFM spikes correspond to a phase transition at nonzero temperatures. According to Fig. 11.21c, the spike magnitude exhibits a sharp maximum at the temperature that could be identified as Curie point of ferromagnetic ordering, $T_C \approx 1.3$ K. At the same time, a hysteresis loop of $\rho_{xx}(B)$ develops when B is swept in two directions below T_C at the precise location of the spike, as presented in Fig. 11.22b. Moreover,

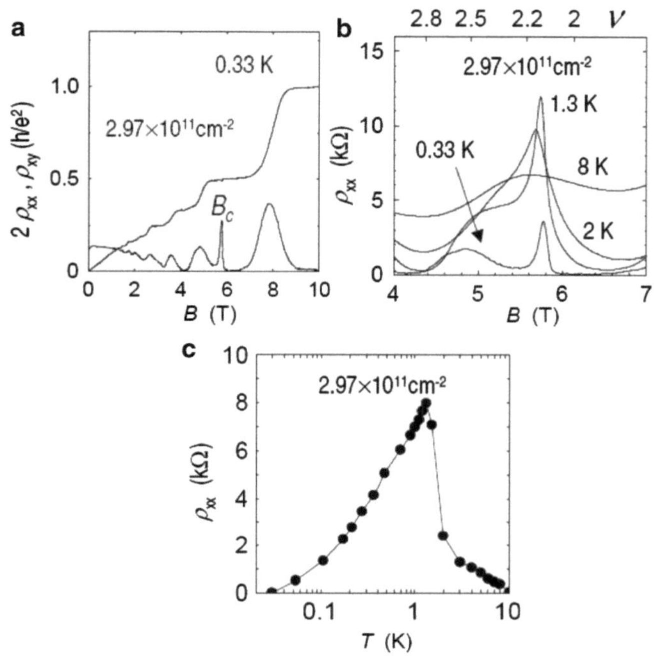

Fig. 11.21 (**a**) Resistances ρ_{xx} and ρ_{xy} at $T = 0.33$ K measured in gated (Cd,Mn)Te quantum well for $n_s = 2.97 \times 10^{11}$ cm^{-2}. Note the presence of a spike in ρ_{xx} at $B_c \simeq 5.8$ T, shown at selected temperatures in (**b**). (**c**) The spike height as a function of T for $n_s = 2.97 \times 10^{11}$ cm^{-2}. Reprinted from [103]

Fig. 11.22c shows that the QHFM spike reaches its maximum when the LLs crossing occurs at exactly integer filling factor.

These results strongly support theory of QHFM [99, 104] predicting that if ν is close to an integer at B_c, a transition to Ising QHF ground state takes place. In this broken symmetry state, all electrons fill up one LL, leaving the other one empty. This is evidenced in Fig. 11.21b, which reveals the absence of a SdH maximum at B_c below T_C. However, depending on a local potential landscape either $0\uparrow$ or $1\downarrow$ LL is filled up in a given space region. Domain walls appearing in this way form 1D conduction channels across the sample. Their presence gives rise to an additional scattering that results in the resistance spike at B_c [99]. Random configurations of the domains below a critical temperature T_C lead to large energy barriers between adjacent domains. This gives rise to metastable states with slow evolution and leads to the observed hysteretic behavior. The interplay of domain walls energy and the entropy of the system results in increasing domain size when T is decreasing to minimize the free energy: $F = W - TS$, where W is the energy of domain walls and S is the entropy. Thus, since the magnitude of the QHFM spike is proportional to the length of domain walls present in the sample, it decreases when $T \to 0$, as experimentally observed.

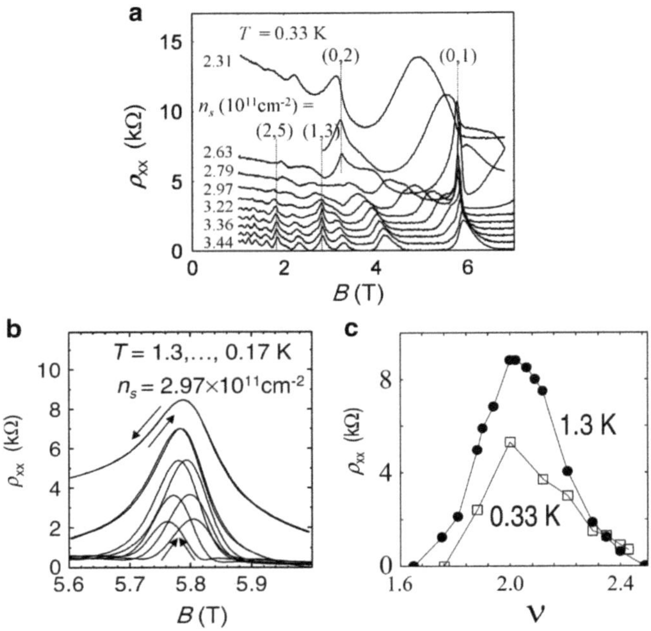

Fig. 11.22 (**a**) Longitudinal resistivity at $T = 0.33$ K at various electron densities $n_s = 2.31 - 3.44 \times 10^{11}$ cm^{-2}. *Dashed lines* mark resistance spikes at the crossing of LLs with the indices (2↑,5↓), (1↑,3↓), (0↑,2↓), and (0↑,1↓). (**b**) Hysteresis loops, where ρ_{xx} is depicted in the region of the QHFM spike for sweeping the magnetic field in two directions. (**c**) The height of the QHFM spike at $B_c \simeq 5.8$ T as a function of the filling factor ν showing that the spike peaks at integer ν, i.e., when at the coincidence two LL are half filled. Reprinted from [103]

It should be noted that although domain origin of the QHFM is widely accepted, there is also a competing theory, which attributes QHFM spikes to the critical spin reversal at LL crossing with no domain picture involved [105] These calculations describe well the above results in (Cd,Mn)Te QW, including resistance spikes, their temperature dependence and hysteretic behavior.

11.7 Summary and Perspectives

Diluted magnetic semiconductors proved themselves as an invaluable laboratory for fundamental studies of the influence of spin degrees of freedom on electric charge transport properties. In particular, the giant $s-d$ spin-splitting of the electron band strongly influences quantum corrections to the conductivity, results in extremely strong magnetoresistance, alters metal-to-insulator transition, constitutes a novel mechanism of the universal conductance fluctuations. DMSs are particularly suitable for the meaningful examination of the spin-glass phase by means of coherent transport and generally very useful for studying an influence of magnetic ions on mesoscopic transport in semiconductors. Phenomena similar to those occurring in the colossal-magnetoresistant materials point out that DMSs may constitute a bridge between nonmagnetic semiconductors and complex electronic materials as manganites, etc. DMSs offer a worthwhile opportunity to examine still not well understood anomalous Hall effect. At the same time, DMSs interface low-dimensional transport phenomena with thin film magnetism.

Magnetically doped 2DES formed in modulation-doped semiconductor heterostructures make it possible to study the interplay between quantum transport, localization, and electron–electron interactions. In particular, it is possible to study quantum Hall effect and related phenomena in highly or completely spin-polarized electron liquids to observe phenomena stemming from coincidence of LLs with opposite real spins. In turn, the narrow gap magnetic heterostructures offer an important test bed to study an influence of both giant spin-splitting and spin-orbit coupling on transport phenomena.

There are still many experiments and DMS devices to be done in the future. Particularly interesting would be a hybrid system consisting of paramagnetic DMS quantum well and a superconducting (SC) film. According to theoretical predictions [106], the local magnetic field of Abrikosov vortices in SC create a strong local spin-splitting in DMS, because of giant effective Landé factor in DMS. This, in turn, leads to spin and charge textures in the semiconductors. Moreover, these textures could be manipulated and controlled by manipulating vortices in SC. This opens the doors to investigate new striking physics phenomena, as unusual quantum Hall effect and to produce devices to manipulate spin and charge in semiconductor as well.

References

1. G. Grabecki, T. Dietl, J. Kossut, W. Zawadzki, Surf. Sci. **142**, 588 (1984)
2. G. Grabecki, T. Dietl, P. Sobkowicz, J. Kossut, W. Zawadzki, Appl. Phys. Lett. **45**, 1214 (1984)
3. L.A. Kolodziejski, T. Sakamoto, R.L. Gunshor, S. Datta, Appl. Phys. Lett. **44**, 799 (1984)
4. R.N. Bicknell, R.W. Yanka, N.C. Giles-Taylor, D.K. Blanks, E.L. Buckland, J.F. Schetzina, Appl. Phys. Lett. **45**, 92 (1984)
5. D.E. Ashenford, B. Lunn, J.J. Davies, J.H.C. Hogg, J.E. Nicholls, C.G. Scott, D. Shaw, H.H. Sutherland, C.P. Hilton, T.J. Gregory, D. Johnston, B.C. Cavenett, G.R. Johnson, M. Haines, J. Cryst. Growth **95**, 557 (1989)
6. D.E. Ashenford, J.H.C. Hogg, D. Johnston, B. Lunn, C.G. Scott, D. Staudte, J. Cryst. Growth **101**, 157 (1990)
7. A. Waag, S. Scholl, K. von Schierstedt, D. Hommel, G. Landwehr, G. Bilger, J. Cryst. Growth **129**, 243 (1993)
8. D. Ashenford, B. Lunn, K.A. Dhese, J.E. Nicholls, P. Devine, G.G. Scott, J. Jaroszyński, J. Cryst. Growth **138**, 443 (1994)
9. I.P. Smorchkova, N. Samarth, J. Appl. Phys. **69**, 1640 (1996)
10. T. Baron, K. Saminadayar, S. Tatarenko, J. Cryst. Growth **159**, 271 (1996)
11. G. Karczewski, J. Jaroszyński, A. Barcz, M. Kutrowski, T. Wojtowicz, J. Kossut, J. Cryst. Growth **184–185**, 814 (1998)

12. T. Yasuda, M.D. Duc, Y. Segawa, J. Cryst. Growth **214–215**, 159 (2000)
13. S. Scholl, H. Schäfer, A. Waag, D. Hommel, K. von Schierstedt, B. Kuhn-Heinrich, G. Landwehr, J. Appl. Phys. **62**, 3010 (1993)
14. H. Ohno, H. Munekata, T. Penney, S. von Molnár, L.L. Chang, Phys. Rev. Lett. **68**, 2664–2667 (1992)
15. H. Ohno, A. Shen, F. Matsukura, A. Oiwa, A. Endo, S. Katsumoto, Y. Iye, App. Phys. Lett. **69**, 363–365 (1996)
16. F. Matsukura, H. Ohno, A. Shen, Y. Sugawara, Phys. Rev. B **57**, R2037–R2040 (1998)
17. A. Van Esch, L. Van Bockstal, J. De Boeck, G. Verbanck, A.S. van Steenbergen, P.J. Wellmann, B. Grietens, R. Bogaerts, F. Herlach, G. Borghs, Phys. Rev. B **56**, 13103–13112 (1997)
18. M. Sawicki, T. Dietl, J. Kossut, J. Igalson, T. Wojtowicz, W. Plesiewicz, Phys. Rev. Lett. **56**, 508 (1986)
19. Y. Shapira, N.F. Oliveira, P. Becla, T.Q. Vu, Phys. Rev. B **41**, 5931–5941 (1990)
20. C. Leighton, I. Terry, P. Becla, Phys. Rev. B **58**, 9773–9782 (1998)
21. D. Ferrand, J. Cibert, A. Wasiela, C. Bourgognon, S. Tatarenko, G. Fishman, T. Andrearczyk, J. Jaroszyński, S. Koleśnik, T. Dietl, B. Barbara, D. Dufeu, Phys. Rev. B **63**, 085201 (2001)
22. J.H. Kim, H. Kim, D. Kim, Y.E. Ihm, W.K. Choo, Physica B **327**, 304 (2003)
23. Q. Xu, L. Hartmann, H. Schmidt, H. Hochmuth, M. Lorenz, R. Schmidt-Grund, C. Sturm, D. Spemann, M. Grundmann, Phys. Rev. B **73**, 205342 (2006)
24. S-J. Han, J.W. Song, C.-H. Yang, S.H. Park, J.-H. Park, Y.H. Jeonga, K.W. Rhie, Appl. Phys. Lett. **81**, 4212 (2002)
25. H. Kimura, T. Fukumura, M. Kawasaki, K. Inaba, T. Hasegawa, H. Koinuma, Appl. Phys. Lett. **80**, 94 (2002)
26. S.R. Shinde, S.B. Ogale, S. Das Sarma, J.R. Simpson, H.D. Drew, S.E. Lofland, C. Lanci, J.P. Buban, N.D. Browning, V.N. Kulkarni, J. Higgins, R.P. Sharma, R.L. Greene, T. Venkatesan, Phys. Rev. B **67**, 115211 (2003)
27. I.P. Smorchkova, N. Samarth, J.M. Kikkawa, D.D. Awschalom, Phys. Rev. Lett. **78**, 3571–3574 (1997)
28. I.P. Smorchkova, N. Samarth, J.M. Kikkawa, D.D. Awschalom, Phys. Rev. B **58**, R4238–R4241 (1998)
29. J. Jaroszyński, J. Wróbel, M. Sawicki, E. Kamińska, T. Skośkiewicz, G. Karczewski, T. Wojtowicz, A. Piotrowska, J. Kossut, T. Dietl, Phys. Rev. Lett. **75**, 3170–3173 (1995)
30. T. Andrearczyk, J. Jaroszyński, G. Karczewski, J. Wróbel, T. Wojtowicz, T. Dietl, E. Papis, E. Kamińska, A. Piotrowska, Physica E **12**, 361 (2002)
31. H. Fukuyama, K. Yoshida, Physica B+C **105**, 132 (1981); H. Fukuyama and K. Yosida, J. Phys. Soc. Jpn. **46**, 1522 (1979)
32. P. Głód, T. Dietl, M. Sawicki, I. Miotkowski, Physica B **194–196**, 995 (1994)
33. T. Dietl, Physica E **35**, 293 (2006)
34. Le Van Khoi, J. Jaroszyński, B. Witkowska, A. Mycielski, R.R. Gałazka, Acta Phys. Polon. A **92**, 833 (1997)
35. T. Andrearczyk, J. Jaroszyński, G. Grabecki, T. Dietl, T. Fukumura, M. Kawasaki, Phys. Rev. B **72**, 121309 (2005)
36. J. Jaroszyński, T. Andrearczyk, G. Karczewski, T. Wojtowicz, D. Popović, T. Dietl, Phys. Rev. B **76**, 045322 (2007)
37. N. Mathur, P. Littlewood, Phys. Today **56(1)**, 25 (2003)
38. E. Dagotto, T. Hotta, A. Moreo, Phys. Rep. **344**, 1 (2001)
39. T. Dietl, H. Ohno, F. Matsukura, J. Cibert, D. Ferrand, Science **287**, 1019 (2000)
40. G. Alvarez, M. Mayr, E. Dagotto, Phys. Rev. Lett **89**, 277202 (2002)
41. M. Mayr, G. Alvarez, E. Dagotto, Phys. Rev. B **65**, 241202(R) (2002)
42. U. Yu, B.I. Min, Phys. Rev. Lett. **94**, 117202 (2005)
43. D. Kechrakos, N. Papanikolaou, K.N. Trohidou, T. Dietl, Phys. Rev. Lett. **94**, 127201 (2005)
44. M.N. Baibich, J.M. Broto, A. Fert, F. Nguyen Van Dau, F. Petroff, P. Eitenne, G. Creuzet, A. Friederich, J. Chazelas, Phys. Rev. Lett. **61**, 2472–2475 (1988)

45. P. Grünberg, R. Schreiber, Y. Pang, M.B. Brodsky, H. Sowers, Phys. Rev. Lett. **57**, 2442–2445 (1986)
46. T. Wojtowicz, T. Dietl, M. Sawicki, W. Plesiewicz, J. Jaroszyński, Phys. Rev. Lett. **56**, 2419–2422 (1986)
47. S.V. Kravchenko, Whitney E. Mason, G.E. Bowker, J.E. Furneaux, V.M. Pudalov, M. D'Iorio, Phys. Rev. B **51**, 7038–7045 (1995)
48. E. Abrahams, S.V. Kravchenko, M.P. Sarachik, Rev. Mod. Phys. **73**, 251 (2001)
49. E. Abrahams, P.W. Anderson, D.C. Licciardello, T.V. Ramakrishnan, Phys. Rev. Lett. **42**, 673–676 (1979)
50. Y. Hanein, D. Shahar, J. Yoon, C.C. Li, D.C. Tsui, H. Shtrikman, Phys. Rev. B **58**, R13338–R13340 (1998)
51. T. Kasuya, Progr. Theor. Phys. **16**, 58 (1956)
52. F. Matsukura, M. Sawicki, T. Dietl, D. Chibab, H. Ohno, Physica E **21**, 1032 (2004)
53. D.V. Baxter, D. Ruzmetov, J. Scherschligt, Y. Sasaki, X. Liu, J.K. Furdyna, C.H. Mielke, Phys. Rev. B **65**, 212407 (2002)
54. C.P. Moca, B.L. Sheu, N. Samarth, P. Schiffer, B. Janko, G. Zarand, Phys. Rev. Lett. **102**, 137203 (2009)
55. J. Honolka, S. Masmanidis, H.X. Tang, D.D. Awschalom, M.L. Roukes, Phys. Rev. B **75**, 245310 (2007)
56. B.L. Al'tshuler, A.G. Aronov, JETP Lett. **37**, 410 (1983)
57. G. Zaránd, C.P. Moca, B. Jankó, Phys. Rev. Lett. **94**, 247202 (2005)
58. J. Jaroszyński, J. Wróbel, M. Sawicki, E. Kamińska, T. Skośkiewicz, G. Karczewski, T. Wojtowicz, A. Piotrowska, J. Kossut, T. Dietl, Phys. Rev. Lett. **75**, 3170 (1995)
59. Y. Shapira, N.F. Oliveira, Phys. Rev. B **35**, 6888 (1987)
60. J. Jaroszyński, T. Dietl, J. Wróbel, G. Karczewski, T. Wojtowicz, J. Kossut, D. Maude, P. van der Linden, J.C. Portal, Acta Phys. Polon. A **92**, 797 (1997)
61. J. Jaroszyński, J. Wróbel, G. Karczewski, T. Wojtowicz, T. Dietl, Phys. Rev. Lett. **80**, 5635 (1998)
62. M.A. Novak, O.G. Symko, D.J. Zheng, S. Oseroff, J. Appl. Phys. **57**, 3418 (1985)
63. K. Wagner, D. Neumaier, M. Reinwald, W. Wegscheider, D. Weiss, Phys. Rev. Lett. **97**, 056803 (2006)
64. L. Vila, R. Giraud, L. Thevenard, A. Lemaitre, F. Pierre, J. Dufouleur, D. Mailly, B. Barbara, G. Faini, Phys. Rev. Lett. **98**, 027204 (2007)
65. S. Lee, A. Trionfi, T. Schallenberg, H. Munekata, D. Natelson, Appl. Phys. Lett. **90**, 032105 (2007)
66. J.E. Hirsch, Phys. Rev. Lett. **83**, 1834 (1999)
67. J. Sinova, D. Culcer, Q. Niu, N.A. Sinitsyn, T. Jungwirth, A.H. MacDonald, Phys. Rev. Lett. **92**, 126603 (2004)
68. C. Day, Phys. Today **58**(2), 17 (2005)
69. R. Karplus, J.M. Luttinger, Phys. Rev. **95**, 1154 (1954)
70. J.M. Luttinger, Phys. Rev. **112**, 739 (1958)
71. M. Berry, Phys. Today **43**(12), 34 (1990)
72. T. Smit, Physica **21**, 877 (1955)
73. L. Berger, Phys. Rev. B **2**, 4559 (1970)
74. T. Jungwirth, Q. Niu, A.H. MacDonald, Phys. Rev. Lett. **88**, 207208 (2002)
75. D. Ruzmetov, J. Scherschligt, David V. Baxter, T. Wojtowicz, X. Liu, Y. Sasaki, J.K. Furdyna, K.M. Yu, W. Walukiewicz, Phys. Rev. B **69**, 155207 (2004)
76. J. Cumings, L.S. Moore, H.T. Chou, K.C. Ku, G. Xiang, S.A. Crooker, N. Samarth, D. Goldhaber-Gordon, Phys. Rev. Lett. **96**, 196404 (2006)
77. J.S. Dyck, C. Drasar, P. Lost'ak, C. Uher, Phys. Rev. B **71**, 115214 (2005)
78. Q. Xu, L. Hartmann, H. Schmidt, H. Hochmuth, M. Lorenz, R. Schmidt-Grund, C. Sturm, D. Spemann, M. Grundmann, J. Appl. Phys. **101**, 063918 (2007)
79. W. Allen, E.G. Gwinn, T.C. Kreutz, A.C. Gossard, Phys. Rev. B **70**, 125320 (2004)
80. A.A. Burkov, L. Balents, Phys. Rev. Lett. **91**, 057202 (2003)

81. S.H. Chun, M.B. Salamon, Y. Lyanda-Geller, P.M. Goldbart, P.D. Han, Phys. Rev. Lett. **84**, 757 (2000)
82. Sh.U. Yuldashev, H.C. Jeon, H.S. Im, T.W. Kang, S.H. Lee, J.K. Furdyna, Phys. Rev. B **70**, 193203 (2004)
83. D. Culcer, A.H. MacDonald, Q. Niu, Phys. Rev. B **68**, 045327 (2003)
84. J.A. Gaj, W. Grieshaber, C. Bodin-Deshayes, J. Cibert, G. Feuillet, Y. Merle d'Aubigné, A. Wasiela, Phys. Rev. B **50**, 5512 (1994)
85. C.Y. Hu, W. Ossau, D.R. Yakovlev, G. Landwehr, T. Wojtowicz, G. Karczewski, J. Kossut, Phys. Rev. B **58**, R1766 (1998)
86. J.A. Gaj, R. Planel, G. Fishman, Solid State Comm. **29**, 435 (1979)
87. T. Dietl, in *Diluted Magnetic Semiconductors*, ed. by T.S. Moss. Handbook on Semiconductors (North-Holland, Amsterdam, 1994), p. 1251
88. G. Grabecki, A. Wittlin, T. Dietl, P.A.A. Teunissen, S.A.J. Wiegers, J.A.A.J. Perenboom, Semicond. Sci. Technol. **8**, S95 (1993)
89. B. Huckestein, Rev. Mod. Phys. **67**, 357 (1995)
90. J. Jaroszyński, G. Karczewski, J. Wróbel, T. Andrearczyk, A. Strycharczuk, T. Wojtowicz, G. Grabecki, E. Papis, E. Kamińska, A. Piotrowska, T. Dietl, Physica E **6**, 790 (2000)
91. R. Knobel, N. Samarth, J.G.E. Harris, D.D. Awschalom, Phys. Rev. B **65**, 235327 (2002)
92. J.G.E. Harris, R. Knobel, K.D. Maranowski, A.C. Gossard, N. Samarth, D.D. Awschalom, Phys. Rev. Lett. **86**, 4644 (2001)
93. F.J. Teran, M. Potemski, D.K. Maude, T. Andrearczyk, J. Jaroszyński, G. Karczewski, Phys. Rev. Lett. **88**, 186903 (2002)
94. Y.S. Gui, C.R. Becker, J. Liu, V. Daumer, V. Hock, H. Buhmann, L.W. Molenkamp, Europhys. Lett. **65**, 393 (2004)
95. F.J. Teran, M. Potemski, D.K. Maude, D. Plantier, A.K. Hassan, A. Sachrajda, Z. Wilamowski, J. Jaroszyński, T. Wojtowicz, G. Karczewski, Phys. Rev. Lett. **91**, 077201 (2003)
96. J. Liu, H. Buhmann, E.G. Novik, Y. Gui, V. Hockand, C.R. Becker, L.W. Molenkamp, Phys. Stat. Sol. (B) **243**, 835 (2006)
97. G.F. Giuliani, J.J. Quinn, Phys. Rev. B **31**, 6228 (1985)
98. S.M. Girvin, Phys. Today **53**(6), 39 (2000)
99. T. Jungwirth, A.H. MacDonald, Phys. Rev. Lett. **87**, 216801 (2001)
100. S. Koch, R.J. Haug, K. von Klitzing, M. Razeghi, Phys. Rev. B **47**, 4048 (1993)
101. V. Piazza, V. Pellegrini, F. Beltram, W. Wegscheider, T. Jungwirth, A.H. MacDonald, Nature **402**, 638 (1999)
102. E.P. De Poortere, E. Tutuc, S.J. Papadakis, M. Shayegan, Science **290**, 1546 (2000)
103. J. Jaroszyński, T. Andrearczyk, G. Karczewski, J. Wróbel, T. Wojtowicz, E. Papis, E. Kamińska, A. Piotrowska, Dragana Popović, T. Dietl, Phys. Rev. Lett. **89**, 266802 (2002)
104. T. Jungwirth, A.H. MacDonald, Phys. Rev. B **63**, 035305 (2001)
105. H.J.P. Freire, J.C. Egues, Phys. Rev. Lett. **99**, 026801 (2004)
106. M. Berciu, T.G. Rappoport, B. Jankó, Nature **435**, 71 (2005)

Chapter 12
Neutron Scattering Studies of Interlayer Magnetic Coupling

T.M. Giebultowicz and H. Kępa

Abstract This chapter contains a review of experimental methods and results of neutron scattering studies performed on AF and FM all-semiconductor superlattices. It demonstrates the efficiency of neutron diffraction methods in experimental studies of the interlayer coupling and contains a description of theoretical efforts to understand physical mechanisms of this coupling in semiconductor structures containing ferromagnetic and antiferromagnetic layers.

12.1 Introduction

At the time we finish writing this chapter – in the closing months of 2007 – the importance of studying interlayer exchange coupling (IEC) in magnetic multilayers has become self-evident. In October, the 2007 Nobel Prize in physics was awarded to Fert and Grünberg for their 1988 discovery of giant magnetoresistance (GMR) in thin-film systems consisting of alternating layers of ferromagnetic (FM) and non-magnetic metals [1, 2]. From the moment of the discovery, it was obvious that IEC plays a crucial role in this phenomenon. It stimulated vigorous experimental and theoretical studies on IEC effects and on physical mechanisms underlying them. After almost 20 years from the original Fert and Grünberg's discovery, this research field still remains very active.

The GMR magnetic field sensors and spin valves, now widely used in computer hard drives, were the first practical devices utilizing spin-dependent transport phenomena. For such electronics, the term "spintronics" was coined. The currently

T.M. Giebultowicz (✉)
Physics Department, Oregon State University, Corvallis, OR 97331, USA
e-mail: tgiebult@physics.oregonstate.edu

H. Kępa
Institute of Experimental Physics, Faculty of Physics, University of Warsaw, Hoża 69 00-681, Warsaw, Poland
and
Physics Department, Oregon State University, Corvallis, OR 97331, USA
e-mail: Henryk.Kepa@fuw.edu.pl

J. Kossut and J.A. Gaj (eds.), *Introduction to the Physics of Diluted Magnetic Semiconductors*, Springer Series in Materials Science 144,
DOI 10.1007/978-3-642-15856-8_12,

existing first-generation spintronics based on all-metallic multilayers surely has a bright future – no doubt, there will be many new developments in this area in the years to come. There is a broad consensus, however, that the most promising long-term objective is to develop *semiconductor spintronics*, which will combine the potential of semiconductors (i.e., their unique transport and optical properties) with the potential of ferromagnetic conductors (i.e., their spin-dependent magneto-transport properties). It can be expected that IEC effects will also be widely utilized in the future semiconductor spintronics. Yet, much research still needs to be done for achieving a good understanding of physical mechanisms that can give rise to interlayer coupling in structures composed partially or exclusively of semiconducting materials.

The origin of IEC phenomena in all-metallic multilayer is now quite well understood, but those theories cannot be applied to semiconductor systems. The transfer of interactions between two metallic FM blocks across a nonmagnetic metallic spacer is maintained by conduction electrons. Several different approaches have been used to describe the details of the electron-assisted mechanism (e.g., the free-electron model, the RKKY model, the "electron confinement" model), and there are still some disputes on which of the proposed interaction schemes offers the best description [3, 4]. However, all those models agree in that it is the conduction electrons which play the crucial role in the interaction transfer – also, they all stress the importance of the existence of nonzero magnetization in the interacting layers. In other words, in the light of those theories, no significant IEC effects could be expected to occur in all-semiconductor multilayers with low concentrations of mobile carriers – and definitely not in systems composed of alternating nonmagnetic and antiferromagnetic (AF) layers.

The first hints that such predictions based on the IEC theory for metallic multilayers might not be correct were obtained from neutron diffraction experiment on some all-semiconductor superlattices. Around 1990, a program of systematic studies of such systems was launched at the NIST Center for Neutron Research (in Gaithersburg, MD, USA). The first specimens investigated in the project were Mn-VI/II–VI superlattices in which the Mn-VI component is antiferromagnetic. At that time, they were the only all-semiconductor magnetic SL structures available for experimentation. A number of samples for the project were prepared by the laboratory of Prof. J.K. Furdyna at the University of Notre Dame.

As described in closer detail in Sect. 12.2, neutron diffractometry is an excellent tool for investigating magnetic superlattices – especially, those containing AF layers. In particular, this technique has a unique capability of detecting magnetic correlations between such layers. The initial experiments on the Mn-VI/II–VI SLs revealed a number of interesting effects in the strongly strained Mn-VI layers, but no detectable evidence of interlayer coupling was found [5, 6]. These negative results seemed to confirm the predictions that IEC effects cannot occur in AF systems with low concentrations of mobile carriers. However, those pessimistic views soon changed when the team started a new collaborative venture with Prof. Günter Bauer's laboratory at J. Kepler University, Linz, Austria. The Linz group had developed a technology of preparing high-quality EuTe/PbTe superlattices. Diffraction experiments on the new samples yielded spectacular spectra with distinct features that are an unmistakable signature of correlations between magnetic layers [7, 8]. Such effects could be still seen in the spectra from samples with the PbTe spacer thickness well exceeding 50 Å, which indicated a surprisingly long range of the interaction forces giving rise to the correlations. Those results proved that the theories constructed specifically for describing the IEC phenomena in metallic multilayers cannot be used as a basis for making prediction or for result interpretation in research on analogous systems composed of semiconductors. The data from the EuTe/PbTe samples prompted new investigations of the II-Mn/II–VI systems – and pronounced interlayer coupling effects were indeed found [9–11] in some of them.[1] The experimental works on both systems were followed by theoretical studies aimed at identifying the physical mechanisms underlying the observed coupling effects.

Although coupling between AF films across a nonmagnetic spacer is surely of considerable interest from the viewpoint of fundamental magnetic studies, there is much more excitement about IEC phenomena in systems with FM layers because of their potential applications in practical spintronics. However, at the time when neutron studies on the Mn-VI/II–VI and EuTe/PbTe systems were already well advanced, research on all-semiconductor SLs with FM layers could not even be started because no such samples were available. Because of a caprice of Mother Nature, there are few semiconducting compounds that are naturally ferromagnetic, and none of those existing is well suited for being prepared in a thin film form using standard deposition techniques. For instance, one well-known "archetypical" FM semiconductor is EuS. Because of the research done on the AF EuTe/PbTe system, studies on similar EuS-based superlattices performed in parallel with that work would have been of great interest – but no such samples existed at that time.

The situation took a positive turn only in the late 1990s when $Ga_{1-x}Mn_xAs$, the first synthetic FM semiconductor emerged [12]. About the same time at Kharkov, Ukraine, a team headed by Dr. A. Yu. Sipatov developed a unique method of fabricating high-quality superlattices of EuS/PbS. Those technological successes finally made it possible to launch neutron scattering studies of coupling between FM layers. An appropriate neutron scattering tool for investigating multilayers with an FM component is not diffractometry, but a technique commonly referred to as *neutron reflectometry*. This method was used at CNR NIST for studying interlayer exchange in $Ga_{1-x}Mn_xAs$/GaAs superlattices [13, 14], and in two EuS-based multilayered systems, EuS/PbS and EuS/YbSe [15–17]. The origin of ferromagnetism in $Ga_{1-x}Mn_xAs$ is well understood (see, e.g., the papers quoted in [12]) – this material is always strongly *p*-type, and it is the holes that maintain FM interactions between the Mn spins via an RKKY-like Zener mechanism. As indicated by the fact that the FM ordering occurs in a lattice in which only a small fraction of sites is occupied

[1] One of the goals in the early studies of the Mn-VI/II–VI was to study the influence of significant lattice strain on the AF spin structures in the Mn-VI layers. To enhance those effects, the nonmagnetic layers were made thick – as it turned out later when IEC effects were finally detected in Mn-VI/II–VI system, the spacer thickness in the samples used in the earlier studies exceeded the range of the interlayer interactions.

by the Mn spins, the range of these spin–spin interactions is quite long. It was therefore not surprising that neutron experiments revealed an FM coupling between the $Ga_{1-x}Mn_xAs$ layers across the pure GaAs spacers. In contrast to $Ga_{1-x}Mn_xAs$, EuS is nearly insulating, and the range of the FM spin–spin interactions in this material is very short. But in both EuS-based systems investigated, the reflectivity measurements revealed pronounced *antiferromagnetic* coupling between the FM layers [15, 16, 18, 19]. The origin of this coupling was the subject of a theoretical study done in parallel to the experimental work [20, 21].

In this chapter, we present a review of experimental results obtained from neutron scattering studies of the aforementioned AF and FM all-semiconductor superlattices. In Sect. 12.2, we present a brief outline of the basic principles of neutron scattering from magnetic solids and from modulated lattices. Experimental neutron scattering techniques used for investigating such systems are also described. Then, in Sects. 12.3 and 12.4 the experimental findings are presented and discussed in the context of theoretical modeling results (if such studies were performed for a given system). Neutron reflectometry research performed on the FM systems (on EuS/PbS and EuS/YbSe, and on $Ga_{1-x}Mn_xAs$/GaAs multilayers) is reviewed first (in Sects. 12.3.1 and 12.3.2, respectively). Diffraction studies of antiferromagnetic EuTe/PbTe superlattices are described in Sect. 12.4.1, and some findings from diffraction experiments on AF multilayered systems based on II–VI compounds are briefly presented in Sect. 12.4.2. A short summary and concluding remarks are given in Sect. 12.5.

12.2 Neutron Scattering Tools

12.2.1 Neutron Diffractometry

Diffractometry is historically the oldest neutron scattering method for investigating atomic and magnetic ordering in condensed matter systems [22]. The first neutron diffraction studies of crystals were performed in late 1940s, when intense thermal neutron beams from reactors became available. A well-known success of those pioneering efforts was the 1949 study of MnO by Shull and Smart [23] that provided the first direct evidence for the existence of antiferromagnetism. The physical effect measured in such experiments is elastic Bragg diffraction, which produces maxima at scattering angles θ satisfying the known Bragg condition:

$$\lambda = 2d_{hkl} \sin\theta, \tag{12.1}$$

where $\lambda = 2\pi\hbar/mv$ is the de Broglie wavelength of incident neutrons, with m – neutron mass, and v – neutron velocity. The d_{hkl} symbol denotes the spacing between crystallographic planes with (hkl) Miller indices. Alternatively, this condition can be written in terms of wavevectors:

$$\mathbf{k}' - \mathbf{k}_0 = \boldsymbol{\tau}_{hkl} \quad \text{with} \quad |\mathbf{k}'| = |\mathbf{k}_0|, \tag{12.2}$$

where $\mathbf{k}_0$ and $\mathbf{k}'$ are the wavevectors of the incident and the scattered neutron waves, respectively, and $\boldsymbol{\tau}_{hkl}$ is the reciprocal lattice vector corresponding to the (hkl) family of crystallographic planes.

From the angular positions of the Bragg reflections, one can determine the basic structural properties of the investigated specimen, such as the unit cell dimensions. Closer insight into the atomic structure can be obtained by analyzing the reflection intensities. In the simplest theoretical approach (the so-called *kinematical theory*[2]), the diffracted wave is obtained by adding up all waves scattered by individual atomic nuclei:

$$\psi_{\text{diff}}^{\text{nucl}} \propto \sum_j b_j \exp(i\mathbf{Q}\cdot\mathbf{r}_j), \tag{12.3}$$

where $\mathbf{r}_j$ is the position of the jth atom, b_j is the scattering amplitude of its nucleus, and $\mathbf{Q} = \mathbf{k}' - \mathbf{k}_0$ is the scattering vector. However, interaction with the nuclear potential is not the only possible scattering mechanism. The neutron has a magnetic moment, and if an atom has a nonzero electronic magnetic moment, the dipole–dipole interaction between the two moments gives rise to additional magnetic scattering. If the incident neutron beam is not polarized, there is no interference between the nuclear and magnetic diffraction. So, the "magnetic" component in the diffracted wave can be obtained in a similar way by adding up all the waves diffracted by individual magnetic atoms. Yet, an individual magnetic scattering process is more complicated than a nuclear one: here, the scattering amplitude is essentially a *vector* quantity depending on the momentum transfer $Q = |\mathbf{Q}|$ and the mutual orientation of the neutron and atomic moments. In most practical cases, however, we deal with situations in which all magnetic atoms in the sample are identical, and their moments have only two orientations, "up" and "down." Then the magnetic scattering amplitude of the jth atom can be written simply as $\kappa_j f(Q)$, with $\kappa_j = +\kappa$ or $\kappa_j = -\kappa$ for the "up" or "down" orientation, respectively. Here, κ is proportional to the atomic magnetic moment, and $f(Q)$ is the normalized single-atom form factor, a decreasing function of Q with $f(0) = 1$. It leads to the following expression for the magnetically diffracted wave:

$$\psi_{\text{diff}}^{\text{mag}} \propto f(Q) \sum_j \kappa_j \exp(i\mathbf{Q}\cdot\mathbf{r}_j) \tag{12.4}$$

The above equation can be now used for calculating theoretical diffraction spectra from magnetic superlattices for several simple model situations. Most often the experiments on such systems are done in symmetric reflection geometry, so that the

[2] The kinematical theory of diffraction is based on the known Born approximation (see, e.g., [24]). In this approach, effects such as the loss of the incident wave intensity on its path through the crystal due to scattering processes, multiple scattering, and interference between the incident wave and diffracted waves, are assumed to be negligible. It is a legitimate approximation for small crystals, and thus, also for thin films.

scattering vector is parallel to the superlattice growth axis z: $Q = (0, 0, Q_z)$, and $\mathbf{Q} \cdot \mathbf{r}_j = Q_z \cdot z_j$. Ergo, only the z component of the atom's position matters, and since all atoms residing in a given monolayer have the same z component, the summation over individual atoms can be replaced by summation over the monolayers. If the spacing between the monolayers is d, then the z coordinate of all atoms in the lth monolayer is $z_l = ld$. The equation simplifies to:

$$\psi_{\text{diff}}^{\text{mag}} \propto f(Q) \sum_l \mathcal{M}_l \exp(iQ_z ld), \tag{12.5}$$

where $\mathcal{M}_l$, the sum of all amplitudes κ_j of atoms in the lth monolayer, is proportional to the monolayer magnetization. Taking advantage of the SL periodicity, one can separate this equation into summation over all monolayers within an SL "elementary cell" – a bilayer (BL) – and over N SL repeats. Suppose that each individual bilayer consists of m magnetic monolayers followed by n nonmagnetic ones (for which $\mathcal{M}_l = 0$). For such a bilayer, one can define the *magnetic structure factor* F_{BL} as

$$F_{\text{BL}}(Q_z) \equiv f(Q) \sum_{\mu=0}^{m-1} \mathcal{M}_\mu \exp(iQ_z \mu d). \tag{12.6}$$

Equation (12.5) can be thus written as a sum over the N bilayers:

$$\psi_{\text{diff}}^{\text{mag}} \propto \sum_{\nu=0}^{N-1} \exp(iQ_z z_\nu)[\xi_\nu F_{\text{BL}}(Q_z)]. \tag{12.7}$$

Here, z_ν is the z coordinate of the first magnetic monolayer in the νth bilayer, and it can be further on replaced by $z_\nu = D\nu$, where $D = (m + n)d$ is the SL period. The ξ_ν coefficient is introduced to describe the monolayer magnetization sequence in each bilayer: in the νth bilayer, this sequence either may be the same as in the first one (then $\xi_\nu = +1$), or may be reversed relative to that in the first bilayer (then $\xi_\nu = -1$). Equation (12.7) takes now the form:

$$\psi_{\text{diff}}^{\text{mag}} \propto F_{\text{BL}}(Q_z) \sum_{\nu=0}^{N-1} \xi_\nu \exp(iQ_z D\nu). \tag{12.8}$$

In the experiments, one measures the *intensity* of the diffracted wave, which is proportional to the squared modulus of the diffracted wave amplitude: $I(Q_z) \propto |\psi_{\text{diff}}|^2$. From (12.8), one obtains:

$$I(Q_z) \propto |F_{\text{BL}}(Q_z)|^2 \left| \sum_{\nu=0}^{N-1} \xi_\nu \exp(iQ_z D\nu) \right|^2, \tag{12.9}$$

or, an equivalent expression:

$$I(Q_z) \propto |F_{\rm BL}(Q_z)|^2 \sum_{\alpha,\beta=0}^{N-1} \xi_\alpha \xi_\beta \exp[iQ_z D(\alpha-\beta)] \tag{12.10}$$

The physical meaning of the right-hand components in the above two equations can be readily explained. Note that for a system consisting only of a *single* bilayer, both equations yield $I(Q_z) \propto |F_{\rm BL}(Q_z)|^2$. So, the squared structure factor modulus simply describes the diffraction spectrum from a *single bilayer*. Calculating the $|F_{\rm BL}(Q_z)|^2$ for some idealized model situations is a straightforward task. In the case of *ferromagnetic* layers, the magnetization of all monolayer is the same, and (12.6) then becomes a sum of a simple geometric series:

$$F_{\rm BL}(Q_z) = f(Q_z)\mathcal{M}\sum_{\mu=0}^{m-1}[\exp(iQ_z d)]^{\mu} = f(Q_z)\mathcal{M}\frac{1-\exp(iQ_z md)}{1-\exp(iQ_z d)} \tag{12.11}$$

from which one obtains:

$$|F_{\rm BL}(Q_z)|^2 = f(Q_z)\mathcal{M}\frac{\sin^2(mQ_z d/2)}{\sin^2(Q_z d/2)}. \tag{12.12}$$

This function, as illustrated in Fig. 12.1a, has broad maxima (with weak subsidiaries on both sides) positioned at regular intervals $Q_z = 2\pi n/d$ (with $n = 0, 1, 2, \ldots$). The width (FWHM) of each maximum is given by the approximate formula: $\Delta Q_z \approx 2\pi/md$, i.e., the maxima become sharper with the increasing number of monolayers in the magnetic layer.

In *antiferromagnetic layers*, the monolayer magnetization alternates, which can be introduced into (12.6) as $\mathcal{M}_\nu = (-1)^\nu \mathcal{M}$. By analogous calculations as above, one obtains:

$$|F_{\rm BL}(Q_z)|^2 \propto \frac{\cos^2(mQ_z d/2)}{\cos^2(Q_z d/2)} \quad \text{for } m \text{ odd, or}$$
$$\frac{\sin^2(mQ_z d/2)}{\cos^2(Q_z d/2)} \quad \text{for } m \text{ even.} \tag{12.13}$$

Both these functions have major maxima centered at $Q_z = \frac{1}{2}(2\pi/d), \frac{3}{2}(2\pi/d), \ldots$, i.e., half-way in between the points at which the maxima occur in the case of ferromagnetic layers (see Fig. 12.2a).

The sum-containing parts of (12.9) and (12.10) describe the effects of interference of the waves diffracted by N single bilayers. Here, a major role is played by the magnetic correlations between the successive layers. In an idealized model description, they may be either "perfectly correlated" (the monolayer magnetization sequence in each layer is identical, so that all coefficients $\xi_\nu = 1$), or "perfectly anticorrelated" (the sequence in the $(\nu + 1)$th layer is reversed relative to that in

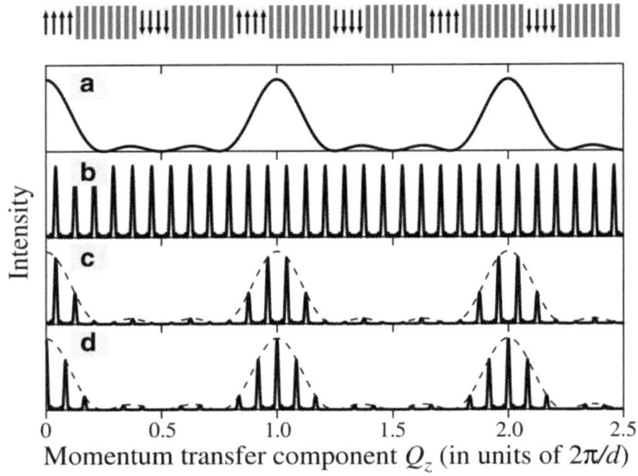

Fig. 12.1 An illustrative example of calculating the diffraction spectrum from a simple SL model with FM layers, depicted at the *top* of the figure. Each FM layers consists of four monolayers, and each nonmagnetic spacer of eight monolayers – all with the same thickness d. There are six bilayers. The FM layers are "anticorrelated" (or, using a popular terminology, the interlayer correlations are *antiferromagnetic*). The plot in panel (**a**) is the squared single bilayer structure factor described by (12.12) (for simplicity, $f(Q_z)\mathcal{M} = \infty$ was used). Panel (**b**) shows a plot of the $\sin^2(NQ_zD/2)/\sin^2(Q_zD/2)$ function in (12.16), and panel (**c**) shows the resultant spectrum, which can be thought of as the function plotted in panel (**b**) being "modulated" by that in panel (**a**). Panel (**d**) shows the spectrum from the same superlattice model, but with all FM layers magnetized in the same direction ("FM interlayer correlations"). As can be seen in the plots, in the latter case the "satellites" are shifted by one-half period compared to the situation in plot (**c**). From the peak positions in the measured spectra, one can thus determine the type of the interlayer correlations in an SL specimen investigated by neutron diffraction

the νth layer, so that $\xi_{\nu+1} = -\xi_n u$). The third model idealization is a "perfectly random" superlattice, in which the $(\nu + 1)$th layer may be either "correlated", or "anticorrelated" with the νth layer with a 50% probability, so that the $\xi_n u$ coefficients form a random sequence of $+1$ and -1 values. For analyzing this latter situation, it is convenient to use (12.10). It is easy to see that in the sum only the terms with $\alpha = \beta$ survive statistical averaging, whereas all terms with $\alpha \neq \beta$ are averaged to zero – and, consequently, for a "random" superlattice the diffraction spectrum shape simply reproduces the shape of the spectrum from a *single* bilayer:

$$I(Q_z) \propto N|F_{\rm BL}(Q_z)|^2. \qquad (12.14)$$

Equation (12.10) can also be used as a basis for calculating the spectra from *partially correlated* superlattices – i.e., such in which the probability that layers ν and $\nu + 1$ are correlated is higher than that they are anticorrelated, or vice versa. Such systems will be discussed in Sect. 12.3.1.

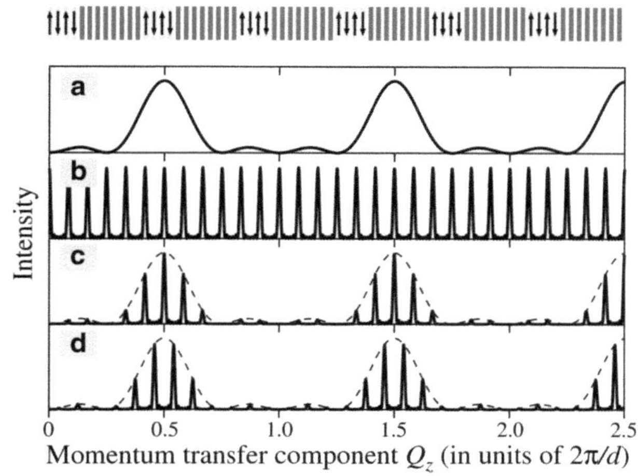

Fig. 12.2 A similar example as in Fig. 12.1, but for a simple SL model with AF layers. As can be seen in panel (**a**), the squared single-bilayer structure factor function now has maxima positioned half-way in between the positions, where the maxima from a superlattice with FM layers occur. Again, the resultant diffraction spectrum calculated from (12.16), shown in panel (**c**), can be thought of as the multipeaked $\sin^2(NQ_zD/2)/\cos^2(Q_zD/2)$ function (panel (**b**)) "modulated" by the squared structure factor function. And again, changing simple interlayer correlations (i.e., identical sequence of monolayer magnetization in each layer) to "anticorrelations" (spin configuration reversed in every second layer) results in a shift of the satellite peaks by one-half period (plot (**d**))

However, (12.9) is a more convenient form for analyzing perfectly correlated or anticorrelated multilayers. For correlated ones, all ξ_ν coefficients are 1. By the same method that is used in (12.11) and (12.12), one obtains:

$$I(Q_z) \propto |F_{\mathrm{BL}}(Q_z)|^2 \frac{\sin^2(NQ_zD/2)}{\sin^2(Q_zD/2)}. \tag{12.15}$$

For anticorrelated superlattices, $\xi_\nu = (-1)^\nu$, which leads to:

$$I(Q_z) \propto |F_{\mathrm{BL}}(Q_z)|^2 \times \begin{cases} \dfrac{\cos^2(NQ_zD/2)}{\cos^2(Q_zD/2)} & \text{for } N \text{ odd} \\ \dfrac{\sin^2(NQ_zD/2)}{\cos^2(Q_zD/2)} & \text{for } N \text{ even} \end{cases} \tag{12.16}$$

The $I(Q_z)$ function describes the diffraction spectrum shape observed in experiments (it has to be additionally corrected for $f(Q_z)$ and for the instrumental resolution – however, these corrections usually do not change its shape in a significant way). Examples of such spectra calculated for simple FM and AF superlattice models are shown, respectively, in Fig. 12.1c, d, and in Fig. 12.2c, d.

As follows from the above brief theory outline, superlattices containing ferromagnetic layers produce neutron diffraction maxima centered at $Q_z = (2\pi/d)$,

$2(2\pi/d), 3(2\pi/d), \ldots$ positions, and those with antiferromagnetic layers, at $Q_z = \frac{1}{2}(2\pi/d), \frac{3}{2}(2\pi/d), \frac{5}{2}(2\pi/d), \ldots$ positions, which are distinctly different than the former ones. Thus, in principle, diffraction studies can be used for identifying the type of spin ordering in the magnetic layers; however, that is usually known beforehand. Rather, the main goal in neutron experiments on magnetic superlattices is to investigate the *profile* of a single diffraction maximum as it carries information about the correlations between the magnetic layers. As shown above, if there are no such correlations, the maximum has the form of a broad smooth curve, reflecting the shape of the squared single bilayer structure factor $|F_{\mathrm{BL}}|^2$ – and if the correlations exist, it splits into a number of sharp peaks, whose intensities are "enveloped" by the $|F_{\mathrm{BL}}|^2$ function. This information is of unique value because there are no other experimental tools capable of detecting such correlations directly.

In principle, diffraction experiments should enable one to investigate interlayer correlations in all types of superlattices, no matter whether the constituent layers are antiferromagnetic, or ferromagnetic. Unfortunately, there is one additional factor that makes studies of ferromagnetic systems very difficult, or even impossible. Namely, all multilayered systems are grown on substrates, and most often they are single crystal wafers, whose structure closely matches the lattice periodicity of the superlattice constituents. In such situation, structural (nuclear) Bragg reflections from the substrate occur close to the positions of the magnetic maxima from the superlattice structure. Since the substrate reflections are usually more intense than the diffraction maxima from the superlattice by several orders of magnitude, those maxima may become completely obscured. In contrast, there is no such problem in the case of superlattices containing antiferromagnetic layers. Here, the superlattice maxima occur at $Q_z = \frac{1}{2}(2\pi/d), \frac{3}{2}(2\pi/d), \frac{5}{2}(2\pi/d)$ positions, which are normally half-way in between the structural Bragg reflections from the substrate.

In view of the above, the "wide-angle" diffraction technique is used primarily for investigating superlattices with antiferromagnetic layers. A tool much better suited for studying ferromagnetic superlattices is the technique commonly referred to as *neutron reflectometry*. In this method, one measures the intensity of neutrons scattered from a flat specimen at very small angles – i.e., corresponding to very small values of Q. As follows from (12.12), (12.15) and (12.16), and is graphically illustrated in Fig. 12.1, the kinematical theory does predict that ferromagnetic superlattices produce diffraction maxima also in the region of Q values close to zero. However, this result is only *qualitatively* correct. As noted, the kinematical theory is based on the Born approximation that ignores the weakening of the incident beam amplitude on its path in the scattering system, and the interference of the incident and the diffracted waves. This is certainly a reasonable assumption for larger scattering angles θ, but in the case of very small θ values the path traveled by the incident and the scattered waves inside the specimen become long even if the specimen is thin. Thus, the Born approximation can no longer be used. Instead, for obtaining a quantitatively valid description of the scattering effects in the small-angle region (usually, referred to as the *reflectivity region*), one has to employ the formalism of *neutron optics*, in which the effects neglected in the kinematical approach are properly taken into account.

12.2.2 Neutron Reflectometry

The formal theory of small-angle neutron scattering from flat multilayered structures (or, in short, of *neutron reflectivity*) is mathematically far more complex than the kinematical theory of diffraction. Therefore, the description in this section is focused primarily on qualitative aspects of reflectometry studies, with the use of mathematics reduced to the necessary minimum. The theory and the experimental reflectometry techniques are presented in greater detail in [25–27].

In close analogy to the theory of light reflectivity, for describing the propagation of a neutron wave in a material medium the neutron reflectivity theory utilizes the *refractive index* n [22,28]. For most condensed matter systems, this index is slightly less than 1 (by $\sim 10^{-6}$), so that neutrons impinging a flat surface at a grazing angle θ lower than the critical angle $\theta_c = \sqrt{2(1-n)}$ are totally reflected ($R = 1$). Since $(1-n)$ is of the order of 10^{-6}, the value of θ_c is normally only of the order of $10'$ of arc. As the angle θ is increased beyond the critical value, the reflectivity decreases extremely rapidly according to the formula:

$$R(\theta) = \left[\frac{\sqrt{n^2-\cos^2\theta}-\sin\theta}{\sqrt{n^2-\cos^2\theta}+\sin\theta}\right]^2 \approx \left[\frac{1-\sqrt{1-\theta_c^2/\theta^2}}{1+\sqrt{1-\theta_c^2/\theta^2}}\right]^2, \tag{12.17}$$

which is obtained from the well-known Fresnel equations, originally derived for describing the reflectivity of light incident at an interface between a transparent medium and vacuum. It should be stressed that the refractive index n for neutron waves does not depend on the crystallographic structure of the medium. For a nonmagnetic material, it is given by:

$$n = 1 - \left(\frac{\rho_{SL}\lambda^2}{2\pi}\right), \tag{12.18}$$

where λ is the neutron's de Broglie wavelength, and $\rho_{SL} = N\langle b\rangle$ is the so-called nuclear "scattering length density" (SLD), with N being the number of nuclei per unit volume, and $\langle b\rangle$ the average nuclear scattering length. The refraction index is also sensitive to the material magnetism, but only if the system has a nonzero net magnetization M. Thus, (12.18) is valid in the same form for antiferromagnets (regardless of their spin state) and ferromagnets above the Curie temperature T_C. However, for a ferromagnet at $T < T_C$ the equation changes to [22,25]:

$$n = 1 - \left(\frac{\rho_{SL}\lambda^2}{2\pi} \pm \mu_N\frac{M}{2E}\right), \tag{12.19}$$

where μ_N is the magnetic moment of the neutron, E is the neutron energy related to its wavelength and mass as $E = (2\pi^2\hbar^2/m)\lambda^{-2}$. The $+$ or $-$ sign depends on whether the neutron spin has, respectively, an opposite or parallel orientation relative to the magnetization vector **M**.

For incident angles $\theta > \theta_c$, only some part of the impinging radiation is reflected from the surface, and the remaining part penetrates the material (usually, it is referred to as the "transmitted wave"). If the reflecting specimen is a multilayered structure made of components with different refractive indices, the penetrating radiation is partially reflected by each layer–layer interface, and all those waves add up to the total reflectivity. In a rigorous treatment of the problem, one has also to consider secondary reflections from the "back side" of the layers that redirect the reflected radiation back to its original propagation direction, tertiary reflections, and so on. The total reflectivity $R(\theta)$ of an arbitrary structure consisting of N layers of thickness D_i and refractive index n_i can be exactly calculated using recurrent procedures in which all multiple reflection effects are properly taken into account. However, if the layers are arranged into a periodic structure – which is the case for a binary superlattice – then the most important features of the $R(\theta)$ spectrum arise from the interference of primary reflections, and – at least in a qualitative description – the multiple reflections can be neglected in the first approximation.

Suppose that a superlattice is made of alternating layers of an FM material and a nonmagnetic material. For $T > T_C$, the refractive indices of both components, denoted as n_1 and n_2, are purely nuclear. If there is an "SLD contrast" between the layers (i.e., $n_1 \neq n_2$), the impinging neutron waves are reflected at each interface. The bilayers can be thought of as "unit cells" that all produce identical reflected waves. In a way similar to that used for deriving the Bragg condition for a system atomic planes, one can readily show that a constructive interference between all these waves occurs when

$$\eta\lambda = 2D_{\mathrm{BL}} \sin\theta, \tag{12.20}$$

where $\eta = 1, 2, 3, \ldots$ (to avoid confusion with the refractive index, the n symbol conventionally used in this equation is replaced by η). This interference produces maxima superimposed on the rapidly falling curve produced by the surface reflection. As in diffraction measurements, those maxima are usually called "nuclear," "structural," or "chemical." The intensity I of these maxima is proportional to $(n_1^2 - n_2^2)^2$.

Below T_C, the spectrum changes depending on how the directions of magnetization vectors $\boldsymbol{M}$ in successive FM layers are correlated. If unpolarized neutrons are used, there is no coherence between the nuclear and magnetic scattering. The latter component can be thought of as arising from an independent "magnetic superlattice." For FM interlayer correlations (all layers magnetized in the same direction), the magnetic and the chemical periodicity is the same, and one observes additional magnetic peaks superimposed on the structural maxima (Fig. 12.3a). For AF correlations (alternating M), the magnetic SL period is effectively doubled, giving rise to extra peaks half-way between the nuclear ones (Fig. 12.3b).

If the incident neutrons are polarized, nuclear and magnetically scattered waves do interfere, and reflectivity effects can be analyzed in terms of the overall refractive index given by (12.19). If all FM layers are magnetized in the same direction,

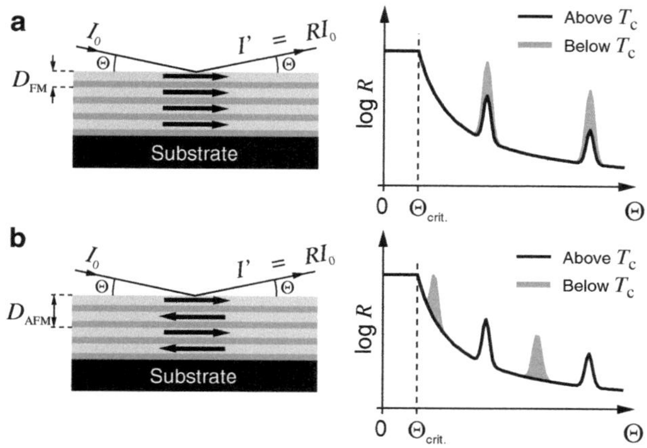

Fig. 12.3 Reflectivity spectra from superlattices with ferromagnetic layers and (**a**) FM, and (**b**) AF interlayer correlations, for unpolarized neutrons. The shape for $T > T_C$ is shown by the *thick solid curve*, and the *shaded* profiles show the extra scattering arising below T_C

with $\boldsymbol{M}$ parallel to the polarization direction $\hat{\boldsymbol{\sigma}}$, then, as follows from (12.19), the n contrast between the FM layers and the nonmagnetic ones increases (relative to the situation for $T > T_C$) for one spin polarization, and decreases for the other. Consequently, for one neutron polarization the intensity of reflectivity peaks increases when the temperature is lowered below T_C, and it decreases for the other polarization.

When the spin polarization of incident neutrons is exactly parallel or antiparallel to $\boldsymbol{M}$, the scattering is non-spin-flip (NSF). A magnetization component *perpendicular* to $\hat{\boldsymbol{\sigma}}$ gives rise to spin-flip (SF) scattering. SF scattering is exclusively magnetic – i.e., it does not occur above T_C. The NSF and SF processes can be distinguished if an analyzer of the reflected beam polarization is applied. A standard experimental practice in polarized neutron reflectometry is to measure $R(\theta)$ spectra for all four combinations of "up" and "down" states of the polarizer and the analyzer. Conventionally, for describing these states one uses "+" and "−" symbols; e.g., "(+ −)" means "polarizer up, analyzer down."

As follows from the above, reflectivity measurements using unpolarized neutrons enable one to detect interlayer magnetic correlations in a ferromagnetic/nonmagnetic superlattice and to determine whether they are of the FM or AF type. By applying polarized neutron reflectometry, one can additionally determine the orientation of M in ferromagnetically correlated layers. The latter technique, as discussed in Sect. 12.3.2, offers valuable insight into magnetism of the $Ga_{1-x}Mn_xAs$ system.

12.3 Studies of Ferromagnetic Semiconductor Superlattices (Primarily, by Neutron Reflectometry)

12.3.1 EuS-Based Multilayers

Eu chalcogenides (EuO, EuS, EuSe, EuTe) are a well-known "prototypical" family of magnetic semiconductors. They have been extensively studied since the 1960s, so their magnetic and electronic properties have been well characterized [29–31]. Their crystallographic and electronic data are given in [32,33]. All four members of the family crystallize in the NaCl structure, and in all of them the Eu–Eu magnetic exchange interactions are very short-ranged – essentially, the only relevant magnetic interactions are those between the nearest- and the next-nearest neighbors. The nearest neighbors are coupled by direct ferromagnetic exchange ($J_1 > 0$), and the interaction between next-nearest neighbors is an antiferromgnetic ($J_2 < 0$) superexchange mediated by the intervening anion. The ratio of $|J_1|/|J_2|$ decreases with the increasing anion atomic number. In EuO and EuS J_1, is the dominant exchange coupling, so both these compounds are ferromagnets – and in EuTe J_2 takes over, making the system antiferromagnetic. In EuSe, the FM and AF interactions are of nearly equal strength, resulting in a complicated low-T phase diagram.

Because of their low Curie temperatures, EuO and EuS ($T_C = 69$ K and 16.5 K, respectively) are certainly not good candidates for practical spintronics applications. However, since their magnetism and electronic properties are well understood, they may serve as good "prototypes" for investigating the fundamental physical mechanisms underlying phenomena that are highly interesting in the context of spintronics studies. Interlayer magnetic coupling in all-semiconductor superlattices is one of such topics because the basic mechanisms of magnetic interactions are either temperature-insensitive (e.g., direct magnetic exchange or superexchange), or vary with temperature in a predictable way (e.g., carrier-mediated interactions). Hence, the information obtained from low-temperature studies can be used for interpreting effects seen in other systems at higher temperatures. For an experimentalist, a considerable advantage of EuS- and EuO-based superlattices is that in such systems the magnetic lattices are fully occupied, which leads to a relatively strong signal.

The fabrication of EuO-based epitaxial structures is not easy because of the very high melting temperature of this material. But EuS can be combined with several other semiconducting compounds into good-quality superlattices. One group that specializes in the growth technology of such systems, and conducts research primarily on EuS/PbS structures, is the team of Dr. A. Yu. Sipatov from Kharkov, Ukraine. Neutron studies of EuS-based superlattices were performed on specimens obtained from that source.

PbS is isostructural with EuS, and the lattice mismatch between the two materials is about 0.5%. PbS is a diamagnetic semiconductor with a narrow band gap ($E_g = 0.3$ eV); its carrier concentration is typically of the order of 10^{17}–10^{18} cm^{-3}. The EuS/PbS multilayers were grown epitaxially on monocrystalline KCl(001) substrates topped with a PbS buffer layer of ~50 nm thickness. An electron beam was used for EuS evaporation, and standard resistive heating for PbS evaporation. Detailed studies of the growth and magnetic properties of EuS/PbS superlattices with thick PbS spacer (magnetically decoupled case) have been reported in [34].

Neutron scattering studies of magnetic interlayer correlations in the EuS/PbS system were performed on a series of samples with relatively thin PbS spacers (beginning from layer thickness as small as 4.5 Å). To check the structural quality, the samples were first examined by neutron diffraction. It may be of interest to mention that those measurements, intended only to do routine tests before starting neutron reflectometry studies, already provided evidence that the EuS layers are "anticorrelated." As noted, neutron diffraction is not the optimal tool for studying ferromagnetic superlattices because the magnetic diffraction signal produced by the thin-film structure is "out shined" by the overwhelmingly stronger Bragg reflections from the substrate that usually are positioned very close to the superlattice peaks. Yet, in the case of EuS/PbS multilayers grown on KCl(001) surfaces, the lattice mismatch between epitaxial structure and the substrate is unusually large (~5.5%), so that the substrate reflections and the superlattice maxima are shifted apart. This is illustrated by a reciprocal space diagram shown in Fig. 12.4. As shown in Fig. 12.5a, c in which data from two samples are displayed, scans along the Q_z axis reveal a maximum in the region where magnetic diffraction from the

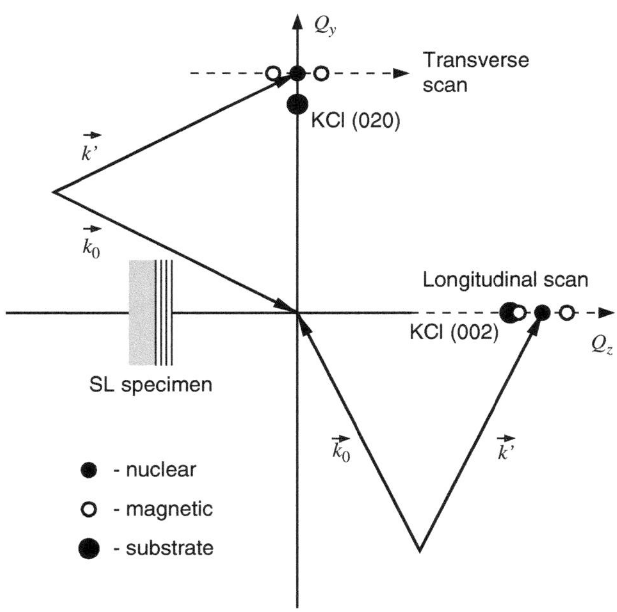

Fig. 12.4 Reciprocal lattice diagram showing the scanning trajectories (longitudinal and transverse) that were used in the search for neutron diffraction maxima arising from AF interlayer magnetic correlations in a EuS/PbS SL specimen. Also are shown (not-to-scale) the positions of the corresponding nuclear and magnetic SL reciprocal lattice points as well as the substrate reflection points

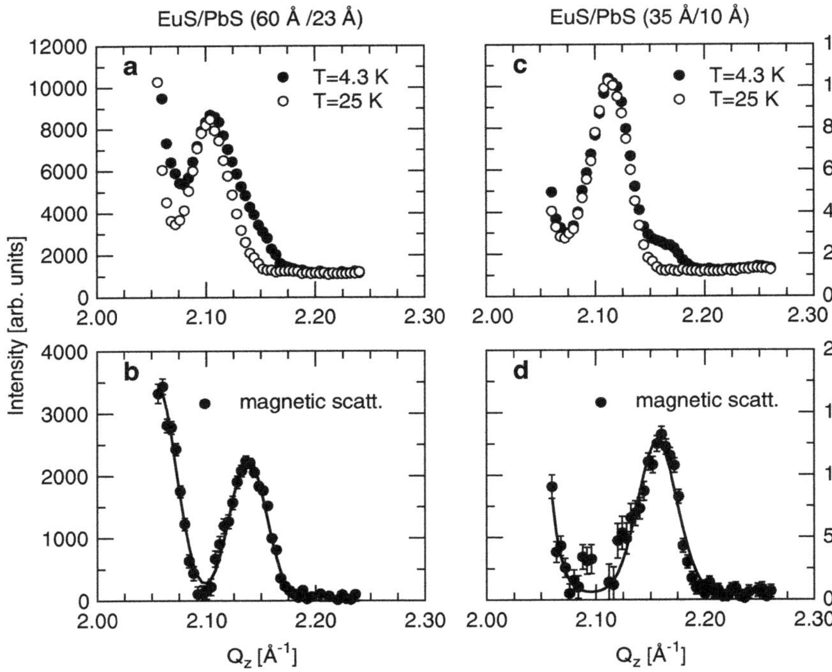

Fig. 12.5 (**a**) and (**c**): Neutron diffraction scans in reflection geometry (longitudinal scans) along the reciprocal lattice axis $(0, 0, Q_z)$ for two EuS/PbS SL specimens taken below and above T_C. (**b**) and (**d**): Purely magnetic scattering spectra for the same two samples obtained by subtracting the data measured above T_C from those measured below T_C

superlattice is expected to occur. A truly magnetic peak should disappear if the sample temperature is raised above the Curie point. But as can be seen in the figures, the spectrum only slightly changes when the sample is heated up from 4.2 K to 25 K. It indicates that the peak is in most part produced by Bragg scattering from the PbS buffer layer. Yet, the purely magnetic signal can be extracted by subtracting the high-T spectrum from the low-T one. Such differential scan data are shown in Fig. 12.5b, d indeed reveal a maximum. However, it is not possible to tell whether this is a single peak, or a fragment of a multipeak pattern, whose remaining part is "hidden" under the substrate Bragg reflection.

A more conclusive information can be obtained by performing a symmetry-equivalent scan in the same direction through a Q-space point located on the Q_y axis. In a "transverse" diffraction scan (i.e., with scanning direction perpendicular to the reciprocal lattice vector), the Q-resolution of a neutron diffractometer is typically about two times better than in a "longitudinal scan" (scanning along the reciprocal lattice vector). As can be seen in Fig. 12.6, in such scans the magnetic components are indeed better resolved from the PbS Bragg line. Here, the differential scans clearly show a symmetric two-peak pattern with a central minimum.

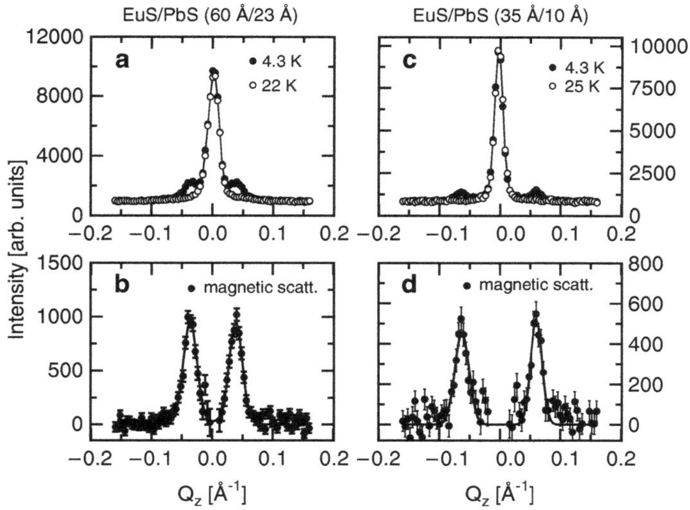

Fig. 12.6 (**a**) and (**c**): Neutron diffraction scans in transmission geometry (transverse scans) along the reciprocal space vector $(0, 2.14, Q_z)$ for two EuS/PbS SL specimens taken above and below T_C. (**b**) and (**d**): Purely magnetic signals from the same two samples obtained by subtracting the data measured above T_C from those measured above T_C

Such a spectrum is consistent with magnetically "anticorrelated" EuS layers (i.e., the magnetization vectors in adjacent EuS layers are aligned in opposite directions). In contrast, a "correlated" system of layers (all magnetization vectors pointing in the same direction) would produce a spectrum with a central maximum and satellite peaks symmetrically located on both sides.

As follows from the above, diffraction experiments may in certain situations provide information about interlayer correlations in ferromagnetic superlattices. But the reason of presenting those data was rather to show that neutron reflectometry is a far superior tool for studying such systems. Here, the troublesome Bragg scattering from the substrate and the buffer layer no longer pose problems – essentially, the scattering process probes only the multilayered structure, but is virtually insensitive to the presence of other specimen components.

Neutron reflectometry studies of the EuS/PbS system led to a spectacular confirmation of the preliminary findings from diffraction experiment. Examples of reflectivity spectra obtained from one of the samples investigated (with 60 Å thick EuS layers, and 23 Å thick PbS spacers) are shown in Fig. 12.7. When the sample temperature is well above the Curie point, the spectrum (open circles) shows a peak at $Q = 0.076\,\text{Å}^{-1}$, which agrees well with the calculated position of the first Bragg maximum in the reflectivity spectrum from a superlattice specimen with a bilayer period of $D = 83\,\text{Å}$ $(2\pi/D = 0.0757\,\text{Å}^{-1})$. So, this peak is obviously a "structural" maximum associated with the difference between the refractive indices of EuS and PbS. However, when the sample is cooled down to 4.2 K, which is well below

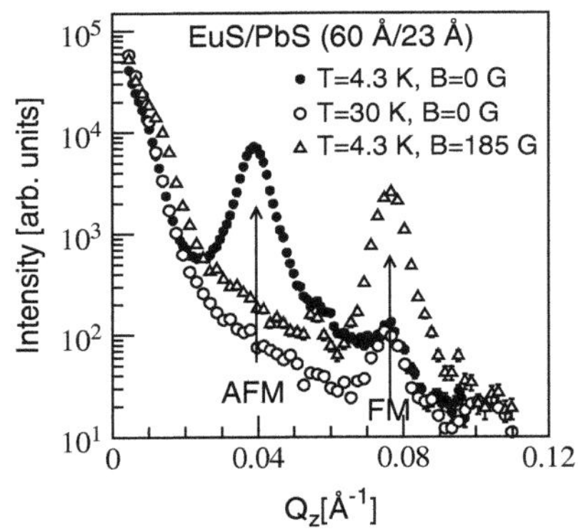

Fig. 12.7 Neutron reflectivity spectra from an EuS/PbS (60 Å/23 Å specimen measured below and above T_C in zero magnetic field (*filled* and *empty circles*, respectively) and a saturating field (*triangles*)

the Curie point ($T_C = 18.5$ K), there is a dramatic change in the spectrum (black circles) – another much stronger maximum appears at $Q = 0.038\,\text{Å}^{-1}$, which is a position corresponding to a *doubled* bilayer period. This maximum clearly indicates that the EuS layers – after acquiring magnetization by being cooled below the Curie point – form a magnetically "anticorellated" sequence. Note that the height of the structural maximum remains unchanged, meaning that there is no observable tendency in the system to form a "correlated" sequence.

As demonstrated by the above, neutron reflectometry is indeed a highly efficient tool for investigating ferromagnetic superlattices. While a sophisticated experimental procedure and elaborate data analysis were needed for detecting the type of interlayer magnetic correlations in the EuS/PbS system by diffraction studies, the reflectivity experiments provided a direct proof for their "anticorrelated" nature based on just two measured spectra! In addition, reflectometry studies make it possible to determine in a straightforward way the *strength* of the coupling between the EuS blocks. By applying an external magnetic field of sufficient magnitude parallel to the superlattice growth plane, one can overcome the forces leading to the antiparallel orientation of the magnetization vectors in adjacent EuS layers, and enforce the magnetization vectors in all individual EuS layers to get aligned with the field. In such a case, the specimen will no longer be "seen" by neutrons as a sequence of "anticorrelated" layers, but, instead, as a "correlated" chain. As one can expect taking into consideration the spectrum schemes illustrated in Fig. 12.3b, the magnetic peak should now shift from the $Q = 2\pi/2\text{D}$ position to the $Q' = 2\pi/D$ position (marked in the plots, respectively, as the "AF" and "FM" positions) and it should overlay the structural maximum that occurs at the latter spot. As can be seen in Fig. 12.7, after applying a field of 185 G the spectrum (triangles) changes exactly as predicted by the above scenario. However, the spectrum does not change abruptly at some threshold value – the process is continuous, as shown in Fig. 12.8, where the intensity of the AF peaks seen in several different samples is plotted versus the applied field. In those measurements, the field was gradually increased to a level that caused a total disappearance of the AF peak (saturation value), and then it was

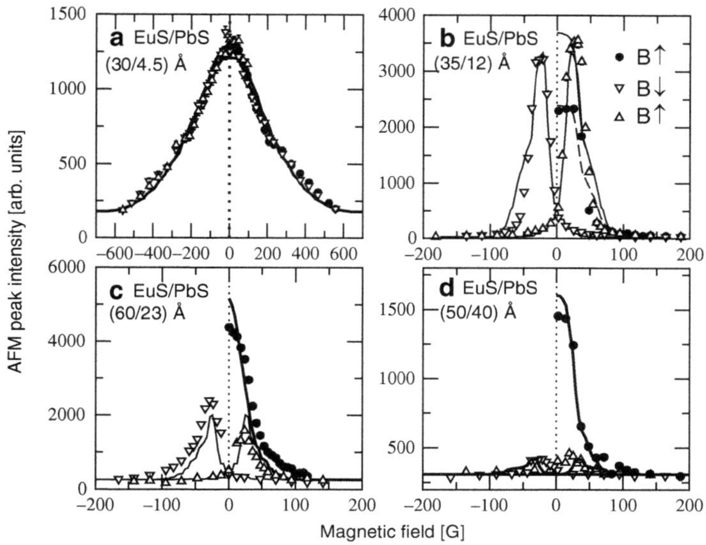

Fig. 12.8 The intensity of the AF Bragg peak vs. magnetic field for four different EuS/PbS SL specimens

gradually decreased back to zero. Then, the entire cycle – going up to the saturation value, and back to zero – was repeated with the field direction reversed.

The "saturation value" – i.e., the applied field magnitude needed for a total suppression of the "anticorrelations" – can be regarded as a measure of the strength of the AF interlayer coupling. The measurements performed on several samples with different PbS spacer thickness (d_{PbS}) show that the coupling strength decreases with increasing d_{PbS} value. This is certainly a behavior one might expect. Surprisingly, however, when d_{PbS} reaches the value of ~15 Å (equivalent to about 5 PbS monolayers), the interlayer coupling strength stabilizes at certain level (Fig. 12.9).

Another important fact revealed by the data displayed in Fig. 12.8 is that for only one of the investigated samples (with $d_{\mathrm{PbS}} = 4.5$ Å) the field characteristics were found to be almost completely reversible. In all other cases, the characteristics exhibit a pronounced hysteresis. It points to a significant role of magnetic anisotropy in the process.

The total magnetic energy of a pair of coupled EuS layers in external magnetic field H is given by [35]:

$$E = E_J + E_{K,1} + E_{K,2} + E_{H,1} + E_{H,2}. \tag{12.21}$$

Here, E_J is the interlayer coupling energy described by the constant J:

$$E_J = -J \frac{\mathbf{M}_{s,1} \cdot \mathbf{M}_{s,2}}{M_{s,1} M_{s,2}} = -J \cos(\theta_1 - \theta_2), \tag{12.22}$$

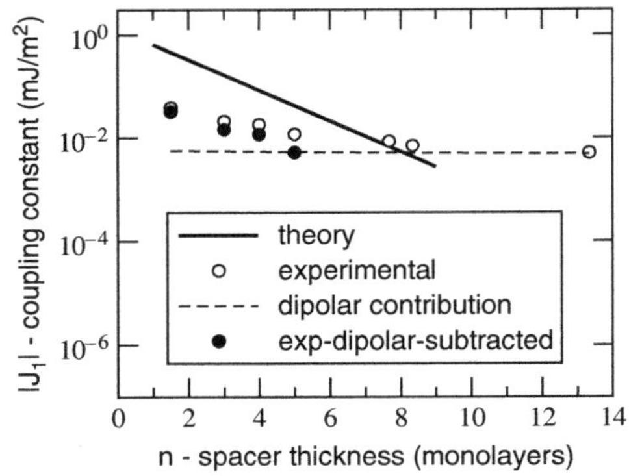

Fig. 12.9 The strength of the interlayer coupling in EuS/PbS superlattices plotted vs. the PbS spacer thickness. *Open circles* represent the mean J values obtained by fitting theoretical functions calculated from the Stoner–Wohlfarth model to the measured data. The *solid line* shows the $J(d)$ dependence obtained from the Blinowski–Kacman model. The *dashed line* represents the estimated dipolar contribution to the interlayer coupling. *Black circles* show the experimental J values corrected for the dipolar interactions

where $\mathbf{M}_{s,1}$ and $\mathbf{M}_{s,2}$ are the magnetization vectors of the layers, and θ_1, θ_2 are the angles between these vectors and the magnetic field vector $\mathbf{H}$. $E_{K,1}$, $E_{K,2}$ describe the cubic anisotropy energy of each layer:

$$E_{K,i} = -tK\cos^4\theta_i, \tag{12.23}$$

where t is the layer thickness, and $E_{H,1}$, $E_{H,2}$ are the Zeeman terms:

$$E_{H,i} = -t\mu_0\mathbf{M}_{s,i}\cdot\mathbf{H} = -t\mu_0 M_s H\cos(\theta_i). \tag{12.24}$$

The magnetic anisotropy in Eu chalcogenides is known to fall to zero before the Curie temperature is reached [36–38]. Therefore, the anisotropy has no influence on the formation of the interlayer correlations when the sample is cooled down through the Curie point. However, at low temperatures the K constant may acquire a value that makes the anisotropy energy comparable with the interlayer coupling energy. If the system is initially in a ↑↓↑↓↑↓ configuration, then a sufficiently strong magnetic field will always enforce a ↑↑↑↑↑↑ sequence. However, the multiple minima in the anisotropy energy will not allow the restoration of the initial sequence after the field is removed – the system may stay in some "intermediate" state. Then, the application of the field in the opposite direction may rather favor a transition to the ↓↓↓↓↓↓ configuration. Thus, the initial configuration, and the initial AF peak intensity may never be restored. Only if the interlayer coupling energy is much stronger than the anisotropy energy, it may be able to push the system through the barriers between the anisotropy energy minima, and return it to the original state. This is apparently the case in the sample with the 4.5 Å PbS spacer thickness. Determining the interlayer coupling energy for this particular sample was therefore a straightforward task – but for the other samples it required a much more complicated deconvolution procedure based on the Stoner-Wohlfarth model [39]. The details of that procedure are discussed in [19]. In this article, we show in Fig. 12.9 the values of J obtained in that work plotted vs. the PbS spacer thickness.

Needless to say, an important matter is to explain the origin of the interlayer coupling in a system with no conduction electrons. For that, a model was proposed by Blinowski and Kacman (B&K) [20, 21] in which the exchange interactions are conveyed across the PbS spacers by *valence band electrons*. The model does not assume any particular interaction mechanism, but attributes the interlayer coupling to the sensitivity of the superlattice electronic energies to the magnetic order in the layers – i.e., it accounts globally for the spin-dependent band structure effects.

According to the Blinowski and Kacman model, the strength of the coupling between the EuS layers decreases exponentially with the PbS layer thickness. This result is indeed in a good qualitative agreement with the experimental results from specimens with the PbS layer thickness up to $n = 5$ monolayers. However, as noted, for thicker PbS spacers the dependence of J on n visibly flattens out, showing that the interlayer correlations are of longer range than predicted by the model. The observed behavior of J clearly suggests that there are two interaction components – one that decays exponentially with increasing n, and the other that is approximately constant.

One possible explanation of the observed effects is that the other interaction component arises from *dipolar* forces. It is known that the interaction energy between two uniformly magnetized infinite planes of magnetic dipoles is zero. However, as shown in [40], if the layers exhibit a domain structure with sufficiently small average domain size (1 μm or less), the coupling by dipolar forces may become significant. A characteristic feature of such coupling is a relatively weak dependence of its strength on the layer–layer separation (see (9) in [40]). On the other hand, this strength depends quite strongly on the lateral dimensions of the domains, and as yet, there is no available information on this subject. So, the contribution of the dipolar coupling mechanism to the total value of J cannot be obtained from pure calculations. However, since the calculations of Blinowski and Kacman indicate that for PbS spacers thicker than $\sim$25 Å the coupling via valence electrons is already very weak, it is reasonable to assume that all interlayer coupling in this region is of dipolar origin. So, one can extrapolate its strength for thinner spacers (see the dashed line in Fig. 12.9), subtract it from the measured total value, and thus obtain the coupling energy corresponding to mechanisms other than dipolar coupling. The J values corrected in this way are shown in Fig. 12.9 as filled circles. As can be seen in the figure, the slope of the experimental $J(d_{\mathrm{PbS}})$ dependence matches well the slope of the theoretical characteristic (shown as the solid line), but the experimental J values are still almost an order of magnitude lower than the theoretical ones. It should be noted, however, that such a discrepancy between the calculated and measured values of interlayer coupling energies is also typical in the case of metallic ferromagnetic superlattices. So, the Blinowski and Kacman model can be regarded as being in a "semiquantitative" agreement with the experiment.

Still, the Blinowski and Kacman explanation of the origin of the interlayer coupling forces in the EuS/PbS system was met with criticism from other researchers who pointed out that PbS cannot be treated as a "good" insulator. In another theoretical study, Zorchenko [41] attributed the interlayer coupling effects to the small population of mobile electrons ($n \sim 10^{17}\,\mathrm{cm}^{-3}$) that exist in PbS even at low T due to its relatively narrow energy gap ($\sim$0.2 eV). In contrast to the AF-only coupling in the B&K model, Zorchenko's theory predicts coupling sign oscillations – in certain analogy to those seen in metallic SLs [3,42], but with a considerably longer (10–20 times) period. According to the Zorchenko's calculations, the coupling between EuS blocks should change to *ferromagnetic* for D_{PbS} larger than $\sim$100 Å.

One possible way of resolving the controversy concerning the coupling mechanism was to carry out experiments on samples with modified spacer material. An ideal situation, of course, would be to vary the electron concentration (n) in the spacers. If Zorchenko's model were correct, adding more electrons would enhance the correlations, whereas lowering n would weaken them. Unfortunately, there is no easy way for manipulating the n value in PbS. However, one can prepare samples similar to the hypothetical EuS/PbS system with $n \rightarrow 0$ by replacing PbS with a truly insulating compound. There are two materials very well suited for that purpose, YbSe and SrS. YbSe is a wide-gap semiconductor ($E_g = 1.6\,\mathrm{eV}$), and SrS is usually referred to as an insulator. Both have the same structure (NaCl type) as EuS and PbS, and are nearly perfectly lattice-matched to PbS (this aspect is important because strain effects in EuS/YbSe and EuS/SrS systems will remain essentially the same as in EuS/PbS).

It was the EuS/YbSe system that became available for experimentation. As shown in Fig. 12.10, the reflectivity spectra from an EuS/YbSe specimen are strikingly similar to those obtained at the same conditions from an EuS/PbS systems of similar parameters. Needless to say, this result strongly argues in favor of the Blinowski and Kacman theory, and against the Zorchenko's interpretation. According

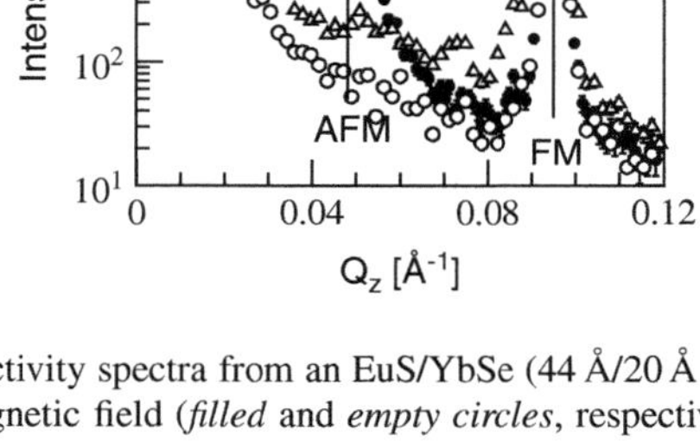

Fig. 12.10 Neutron reflectivity spectra from an EuS/YbSe (44 Å/20 Å specimen measured below and above T_C in zero-magnetic field (*filled* and *empty circles*, respectively) and a saturating field (*triangles*)

to that latter model, there should be no detectable interlayer coupling in the EuS/YbSe system because of the nearly total absence of conducting electrons in the YbSe spacers – whereas the Blinowski and Kacman theory predicts only a faster decrease of the coupling strength with the YbSe spacer thickness than in the case of the PbS spacers, but essentially the same qualitative behavior of both systems.

12.3.2 Neutron Reflectometry Studies of $Ga_{1-x}Mn_xAs$/GaAs Superlattices

$Ga_{1-x}Mn_xAs$ is surely the best known prototypical spintronics material. The technology of preparing thin epitaxial $Ga_{1-x}Mn_xAs$ films, and research on this system were pioneered by the team of Prof. Hideo Ohno from Sendai, Japan [12]. The question of whether two ferromagnetic $Ga_{1-x}Mn_xAs$ layers would interact magnetically across an intervening nonmagnetic spacer already emerged at the early stage of that research. Prof. Ohno's team obtained insight into that problem using an ingenious experimental technique [12]. From prior experiments, it was known that $Ga_{1-x}Mn_xAs$ has a nearly rectangular magnetic hysteresis loop, and that the coercive field and the remanence values are approximately proportional to x. So, they prepared a $Ga_{1-x}Mn_xAs$/GaAs/$Ga_{1-y}Mn_yAs$ trilayer with different Mn concentration values of the two FM layers ($x \neq y$), and examined the hysteresis loop of such a heterostructure. If there were no "magnetic communication" between the two FM blocks across the intervening layer of pure GaAs, measurements of magnetization vs. H in such a system would produce a figure that would be a simple sum of two different hysteresis loops, exhibiting characteristic "step-like" features. In fact, the observed figure showed no such features, thus proving that the two FM layers react coherently to the applied magnetic field – ergo, the two spin systems are coupled across the magnetically neutral GaAs layer by forces that tend to align them *parallel*. The coupling is thus ferromagnetic.

However, this experimental technique can be used only for investigating trilayers of special composition, but not true superlattices. Furthermore, the method requires using relatively strong magnetic fields – and it cannot be ruled out that the observed ferromagnetic coupling is enhanced by the field, or even that the coupling is a field-induced effect that completely disappears at $H = 0$. Therefore, the most conclusive insight into the phenomenon can be obtained from a method that does not necessarily require using an external magnetic field. Neutron reflectometry certainly meets such criteria. The probe used in such measurement – neutron radiation – does not interfere in any way with the physical mechanism responsible for the correlation formation. So, reflectometry enables one to detect interlayer correlations in their "virgin state", as they form in the system in the process of cooling through the Curie point at zero external field.

Yet, reflectometry studies of $Ga_{1-x}Mn_xAs$-based systems are much more challenging than the studies of EuS-based systems described in the preceding section. In EuS/PbS multilayers, the nuclear scattering length density (SLD) values for the constituent materials are $1.913 \times 10^{-8}\,\text{nm}^{-2}$ for EuS, and $2.326 \times 10^{-8}\,\text{nm}^{-2}$ for PbS. Because of the full occupancy of magnetic sites in the EuS lattice, and the large spin value of the Eu^{++} ion ($S = 7/2$), the magnetic SLD of EuS has a relatively high value of $3.584 \times 10^{-8}\,\text{nm}^{-2}$ (whereas for PbS the magnetic SLD is zero, of course). As shown by these numbers, in the EuS/PbS system the "magnetic SLD contrast" between the constituent layers is significant, and is much stronger than the "nuclear SLD contrast." Consequently, the magnetic peaks in the reflectivity spectra from EuS/PbS superlattices are quite intense, and are much stronger than the nuclear peaks – which is certainly a favorable situation from the experimenter's viewpoint.

The situation in $Ga_{1-x}Mn_xAs$/GaAs superlattice is much less favorable. The difference between the nuclear scattering lengths of Ga and Mn (7.88 fm, and −3.73 fm, respectively[3]) is relatively large – however, in $Ga_{1-x}Mn_xAs$ only a few percent of Ga atoms is substituted by Mn atoms, so the effective nuclear SLD contrast between the constituent layer is rather weak (for instance, for $Ga_{1-x}Mn_xAs$ with $x = 0.06$ the SLD value is $2.890 \times 10^{-8}\,\text{nm}^{-2}$, and for pure GaAs it is $3.070 \times 10^{-8}\,\text{nm}^{-2}$. The magnetic scattering length of the $Mn^{+}+$ ion is 13.7 fm – but, again because of the dilution, the purely magnetic SLD value is only a small fraction of that for a fully occupied magnetic lattice (for $Ga_{1-x}Mn_xAs$ with $x = 0.06$, it is only $0.177 \times 10^{-8}\,\text{nm}^{-2}$, 17 times less than it would be for a "hypothetical" system with $x = 1$).

As indicated by the above numbers, in the $Ga_{1-x}Mn_xAs$/GaAs system the nuclear and the magnetic SLD contrasts are of similar magnitude, and both are by far weaker than the magnetic SLD contrast in the EuS/PbS system that yields spectacular reflectivity peaks. Therefore, the peaks in the reflectivity spectrum from $Ga_{1-x}Mn_xAs$/GaAs multilayers cannot be expected to be very intense. In fact, measurements performed on a series of SL samples with $x = 0.06$ well above the Curie point (known from magnetometric measurements) showed in all cases weak peaks that could be identified as the first-order nuclear Bragg reflectivity maximum (based on the superlattice periodicity known from earlier X-ray reflectivity studies of the specimens). When the samples were cooled in zero magnetic field to 6 K (well below the Curie point of ∼30 K), the measurements [13] revealed no detectable peaks at new positions – however, the nuclear maxima visibly increased in intensity (roughly by ∼20%). The results show therefore that the nuclear and magnetic reflectivity maxima occur at the same spots, which is a clear evidence of *ferromagnetic* coupling between the $Ga_{1-x}Mn_xAs$ blocks. Thus, neutron reflectivity data confirm the earlier findings from trilayer studies. In addition, they show that the FM correlations form spontaneously, without any assistance from external magnetic field.

[3] Note that the scattering length of Mn is *negative*. The scattering of a neutron wave from a nucleus is associated with a phase shift that may be close to zero or to π. The latter is the case for a large majority of known nuclei. A convention has been adopted to regard their scattering lengths as positive. For a small number of elements (e.g., H, Mn, Ti), the phase shift is zero, and then, by the same convention, the scattering length is negative. In a material composed of different elements, the effective scattering length is a weighted average of the scattering lengths of the constituent nuclei.

Standard reflectivity measurements cannot provide much more information about the observed correlations. As described in the preceding section, in studies of EuS-based superlattices it was possible to determine the strength of the interlayer coupling by investigating changes in the reflectivity spectrum caused by the application of external magnetic field. However, this experimental technique can be used only for systems with *antiferromagnetic* interlayer correlations. In systems with FM coupling, an applied magnetic field does not perturb the correlations between individual layers – it may only cause a global rotation of an entire assemble of coupled layers. The relevant factor in this process is magnetic anisotropy, but the magnitude of the interlayer coupling does not play any meaningful role in it. So, the interlayer coupling constants for the $Ga_{1-x}Mn_xAs$/GaAs superlattices cannot be extracted from the reflectivity data, as it was done for the EuS-based multilayers.

However, the type of the interlayer correlations in $Ga_{1-x}Mn_xAs$/GaAs superlattices is not the only piece of information that neutron reflectivity studies can provide. More insight into magnetism of this system can be obtained by employing an advanced reflectometry technique in which a spin-polarized incident neutron beam is used, and the spin polarization of the reflected beam is analyzed. As described by (12.19), the refractive index of a magnetized material depends on whether the neutron spin is parallel, or antiparallel to the magnetization vector **M**. This sensitivity can be taken advantage of when investigating phenomena in which the orientation of **M** is relevant – e.g., in studies of magnetic domains in buried layers.

A scheme of a neutron reflectometer with polarization analysis capability is shown in Fig. 12.11. An incident monochromatic beam is first produced by a Bragg reflection from a crystal. The beam is then polarized using a "supermirror," a sophisticated thin film structure consisting of alternating layers of silicon and magnetized iron. The reflectivity of such a multilayer for neutrons of one spin orientation is approximately 100%, and almost zero for the other. Normally, the supermirror is used to reflect neutrons of one spin orientation *out of the beam*, thus making the transmitted beam totally polarized. The beam passes then through a "flipper" – an electromagnetic device that utilizes Larmor precession for reversing the spin

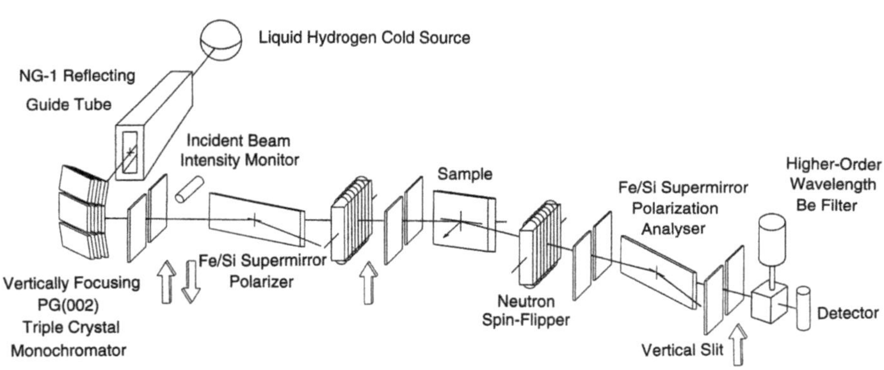

Fig. 12.11 A scheme of a neutron reflectometer with polarization analysis capability (see text)

orientation when it is activated. If the flipper is turned off, the beam passes with its polarization unchanged. Thus, the experimenter has the choice of two polarization states of the incident beam – e.g., "spin up" in the flipper's "on" or "+"state, and "spin down" in its "off," or "–" state.

After being reflected from the sample, the beam passes through a polarization analyzer that consists of another flipper and another supermirror (transmitting neutrons of the same polarization as the first one) behind it, and finally reaches the detector. So, the first flipper determines the incident beam polarization. The beam will then pass through the analyzer only if the second flipper is in the *same* state as the first one. If the second flipper were in the *opposite* state than the first one, the analyzer would transmit the beam only if its polarization were reversed. Thus, in the "off-off" and the "on-on" flipper configurations, one would measure the reflectivity spectrum corresponding to scattering processes in which neutron spin polarization *does not change* (usually referred to as "nonspin flip", or "NSF" scattering), and in the "on-off" and "off-on" configurations the spectrum corresponding to processes in which the spin polarization *is reversed* ("spin flip," or "SF" scattering).

The general theory of polarized neutron scattering in magnetic solids, encompassing all SF and NSF processes, is given in specialized literature (see, e.g., [25–27]). However, when discussing the experiments on $Ga_{1-x}Mn_xAs$/GaAs systems, the consideration can be limited to the following special situations:

(a) For nonmagnetic systems and FM systems above the Curie point, all scattering is purely nuclear, and NSF only.
(b) When a polarized beam is reflected from a magnetized film with the magnetization vector **M** lying in the film plane, and parallel/antiparallel to the beam polarization vector, all scattering is NSF. The magnetic and nuclear scattering are then coherent, with the effective scattering amplitude being $b \pm p$, and the refractive index given by (12.19) (with "+" or "−", respectively, for **M** parallel or antiparallel to the polarization vector).
(c) When the in-plane film magnetization vector is *perpendicular* to the polarization vector, the nuclear and magnetic scattering processes are totally decoupled. All magnetic scattering is then of SF type, and all scattering occurring without a spin flip is purely nuclear.

In these described experiments of $Ga_{1-x}Mn_xAs$/GaAs superlattices, the neutron polarization vector was vertically oriented. The SL specimens were grown on [100] GaAs substrates. In studies conducted by others, it had been established that in such systems the magnetization vector usually lies in the SL growth plane, and the [011] and $[01\bar{1}]$ are the easy axes. So, the samples were mounted on the reflectometer with the [011] axis vertical. It could be therefore expected that in the ferromagnetic domains forming in the specimens the magnetization vectors be either exactly parallel, or exactly perpendicular to the neutron polarization vector.

In Figs. 12.12–12.14, it is explained how different domain types that may form in the specimens would reflect the polarized beam. The first one (Fig. 12.12) depicts the situation above the Curie point. Here, only the nuclear scattering matters. The blue blocks in the plot symbolize the $Ga_{1-x}Mn_xAs$ layers, and the yellow ones – the

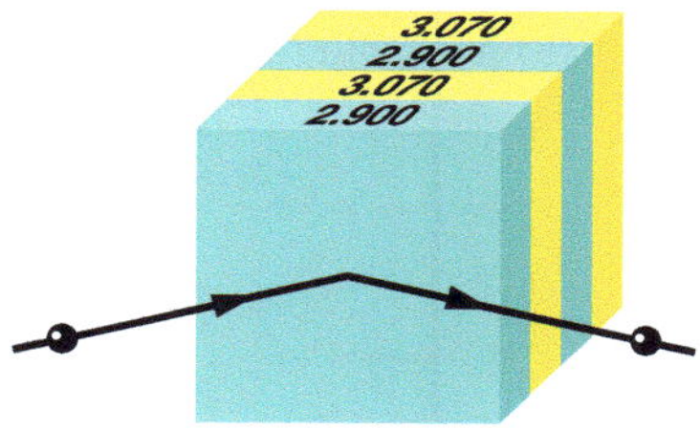

Fig. 12.12 A diagram illustrating neutron reflection from a $Ga_{0.94}Mn_{0.06}As$/GaAs sample above the Curie temperature. The numbers at the layer *tops* are their nuclear SLD values in the units of $10^{-8}\,nm^{-2}$. Because the magnetization of the $Ga_{0.94}Mn_{0.06}As$ layers is zero, neutron spin polarization is irrelevant

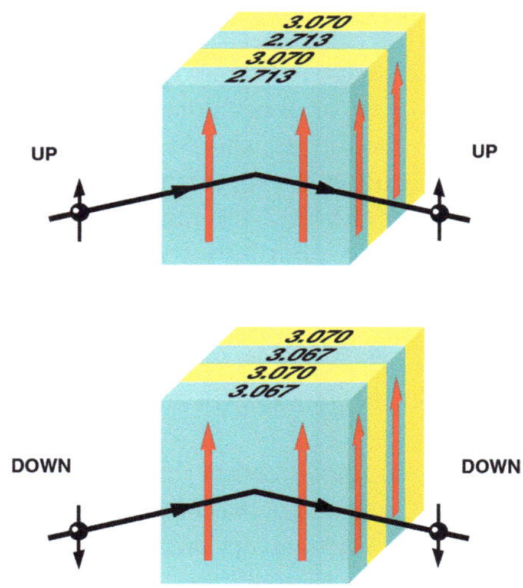

Fig. 12.13 The $Ga_{0.94}Mn_{0.06}As$/GaAs sample shown in Fig. 12.12, now below T_C and magnetized in the vertical direction, "as seen" by polarized spin-up and spin-down neutrons. For the "up" polarization, the magnetic SLD of the $Ga_{0.94}Mn_{0.06}As$ layers is subtracted from the nuclear one, thus increasing the SLD contrast between the layers. For the "down" polarization, the magnetic SLD adds up to the nuclear one. In effect, the SLD contrast between the magnetic and nonmagnetic layers almost completely disappears

GaAs layers. The numbers placed on the top of each block is the nuclear scattering length density (SLD) of the layer material. As noted, above the Curie temperature only nuclear scattering contributes to the reflectivity, and only NSF scattering occurs. The orientation of the incident beam polarization vector is thus irrelevant. The spectra obtained from measurements with the "off-off" and "on-on" flipper settings should exhibit Bragg peaks of the same intensity, and there should be no peaks for the "off-on" and "on-off" settings.

Figures 12.13 and 12.14 illustrate the situations when the system is cooled well below the Curie point. The magnetization in the $Ga_{1-x}Mn_xAs$ layers is now near

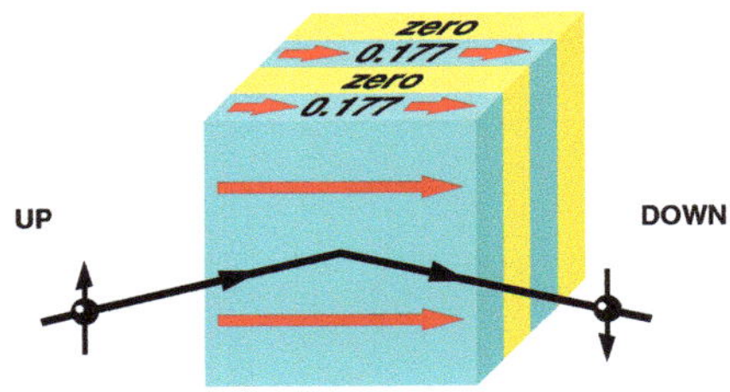

Fig. 12.14 The same sample as in the preceding two figures, now magnetized horizontally, "as seen" by impinging polarized neutrons. In such situation, the neutron spin is flipped in the scattering process. The only SLD component that matters in the process is the magnetic one – the effective SLD value for the nonmagnetic layers is now zero. The scattering is thus "purely magnetic"

its saturation value. If the **M** vector is oriented "up" (Fig. 12.13a, b), there is no spin-flip in the scattering process for either polarization of the incident beam (so, for observing these processes the flippers should be both "off", or both "on"). Yet, for each incident beam polarization the spectrum shape will be quite different. If the neutron spins are "up" (Fig. 12.13a), the magnetic SLD in the $Ga_{1-x}Mn_xAs$ layers *adds up* to the nuclear SLD. Fortuitously, the amount by which the SLD of $Ga_{1-x}Mn_xAs$ increases is nearly the same as the difference between the pure nuclear SLD of the constituent materials. Hence, for this spin polarization the SLD contrast between the layers almost completely disappears, making the Bragg peak in the reflectivity spectrum extremely weak (in practical experimental conditions, it is virtually invisible).

For the "down" incident beam polarization (Fig. 12.13b), on the other hand, the magnetic SLD is *subtracted* from the nuclear one. Consequently, the SLD contrast between the $Ga_{1-x}Mn_xAs$ and GaAs layers increases about twofold as compared with the situation above the Curie temperature, resulting in a much more intense Bragg peak than that seen in the paramagnetic state. So, in summary – for the "up" orientation of the **M** vector, essentially no peak should be seen in the "off-off" flipper configuration, and a sizable maximum should occur for the "on-on" configuration. Using the same reasoning, it can be readily concluded that in the case of the **M** vector pointing "down", the maximum occurs when the flippers are "off-off," and no maximum when they are "on-on."

In the case of *horizontal* **M** orientation (Fig. 12.14), the magnetic scattering is exclusively of the "spin-flip" type. For measuring this mode of scattering, the flipper setting should be "off-on" or "on-off" (because of the symmetry, the measured intensity should be the same for both these settings). Now only the magnetic scattering length of the Mn^{++} ions contribute to the effective SLD values of the layers (accordingly, for the GaAs blocks SLD=0). The Bragg peaks seen in the reflectivity spectrum are now *purely magnetic*. If, however, the flipper setting were changed to "on-on" or "off-off," only neutrons that underwent NSF scattering would get through the analyzer, and the spectrum recorded would then be *purely nuclear*.

As follows from the above, the magnetic and nuclear scattering can be totally separated only if the sample magnetization is exactly perpendicular to the neutron beam polarization. If these two vectors are parallel, the measured effect is always the result of interference of these two modes of scattering.

When a ferromagnetic sample is cooled down from a paramagnetic state through the Curie point, the system tends to minimize its net macroscopic magnetic moment. The final FM state that forms in such situations most often consists of a large number of microdomains. If there are no additional factors that make some of the domain orientations "privileged" (for instance, anisotropic crystal lattice strain may introduce additional uniaxial anisotropy, and thus favor domains with magnetization along such axis), all possible domain states in the low-T phase should be equally populated, which would effectively cancel out their magnetic moments.

In $Ga_{1-x}Mn_xAs$ films grown on (100) GaAs wafers, there are two orthogonal in-plane easy axes, which correspond to four symmetry-equivalent in-plane domain types. The SL specimens prepared for these discussed neutron reflectometry studies were all grown on high-quality substrates. So, it was reasonable to expect that cooling down through T_C in zero-magnetic field those samples would form multidomain phases with nearly equal populations of the four domain states. With the [011] sample axis vertically mounted on the instrument, in two of those states **M** is along the polarization vector, which corresponds to NSF scattering. Such domains should produce peaks in measurements with "off-off" and "on-on" flipper settings. In the other two domain types, **M** is orthogonal to the polarization vector, giving rise to SF magnetic scattering. Peaks produced by those domains should be seen in the spectra measured with the flippers set "on-off" or "off-on."

In view of the above, it was expected that reflectivity data from $Ga_{1-x}Mn_xAs$/GaAs multilayers cooled in zero-magnetic field would exhibit peaks of similar intensity for all four combinations of the flipper settings. Quite surprisingly, the experiments revealed that the system is not as "well behaved" as it was assumed in that idealized scenario. In fact, the data clearly show that in the system there is an overwhelming tendency of forming *single-domain* states. A set of measured spectra from one of the investigated samples is shown in Fig. 12.15, as a representative example of the results obtained in experiments on eight specimens. The plots in Fig. 12.15c display the spectra from the sample in the paramagnetic state (at $T = 90$ K, well above the 40 K Curie point). When the instrument is configured to record the NSF scattering component – flippers "off-off," or "on-on" – in each case the measurements yield an identical nuclear Bragg peak. In contrast, the two SF scattering spectra are completely flat. Such an outcome is, of course, in perfect agreement with the basic reflectivity theory – in addition, it proves that the instrument is perfectly "tuned up."

The data in Fig. 12.15a were measured after the sample was cooled down to 7.8 K. In the spectra obtained in the spin-flip modes ("on-off" and "off-on"), one can now see maxima – however, they are extremely weak, indicating that only a small fraction of the Mn^{++} spins form domains with a horizontal orientation of the **M** vector. In the NSF scattering, in contrast, there is a sizable Bragg maximum – but only for the "on-on" flipper setting, and only a trace of a peak is seen in the data taken in the "off-off" configuration. As discussed above, such a disproportion in the NSF spectral line intensities – a sizable maximum for one incident beam

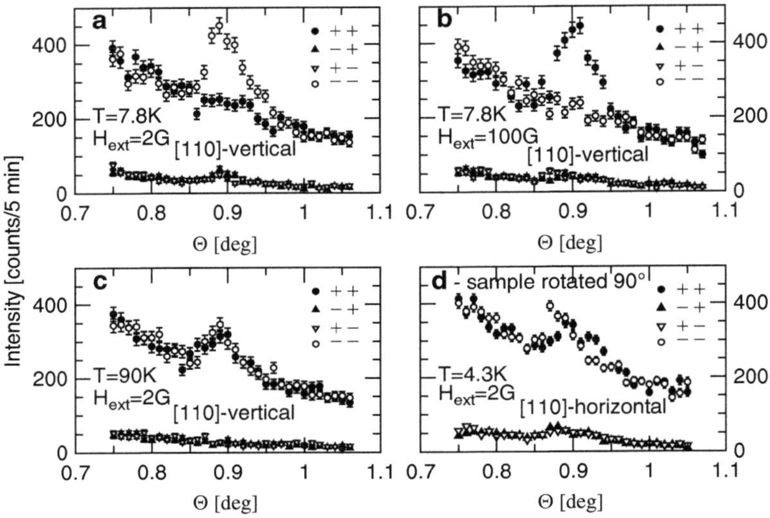

Fig. 12.15 Polarized neutron reflectivity profiles in the vicinity of the first-order reflectivity Bragg peak for a (50ML/6ML)×50 $Ga_{0.94}Mn_{0.06}As$/GaAs SL specimen. The intensity of non-spin-flip (++) and (− −) scattering and of spin-flip (+−) and (−+) scattering is shown. No peak in the spin-flip scattering indicates the absence of any horizontal magnetization component in the sample. Note the swap in the (++) and (− −) scattering after applying an external magnetic field of 100 G

polarization, and an almost flat spectrum for the other – occurs when the sample is uniformly magnetized parallel to the neutron polarization vector. In particular, the strong effect occurring in the "on-on" spectrum is consistent with a *downward* orientation of the **M** vector. Thus, the measured results indicate that in the investigated sample a large majority of the Mn^{++} spins form a domain with **M** oriented downward, and only a tiny fraction belongs to domains in which **M** points upward (if a significant fraction of spins formed such domains, it would be reflected by a much stronger line seen in the "off-off" spectrum).

The validity of the above interpretation can be tested by performing analogous experiments on the same sample after enforcing it to be in a truly single-domain state, which can be done by applying an external magnetic field. Fig. 12.15b shows the data from the specimen kept in a field of 100 G oriented up. With all the magnetic spins pointing in the field direction, a Bragg maximum should appear only in the "off-off" reflectivity scan. Because of the total absence of domains with other spin orientations, the other three spectra should be completely flat. This is indeed what the results in Fig. 12.15b show. Taken together, the data thus prove that the magnetic phase forming spontaneously in the sample in the process of zero-field cooling is a nearly single-domain state, with only a marginal fraction of spins belonging to domains with other orientations of **M** than that in the main domain.

The experiment performed on other $Ga_{1-x}Mn_xAs$/GaAs SL samples in all cases yielded very similar results as those shown in Fig. 12.15 with the exception of a single specimen. Reflectivity measurements on that one sample revealed distinct

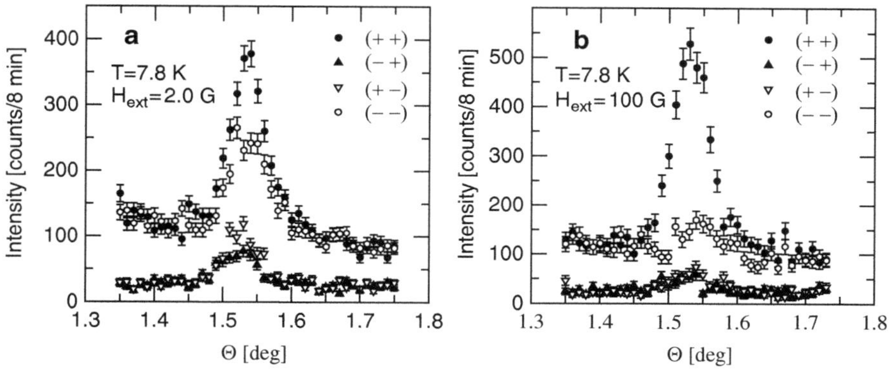

Fig. 12.16 (**a**) Reflectivity data from a sample cooled in zero magnetic field. Here peaks are seen in all four scattering modes, indicating that in this particular sample all four domain states are populated. (**b**) After applying a saturating magnetic field, only one strong peak is seen

Bragg peaks at all four combinations of flipper states (Fig. 12.16), clearly indicating the formation of a multidomain structure.

In summary, neutron reflectivity studies of the $Ga_{1-x}Mn_xAs/GaAs$ SL system confirmed the prior results pointing out that the interlayer magnetic coupling in this system is of the FM type. The study was done on samples with GaAs spacer thickness up to eight monolayers, so that the results do not exclude the possibility that in systems with thicker spacers the sign of the coupling changes to antiferromagnetic. Experiments with polarization analysis provided insight into the domain structure in the low-T FM phase. They revealed that in this multilayered system there exists a strong tendency toward forming single-domain FM states, but occasionally a sample disobeys that tendency and forms a multidomain structure. This information may be of some value from the viewpoint of practically oriented research on the $Ga_{1-x}Mn_xAs/GaAs$ system. However, as far as the physical mechanisms responsible for the observed peculiarities in the domain formation are concerned, neutron reflectivity studies alone do not provide enough insight into details that need to be known for their precise identification.

12.4 Neutron Diffraction Studies of Antiferromagnetic Multilayered Systems

12.4.1 EuTe/PbTe Superlattices

As noted in Sect. 12.1, EuTe/PbTe is an *antiferromagnetic* system. The magnetic component, EuTe, is a well-known "prototypical" antiferromagnet. It has been the subject of many studies (most notably, in the 1970s [29, 43]), and its magnetic properties are well understood. The material forms crystals of NaCl structure

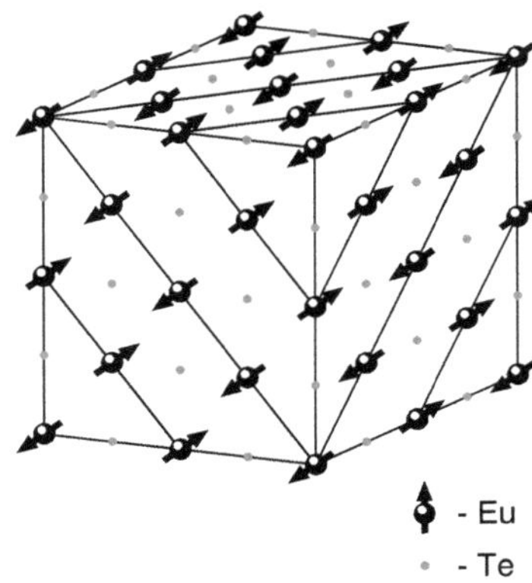

Fig. 12.17 The AF structure of EuTe consisting of antiferromagnetically coupled FM spin "sheets" on (111)-type planes

(with $a_{\mathrm{EuTe}} = 6.598\,\text{Å}$) in which Eu cations and Te anions occupy sites in two interpenetrating FCC sublattices. Tellurium is the magnetically inactive component. The Eu^{++} ions (f-state, with $S = \frac{7}{2}$) form a lattice of magnetic moments coupled via short-range exchange forces. The only relevant exchange interactions are those between the nearest neighbors (J_1, ferromagnetic) and the next-nearest neighbors (J_2, antiferromagnetic). The latter interaction is the dominant one. Below the Néel point of 9.6 K, the Eu^{++} spins form an ordered structure that is usually referred to as the "Type II AF ordering" or "AF ordering on the second kind" on an FCC lattice. The Eu spins are arranged in FM sheets on (111)-type planes, and these sheets are antiferromagnetically coupled to one another (Fig. 12.17).

PbTe, a nonmagnetic semiconductor, is isostructural with EuTe, with a lattice parameter $a_{\mathrm{PbTe}} = 6.462\,\text{Å}$. The mismatch between the two lattices is thus only about 2%, which is a favorable situation for MBE superlattice growth. An extensive program of research on such structures was launched in the early 1990s by the team of Prof. Günther Bauer at Johannes Kepler University in Linz, Austria. The EuTe/PbTe superlattices prepared by the Linz group on (111) BaF_2 substrates appeared to be of remarkably good crystalline quality. Those specimens were used in a variety of research projects conducted at Linz and in many collaborating laboratories. Neutron diffraction studies, in particular, were done at the Center for Neutron Research at the National Institute of Standards and Technology at Gaithersburg, MD in USA.

One question that naturally emerges in studies of (111) EuTe/PbTe superlattices is whether the spins in the EuTe layers form an ordered structure, and how that spin structure is accommodated in its very thin "host"? In bulk EuTe, as noted, the spins are organized into FM (111)-type planes. There are sets of such planes, and in a bulk crystal they are all symmetry equivalent. In a EuTe single-crystal film with a [111] growth axis, however, this symmetry is perturbed: one of the (111) planes is perpendicular to the growth axis, while the other three have an "oblique" orientation. On which set of planes would then the FM sheets form? Or, perhaps, would it be different type of spin ordering than that occurring in bulk crystals? Answers to such questions can be obtained only by performing neutron scattering experiments.

As noted in Sect. 12.2, the appropriate neutron scattering technique to be used in studies of superlattices containing layers of antiferromagnetic materials is conventional ("wide angle") diffraction. The net magnetization of an ordered AF lattice is zero, so that there is no magnetic SLD component, and the reflectivity spectra from such superlattices do not change when they are cooled below the AF transition point. Therefore, small-angle neutron reflectivity techniques are not particularly useful for investigating AF multilayers. However, the AF ordering of the spins gives rise to additional purely magnetic Bragg *diffraction* maxima at larger scattering angles. Their positions in Q-space coincide neither with the positions of nuclear Bragg reflections from the multilayered structure, nor (in most cases) with those from the substrate. It automatically solves the problem of separating the magnetic scattering components from the nuclear ones.

The first nuclear Bragg reflection (i.e., the one for which the diffraction angle θ is the lowest) from a EuTe crystal occurs at $\mathbf{Q} = (1,1,1)$ (in Q-space coordinates corresponding to the conventional description of an FCC system in terms of a cubic lattice), and at seven other symmetry-equivalent spots. The first *magnetic* reflection is positioned half-way between the origin and the first nuclear peak – but only on the axis perpendicular to the FM spin planes. For example, if the FM sheets have a $(1,1,1)$ designation, then magnetic peaks occur at the $(\frac{1}{2}, \frac{1}{2}, \frac{1}{2})$ and $(\frac{\bar{1}}{2}, \frac{\bar{1}}{2}, \frac{\bar{1}}{2})$ points, but not at the six other symmetry-related spots. Accordingly, to determine on which particular set of (111)-type planes the FM sheets form, it is enough to carry out diffraction scans through four Q-space points, namely $(\frac{1}{2}, \frac{1}{2}, \frac{1}{2})$, $(\frac{\bar{1}}{2}, \frac{1}{2}, \frac{1}{2})$, $(\frac{1}{2}, \frac{\bar{1}}{2}, \frac{1}{2})$, and $(\frac{1}{2}, \frac{1}{2}, \frac{\bar{1}}{2})$. As a matter of convenience, the superlattice growth axis is always labeled as [111], so the $(\frac{1}{2}, \frac{1}{2}, \frac{1}{2})$ magnetic reflection corresponds to the FM sheets lying in the growth plane, and the three other to an "oblique" sheet orientation.

To identify the AF arrangement in the EuTe layers, a standard procedure was used in the NIST experiments. Diffraction data were taken first from a sample kept at $T = 78$ K, and then repeated after cooling it through the Néel point of bulk EuTe (9.6 K) down to 5 K. No detectable maxima were ever found in the in the low-T data at the $(\frac{\bar{1}}{2}, \frac{1}{2}, \frac{1}{2})$, $(\frac{1}{2}, \frac{\bar{1}}{2}, \frac{1}{2})$, and $(\frac{1}{2}, \frac{1}{2}, \frac{\bar{1}}{2})$ Q-space points, whereas scans through the $(\frac{1}{2}, \frac{1}{2}, \frac{1}{2})$ point revealed a pronounced magnetic diffraction component. The experiments thus resolved the dilemma concerning the spin ordering in the EuTe layers showing that the system always chooses the arrangement with the FM planes parallel to the growth plane. In addition to that, however, the experiments revealed a highly intriguing fact: namely, in most cases the diffraction spectra obtained by scanning along the [111] direction exhibited a distinct pattern of narrow satellite lines emerging at regular intervals ΔQ. Examples of measured spectra from several different samples, in which the effect is clearly seen, are displayed in Fig. 12.18. In each investigated specimen, the observed satellite spacing was consistent with the SL periodicity, showing a good agreement with the theoretical formula $\Delta Q = \pi / D_{SL}$, where $D_{SL} = D_{EuTe} + D_{PbTe}$ is the thickness of single bilayer. The occurrence of such satellites clearly indicates the formation of magnetic interlayer correlations across the PbTe spacers.

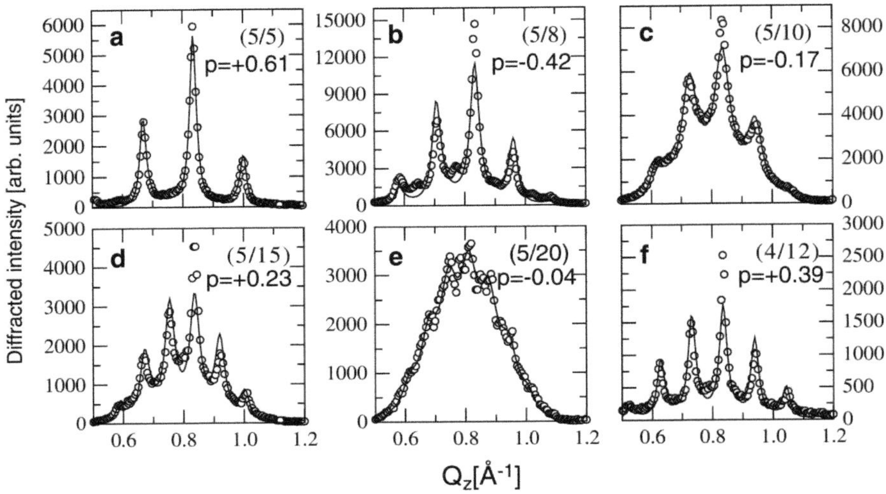

Fig. 12.18 Magnetic diffraction patterns from several EuTe/PbTe SL specimens with different thickness of the constituent layers, at $T = 4.2$ K. The *solid curves* are fits of (12.25) to the data points

As the FM sheets in the EuTe layers appear to be parallel to SL growth plane, and the net magnetization vector directions in successive FM sheets alternate, a single layer can be thought of as a quasi-one-dimensional chain, and schematically illustrated as $\uparrow\downarrow\uparrow\downarrow\uparrow$, where each $\uparrow$ and the $\downarrow$ symbol represent a single EuTe monolayer. The entire superlattice can be thus plotted as $\cdots \uparrow\downarrow\uparrow\downarrow\uparrow$ 000 $\uparrow\downarrow\uparrow\downarrow\uparrow$ 000$\cdots$, with the 0-s symbolizing magnetically neutral PbTe monolayers. Such a chain may have two ideally ordered states: a "correlated" one in which the magnetization sequence in each layer is identical: $\cdots \uparrow\downarrow\uparrow\downarrow$ 000 $\uparrow\downarrow\uparrow\downarrow$ 000 $\uparrow\downarrow\uparrow\downarrow$ $\cdots$, or an "anticorrelated" state in which the sequence in the $(i + 1)$th layer is reversed relative to that in the ith layer: $\cdots \uparrow\downarrow\uparrow\downarrow$ 000 $\downarrow\uparrow\downarrow\uparrow$ 000 $\uparrow\downarrow\uparrow\downarrow$ $\cdots$. In an ideally disordered state, there is a 50% probability that the magnetization sequence in the $(i + 1)$th layer is the same as in the ith layer, and 50% probability that it is reversed. This "Ising-like model" may be helpful in some considerations, but, in fact, it is oversimplified. Because of the hexagonal symmetry of the system, it is realistic to assume that in each EuTe monolayer there are three equivalent magnetic easy axes, and thus there are six equivalent magnetization vector orientations with a 60° spacing between them. A better graphical scheme is therefore that shown in Fig. 12.19, in which each EuTe monolayer is represented by a hexagon, with its three main diagonals symbolizing the magnetic easy axes. The arrow representing the magnetization vector lies along one of the easy axes, and can thus take six different orientations. In an ideally correlated superlattice (Fig. 12.19a), all spins in all layers are oriented "up" or "down" along the same easy axis. In an ideally disordered chain (Fig. 12.19b), only the spins within individual EuTe layer are antiferromagnetically correlated, but each layer independently chooses a configuration out of the possible six.

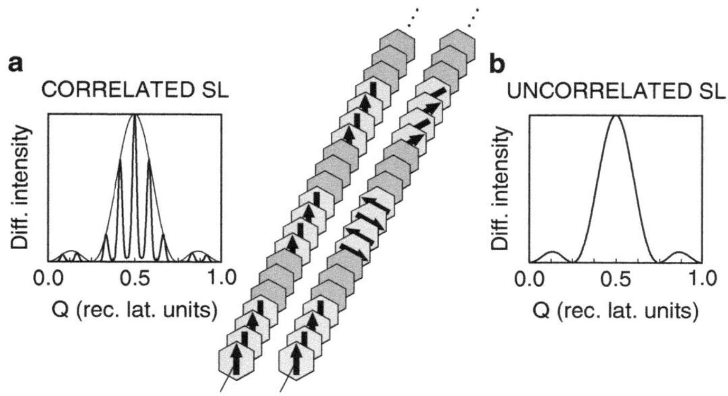

Fig. 12.19 *Middle*: schematic illustration of monolayer magetization orientation in a multilayer with hexagonal in-plane symmetry, in the case of ideal interlayer correlations (*left*) and ideally uncorrelated layer chain (*right*). Neutron diffraction profiles corresponding to such idealized situations are shown in boxes (**a**) and (**b**), respectively

The disordered chain produces a single broad maximum with weak "sidebands" in the diffraction spectrum (Fig. 12.19b, right panel) As noted in Sect. 12.2, due to lack of coherence between the waves scattered by individual layers, this profile essentially reproduces the shape of the $|F_{\mathrm{BL}}(Q_z)|^2$ function (or the lineshape one would observe if the specimen consisted only of a single layer). The diffraction pattern from the correlated chain, in contrast, consists of a number of narrow satellites, whose intensities are "modulated" by the $|F_{\mathrm{BL}}(Q_z)|^2$ function.

However, a look at Fig. 12.18 makes it clear that neither of the data sets displayed in the panels is accurately described by any of the two idealized spectrum spectrum shapes. Rather, each pattern looks like a combination of the two shapes taken in various proportions. The data set in panel (a) is the one that most closely resembles the spectrum from a fully correlated system – however, one can still see that there is a weak broadened component under the three narrow lines. The spectrum in the (e) panel, on the other hand, looks much like one from an ideally random chain – but still one can see evidence of weak satellite peaks forming near its maximum.

The data show therefore that the interlayer correlations in the EuTe/PbTe system are never perfect. In addition to that, there is an obvious trend in the data: in the spectra from samples with thin PbTe spacers the satellites are the dominating feature, but when the spacer thickness increases, the peaks become less pronounced, and spectrum visibly evolves toward the shape characterizing a disordered chain. Evidently (and not surprisingly), the ordered chain structure gradually deteriorates when the distance between the EuTe blocks grows larger. A question then arises – could the deviation from a perfectly correlated sequence be described in quantitative terms? Necessarily, such a description should be based on a model. Perhaps, the simplest conceivable model of "imperfect correlations" between the EuTe layers can be obtained by "randomizing" the aforementioned Ising-like chain. In such a system, if one selects at random pairs of adjacent layers – (i.e., with numbers i and

$i + 1$), then the *probability* that in a given pair the magnetization sequence is the same is P, and that the sequence in the $(i + 1)$th layer is reversed compared to that in the ith layer is $(1 - P)$. Note that for $P = 1$ the chain becomes "perfectly correlated," and for $P = 0$ – "perfectly anticorrelated," whereas $P = \frac{1}{2}$ corresponds to the "perfectly random" situation. In practice, it is more convenient to introduce another parameter: $p = 2P - 1$ that will be called below "the partial correlation coefficient." It takes the value of +1 or −1, respectively, for a perfectly correlated or anticorrelated chain, and is 0 for an ideally random chain. The advantage of this simple model is that neutron diffraction spectrum from a system with any p value can be analytically calculated. As shown in Appendix A in [8], the spectrum profile is given by:

$$I(Q_z) \propto |F_{\mathrm{BL}}(Q_z)|^2 \frac{1 - p^2}{1 - 2p\cos(Q_z D) + p^2}. \qquad (12.25)$$

This formula can be fitted to the experimental spectra, with p being the only adjustable parameter. The solid curves in Fig. 12.18 are the fits of (12.25) to the data points, and it can be seen that the agreement between the measured data and the calculated function is indeed very good.

In a more realistic modeling of imperfect correlations in the superlattice chain, one has to take into account that there are three easy axes in the EuTe films (the Ising-like model is, in fact, "a single easy axis approximation"). The angles between spins in adjacent layers may thus take not only the values of $\alpha = 0$ or $\alpha = 180°$, but also of $\alpha = \pm 60°$ or $\alpha = 120°$. The model becomes much more complicated. Essentially, the situation in such chain cannot be described by a single parameter because the probability of finding a pair with spins making an α angle should be specified for all possible angle values. There is no longer an analytical solution. However, the problem can be approached numerically – e.g., by using the Monte–Carlo simulation method. In a study performed by this technique [44, 45], imperfectly correlated chains were numerically generated, taking different probability distributions for the α angles. Next, the diffraction spectrum from the chain was calculated. An interesting finding from that work was that if a "partial correlation coefficient" was calculated as the $\langle \cos\alpha \rangle$ average for the entire chain, than, regardless of the probability distribution function $P(\alpha)$ that had been used for the chain generation, the spectrum shape obtained by putting this value into (12.25) was essentially identical with the numerically calculated shape.

However, there is still one question that needs to be answered – namely, what physical mechanism is responsible for the observed correlations? Surely, it cannot be the same conduction-electron-mediated process that gives rise to the interlayer exchange coupling seen in all-metallic multilayers. Looking for a mechanism that could lead to analogous effects in systems where mobile electrons are scarce (i.e., in superlattices made of insulating or semiconducting materials), Blinowski and Kacman, as already mentioned, cf. p. 439, considered another possible scenario – namely, exchange interaction transfer by valence electrons. The B&K theory is based on energy calculations using the tight-binding approach. The total energy calculations do not focus on a particular interaction mechanism, but account globally for the spin-dependent structure of the valence bands. The crystal structure and the structure of the electronic bands in the constituent materials are taken into account. The energy is calculated for two different configurations, with the spin structure in adjacent EuTe layers being either in-phase, or out-of-phase. The difference ΔE between these two energies per unit surface area can be regarded as a measure of the strength of the interlayer coupling. It is related to the commonly used coupling constant J_1 as $\Delta E = 4|J_1|$ (the factor of 4, instead of 2, accounts for the fact that in a real SL system each magnetic layer is coupled to two adjacent layers).

The calculations performed by Blinowski and Kacman for the EuTe/PbTe system yielded different energies for the "in-phase" and "out-of-phase" configuration, thus offering an explanation why interlayer coupling occurs in these superlattices despite the absence of mobile carriers. The calculations also predicted that the coupling strength exponentially decreases with increasing thickness of the PbTe spacer. An ideal test for the model would be a comparison of the calculated J_1 results with experimentally determined values. Unfortunately, neutron diffraction methods do not offer such a possibility. As was discussed in Sect. 12.3.1, in studies of the EuS/PbS and EuS/YbSe the strength of the AF interlayer coupling in these multilayers could be determined by applying an external magnetic field. A sufficiently strong magnetic field overcomes the AF coupling between the ferromagnetic EuS layers, so that the magnetization vectors in all the layers get aligned with the field. This process can be monitored by measuring the reflectivity spectra while gradually increasing the applied field. From the field magnitude needed for obtaining a fully aligned configuration, one can next determine the strength of the interlayer coupling. However, this simple method does not work in systems, where the layers are antiferromagnetic – their net magnetization is zero, so they do not react to an external field in the same way as the FM layers do.

In the present case of EuTe/PbTe multilayers, the only measured parameter that can offer some idea of how the interlayer coupling strength depends on the PbTe spacer thickness is the "partial correlation coefficient" p. This parameter can be thought of as an analog of the "correlation range" coefficient used for describing the dependence of spin–spin correlations on the interatomic distance in short-range ordered magnetic systems (such as, e.g., spin glasses). The interlayer exchange in the EuTe/PbTe system is a factor that tends to produce a correlated sequence of the layers. If the system were absolutely perfect (i.e., if all individual EuTe layers ideally obeyed the hexagonal symmetry), the interlayer correlations would also be perfect, regardless of the interlayer exchange strength. In other words, in an ideal case the interlayer correlations should not significantly deteriorate when the coupling constant J_1 decreases due to the increasing thickness of the PbTe spacer. However, the experiments clearly show that they *do* deteriorate. It leads to the conclusion that in the system there exist a "randomizing factor," which opposes the formation of an ordered chain of layers. What underlies this randomizing tendency is not entirely clear – one can only speculate on the physical mechanism that is involved. One conceivable explanation might be that in individual EuTe layers there exist "preferred" easy axes. Because of the hexagonal symmetry, in each layer there should be three equivalent in-plane easy axes. However, a nonuniform in-plane strain may give rise to additional single-axis magnetic anisotropy. Further, one can speculate that both the magnitude of this "extra" anisotropy and the orientation of its axis may change from layer to layer in a random fashion. Then, if the interlayer exchange is strong, in most cases it will be able to overcome those random anisotropies, and to correlate the spin orientation in consecutive layers. However, if the interlayer coupling becomes weaker due to the increased thickness of the PbTe spacers, then more often the spins in the $(i + 1)$th layer will choose an orientation along the "random anisotropy" axis rather than an orientation correlated with that in the ith layer. In other words, the weaker the interlayer exchange is, the larger becomes the number of "random phase shifts" in the layer chain. Note now that the p coefficient can be thought of as a measure of the average correlation length in layer chain. The above simple model of "random anisotropies" thus enables one to assume that the value of p is roughly proportional to the $J_1/\langle A \rangle$ ratio, where $\langle A \rangle$ is the average magnitude of the random anisotropy fields. Assuming, in addition, that the "random anisotropy effects" in all EuTe/PbTe multilayer specimens are similar, one reaches the final conclusion that the value of p can be taken as a relative measure of the strength of the interlayer coupling forces in a given SL specimen.

The above model is surely only heuristic – and most likely one can think of other physical scenarios that might explain the fact that the interlayer correlations gradually deteriorate with the increasing PbTe spacer thickness. However, the conclusion that $J_1 \propto p$ is certainly consistent with one's intuitive judgment. Given that there is no way of obtaining the information of the dependence of the J_1 parameter on the PbTe spacer thickness from direct measurements, using the $J_1 \propto p$ assumption when comparing the results of the B&K model with the measured data seems to be a reasonable option. A comparison of the theoretical $J_1(D_{\mathrm{PbTe}})$ dependence with the $p(D_{\mathrm{PbTe}})$ data obtained from the analysis of experimental spectra is shown in Fig. 12.20. Even though the fluctuations in the p data are quite large, these results are certainly in favorable agreement with an exponential dependence predicted by the model. Surprisingly, however, the observed $p(D_{\mathrm{PbTe}})$ characteristic clearly suggests

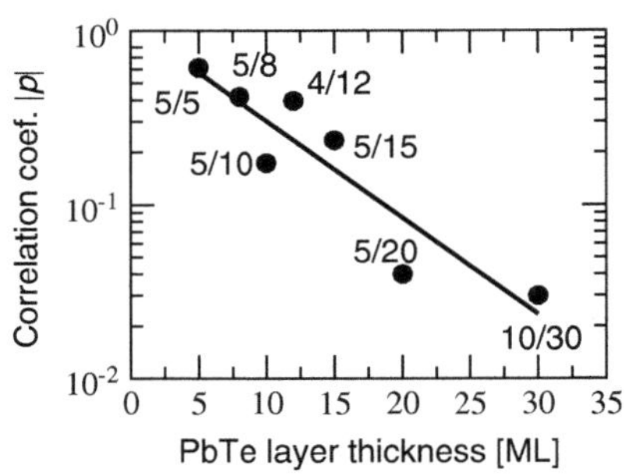

Fig. 12.20 Dependence of the partial interlayer correlation parameter p on the PbTe spacer thickness, as determined by fitting (12.25) to the measured data obtained from several EuTe/PbTe SL specimens. The *line* represents the best fit of an exponential dependence to the data points

that the decrease of interlayer coupling strength with increasing spacer thickness is much slower than predicted by the model. The model appears to be qualitatively correct, since the exponential dependence of J_1 on D_{PbTe}) is indeed consistent with the observed system behavior. It should be noted that the very same theoretical approach was used by B&K for modeling the exchange coupling between FM EuS films in the EuS/PbS and EuS/YbSe structures. Here, the slope of the calculated $J_1(D_{PbS})$ and $J_1(D_{YbSe})$ characteristics was found to be in a reasonably good agreement with the slope of the experimental characteristics in the region of D values up to $\sim$20 Å. Those results certainly argue in favor of the model trustworthiness. Therefore, the fact that in the case of the EuTe/PbTe system the observed range of interlayer correlations much exceeds the range predicted by the same theory is somewhat puzzling. It may suggest that in these antiferromagnetic superlattices the transfer interactions by the valence electrons is not the *only* physical mechanism giving rise to interlayer coupling – and not even the dominant one.

As discussed in Sect. 12.3.1, the experimental data from the ferromagnetic systems also indicated the existence of two coupling mechanisms. The initial part of the $J_1(D_{PbS})$ characteristic for EuS/PbSe is consistent with the B&K model prediction – but beginning from the $\sim$25 Å spacer thickness, the coupling strength apparently stabilizes, and remains approximately constant up to PbS layer thickness as large as 150 Å (i.e., the largest value investigated). A coupling strength weakly depending on the interlayer distance is indicative for dipole–dipole interaction. It can be therefore concluded that for thinner PbS spacers the electronic coupling mechanism plays a dominant role, and then it is the dipole–dipole interaction that takes over. Therefore, it is conceivable that in the EuTe/PbTe the dipolar coupling mechanism also plays a significant role – and perhaps even a dominant role in the entire range of D_{PbTe}) values investigated. It might explain the discrepancy between the calculated characteristics and the results of observations in the case of this system. However, quantitative calculations of the J_1 component arising from dipole–dipole interactions would require a detailed knowledge of the domain structure of the layers, and such information cannot be extracted from neutron diffraction data. In this situation, when neither the theoretical nor the experimental $J_1(D_{PbTe})$ dependencies are accurately known, the above interpretation is merely a conjecture. Even though the experiment described in this section revealed much highly interesting information about the interlayer coupling in the EuTe/PbTe system, more research would still be needed for determining the exact nature of physical mechanisms responsible for this coupling.

12.4.2 II–VI-Based Systems

II–VI/Mn-VI multilayers were the first all-semiconductor magnetic superlattices investigated by neutron diffraction. In these epitaxial structures their magnetic constituents, MnSe and MnTe, crystallize in the metastable zinc-blende (ZB) phase that does not exist in bulk. In these ZB modifications, the Mn^{++} cations form an

FCC spin lattice with dominant AF superexchange interactions between the nearest-neighbor (NN) spins, and weak AF coupling between more distant neighbors. The magnetic ground state of such a spin system is the so-called Type III AF ordering on an FCC lattice, which has a tetragonal unit cell of a, a, $2a$ dimensions along the principal cubic axes (where a is the lattice constant of the "chemical" FCC cell).

Pronounced interlayer coupling effects were observed in two such SL structures, MnTe/CdTe and MnTe/ZnTe with [001] growth axis. Although their chemical formulae are very similar, these two systems are magnetically quite different. This comes from the fact that the Type III AF spin lattice is strongly frustrated. Because of the lattice geometry, only 2/3 of all AF bonds between NN spin pairs can be "satisfied" (i.e., the two spins are antiparallel), while the remaining 1/3 bonds are "frustrated" (both spins have the same orientation). The ground state has an infinite number of degenerate configurations. The system is very sensitive to symmetry-breaking perturbations (such as anisotropic lattice strain), which can lift the degeneracy and either stabilize one of the configurations, or cause a transition to an ordering of different symmetry.

The unstrained lattice parameters of CdTe and MnTe are $a_{\mathrm{CdTe}} = 6.48\,\text{Å}$ and $a_{\mathrm{MnTe}} = 6.34\,\text{Å}$. Hence, in the MnTe/CdTe SL structures, the MnTe layers are *stretched* – i.e., the in-plane lattice constants are elongated ($a_x = a_y \equiv a_{x,y} > a_{\mathrm{MnTe}}$), whereas the period in the growth direction is shortened ($a_z < a_{\mathrm{MnTe}}$). Such a lattice distortion leads to a transition from the Type III order to a new *incommensurate* structure, in which the Mn spins are arranged in a helical fashion. The spin lattice periodicity along the principal crystallographic axes is now $(2+\delta)a_{x,y}$, $a_{x,y}$, a_z (actually, two symmetry-equivalent in-plane domain types may form, with the lattice periodicity in the other one $a_{x,y}$, $(2+\delta)a_{x,y}$, a_z). The $(2+\delta)$ coefficient can be thought of as the spin helix "pitch." The fact that such a spin structure forms in the stretched layers can be explained by relatively simple energy considerations, making it possible to obtain a theoretical relation between the helix pitch and the strain magnitude. This model is presented in [46, 47], which describe the experiments on MnSe/ZnTe SL system, where such "strain-engineered" helical ordering effects are strongly pronounced (however, interlayer correlations effects in MnSe/ZnTe have not been investigated).

Neutron diffraction studies [11, 48] of the MnTe/CdTe system were performed on a number of specimens prepared in Prof. G. Bauer's laboratory at Linz, Austria. The thickness of the magnetic MnTe layers in all samples was 10 monolayers (Mls), and the CdTe spacer thickness was changed from 2 Mls to 10 Mls. As expected, the pitch $(2 + \delta)$ of the helical order seen in the samples gradually increased with increasing spacer thickness, reflecting the increase of strain in the MnTe layers. In addition, the measurements revealed distinct patterns of interference fringes, showing that the spin helices forming in successive layers are phase-synchronized. Such an effect clearly indicates the existence of some kind of "magnetic communication" between the MnTe layers across the nonmagnetic CdTe spacers. An intriguing question becomes what physical mechanism is involved in this process?

The B&K model is not applicable to the MnTe/CdTe system because the energy of the valence electrons is not sensitive to the phase shift between spin helices in successive layers. A different theoretical model that would be able to explain the observed correlation effects was proposed by Rusin [49]. He pointed out that even though in the system there are no *mobile* carriers which might give rise to an RKKY-like interlayer coupling, epitaxial CdTe layers usually do contain carriers that are bound by impurities or defects, forming "hydrogenic centers" with the Bohr radius of several tens of Å. In diluted magnetic semiconductors such centers are known to polarize the magnetic spins within the Bohr orbit, which results in an effect usually referred to as "bound magnetic polaron" (BMP). As reasoned by Rusin, such centers located in the spacers might act in a similar way on the Mn spins in the two adjacent MnTe blocks, thus "synchronizing" their polarization and effectively introducing magnetic coupling between the spin helices.

The other Mn-VI/II–VI SL system in which neutron diffraction experiments reveal pronounced interlayer correlation effects is MnTe/ZnTe [9, 10]. In contrast to the situation in MnTe/CdTe multilayers, here the strain in the MnTe layers is *compressive* because $a_{\mathrm{ZnTe}} = 6.10\,\text{Å}$. Now the parameters of the distorted MnTe lattice are $a_{xy} < a_{\mathrm{MnTe}}$ and $a_z > a_{\mathrm{MnTe}}$. Such a distortion of the FCC lattice does not change the Type III AF structure but only selects an energy minimizing configuration with the unit cell doubling direction along the SL growth axis (i.e., the magnetic unit cell parameters are a_{xy}, a_{xy}, $2a_z$). Neutron diffraction experiment was done on samples prepared in the laboratory of Prof. J. K. Furdyna at the University of Notre Dame, USA. The data from specimens with thin spacers (up to 15 Å) revealed pronounced interlayer coupling effects. In addition, this coupling shows a peculiar temperature behavior, never seen in any other magnetic semiconductor SL system investigated by neutron scattering. This is illustrated by Fig. 12.21, which

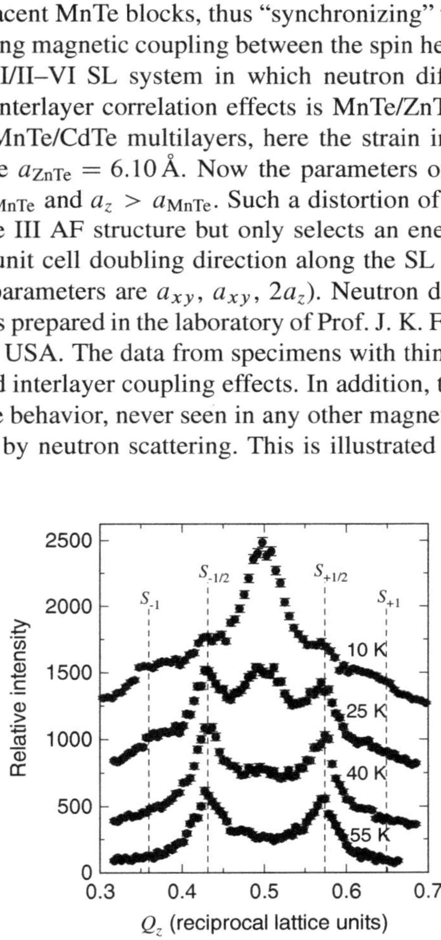

Fig. 12.21 Magnetic diffraction spectra from a $[(\mathrm{MnTe})_{10})|(\mathrm{ZnTe}_5)]_{400}$ specimen measured at different temperatures. The data measured at 10 K show a central peak and weak satellites at S_{-1} and S_{+1} positions, and additional weak satellites at $S_{-\frac{1}{2}}$ and $S_{+\frac{1}{2}}$ positions. As the temperature is increased, the central peak gradually disappears, whereas the $S_{-\frac{1}{2}}$ and $S_{+\frac{1}{2}}$ peaks increase in intensity and eventually become the dominant spectrum features, indicating a change in the sign of the interlayer magnetic interactions. The spectra were vertically offset for clarity

shows the diffraction data from a $[(MnTe)_{10}|(ZnTe_5)]_{400}$ specimen measured at several different temperatures. The spectrum measured at $T = 10$ K shows a central peak accompanied by two weak maxima at the calculated first-order satellite positions S_{-1} and S_{+1} – but one can also recognize another pair of weak peaks emerging at "half-integer" satellite positions $S_{-1/2}$ and $S_{+1/2}$. This spectrum indicates that in most of the specimen volume the Type III order in successive layers is correlated, but there are also small regions in which the sequence is "anticorrelated" (i.e., the spin structure in the $(i + 1)$th layer is reversed compared to that in the ith one). When the temperature is raised, the central peak and the S_{-1} and S_{+1} satellites gradually disappear while the peaks at $S_{-1/2}$ and $S_{+1/2}$ grow in intensity and become the dominant spectrum features. Such a behavior clearly indicates that with increasing T the *sign* of the interlayer interaction changes. Similar experiments were performed somewhat later on a specimen of the same composition doped with Cl, which introduces deep electronic levels in ZnTe [10]. The doping was found to enhance the low-T correlations, but not to significantly influence the system behavior at $T > 40$ K. Regretfully, after that the studies of the MnTe/ZnTe system were not further pursued, so the mechanism responsible for that extremely intriguing change of sign of the interlayer coupling still remains unexplained.

12.5 Closing Remarks

The experimental results presented in this chapter certainly demonstrate that neutron scattering is an excellent tool for investigating the properties of thin magnetic films buried in multilayered structures. This review was focused specifically on interlayer magnetic coupling phenomena. Here, the advantages of neutron techniques are unquestionable. There are no other experimental tools capable of directly probing magnetically correlated states in multilayered structures – especially, in the case of antiferromagnetic coupling between the layers. However, neutron scattering measurements also offer insight into other important aspects of such systems. One example is the domain structure in the magnetic layers – as illustrated by the results of studies of the $Ga_{1-x}Mn_xAs$/GaAs superlattices, in Sect. 12.4, much information about the domains can be obtained using polarized neutron reflectometry. It is worth mentioning that similar studies (not included in the present review) performed on the EuS-based systems revealed intriguing anomalies in the population of the domain states in the FM layers [50–52].

Diffraction and reflectivity measurements performed on the EuTe and EuS-based superlattices, and on the II–VI-based multilayers, have shown that the phenomenon of interlayer exchange coupling is not restricted to structures composed of ferromagnetic/nonmagnetic metals in which mobile electrons are abundant. Therefore, new theoretical models, based on mechanisms other than mobile electron-assisted transfer of exchange interactions, have to be developed for all-semiconductor multilayered systems. The Blinowski and Kacman theory, in which the principal role in the interlayer interaction transfer is ascribed to the valence band electrons, emerges here as an elegant solution option. One considerable advantage of the B&K approach is it soundness: it does not introduce any special assumptions, everything is based on energy calculations in the framework of existing well-established models in semiconductor physics. The good qualitative agreement between the calculations and the measured results in the case of the EuS-based system argues in the model's favor. Another advantage of the B&K theory is that it explains the origin of coupling between *antiferromagnetic* layers by showing that total magnetic energy *does* depend on the mutual orientation of spins in two such objects. However, the experimental results from the EuTe/PbTe SLs may suggest this coupling mechanism is not solely responsible for the interlayer correlations seen in this system.

As far as the IEC phenomena are concerned, the scientific profit from the studies described in this chapter may be summarized as follows. The most important piece of information provided by the experiments is that such effects may occur in all-semiconductor multilayered structures of various types – in those containing FM layers as well as in those with AF ones. In addition to that, a large volume of experimental information was collected by performing systematic studies of interlayer correlations in specimens with different thickness of the nonmagnetic spacer, and of the effects of applied magnetic fields on the correlations. Some of the investigated systems were the subjects of theoretical studies. A comparison of the experimental results with the theoretical predictions shed much light on the physical mechanisms underlying the IEC effects seen in those systems. From the standpoint of fundamental spintronic physics, all those results are certainly of considerable interest.

However, the low Curie and Néel temperatures of the magnetic constituents obviously rule out any practical applications of the systems described in this review. Practical semiconductor spintronics will become possible only when new room-temperature FM semiconductors emerge. Practice-oriented research on IEC phenomena in superlattices and heterostructures composed of the new materials may then be needed – and neutron scattering tools will surely be employed for this purpose. Reports from prior studies performed on low-T_C systems such as EuS/PbS or $Ga_{1-x}Mn_xAs$/GaAs may offer guidance and helpful hints for these future investigations.

Acknowledgements

All neutron scattering experiments reviewed in this chapter were performed at the Center for Neutron Research at National Institute of Standards and Technology (CNR NIST) in Gaithersburg, Maryland, USA. The Authors express their sincere gratitude to the CNR leaders and staff members for their hospitality and assistance in carrying out the projects – and especially, to Dr. Charles F. Majkrzak, who was an invaluable partner and mentor in neutron reflectometry studies of ferromagnetic multilayered systems.

Special thanks are due to collaborators who supplied the Authors with samples for neutron scattering studies – namely, to (alphabetically): G. Bauer, W. Faschinger,

J. K. Furdyna, J. Sadowski, H. Sitter, A. Yu. Sipatov, and V. Volobuev. The Authors are also thankful to their colleagues who took part, directly or indirectly, in various phases of the projects (alphabetically): J. Blinowski, J. Chen, W. J. M. de Jonge, M. S. Dresselhaus, R. R. Galazka, K. I. Goldman, K. Ha, S. Hall, P. Kacman, P. Klosowski, J. Kutner-Pielaszek, H. Krenn, V. Nunez, J. J. Rhyne, P. Sankowski, S. K. Sinha, G. Springholz, T. Story, H. J. M. Swagten, A. Twardowski, and Z. Q. Wiren.

Last, but not least, the authors would like to gratefully acknowledge the generous support from the following agencies and institutions: National Science Foundation (DMR Grants 0508478, 0204105, 9972586, 9510434, and 9121353), Civilian Research and Development Foundation (Grant CRDF UP2-2444-KH-02), and NATO (Grant PST.CLG 975228).

References

1. M.N. Baibich, J.M. Broto, A. Fert, F.N.V. Dau, F. Petroff, P. Eitenne, G. Creuzet, A. Friederich, J. Chazelas, Phys. Rev. Lett. **61**, 2472 (1988)
2. G. Binash, P. Grünberg, F. Saurenbach, W. Zinn, Phys. Rev. B **39**, 4828 (1989)
3. P. Bruno, Phys. Rev. B **52**, 411 (1995)
4. P. Bruno, J. Phys. Condens. Matter **11**, 9403 (1999)
5. T.M. Giebultowicz, P. Klosowski, N. Samarth, H. Luo, J.K. Furdyna, J.J. Rhyne, Phys. Rev. B **48**, 12817 (1993)
6. N. Samarth, P. Klosowski, H. Luo, T.M. Giebultowicz, J.K. Furdyna, B. Larson, N. Otsuka, Phys. Rev. B **44**, 4701 (1991)
7. T.M. Giebultowicz, V. Nunez, G. Springholz, G. Bauer, J. Chen, M.S. Dresselhaus, J.K. Furdyna, J. Magn. Magn. Mater. **140–144**, 635 (1995)
8. H. Kępa, G. Springholz, T.M. Giebultowicz, K.I. Goldman, C.F. Majkrzak, P. Kacman, J. Blinowski, S. Holl, H. Krenn, G. Bauer, Phys. Rev. B **64**, 024419 (2003)
9. J.J. Rhyne, J. Lin, J.K. Furdyna, T.M. Giebultowicz, J. Magn. Magn. Mater. **177–181**, 1195 (1998)
10. J. Lin, J.J. Rhyne, J.K. Furdyna, T.M. Giebultowicz, J. Appl. Phys. **83**, 6554 (1998)
11. V. Nunez, T.M. Giebultowicz, W. Faschinger, G. Bauer, H. Sitter, J.K. Furdyna, J. Magn. Magn. Mater. **140–144**, 633 (1995)
12. H. Ohno, Science **281**, 951 (1998)
13. H. Kępa, J. Kutner-Pielaszek, A. Twardowski, J. Sadowski, T. Story, T.M. Giebultowicz, Phys. Rev. B **64**, 121302(R) (2001)
14. H. Kępa, J. Kutner-Pielaszek, A. Twardowski, C.F. Majkrzak, T. Story, J. Sadowski, T.M. Giebultowicz, Appl. Phys. A **74**, S1526 (2001)
15. H. Kępa, C.F. Majkrzak, P. Sankowski, A.Y. Sipatov, T.M. Giebultowicz, J. Alloy. Comp. **401**, 238 (2006)
16. H. Kępa, J. Kutner-Pielaszek, J. Blinowski, A. Twardowski, C.F. Majkrzak, T. Story, P. Kacman, R.R. Gałązka, K. Ha, H.J.M. Swagten, W.J.M. de Jonge, A.Y. Sipatov, V. Volobuev, T.M. Giebultowicz, Europhys. Lett. **56**, 54 (2001)
17. H. Kępa, C.F. Majkrzak, A.Y. Sipatov, T.M. Giebultowicz, Physica B **345**, 193 (2004)
18. H. Kępa, P. Sankowski, P. Kacman, A.Y. Sipatov, C.F. Majkrzak, T.M. Giebultowicz, J. Magn. Magn. Mater. **272–276**, 323 (2004)
19. P. Sankowski, H. Kępa, P. Kacman, A.Y. Sipatov, C.F. Majkrzak, T.M. Giebultowicz, Acta Phys. Polon. A **105**, 607 (2004)
20. J. Blinowski, P. Kacman, Acta Phys. Polon. A **92**, 719 (1997)

21. J. Blinowski, P. Kacman, Phys. Rev. B **64**, 045302 (2001)
22. G.E. Bacon, *Neutron Diffraction* (Clarendon Press, Oxford, 1975)
23. C.G. Shull, J.S. Smart, Phys. Rev. **76**, 1256 (1949)
24. S.W. Lovesey, *Theory of Neutron Scattering from Condensed Matter* (Oxford University Press, London, 1986)
25. W.G. Williams, *Polarized Neutrons* (Clarendon Press, Oxford, 1987)
26. C.F. Majkrzak, Physica B **173**, 75 (1991)
27. C.F. Majkrzak, J. Kwo, M. Hong, Y. Yafet, D. Gibbs, C.L. Chien, J. Bohr, Adv. Phys. **40**, 99 (1991)
28. J. Dailaut, A. Gibaud (eds.), *X-Ray and Neutron Reflectivity: Principles and Applications* (Springer, Berlin, 1999)
29. T. Kasuya, IBM J. Res. Dev. **14**, 214 (1970)
30. L. Passell, O.W. Dietrich, J. Als-Nielsen, Phys. Rev. B **14**, 4897 (1976)
31. J. Kunes, W. Ku, W.E. Pickett, J. Phys. Soc. Jpn. **74**, 1408 (2005)
32. P. Wachter, in *Handbook on the Physics and Chemistry of Rare Earth*, vol. 2, ed. by K.A. Gschneider, L.R. Eyring (North Holland, Amsterdam, 1979), pp. 507–574
33. A. Mauger, C. Godart, Phys. Rep. **41**, 51 (1986)
34. A. Stachow-Wójcik, T. Story, W. Dobrowolski, M. Arciszewska, R.R. Gałązka, M.W. Kreijveld, C.H.W. Swüste, H.J.M. Swagten, W.J.M. de Jonge, A. Twardowski, A.Y. Sipatov, Phys. Rev. B **60**, 15220 (1999)
35. W. Folkerts, J. Magn. Magn. Mater. **94**, 302 (1991)
36. G.E. Everett, R.A. Ketcham, J. Phys. Colloq. **32**, C1 (1971)
37. G.E. Everett, R.C. Jones, Phys. Rev. B **4**, 1561 (1971)
38. R.S. Hughes, G.E. Everett, A.W. Lawson, Phys. Rev. B **9**, 2394 (1974)
39. E.C. Stoner, E.P. Wohlfarth, Philos. Trans. R. Soc. Lond. Ser. A **240**, 599 (1948)
40. J.A. Borchers, P.M. Gehring, R.W. Erwin, J.F. Ankner, C.F. Majkrzak, T.L. Hylton, K.R. Coffey, M.A. Parker, J.K. Howard, Phys. Rev. B **54**, 9870 (1996)
41. V. Zorchenko, A.Y. Sipatov, W.W. Wolobuev, Fiz. Nizk. Temp. **29**, 1208 (2003)
42. B.A. Jones, IBM J. Res. Develop. **42**, 25 (1998)
43. N.F. Oliveira, Jr., S. Foner, Y. Shapira, T.B. Reed, Phys. Rev. B **5**, 2634 (1972)
44. K.I. Goldman, H. Kepa, T.M. Giebultowicz, S.K. Sinha, J. Appl. Phys. **81**, 5309 (1997)
45. K.I. Goldman, H. Kepa, T.M. Giebultowicz, S.K. Sinha, Physica B **241–243**, 717 (1998)
46. T.M. Giebultowicz, N. Samarth, H. Luo, J.K. Furdyna, P. Klosowski, J.J. Rhyne, Phys. Rev. B **46**, 12076 (1992)
47. T.M. Giebultowicz, J.K. Furdyna, J. Rhyne, IEEE Trans. Magn. **6**, 3382 (1993)
48. T.M. Giebultowicz, W. Faschinger, V. Nunez, P. Klosowski, G. Bauer, H. Sitter, J.K. Furdyna, J. Cryst. Growth **138**, 877 (1994)
49. T.M. Rusin, Phys. Rev. B **58**, 2107 (1998)
50. H. Kępa, C.F. Majkrzak, A.Y. Sipatov, T.M. Giebultowicz, Physica B **397**, 36 (2007)
51. H. Kępa, C.F. Majkrzak, A.Y. Sipatov, Z.Q. Wiren, T.M. Giebultowicz, J. Magn. Magn. Mater. **310**, 2280 (2007)
52. H. Kępa, C.F. Majkrzak, A.Y. Sipatov, T.M. Giebultowicz, Physica B **345**, 193 (2004)

Index

J. Kossut and J.A. Gaj (eds.), *Introduction to the Physics of Diluted Magnetic Semiconductors*, Springer Series in Materials Science 144,
DOI 10.1007/978-3-642-15856-8,

MIX
Papier aus verantwortungsvollen Quellen
Paper from responsible sources
FSC® C105338

If you have any concerns about our products,
you can contact us on
ProductSafety@springernature.com

In case Publisher is established outside the EU,
the EU authorized representative is:
Springer Nature Customer Service Center GmbH
Europaplatz 3, 69115 Heidelberg, Germany

Printed by Libri Plureos GmbH
in Hamburg, Germany